Thermal Analysis

Thermal Analysis

Proceedings of the Seventh International Conference on Thermal Analysis

VOLUME II

Edited by **Bernard Miller**
Textile Research Institute,
Princeton, New Jersey

A Wiley Heyden Publication

JOHN WILEY & SONS
Chichester · New York · Brisbane · Toronto · Singapore

Copyright © 1982 by Wiley Heyden Ltd.

All rights reserved.

No part of this book may be reproduced by any means, nor transmitted, nor translated into a machine language without the written permission of the publisher.

British Library Cataloguing in Publication Data:

International Conference on Thermal Analysis
  (7th: 1982: Ontario)
  Thermal analysis.
  1. Thermal analysis—Congresses
  I. Title   II. Miller, Bernard
  543'.086   QD79.T38
  ISBN 0 471 26243 9
    ISBN 0 471 26244 7 (v. 1)
    ISBN 0 471 26245 5 (v. 2)

Printed in Great Britain

CONTENTS OF VOLUME 2

CONTENTS OF VOLUME 1 xv

FOREWORD xxvi

PREFACE xxvii

Section III   Organic Chemistry, Biological and Medical Sciences

Plenary Lecture

Microcalorimetric Studies on Native Human Plasma
Protein Sorption. 793
    E. Nyilas and T.-H. Chiu

Phase Transitions in the Compounds p-n-Alkoxybenzylidene-
p-Amino Benzoic Acids (nBABA, n=6 to 9) 796
    K. U. Deniz, E. B. Mirza, U. R. K. Rao, P. S. Parvathanathan,
    A. S. Paranjpe, and A. V. Patankar

The Application of Thermal Analysis to Investigation of the
Interaction of Boric Acid With Salicylic Acid Derivatives. . 803
    E. M. Schwartz, V. V. Grundstein, and I. M. Vitol

Thermal Studies of Proteins, Peptides, and Amino Acids II . 809
    J. S. Crighton and P. N. Hole

Thermodynamic Properties of the Polyuronates-Ca^{2+} Ions
Interaction in Aqueous Solutions. 815
    A. Cesaro, F. Delben, and S. Paoletti

Role of "Intermediate Water" for the Phase Transition
Phenomena in Surfactant-Water Systems as Revealed by DSC. . 822
    M. Kodama, M. Kuwabara, and S. Seki

Specific Heat Anomalies in Some Organic Compounds 829
    C. A. Martín

Thermoanalytical Investigations on Release of Oxygen
From A Heme-Model   836
    O. Schneider, J. Moll, M. Hanack, and D. Krug

Thermogravimetric Analysis of Powdered Activated Carbon
in Biological Systems   843
    W. K. Bedford and T. R. Bridle

Microcalorimetric Investigations on Mixed Microbial
Cultures.   849
    I. Lamprecht and B. Schaarschmidt

Microcalorimetric Experiments on Skin Cells. Influence
of Drugs on Heat Production   857
    M. Patel, B. Schaarschmidt, and I. Lamprecht

Testing the Effect of Three Chemical Substances Formed
in Livestock Facilities on E. Coli Broth Cultures by
Flow Microcalorimetry   863
    J. Hartung

Analysis of the Thermal Behavior of Human Finger Nail
Cuttings.   870
    B. Marjit and D. Marjit

Applications of Thermal Methods in the Pharmaceutical
Industry - Part I   876
    M. J. Hardy

Applications of Thermal Methods in the Pharmaceutical
Industry - Part II.   887
    M. J. Hardy

Application of Thermal Analysis to Analytical and
Process Research in the Pharmaceutical Industry   893
    J. A. McCauley

Dynamic Purity Measurements   899
    P. D. Garn, B. Kawalec, J. J. Houser, and T. F. Habash

Enthalpy Determination by DSC and DTA   904
    E. Marti, O. Heiber, and A. Geoffroy

Optimization of Accuracy and Precision in the Dynamic
Purity Method   908
    J. E. Hunter, III and R. L. Blaine

Preparation and Thermal Properties of Some L-Proline
Complexes of Nickel, Copper and Zinc   917
    S. Z. Haider, K. M. A. Malik, M. M. Rahman, and
    T. Wadsten

Section IV   Polymer Science

Plenary Lecture

Recent Developments in the Area of Thermotropic Liquid
Crystalline Polymers and Their Thermal Analysis   927
    M. Takeda

The ATHAS Data Bank of Heat Capacity of Linear Macro-
molecules   950
    B. Wunderlich, S.-F. Lau, and U. Gaur

Precision Automated Differential Scanning Calorimetry
of Polymers   954
    R. W. Connelly and J. M. O'Reilly

On Stretching Calorimetry and Thermoelasticity of Rubbers .   955
    G. W. H. Höhne, H. G. Kilian, and P. Trögele

Polymer Orientation Measured from Linear Expansion by
Thermal Analysis.   964
    L. Wang, C. L. Choy, and R. S. Porter

Structural Relaxation and Recovery of Glassy Polymers:
Theory.   966
    T. S. Chow and W. M. Prest, Jr.

Aging Processes in Partially Compatible Polymer Blends. . .   973
    W. M. Prest, Jr and F. J. Roberts, Jr.

Analysis of Polymer Blends and Copolymers by Coupled
Thermogravimetry and Automatic Titration.   979
    S. G. Fischer and J. Chiu

Creep Compliance Measurements by Thermomechanical
Analysis - Parallel Plate Rheometry (TMA-PPR)   984
    R. B. Prime

Recent Developments in Characterization of Polymers by
Dynamic Mechanical Analysis   994
    P. S. Gill and R. L. Blaine

Phenomenological Behavior of Polymers in Stress-
field Diffusion 999
    B. B. Chowdhury

The Role of Zinc Stearate Crystallization in the Reinforcement of Sulfonated EPDM. 1010
    R. A. Weiss

Multiple Melting in Nylon 1010. 1016
    S. Fu and T. Chen

A Study of Isothermal Crystallization and Lamellar
Thickening in Poly (Ethylene-Terephthalate) Using
Differential Scanning Calorimetry 1024
    Y. S. Yadav, P. C. Jain, and V. S. Nanda

Crystallinity as a Selection Criterion for Engineering
Properties of High Density Polyethylene 1029
    D. M. Hoffman

The Effect of Carrier Gas on Rates of Crystallization
of Isotactic Polypropylene Obtained by Differential
Scanning Calorimetry. 1030
    J. M. Willis, G. R. Brown, and L. E. St-Pierre

Analysis of Microstructure and Domain Formation in Sulfonated Ethylene-Propylene-Ethylidene Norbornene Ionomers . 1040
    J. J. Maurer

Rapid Quantitative Characterization of Elastomers by
Dynamic Load Thermomechanical Analysis. 1050
    R. Riesen and W. Bartels

Thermal Behaviors of Long-Chain Monomers and Comblike
Polymers 1056
    Yoshio Shibasaki and Kiyoshige Fukuda

Thermal Characterization of an Ethylene-Vinyl Alcohol
Copolymer 1057
    W. R. R. Park

Polyurethane/Polymethylmethacrylate Simultaneous Interpenetrating Networks (SINs) - Synthesis and Thermal)
Behavior. 1064
    S. K. Srivastava, A. B. Mathur, and G. N. Mathur

Synthesis and Characterization of Polyimides by
Thermo Analytical Techniques. 1071

    A. B. Mathur, S. K. Srivastava, P. K. Singh and G. N. Mathur

The Investigation of the Compatibility and Stability of
Poly (Chloro-co-fluorostyrene) and PPO Blends by DSC and
TG Analysis 1078

    R. Vuković, V. Kurešević, F. E. Karasz, and
    W. J. MacKnight

The Glass Transition of Mesophase Macromolecules. 1084

    B. Wunderlich and J. Grebowicz

Influence of Inter-/Intramolecular Energy Ratio and Chain
Segment Density on the Gibbs-Dimarzio Theory of the
Glass Transition. 1092

    A. R. Greenberg and R. P. Kusy

The Influence of Crosslink Density on Thermal Properties
of Some Epoxy Resins Based on 4,4'-Diaminophenylmethane . . 1099

    D. Argyropoylos, K. A. Hodd, and W. W. Wright

Glass-Transitions and Melt-Flow of Homo- and Copolymers
of Electro-Donor and Electro-Acceptor Monomer Pairs 1106

    H.-A. Scheider

Kinetic Studies on the Oxidation Characteristics of
Carbons Prepared by the Carbonisation of Various Polymers . 1111

    D. Dollimore and P. Campion

The Thermal Study of Cross-Linked Polyelectrolytes
(Ion Exchange Membranes). 1118

    R. Dabek and R. Morales

DSC Study of Unsaturated Polyester Resin Stability. 1125

    S. P. Molnar

Thermal Degradation of Phenol-Formaldehyde Polycondensate . 1137

    G. Camino, M. P. Luda di Cortemiglia, L. Costa, and
    L. Trossarelli

Thermal Degradation of Urea-Formaldehyde Polycondensate . . 1144

    G. Camino, L. Operti, L. Costa, and L. Trossarelli

Curing Kinetics of Poly (Ethylene Adipate) and Toluene
Diisocyanates 1150

    R. G. Ferrillo, A. Granzow, and V. D. Arendt

Quantitative DTA of Epoxy Adhesives and Prepregs.   1151
    A. Schiraldi and P. Rossi

Curing of a Polyimide Resin   1155
    G. A. Pasteur, H. E. Bair, and F. Vratny

Section V   Applied Science and Industrial Applications

Plenary Lecture

Coal, Oil Shale and Thermal Analysis.   1161
    S. St. J. Warne

Determination of Reaction Rate and Kinetics of the Nickel
Catalysed Methanation of CO, Measured by DSC.   1175
    G. Hakvoort and L. L. van Reijen

Thermoanalytical Interpretation of the Reaction Mechanism
of Propene Oxidation on Thallium (III) Oxides   1182
    C. Mazzocchia, P. Cardillo, F. Di Renzo, R. Del Rosso,
    and P. Centola

Calorimetric Study of the Interaction of Hydrogen with
CoMo Hydrodisulphurisation Catalysts.   1189
    M. B. Poleski, A. Auroux, and P. C. Gravelle

Effect of Acidic and Basic Catalysts on Pyrolytic Decomposition of Cellulose Studied by Computer Curvefitting of
Thermogravimetric Data.   1190
    V. G. Randall, M. S. Masri, and A. E. Pavlath

A Thermogravimetric Study of the Catalytic Decomposition
of Non-Edible Vegetable Oils.   1197
    K. N. Ninan, K. Krishnan, and K. V. C. Rao

A TG-DTA Study of the Adsorption of Small Hydrocarbon
Molecules by Various Modified ZSM-5 Zeolites.   1203
    Z. Gabelica, J.-P. Gilson, G. Debras and E. G. Derouane

The Effect of Different Types of Distributions on Thermodesorption Kinetics   1209
    V. Dondur, D. Fidler, and D. Vučelić

Thermal Programmed Reduction (TPR): A Novel Thermal
Analysis Technique for the Investigation of Catalysts . . .   1217

    A. Bossi, A. Cattalani, and N. Pernicone

Characterization of Supported Metal Catalysts by Thermal
Programmed Reduction.   1224

    R. Bertè, A. Bossi, F. Garbassi, and G. Petrini

Strong Metal-Support Interaction on Nickel Catalysts. . . .   1230

    L. Y. Chen and C. Y. Lin

Catalyst Change Investigation of the System $ZnO-CuO-Cr_2O_3$
During Its Regeneration in the Industrial Reactor   1237

    G. Rasulić, L. Milanović, and S. Jovanović

Calorimetric Study of Water Vapour Interaction with
Bismuth Molybdate (2:1) Reduced and Reoxidized.   1244

    L. Stradella and G. Venturello

The Behavior of Iron Oxides in Reducing Atmospheres   1249

    S. Soled, M. Richard, R. Fiato, and B. DeRites

The Evaluation of Fossil Fuels by the DTA and PY-GC
Methods   1254

    V. Dobal, P. Šebesta, and V. Káš

A Thermogravimetric Method for the Rapid Proximate and
Calorific Analysis of Coals and Coal Products   1260

    C. M. Earnest and R. L. Fyans

Thermo-Magneto-Gravimetric Analysis of Pyrite in Coal
and Lignite   1270

    D. Aylmer and M. W. Rowe

Thermal Analysis of Mixtures of Polyvinylchloride with
Various Coke Chemical and Petrol Products   1276

    D. Rustchev, T. Gantcheva, and O. Atanassov

Investigation of Calcium Carbonate-Sulfur Trioxide
Reaction by Thermogravimetry.   1280

    R. K. Chan and K. Ser

Temperature-Adaptable Hollow Fibers Containing Inorganic
and Organic Phase Change Materials.   1286

    T. L. Vigo and C. E. Frost

Investigation of the Role of Chemical Admixtures in
Cements - A Differential Thermal Approach   1296
    V. S. Ramachandran

The Application of DTA and TGA to Trouble-shooting of the
Portland Cement Burning Process   1303
    H. Chen

Mechanism of Hydration and the Role of an Admixture in
Cement - A Thermal Analytical Approach.   1310
    A. A. Rahman

Some Aspects of the Application of TMA in the Selection
of Binding and Construction Materials   1318
    H. G. Wiedemann and M. Roessler

Flame Retardancy Effects on the Thermal Degradation of
Poly (Ethylene Terephthalate) Fabrics as Studied by
Thermogravimetric Analysis.   1325
    J. D. Cooney, M. Day, and D. M. Wiles

Thermal Characterization and Stability Studies of a
Polyester-based Polyurethane Laminate   1332
    J. T. Stapler and F. H. Bisset

A Thermal Analytical Technique for Evaluating Bromine
Containing Flame Retardants   1342
    M. Day and D. M. Wiles

Thermal Analysis of Brominated Fire Retardants Using
Pyrolysis-Mass Spectrometry   1349
    R. M. Lum, R. P. Jones, and X. Quan

TG/MS of Some Phosphate Systems   1350
    H. G. Langer, J. D. Fellmann, and C. D. Wood

Thermal Investigation of Electric Insulating Materials
Polyolefin types.   1356
    G. Liptay, M. Laczkó, L. Ligethy, and E. Petrik-Brandt

Thermal-Oxidative Degration of Poly(Phenyl Sulfide)   1362
    C. L. Markert, H. E. Bair, and P. G. Kelleher

Oxidative Stability Applications for the Du Pont 1090
Thermal Analysis Program.   1368
    T. A. Blazer and R. L. Blaine

Thermogravimetric Profile of Photocured Acrylate Systems. . 1373
    D. LaPerriere, J. A. Ors and F. R. Wight

Use of Thermal Analysis to Study the Degree of Cure of
Epoxy Resins. 1380
    T. R. Manley and G. Scurr

The Study of the Cure of Epoxy Compounds by Thermoelectrometry. 1386
    L. Fu

Application of Thermal X-Ray Diffraction in Electronic
Materials Research. 1392
    D. D. L. Chung

Industrial Applications of Curie Temperature Determination of Alloys. 1402
    B. O. Haglund

Effects of Powder Processing Upon the Thermal Behavior
of the TiC-Ni-Mo$_2$C System 1409
    M. A. Tindyala and R. A. McCauley

Thermal Analysis in Manganese Metallurgy and the Prospects
of Its Development in Combination With Other Investigation
Methods 1419
    D. V. Mosia, L. K. Svanidze, T. N. Zagu, and T. I. Sigua

Thermochemistry of Mixed Explosives 1426
    J. L. Janney and R. N. Rogers

Thermochemical Evaluation of Zero-order Processes
Involving Explosives. 1434
    R. N. Rogers and J. L. Janney

An Investigation of the Influence of Organic Binders on
the Reaction of Pyrotechnic Systems Using Thermal Analysis
and Related Techniques. 1440
    E. L. Charsley, J. A. Rumsey, T. J. Barton, and T. Griffiths

Integrated Testing For the Evaluation of Thermal Hazards. . 1447
    T. F. Hoppe and E. D. Weir

Thermal Hazard Evaluation of Styrene Polymerization by
Accelerating Rate Calorimetry 1456
    L. F. Whiting and J. C. Tou

Geological Dating by Thermal Analysis 1463
    G. Szöőr

Thermoanalytical Measurements in Archaeometry 1470
    G. Bayer and H. G. Wiedemann

Thermoluminescence of Quartz, Its Use in Archaeology. . . . 1477
    D. B. Nuzzio

Investigation of the Thermal Degradation of Paints
Used in the Body of Automobiles 1483
    D. Marjit

Thermogravimetric Signatures of Complex Solid Phase
Pyrolysis Mechanisms and Kinetics 1490
    M. J. Antal, Jr.

Characterization of Technical Products by Automated
Thermal Analysis. 1497
    H. G. Wiedemann and G. Widmann

Crystal Transformation IV-III Kinetics of Ammonium
Nitrate in the Lime Ammonium Nitrate. 1504
    G. Rasulic

DTA Studies on Chalcopyrite-Copper Sulfate Conversion . . . 1510
    M. Aneesuddin, P. N. Char, and E. R. Saxena

Thermal Behavior of Long-Chain Monomers and Comblike
Polymers. 1517
    Y. Shibasaki and K. Fukuda

A New EGA for the Kinetical Study of a Complex Thermolysis
(Kerogens). 1524
    G. Therand, F. Rouquerol and J Rouquerol

APPENDIX : Report of the Committee on Standardization . . 1530

AUTHOR INDEX xxviii

CONTENTS OF VOLUME 1

CONTENTS OF VOLUME 2 xvi

FOREWORD xxvi

PREFACE xxvii

Award Papers

Du Pont - ICTA

Some Recent Applications of Thermal Analysis to the Study of Electronic Materials and Processes 1
    P. K. Gallagher

Netzsch-GEFTA

Down-to-Earth Thermal Analysis. 25
    R. C. MacKenzie

Mettler-NATAS

Enhanced Capability For Materials Characterization By Combined Thermogravimetry-Evolved Gas Analysis. 37
    Jen Chiu

Section I   Theory and Instrumentation

Plenary Lecture

The Interpretation of Solid State Kinetics. 38
    A. K. Galwey

Calculation of the Kinetic Compensation Effect. 54
    J. R. MacCallum

Kinetics of Solid-State Reactions from Isothermal Rate-Time Curves 58
    M. E. Brown and A. K. Galwey

Computation of Rate Parameters from Non-Isothermal Diffusion Experiments 65
    W. K. Rudloff

Kinetical Study of a Complex Thermolysis by Controlling
the Rate of Production of a Chosen Gaseous Species.   70
(Complete text appears on p. 1524, vol. 2.)

    G. Thevand, F. Rouquerol, and J. Rouquerol

Interpretation of Complex Non-Isothermal Rate Signals
by the Preliminary Assumption of an Elementary Process. . .   71

    E. Koch

Effects of Operational Factors on Kinetic Parameters
Determined with DSC   80

    A. A. Van Dooren

Kinetic Data from DDTA Curves   85

    A. Marotta, S. Saiello and A. Buri

Kinetic Data of the Pyrolysis of Cellulose Obtained
Using Differential Scanning Calorimetry and Isothermal
Thermogravimetry.   90

    K. S. Gregorski and A. E. Pavlath

The Degradation of Polyurethanes by Thermogravimetry. . . .   97

    J. H. Flynn

Calorimetric Evaluation of Cyclization Kinetics of
Experimental and Commercial Polyacrylonitrile Polymers
Via a Modified Statistical Arrhenius Treatment.   98

    S. P. Sponseller, P. M. D. Benoit and D. W. Behnken

Determination of the Mechanism of Thermal Decomposition
of $MnCO_3$, $CdCO_3$, and $PbCO_3$ by Using Both the TG and
the Cyclic and Constant Decomposition Rate of Thermal
Analysis.   99

    · J. M. Criado

Kinetics and Topochemistry of the Endothermic Solid-State
Reaction of 2-(5-Cyanotetrazolato) Pentammine Cobalt (III)
Perchlorate 106

    J. M. Pickard, P. S. Back, and R. R. Walters

Kinetic Studies on the Oxidation of V_2S_3. 113

    M. Taniguchi and S. Ohara

Thermal Analysis of Oxidation of Nickel-Chromium Alloys . . 120

    G. Baran and A. R. McGhie

Kinetic Predictions of Glass-Formation in GeSe$_2$-Sb$_2$Te$_3$-GeTe Molten Alloys 127

    S. Suriñach, M. D. Baró, M. T. Clavaguera-Mora, and N. Clavaguera

Free Volume Driven Crystallization in Metallic Glasses . . . 134

    J. M. Barandiarán, J. Colmenero, I. Telleria and A. Rivacoba

Further Study on Possible Chemical Effects of a Strong Magnetic Field 141

    D. Aylmer and M. W. Rowe

A Kinetical Model Based on Simultaneous Isothermal Calorimetry-Conductimetry and Microscopical Observations for CaSO$_4 \cdot \frac{1}{2}$ H$_2$O Hydration Process 148

    E. Karmazsin, C. Comel, and M. Murat

Enthalpy Calibration of a Differential Heat Flux Calorimeter 156

    W. Hemminger and K.-H. Schonborn

A New Radiation-Heated Differential Calorimeter 163

    W.-D. Emmerich, E. Kaisersberger, and H. Pfaffenberger

The Calorimetric Calibration of DSC-Cells 169

    W. Eysel and K.-H. Breuer

Microcomputer-Assisted Heat Exchange Calorimetry 176

    S. Fujieda and M. Nakanishi

A New High Sensitivity Calorimeter for Investigating Biological Solutions 183

    F. Pithon and P. LeParlouër

Recent Progress in Mixing Calorimetry 190

    P. LeParlouër

Dynamic Aspects of Isothermal Calorimetry 196

    S. L. Sparks

Simple and Inexpensive Apparatus to Study Anomalies in the Specific Heat 205

    C. A. Martin

Influence of the Special Experimental Conditions Established by Quasi-Isothermal Quasi-Isobaric Thermogravimetric Technique on the Kinetics and Mechanism of Thermal Decomposition Reactions. 214
    F. Paulik, J. Paulik, and M. Arnold

Role of Sample Size and Gaseous Atmosphere in the Interpretation of Thermogravimetric Curves 220
    M. Shyamala, S. R. Dharwadkar and M. S. Chandrasekharaiah

A Study of the Effect of Simultaneous Variation of Sample Mass and Heating Rate in Thermogravimetry 226
    P. M. Madhusudanan, K. Krishnan, and K. N. Ninan

The Effect of the Starting Temperature of the Furnace on the Basic Elements of the DTA Curve. 233
    Ž. D. Živković and D. Blečić

Interpretation of DTA Curves on the Base of Heat Transfer Analysis 240
    E. Campero, Z. Kolenda, L. Martínez-Báez, and J. Norwisz

Mettler TA3000: A New Way to Automated Thermal Analysis. . 246
    K. Vogel

New Data Analysis Applications of the Du Pont 1090 Thermal Analysis System 254
    P. F. Levy and R. L. Blaine

Advantages and Applications of Computer Graphics in Thermal Analysis. 255
    W. P. Brennan, M. P. DiVito, R. L. Fyans, and D. W. Breakey

Calorimetric Thermobalance and Its Application. 258
    K. Saito, M. Kosaka, A. Kishi, M. Ichihashi, and A. Maesono

Thermomagnetic Analysis as a Means for Following Corrosion Reactions in Sealed Systems 264
    R. G. Charles

Coupling a PE-TGS-2 to the SCIEX TAGA 3000 for Evolved Gas Analysis. 272
    S. M. Dyszel

Theoretical and Instrumental Studies for the Coupling
of TA Systems and Quadrupole Mass Spectrometers   279

    E. Kaisersberger and W.-D. Emmerich

Interaction of an Evolved Gas Detector Integrated with
a Computerized 2400°C TGA-DTA System.   284

    D. H. Filsinger and D. B. Bourrie

Determination of Iron (III) by Automatic Continuous
Flow Enthalpiometry After Separation of the Iron on an
Ion Exchange Resin Column   291

    K. A. Kwakye

Thermosonimetry of Glass.   300

    G. M. Clark

Investigation of Some European and South American
Bauxites by Thermosonimetry (TS).   306

    J. L. Holm and K. Lønvik

A Computerised Technique for Producing W.L.F. Shifted
Data from a Single Frequency Temperature Scan Without
Graphical Manipulation.   313

    D. J. Townend

Thermovoltaic Detection (TVD)   320

    W. W. Wendlandt

Preliminary Observations on the Combined Use of Photo-
acoustic Spectroscopy and DTA in Mineralogical Analysis . .   325

    N. T. Livesey and M. S. Cresser

Thermal Stress/Strain Analyzer and Its Applications   331

    T. Okino, M. Maruta, and K. Ito

Advantages of Optoelectronics in Advanced Thermodilato-
metry   337

    E. Karmazsin, P. Satre, and M. Romand

The Use of Quasi-Isothermal Dilatometry in Evaluation of
the Initial Stage of Sintering UO_2 Powder Compacts.   344

    M. El Sayed Ali, O. T. Sørensen, and L. Halldahl

Application of an Oxygen Pump with Stabilized Zirconia
Electrolyte to Thermal Analysis of Oxides   351

    Y. Saito and T. Maruyama

Thermal Analysis With a Galvanic Cell for Solid
Chloride-Electrolytes   358
    H.-J. Seifert, G. Thiel, and J. Warczewski

Evaluation of the Standard Free Energy Change on Isotypic
Decomposition of Alkaline Earth Oxides.   365
    T. Nakamura

Section II   Inorganic Chemistry, Metallurgy, Earth Science, Ceramics

Plenary Lecture

Emanation Thermal Analysis and Its Application in Inorganic
Chemistry, Ceramics, Metallurgy, and Earth Science.   371
    V. Balek

Calorimetric Studies of Dissolution Kinetics in Aluminum
Alloy 2219.   385
    J. M. Papazian

Thermal Analysis of Precipitation Process of Carbide
from Martensitic Fe-C Alloy   392
    S. B. Won, T. Ohshima, and K. Hirano

Thermal Analysis of Precipitation Process from Super-
saturated α Solid Solution in FE-0.023 mass % C Alloy . .   399
    S. B. Won, T. Ohshima, and K. Hirano

The Effects of Melt-Quenching Rate on the Glass Transition
Temperatures of Metallic Glasses.   406
    Q. Tan, X. Liu, F. Zheng, and Y. Gao

The Degree of Devitrification of Metallic Glasses and
Its Application   413
    Q. Tan, Y. Gao, Z. Lu, and X. Liu

Thermal Analysis of Graphite Intercalated with Bromine. . .   418
    D. D. L. Chung

Thermal Behavior of Intercalation Products of Butylamine
in Anhydrous Crystalline Zirconium Phosphate.   429
    A. La Ginestra, C. Ferragina, R. Di Rocco, and P. Patrono

Thermal Decomposition of $Ni(Pyridine)_4 (NCS)_2$ 436
    P. D. Garn and A. Alamolhoda

The Complementary Use of Thermal Analysis, Electron Microscopy and Product Yield-Time Data to Investigate Possible Melting During the Decomposition of Copper (II) Malonate 443

    N. J. Carr, A. K. Galwey, and W. J. Swindall

Thermal Decomposition of Aluminum Salts of Organic Acids . . 450

    T. Sato, S. Ikoma, and F. Ozawa

The Thermal Decomposition of Complexes of Planar Dithio Oxamides 457

    H. Hofmans, H. O. Desseyn, A. J. Aartsen, and M. A. Herman

Thermal Decomposition of Rare Earth Salicylates 464

    M. A. Nabar and S. D. Barve

A Study of Iron Tannates with Thermogravimetric Analysis . . 470

    J. Argo

Mossbauer Study of Thermal Decomposition of Alkali Bis (Citrato) Ferrates (III) 476

    A. S. Brar and B. S. Randhawa

Thermal Studies on Palladium (II) and Platinum (II) Dithiocarbamates 482

    H. S. Sangari, N. K. Kaushik and R. P. Singh

Thermal Studies of Some Metal Thiocarbonates and Sulphides 488

    K. Singh and M. A. Bappa

Synthesis, Characterization and Thermogravimetric Studies on Thallium (I) Dithiocarbamates 492

    K. C. Jain and N. K. Kaushik

Some Studies on Thallium Oxalates IX: Thermal Decomposition of Thallium (I) Trihydro Dioxalato Dihydrate 499

    S. R. Sagi, M. S. Prasada Rao, and K. V. Ramana

Some Studies on Thallium Oxalates X: Thermal Decomposition of Thallium (III) Bisoxalato Diaquothallate (III) 506

    S. R. Sagi, K. V. Ramana, and M. S. Prasada Rao

DTA as a Significant Means for the Characterization of Ceramic Clays 513

    H. Kromer and K. H. Schüller

Improved DTA Determination of Clay Minerals: Use of
Exothermic Effects and a-Values and Their Dependence on
Crystal Chemistry 518
    W. Smykatz-Kloss

Dilatometric Study of the Sintering Behavior of Kaolins
and Kaolinitic Clays. 526
    K. H. Schüller and H. Kromer

The Study of the Interaction Between Cesium Chloride and
Kaolinite by Thermal Methods. 533
    S. Yariv, E. Mendelovici, and R. Villalba

The Influence of Microstructure on Thermal Behavior of
Bentonite 541
    E. T. Stepkowska and S. A. Jefferis

Some Remarks of the Thermal Study of Pharmaceutical
and Greek Islands Bentonites. 551
    G. Margomenou-Leonidopoulou, and M. Laskou

Thermal Characteristics of Mordenite Type Zeolite 558
    P. K. Bajpai, M. S. Rao, K. V. G. K. Gokhale

Thermal Properties of Modified Zeolites 565
    G. V. Tsitsishvili, G. O. Piloyan, L. K. Kvantaliani,
    and D. S. Chipashvili

The Hydration of 12-(Calcium Oxide)-7-(Aluminium Oxide) . . . 571
    B. E. I. Abdel Razig, K. M. Parker and J. H. Sharp

Thermal Transformation of Gelatinous Aluminium Hydroxides
to Alumina. 578
    T. Sato, S. Ikoma, and F. Ozawa

Thermal Analysis of Naβ and Naβ'' Alumina. 585
    B. Mani, A. R. McGhie, and G. C. Farrington

Thermal Stability Studies on Hydronium β'' Alumina 593
    K. G. Frase, A. R. McGhie, and G. C. Farrington

Thermal Study of Synthesis of Topaz 600
    A. M. Abdel Rehim

Measurement of the Degree of Weathering of Granite Rocks
by Means of DTA. 608
    W. Smykatz-Kloss and J. Goebelbecker

Thermal Analysis of Phosphated Chrysotile Fibers. 614
    F. M. Kimmerle, J. Khorami, and D. Choquette

Determination of Metal Carbonate, Metal Sulphate, Pyrite,
and Organic Substance Contaminants in Mineral Substances
by Simultaneous TG, DTG, DTA, and EGA 621
    J. Paulik, F. Paulik, and M. Arnold

Carbonate Minerals of Ca-Mg-Fe-Mn Series and DTA Characteristic of Siderite and Sideroplesite of Different
Origins 629
    G. Chen

A Comparative Thermal Analysis Study of the Decomposition
of Copper Carbonate 636
    D. Dollimore and T. J. Taylor

Thermal Investigations of Members of the Strontianite-
Calciostrontianite Series 642
    D. J. Morgan and A. E. Milodowski

Thermal Analysis of Borate Minerals 650
    J. B. Farmer, A. J. D. Gilbert, and P. J. Haines

Thermal Analysis of Hectorite 657
    C. M. Earnest

Thermoanalysis of Glass-Fibers' Crystallization 660
    S. Meriani, G. Fagherazzi, B. Locardi, and G. Soraru

Thermal Analysis of the Heterogeneous System Cobalt Oxide-
Oxygen as Function of Defect Structure. 667
    W. K. Rudloff and E. S. Freeman

Isothermal Decomposition of Zinc Azide. 680
    R. K. Sood, A. E. Nya and S. R. Yoganarasimhan

Non-Isothermal Kinetic Studies on the Effects of Semiconductive Oxides on the Thermal Decomposition of Barium
Perchlorate Trihydrate. 687
    F. Jasim, M. M. Barbooti, and K. I. Hussain

Thermal Decomposition of Irradiated Lead Bromate Single
Crystal 693
    A. M. Gavande and M. N. Ray

An Investigation of the Phase Transition PCC to FCC in
CsCl by DSC 699
    J. L. Holm and H. Raeder

Phase Transition in Fe_2MoO_4 706
    J. Ghose

High Temperature X-Ray, Thermal and Infrared Spectrum
Analyses of Johannite and Its Deuteroanalogue 713
    J. Cejka, Z. Mrazek, Z. Urbanec, and S. Vasickova

The Solid Phase Reaction of Ammonium Nitrate With the
Oxides of Ni, Cu, and Zn Studied by Energy Dispersive
X-Ray Diffraction 719
    N. Eisenreich and W. Engel

Thermal Decomposition of Zinc Perchlorate Hexahydrate
in Presence of Chromium (III) Oxide 724
    T. P. George and M. R. Udupa

Phase Transitions in $NaNO_2$-$NaNO_3$ Mixed Crystals 731
    R. N. Phatak and V. S. Darshane

Thermal Dehydration Processes of $BaCl_2 \cdot 2H_2O$ and its
Deuterium Analog. 737
    H. Tanaka, S. Shimada and H. Negita

On the Dehydration of Ferrous Chloride Tetrahydrate
Studied By Thermogravimetric Analysis 744
    J. Argo

Thermophysical Properties of Solvates: DMF Solvates of
$FeCl_3$ 751
    U. R. K. Rao, K. S. Venkateswarlu, and B. R. Wani

DTA and DSC Studies on the III-I Transition in Na_2SO_4 . . . 758
    Y. Saito and K. Kobayashi

Phase Equilibrium Diagram Compilation by Reduction
Thermogravimetry. 765
    H. N. Tran, C. So, D. J. Miller, and D. Barham

Thermal Analysis of Basic Ferric Sulfate and Its Formation
During Oxidation of Iron Pyrite 769
    A. C. Banerjee and S. Sood

Thermogravimetric Studies of Morpholinium and Iron
(III), Iron (II), Cobalt (II), and Nickel (II) Double
Sulphates   775
    M. R. Udupa

Thermogravimetric Analysis of Mixed Crystals of $NiSO_4 \cdot 6H_2O$
and $CuSO_4 \cdot 5H_2O$.   781
    D. A. Deshpande, G. T. Deshpande, A. Pal, N. D. Deshpande,
    and V. G. Kher

Isothermal Thermogravimetric Study of Copper Sulfate
Pentahydrate.   785
    D. A. Deshpande, N. D. Deshpande, and V. G. Kher

APPENDIX: Report of the Committee on Standardization. .   792

AUTHOR INDEX   xxviii

FOREWORD

These Proceedings are the product of the work of scientists from all over the world and represent the state of the art in many and varied fields of application of thermal analysis. As one of the few who has participated in all International Conferences on Thermal Analysis, I am particularly pleased to see how much we have progressed from the original workbook of the 1st International Conference on Thermal Analysis of 1965 at Aberdeen which, in fact, represented the first collection of abstracts dealing with thermal analysis only. At Aberdeen, and then at the 2nd ICTA in Worcester, Massachusetts, in 1968, it was realized that a large living space did exist in the scientific world for our interests. From then on, with the birth of the Journal of Thermal Analysis, of Thermochimica Acta and of Thermal Analysis Abstracts, our specific literature became stronger and stronger.

However, the Proceedings of the International Conferences are still a unique chance to have an updated, comprehensive image of what is going on in thermal analysis, as seen from contributions of wide international character. Among the authors, side by side, are old prophets and new disciples, researchers from industry and academia, applied and theoretical scientists, all representing a true cross-section of the various aspects of thermal analysis in the world.

Parallel to this growth of thermal analysis as an interdisciplinary, yet well-defined, field of scientific activity, ICTA has moved forward, adding to its institutional work on nomenclature, standards, publications, awards, the coordination of the international activity of the many National and Regional Groups. Most of these were initially activated by ICTA members and now play a leading role in communication between scientists interested in thermal analysis as well as, in many countries, calorimetry. From many of the meetings of these Groups other proceedings in the local languages contribute to the overall picture of the state of thermal analysis and calorimetry.

Thermal analysis is indeed growing faster and faster and these Proceedings are undoubtedly a sign of its strength.

G. Lombardi
President, ICTA

Instituto di Mineralogia e Petrografia
Universita degli Studi di Roma
Rome, Italy

PREFACE

Waiting for the convergence at a small town in New Jersey of more than 200 manuscripts from five continents can be a memorable experience. These volumes are proof that it really happened, mainly as a result of the cooperation and diligence of the entire group of authors represented in these pages. Credit should also be given to the many postal systems which functioned quite well (with a few bizarre exceptions) throughout. In a few cases, where for reasons either legal or temporal, authors were not able to supply a complete manuscript, we have included their abstract. In the course of doing what an editor of such proceedings must do, the impressive breadth and quality of the thermal analysis work represented here made the task more stimulating and less arduous than expected. In addition, the assistance of Professor Bernard Wunderlich and Dr. Harriet Heilweil is gratefully acknowledged.

Bernard Miller

Textile Research Institute
Princeton, New Jersey, U.S.A.
May 3, 1982

Section III Organic Chemistry, Biological and Medical Sciences

Plenary Lecture

MICROCALORIMETRIC STUDIES ON NATIVE HUMAN PLASMA PROTEIN SORPTION

E. Nyilas and T-H. Chiu
Sensorlab Division, Instrumentation Laboratory Inc., Andover MA 01810

With respect to its nature and its recognized inherent limitations, the microcalorimetric investigation of the sorption energetics of native human plasma proteins could be considered a rather specialized and hence, an isolated branch of inquiry by thermal analysis. In terms of its broader perspectives, however, the energetics of native human plasma protein sorption is related to the physico-chemical fundamentals of the first phase of an interaction which unavoidably occurs between blood and surfaces foreign to it. The significance of this phase which encompasses the adsorption of native plasma proteins, is that the sorbed protein layer, depending on its state, can either passivate a foreign surface leading to the absence of major thromboembolic phenomena or, the sorbed protein layer can set the stage for ensuing cellular interactions resulting in detrimental physiologic effects. The therapeutic efficacy of modern life supporting devices that are in contact with blood (artificial kidneys, blood oxygenators or heart/lung machines, blood pumps and implantable circulatory assist devices), depends on, and is currently still limited by, the incidence of thromboembolic phenomena, despite the increasing clinical application of these devices.

On any surface freshly exposed to blood, the initial phase of native plasma protein adsorption is usually controlled by a complexity of factors. In the near vicinity of the contact surface, however, the adsorption energetics, a calorimetrically accessible quantity of a particular protein has been postulated to determine its adsorptivity under the competitive conditions represented by the flux of a large number of native plasma proteins impinging onto the surface. In the course of its attachment to an adsorbent surface, a native plasma protein which is a flexible macromolecule, may or may not undergo adsorption-induced conformational alterations, depending on the particulars of the force field the adsorbing protein encounters in the near vicinity of the surface. The incidence, extent and reversibility of these conformational alterations can also be taken as a function of the energetics and the energetics-dependent mechanisms involved. To obtain some insight into this initial but apparently critical phase of blood/foreign surface interactions, which includes selectivity in protein adsorption as well as the formation and state of sorbed plasma proteins, the adsorptive properties of these molecules were studied, in model systems, in terms of the sorption energetics and mechanisms arising under both competitive and noncompetitive conditions.

All studies on native protein sorption energetics were performed microcalorimetrically at 25° and 37°C, using a custom built isothermal-jacketed instrument capable of resolving ± 1×10^{-6}°C in 100 ml of aqueous medium. The calorimetric method was chosen because the data obtained is independent of the type and nature of different forces which can arise between proteins and adsorbent surfaces. In noncompetitive model systems which contained a single protein solute in a buffer simulating, in part, physiologic conditions (pH = 7.2, I=0.05), the sorption energetics of three of the highest ranking proteins in plasma, viz., human serum albumin (HSA), γ-(7S)-globulin (γ-GLB) and fibrinogen (FGN) was investigated. For model systems involving competitive protein adsorption, human platelet poor plasma (PPP) and serum (SER) were utilized. As model adsorbent surfaces, glass, a known strong procoagulant, and pure and silicone-alloyed low temperature isotropic carbons (LTIC and SLTIC, respectively) which are relevant biomaterials, were used. As a result of the inherent limitations in experimental conditions, all adsorbents were employed in a microparticulate form to provide a sufficiently large specific surface area which was determined by standard methods. Additionally, each of these adsorbents was characterized by relevant parameters, including their heats of immersion into water and the selected buffer, as well as the standard thermodynamic functions of water adsorbed from the vapor on their surfaces.

For the quantitation of calorimetric data obtained at 25° and 37°C corresponding protein adsorption isotherms were determined for each of the model systems studied. To complement the sorption energetics data with information pertaining to the state of sorbed protein layers, the electrophoretic mobility of adsorbent particles having known amounts of protein coverage was also determined.

The overall enthalpy change, $h_I(SLP)_T$, a quantity calorimetrically measured upon the adsorption of a known amount of protein at a fixed T temperature, is essentially a sum total of several energy terms including the endothermic displacement of interfacial water and ions from the adsorbent surface. The energy terms which have been used in this work to characterize the "intensity" of interaction between a particular native plasma protein and a given surface, are the mean integral and differential net heats of sorption per mole of sorbed protein. The molar integral net heat, $\Delta H_T = [h_I(SLP)_T - h_I(SLB)_T]/\sigma$ where $h_I(SLB)_T$ is the measured heat of immersion of the adsorbent into the protein-free buffer, and σ represents the moles of protein sorbed. As a result of this definition, ΔH_T is commensurate to the interaction "intensity" because it includes only energy terms which are associated with (a) the exothermic establishment of protein-to-surface bonds, and (b) the potential endothermic breakage of noncovalent intramolecular protein bonds leading to conformational changes, as well as the potential exothermic formation of additional bonds between surface and the unraveling protein until these molecules reach a steady state. The molar differential net heat, $\Delta H_T^\dagger = [h_I(SLP)_T - h_I(SLB)_T]/\Delta\sigma$.

Described in terms of ΔH_T amd $\Delta H_T^\dagger$, the different proteins studied under noncompetitive conditions were found to display clearly distinguishable adsorptive properties on the same microparticulate adsorbent. At submonolayer coverages on glass, the differential heat values obtained for γ-GLB do not decrease rapidly with increasing protein coverage, implying that the number of protein-to-surface bonds formed per sorbed γ-GLB molecule is fairly constant within a certain range and these protein molecules do not tend to maximize by spreading the area they occupy. Also at submonolayer coverages on glass but in contrast to γ-GLB, the heats obtained for FGN (a plasma protein strongly implicated in thromboembolic phenomena) rapidly decrease indicating a sharp decrease of the number of protein-to-surface bonds per sorbed FGN molecule with increasing surface coverages. This adsorptive property implies that, depending on the contact surface and conditions permitting, FGN molecules tend to maximize their interaction.

Under noncompetitive conditions, large differences were found, as well, between the adsorptive properties which were displayed on different surfaces by the same protein. At comparable surface coverages, i.e., at the completion of the first sorbed monolayer, the ΔH_T of FGN on glass is greater, by about a factor of 10, than on LTIC. In addition, the net heat values obtained for the FGN/LTIC system are relatively constant, implying that, in contrast to glass, this protein does not tend to spread on the carbon surface. These calorimetric results, corroborated by the results of electrophoretic mobility measurements performed on FGN-coated glass and LTIC adsorbent particles, are indicative of relatively less intense interactions occurring on the carbon surface, and are implying the incidence of a relatively smaller degree of adsorption-induced conformational changes in FGN molecules sorbed onto LTIC. These calorimetric and the confirming electrophoretic mobility results are also consistent with the known macroscopic procoagulant properties of glass and the established minimal effects of LTIC on blood.

The energetics of protein sorption and especially, of fibrinogen sorption under competitive conditions were studied, using the same microparticulate adsorbents in PPP, PPP diluted with Ringer's lactate, and SER that was prepared from PPP by the elimination of FGN. An analysis of microcalorimetric results obtained indicates that the net heats of protein sorption from PPP and SER on glass are greater, by about a factor of 4, than on either of the carbon-powder adsorbents. Using ^{125}I-labeled FGN for monitoring the adsorbance of this protein from PPP, it was found that about 90% of the total net heat of protein sorption from this medium onto glass is attributable to the sorption of FGN. In addition to the fact that the magnitude of the net heats of protein sorption from PPP is much smaller on LTIC and SLTIC than on glass, the fraction of these heats that is attributable to FGN sorption is also smaller on the carbon adsorbents and amounts only to about 30% to 40%. Thus, as investigated chiefly by microcalorimetric techniques, the relatively selective binding of FGN by otherwise thrombogenic surfaces is not accidental but has a rational basis in the types of sorption energetics displayed by this protein.

The research was performed with the support of the Devices and Technology Branch, Division of Heart and Vascular Diseases, National Heart, Lung, and Blood Institute, under Contract NIH-N01-HV-3-2917.

PHASE TRANSITIONS IN THE COMPOUNDS p-n-ALKOXYBENZYLIDENE-
p-AMINO BENZOIC ACIDS (nBABA, n=6 to 9)

K. Usha Deniz, E.B. Mirza, U.R.K. Rao, P.S. Parvathanathan, A.S. Paranjpe and A.V. Patankar, Bhabha Atomic
Research Centre, Trombay, Bombay 400 085, India.

INTRODUCTION

Phase transitions exhibited by the homologous series of compounds with the general formula (nBABA),

$$C_nH_{2n+1}O-\bigcirc-CH=N-\bigcirc-COOH$$

with n = 6,7,8 and 9, have been studied using the DSC technique, in the temperature range, $-100°C < T < 280°C$. The temperatures, energetics as well as the peculiar features of the heating and cooling cycles, are presented in this communication.

EXPERIMENTAL DETAILS

A Perkin-Elmer DSC-1B was used in most of the experiments. In the case of 7BABA, a Perkin-Elmer DSC-2C was also used. The scanning rates, β, generally used were 16, 8 and $4°C\ min^{-1}$, and the reported transition temperatures were obtained by extrapolating to zero scanning rates. The samples were purified by repeated recrystallisation from ethanol, before being used.

RESULTS AND DISCUSSION

DSC Scans and Transition Schemes. Fig.1 shows typical DSC scans of the four compounds. The nomenclature for the phases is as follows: C = crystalline, S_C = smectic C, N = nematic and I = isotropic liquid phases. Phases designated P are those whose nature has not been determined. Based on our study, the following schemes of transitions are arrived at for the compounds (Fig.2). The salient features of the transitions for the four compounds are discussed below.

6BABA. The phase transitions in 6BABA have been reported by us earlier [1] and only a few additional features have been observed in the present detailed study. During first heating, three peaks, attributed to $C \rightarrow S_C$, $S_C \rightarrow N$ and $N \rightarrow I$ were observed. The small hump which is seen on the low temperature side of the first

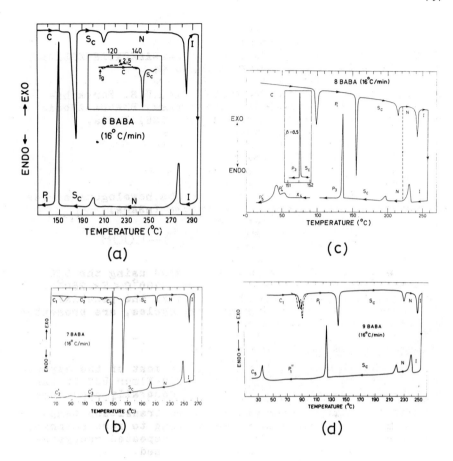

Fig.1. DSC scans of nBABAs. (a) 6BABA: Dashed line curve in the inset shows glassy transition. (b) 7BABA: The dotted line and dashed line curves indicate the $C_1 \rightarrow C_2$ and $C'_1 \rightarrow C_2$ transitions respectively. (c) 8BABA (d) 9BABA: $C_2 \rightarrow P'_1$ and $C_3 \rightarrow P'_1$ transitions are shown by the dashed line curve.

transition is not reproducible and hence is not considered. On cooling from the S_C phase or above, one encounters an intermediate disordered phase P_1, whose disorder freezes on cooling to $-100°C$ (or $0°C$). On reheating, a glassy transition (Fig.1a), which can be clearly seen for $\beta = 8°C\ min^{-1}$, occurs. The glassy transition depends sensitively on the thermal history of the sample. An interesting feature that has been observed is that on cooling the S_C phase, there occurs an extremely narrow tran-

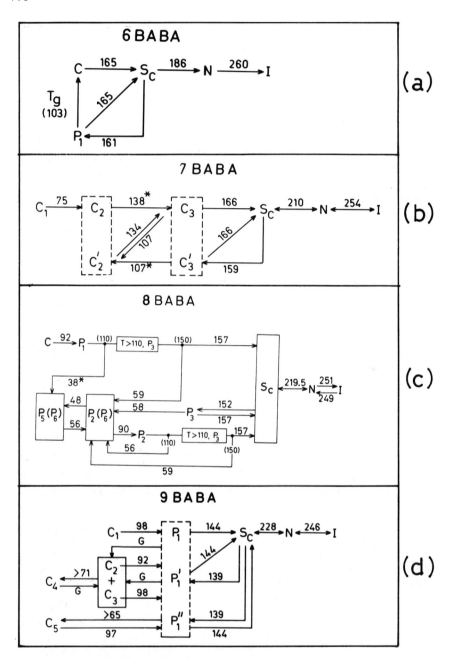

Fig.2 Transition schemes for nBABAs. Temperatures are in °C. (a) G: gradual (b) * : Transition temperature for $\beta = 8°C\ min^{-1}$ (c) T>110 C, P_3 means $P_1(P_2) \xrightarrow[G]{} P_3$ for T>110°C.

sition, $(S_C \rightarrow P_1)$, its sharpness being limited only by the instrumental resolution of $\approx 0.05°C$. Such narrow transitions also occur when the S_C phases of the other three samples are cooled.

7BABA. The first part of the transition scheme is different from that of the other three compounds, because of the two crystal to crystal transitions, $C_1 \rightarrow C_2$ and $C_2 \rightarrow C_3$ (Fig.1b). $T_{C_1 C_2}$ can vary between 65°C and 90°C, the transition can be broad or well-defined, strong or weak or non-existent and reversible or irreversible, depending on the preparative history of the sample. Our scheme is given for a sample having a well defined $C_1 \rightarrow C_2$ transformation not found on subsequent cooling and heating. Both C_2 and C_2' phases supercool to -100°C and they both transform to the C_3 phase on heating. However, $C_2 \rightarrow C_3$ is broader than the $C_2' \rightarrow C_3$ transition (Fig.1b). C_3 and C_3' phases have similar heating transitions but their cooling transitions to the C_2 phase are somewhat different. These strongly resembling phases are boxed in by dashed lines in Fig.2b. The rest of the transition scheme has no noteworthy features except for the sharp $S_C \rightarrow C_3$ transition.

8BABA. The phase transitions in 8BABA have been dealt with in detail elsewhere [2]. The scheme of transitions (Fig.2c) is extremely complex for $T < 100°C$. The notable features in this scheme are: (1) mixed phases $P_2(P_6)$ and $P_5(P_6)$, (2) gradual transformation of P_1 and P_2 phases to P_3, and (3) its extreme sensitivity to thermal history for $T < 100°C$. The $S_C \rightarrow P_3$ transition, observed while cooling the S_C phase, is extremely narrow (Fig.1c) as expected.

9BABA. The earlier paper [1] is incomplete in its description of phase transitions in 9BABA. The transition scheme (Fig.2d) for $T < 100°C$ is very complicated. The interesting features in this scheme are: (1) mixed phase (C_2, C_3), (2) gradual transformation of P_1 and C_4 phases to the mixed (C_2, C_3) phase, evidenced only by the heating transitions, $C_2 \rightarrow P_1'$ and $C_3 \rightarrow P_1'$, (4) the strong dependence of $(C_2, C_3) \rightarrow C_4$ and $P_1'' \rightarrow C_5$ transitions on the thermal history of the sample, specially on the cooling rate (the corresponding transition temperatures are given as $> 71°C$ and $> 65°C$, because extrapolation to zero cooling rates was not possible in these cases), (5) similarity of P_1, P_1', P_1'' phases (boxed in by dashed lines) with identical heating transitions but different cooling transitions.

The starting crystalline phase in all the compounds except 6BABA, once transferred into a high temperature phase, could never be obtained back by any path that could be taken in our experiments. The extremely complicated transition schemes for $T < 100°C$, for 8BABA and 9BABA could be due to their end hydrocarbon chains, which can take up a variety of conformations, the corresponding free energies being very similar. The very sharp transition observed in all four compounds when the S_C phase is cooled, cannot be due to a nucleation process. It must be due to a cooperative phenomenon involving several molecules. Such transitions have also been observed while cooling the N phase of nBABA, n = 1 to 4, in all of which the cooling transition is not to an S_C phase. The peculiar character of the C_1 phase and $C_1 \rightarrow C_2$ transition in 7BABA could be due to traces of impurities, such as the recrystallising solvent.

Heats of Transformation. Table I gives the heats of transformation, ΔH, and the corresponding transition entropies, $\Delta S (= \Delta H/T(K))$, for transitions obtained on first heating up to the S_C phase. The total transition entropy from the starting C or C_1 phase to the S_C phase, increases with increasing n. This behaviour is related to the fact that the number of possible chain conformations increases considerably with increasing n.

The transition temperatures, T_{CN} and T_{NI} and the corresponding transition entropies, ΔS_{CN} and ΔS_{NI} for the $S_C \rightarrow N$ and $N \rightarrow I$ transitions are plotted as a function of n in Fig.3. Both T_{CN} and ΔS_{CN} increase with increasing n. This is accounted for by the fact that interactions between the hydrocarbon chains play an important role in the formation of smectic phases (no smectic phase is seen for nBABA with $n < 6$). ΔS_{CN} varies almost linearly with n and extrapolates to zero for $n \simeq 5.10$. It should be interesting to check whether in such a mixture of 5BABA and 6BABA corresponding to n = 5.1, (a) an S_C phase exists and (b) the $S_C \rightarrow N$ transition would be of second order. T_{NI} decreases with increasing n, as predicted by Marcelja's theory (MT)[3]. However values of ΔS_{NI}, calculated from this theory are three to four times smaller than those observed in these compounds. This is due to the presence of

TABLE I

Compound	Transition	Temp. (°C)	ΔH (KCal/Mole)	ΔS/R_o
6BABA	C ⟶ S_C	165	4.96	5.70
7BABA	C_1 ⟶ C_2	75	1.10	1.60
	C_2 ⟶ C_3	138	0.31	0.37
	C_3 ⟶ S_C	166	3.51	4.03
	C_1 ⟶ S_C			6.00
8BABA	C ⟶ P_1	92	2.47	3.41
	P_1 ⟶ S_C	157	3.32	3.89
	C ⟶ S_C			7.30
9BABA	C_1 ⟶ P_1	98	4.51	6.12
	P_1 ⟶ S_C	144	2.73	3.30
	C_1 ⟶ S_C			9.42

* R_o = gas constant = 1.986 cal mole^{-1} K^{-1}.

S_C-type short range order and 1-d correlation in the nematic phase of nBABAs, not considered in MT. The break-

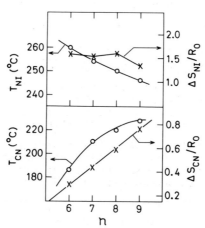

Figure 3 T_{CN} and T_{NI} and reduced transition entropies, $\Delta S_{NI}/R_o$ and $\Delta S_{CN}/R_o$ as a function of n.

down of these structures at T_{NI}, leads to the observed high values for ΔS_{NI}.

REFERENCES

1. K. Usha Deniz, A.S. Paranjpe, E.B. Mirza and P.S. Parvathanathan, J. Physique Colloq. 1979, 40, C3-136.
2. K. Usha Deniz, E.B. Mirza, U.R.K. Rao, P.S. Parvathanathan, A.S. Paranjpe and A.V. Patankar, Thermochim. Acta, 1981, 51, 141.
3. S. Marcelja, J. Chem. Phys. 1974, 60, 3599.

THE APPLICATION OF THERMAL ANALYSIS TO
INVESTIGATION OF THE INTERACTION OF BORIC
ACID WITH SALICYLIC ACID DERIVATIVES
E.M.Schwartz, V.V.Grundstein, I.M.Vitol, USSR

INTRODUCTION

This work deals with the interaction of boric acid with salicylic, 4-amino- and 5-aminosalicylic acids by heating.
Although the range of salts of the di- and monosalicylboric acids, precipitated from aqueous solutions [1-3] is known, the above mentioned acids are not isolated from aqueous solutions. We have established, that in "dry residues", which are formed by evaporation of water solutions of mixtures of boric and salicylic acid at room temperature no new compounds are found. It is known, that by azeotropic destillation of water from the mixture of boric and salicylic acid in toluene, the compounds are isolated, which the authors [1,4] consider to be the disalicylboric acid (m.p. 538-543 K) and its monohydrate. But the structure of these compounds has not been investigated yet, and it is not clear, whether they are acids or esters.
The range of salts of bis(p-aminosalicyl)boric acid [5,6], the free acid [7] and its monohydrate [8] were isolated by precipitation from aqueous ethanol solutions. But the interaction of boric acid with p-aminosalicylic and 5-aminosalicylic acid when heated has not been investigated.

EXPERIMENTAL

The DTA, DTg, Tg-curves of each component and the mixtures of the components in molar ratios from 2:1 to 1:4 were obtained in the range of temperatures 293-773 K. (A derivathograph OD-1). The rate of heating varied from 0,6 to 10 K/min. The reffering substance - Al_2O_3, the atmosphere - the air. The DTA-curves with simultaneous measuring of electric conductivity have also been obtained. The intermediate and end solid products of the interaction were detected by means of crystalloptic, chemical and x-ray analysis and the IR-spectroscopy, the gaseous ones - by means of gas-chromatography (Chromatograph ПАХ-B 07 with the combined column, half of which was filled with zeolite NaX fraction 0,2-0,5 mm and the other half - with polysorb-I).
The fact that salicylic acid, its derivatives and reaction products may be lost by sublimation is taken into account. The loss of weight (Tg) resulting from sublimation, is changed with the change of rate of heating, in contrary to the loss of weight, which is the result of a chemical interaction. So the registration of the Tg and DTg-curves of a cample with various rates of heating enables to detect the effects, connected with sublimation. For example the minimum on the DTg-curve of salicylic acid at 403-423K is connected with

the loss of weight (Tg) - 3% at the rate of heating 10 K/min and 4,7% at 2,5 K/min.

RESULTS AND DISCUSSION

Some data of thermal decomposition of salicylic, 4- and 5-salicylic acids are mentioned in [9-11].
According to ouer data the salicylic acid sublimes at 403-432 K, melts at 432 K with decomposition, the endothermic peak of melting, corresponds to the rise of electric conductivity. The CO_2 is found in the gaseous products of decomposition at 432-493 K. The liquid products of decomposition contain not only phenol, because the boiling point of phenol is 455 K, but endothermic peak on the DTA-curve - at 493 K. All the products of decomposition volatilize by 593 K.

The first endothermic peak on the DTA-curve of p-aminosalicylic acid corresponds to incongruent melting. CO_2 is evolved and M-aminophenol is formed (the loss of weight is 29,5%, calculated according to equation (1)-28,78%).

$$p-NH_2C_6H_3OHCOOH \longrightarrow m\ NH_2C_6H_4OH + CO_2 \quad (1)$$

All the decomposition products volatilize by 613 K.

The endothermic peak (540-543 K) on the DTA-curve of 5-aminosalicylic acid corresponds to incongruent melting. But CO_2 is not detected among the products of decomposition by 583 K.

The new endothermic peaks are found on the DTA-curves of mixtures of boric and salicylic acids (fig.1). As it is seen from table 1, only one substance 1:1(I) is formed at the interaction at 428 K. When heating is continued I is dimerized with splitting off one mole of water. When the mixture contains an excess of boric acid the interaction is acompanied by dehydration of the excess of boric acid into B_2O_3. When the mixture the excess of salicylic acid, the latter is sublimed and decomposed. (The weight loss at 493 K at the heating rate 10 K/min is 22,95%, at 0,6 K/min - 31,5%.).

The compound I is isolated by long heating the equimolar mixture of boric and salicylic acid at 373 K.

Obtained, %:C 51,50; H 3,10; B_2O_3 20,90; calculated according $C_7H_5O_6B(I)$, %:C 51,28; H 3,05; B_2O_3 21,18.
The IR-absorption spectrum of I shows, that the absorption bands are absent in the region 950-1000 cm^{-1}, there is a range of bands at 1350-1500 cm^{-1}, which could be attributed to ν_{B-O} in the BO_3 triangle. There are also bands of $\nu_{C=O}$ - 1650, δ_{B-OH} - 1190 cm^{-1}. δ_{OH} of phenolgroup is absent. ν_{C-O} of carboxylgroup is shifted to 1330 cm^{-1}. All this leads to the conclusion, that I is an ester.

By heating I condensation according 2^2 takes plase. By further heating the compound is decomposed without the evolution of CO_2. The compound II of molar ratio H_3BO_3:HSal=1:2 exists too. It was synthesized by long heating at 373-403 K of the mixture with molar ratio of H_3BO_3:HSal=1:2. Obtained, %:C 56,77; H 3,36; B_2O_3 11,07; calculated according $C_{14}H_{11}O_7B$ (II), C 55,67; H 3,65;

I

TABLE 1. WEIGHT LOSS AT THE INTERACTION OF BORIC ACID WITH SALICYLIC ACID

Molar ratio H_3BO_3:HSal	Weight loss till 428 K, %		Weight loss till 493 K, %	
	obtained	calculated	obtained	calculated
1:0,5	-	-	27,03	27,53 [3]
1:075	-	-	24,56	24,56 [3]
1:1	18:70	18,02 [1]	22,6	22,53 [2]
1:2	10,55	10,66 [1]	The rest of salicylic acid is sublimed and decomposed.	
1:3	7,03	7,57 [1]		
1:4	5,76	5,87 [1]		

[1] and [2] is calculated according the equation 2:

2) $2H_3BO_3 + 2C_6H_4OHCOOH \xrightarrow{428\ K} 2\ C_6H_4OCOOBOH + 4H_2O.-$
   $(18,02\%) \xrightarrow{493\ K} (C_6H_4OCOO)BOB(OOCOC_6H_4) + 5H_2O(22,33\%^2)$

[3] is calculated according 2^2 taking into account the decomposition of boric acid into B_2O_3.

B_2O_3 11,53. Ir-spectrum of II leads to the conclusion, that II also is an ester. The bond of boron with a carboxylgroup and free phenol groups are absent. The probable structure of II is:

$HOOC-C_6H_4-O-B-OC_6H_4-COOH$
                |
               OH   II

When heated over 428 K II loses one mole of water, by further heating CO_2 is evolved.

Table 2 and fig.2 show the weight loss and thermal effects by the interaction of boric acid with p-aminosalicylic acid. Only one compound with molar ratio 1:2 is formed at the interaction. Unfortunately we could not isolate this compound, because by long heating at 373 K it loses 1,5 mole of water more and the compound $C_{28}H_{20}O_{11}N_4B_2$ is formed. The latter is hygroscopic and is turned quiqly to the well known bis(p-aminosalicyl) boric acid, monohydrate [8].

TABLE 2. WEIGHT LOSS AT THE INTERACTION OF BORIC ACID WITH P-AMINOSALICYLIC ACID.

Molar ratio H_3BO_3:PAS	Weight loss by heating till 438 K, %		Weight loss by heating till 473 K, %	
	obtained	calculated	obtained	calculated
1:0,5	–	–	26,77	26,05 [3]
1:1	12,24	12,58 [4]	20,97	20,90 [3]
1:2	9,6	9,79 [1]	17,41	17,13 [2]
1:4	6,03	5,35 [1]	22,32	22,42 [5]

[1)] and [2)] are calculated to the equation 3

3. $H_3BO_3 + (p-NH_2C_6H_3OHCOOH) \xrightarrow{438\ K} C_{14}H_{13}O_7N_2B + 2H_2O\ (9,78\%)$ [1]

$\xrightarrow{473\ K} 1/2\ C_{28}H_{20}O_{11}N_4B_2 + 3,5H_2O\ (17,18\%)$ [2]

[3)] according 3[1] taking into account the dehydration of the excess of boric acid into B_2O_3.
[4)] The same, taking into account the dehydration of the excess of boric acid into HBO_2.
[5)] according 3[2] with account to the evolution of CO_2 from the excess of p-aminosalicylic acid.

TABLE 3. WEIGHT LOSS AT THE INTERACTION OF BORIC ACID WITH 5-AMINOSALICYLIC ACID

Molar ratio H_3BO_3:5-aminosalicylic acid	Weight loss by heating till 473 K	
	obtained	calculated
1:0,5	19,75	19,53 [2]
1:1	12,64	12,57 [2]
1:2	9,20	9,78 [1]
1:4	5,96	5,34 [1]

[1)] Calculated according to (4).

4. $H_3BO_3 + 2(5-NH_2C_6H_3OHCOOH) \xrightarrow{473\ K} C_{14}H_{13}O_7N_2B + 2H_2O\ (9,78\%)$

[2)] Calculated according to 4 with account of the dehydration of the excess of boric acid into B_2O_3.

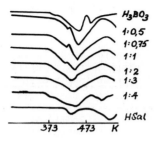

Fig.1. The DTA-curves in the system of boric acid - salicylic acid.

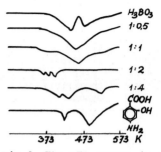

Fig.2. The DTA-curves in the system of boric acid - p-aminosalicylic acid

Only one compound $C_{14}H_{13}O_7N_2B$ is formed at the interaction of boric acid with 5-aminosalicylic acid (see table 3 and fig. 3).

This compound is hygroscopic and forms during the isolation a stable tetrahydrate III. III is a crystalline violate substance. It lose one mole water by heating to 523 K and three moles of water by heating to 573 K. The structure of III is proposed on the basis of IR-absorption and PMR-spetra.

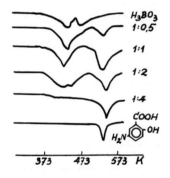

Fig.3. The DTA-curves in the system of boric acid - 5-aminosalicylic acid

SUMMARY

Thermal analysis shows, that the interaction of boric acid with salicylic acid on heating leads to the formation of an ester-type compound 1:1 independently of the molar ratio of the reacting compounds. The other ester 1:2 is obtained only on long heating of the corresponding mixture at 373-403 K.

Only compounds 1:2 are formed on heating mixtures of boric acid and 4- or 5-aminosalicylic acids. In the case of 5-aminosalicylic acid it is an ester. We could not isolate and investigate the nature of the compound with 4-aminosalicylic acid.

REFERENCES

1. H.Steinberg. Organoboron Chemistry. Vol.1, Willey and Son inc., N.-Y, London, Sydney, 1964.
2. V.K.Mardanenko, E.M.Schwartz, N.A.Kostromina. Koordin.Chim. 1980, 6, 1193.
3. V.K.Mardanenko, E.M.Schwartz, N.H.Kostromina. Zurn.neorg.Chim. 1979, 24, 2941.
4. R.S.Mehrotra, R.Srivastava. J. Indian Chem.Soc. 1961, 38, 1.
5. V.K.Mardanenko, E.M.Schwartz. Izv.akad.nauk LSSR, ser.chim., 1978, N3, 259.
6. E.M.Schwartz, V.K.Mardanenko. Zurn.neorg.chim. 1980, 26, 2374.
7. E.M.Schwartz, V.K.Mardanenko. Izv.akad.nauk LSSR,ser.chim. 1977, N3, 276.
8. E.M.Schwartz, V.K.Mardanenko, V.V.Grundstein. Izv.akad.nauk LSSR, ser.chim. 1978, N2, 140.
9. A.Radecki, M.Wesolowski. J.Therm.Anal. 1976, 9, 1367.
10. W.W.Wendlandt, I.A.Noiberg. Anal.Chim.Acta, 1963, 29, 539.
11. F.Wessely, K.Benedikt, H.Genger. Monatsch. Chem. 1948, 79, 185.

THERMAL STUDIES OF PROTEINS, PEPTIDES AND AMINO ACIDS II

J. S. Crighton and P. N. Hole

Schools of Textiles, University of Bradford,
West Yorkshire BD7 1DP, United Kingdom

INTRODUCTION

The influence of heat on proteins, particularly in the form of the commercially important wool (keratin) fibres has been the subject of many investigations [1 - 7]. Any attempt to identify the multiplicity of changes, both physical and chemical, that have been noted can be simplified by the elimination of one of the "additional" experimental variables, namely that of moisture. Thermally induced changes in proteins are significantly affected, in both their extent and their nature, by the prevailing moisture levels [8,9]. Observation of reproducible thermal behaviour from fibrous proteins requires not only the use of a rigorously dry sample but also the removal of those anisotropic characteristics which are inherent in fibrous structures.

Even with the application of these requirements the observed DTA and TG curves reflect the complexity of detailed changes which occur within the temperature range 150° to 450°C [4,10]. The substrate present at any of the prevailing temperatures within this range can still be directly related to the original sample characteristics.

Wool fibre is not a homogeneous structure but is composed of variable sizes of histological regions each with different chemical and physical properties [11]. Variations in the morphological and chemical features of fibrous (keratin) proteins are reflected in the observed thermoanalytical behaviours. With the use of DTA as the reference thermal method, it has proved possible to provide a correlation of fibre structure with the identified thermally induced physical transitions and with some chemical changes [1,12]. To extend further the identification of these induced changes and the characterisation of chemical entities two procedures suggest themselves: (a) examination of the influence, on the observed DTA and TG curves, of chemical modifications to the basic protein structures, (b) a comparative DTA and TG examination of different substrates which possess a range of relevant related and known chemical features. With the use of the former procedure complications arise from the difficulty of achieving any modification which is specific to one chemical entity alone and without disturbing the basic morphology. The aim of this presentation is to report an attempt at further characterisation through the latter alternative procedure.

A range of amino acids, peptides and other proteins which are related in composition to features known to be present in wool were examined. As the emphasis was concerned with the recognition of chemical changes, observations were restricted to the use of thermogravimetry as the primary thermoanalytical method.

EXPERIMENTAL

The range of substrates examined are recorded in Table 1.

TABLE 1. SUBSTRATES EXAMINED BY VACUUM THERMOGRAVIMETRY

PROTEINS	Clupeine Salmine Bovine Insulin
POLYPEPTIDES	poly L-aspartic acid poly L-glutamic acid polyglycine poly L-lysine hydrobromide
PEPTIDES	Glutathione (γ-L-glutamyl-L-cysteinylglycine) Oxytocine Vasopressin Bradykinin Kallidin Gramacin Glycylglycine Triglycine Tetraglycine Hexaglycine Glycyl-L-alanine Glycyl-L-phenylalanine L-leucyl-L-alanine L-leucylglycine L-valyl-L-leucine
AMINO ACIDS	Purified samples of those amino acids identified as present in the acid hydrolysate from keratin proteins

All samples for thermogravimetric study were purified and preprepared to reduce as far as is possible any potential anisotropic character. All were rigorously dried by heating to and by holding at 140°C for 120 minutes under continuous evacuation at a pressure of 0.01 torr. Approximately 5 mg lots of finely spread substrate were so dried on an open 1 cm diameter platinum pan suspended on a fine quartz thread from the arm of a CI Electronics Microforce Mark IIB electrobalance. The balance - sample chamber - furnace assembly has been described previously [13], while the efficiency of this drying procedure with minimal chemical damage has also been demonstrated [1]. After this drying, the samples were cooled to ambient temperatures while still maintaining continuous evacuation. From ambient the sample furnace temperature was programmed to 500°C at a linear rate of 4°C min.$^{-1}$ with the use of a Eurotherm PID controller with thyristor drive. The sample temperature was continuously monitored by a fast response stainless steel sheathed chromel-alumel thermocouple sited immediately above the sample surface. The balance analogue signal, reflecting the sample mass, and the sample thermocouple voltage were monitored on an X-Y recorder. Both sample mass and temperature analogue voltage signals were also amplified and sequentially recorded in digital form through A DART data logger. The logger output was received as a punched tape and as a printed text. This latter text enabled a facile check to be made of the quality of the logging operations.

Concurrent with this vacuum thermogravimetry, evolved gas detection and analysis were attempted. Evacuation of the sample environment was normally achieved through two routes. The major evacuation, and the primary attainment of the operational 0.01 torr pressure, was achieved through the balance head (see Figures 1 & 2 ref.[6]). A second slower speed pumping was achieved via a 5 mm diameter pyrex tube with access internally to close to the sample surface. The external end of this tube was coupled by means of a 0.25 inch diameter stainless steel connection to the basic diffusion and rotary pumps. Sited in this steel connection line was a Hoke "millimite" cross pattern valve fitted with a 10 turn micrometer control. This cross pattern valve was also connected via a Hoke "on-off" valve to the source of a VG Q801 quadropole mass spectrometer. The relative relationships of the components in this 'secondary' pumping line are identified in Figure 1.

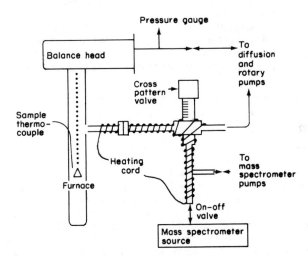

Figure 1. The Vacuum Microbalance-Mass Spectrometer Interface. (a representation of the evacuation and coupling arrangements).

The steel lines are heated to 250°C by Electrothermal heating cord activated by a Eurotherm Power Controller. The cross pattern valve has a maximum operating temperature of 300°C. With the micrometer control this valve can be used as a variable but reproducible splitter of the effluent feed into the mass spectrometer source. At the start of each experiment the needle valve position was set such that the source pressure never exceeded 10^{-5} torr. 1-100 amu mass spectrometer scans were recorded at fixed time intervals across the period of the thermogravimetric experiment; the initiation of each scan was also recorded with the thermogravimetric data.

RESULTS

TG Studies with wool fibre had demonstrated that significant gains in the "visualisation of individual processes within the overall mass loss profile could be achieved by the presentation of the data in the form of a "normalised rate of mass loss as a function of temperature" ($-1/W_o \times dW/dT$) against sample temperature (T_s). The raw TG digital data as received on tape from the logger was processed on the University ICL 1904S Computer. The following sequential operations were involved. The sample temperatures (thermocouple voltages) were linearly interpolated to provide values coincident in time with the recorded masses; both sets of data were then transformed to provide normalised sample mass and temperature (°C) information. Best fit functions were identified to groups of three (mass, temperature) datasets taken sequentially across the experimental temperature range. These functions, and their mathematical differentials, were evaluated at integer sample temperatures over the scanned temperatures. Plots of the resultant differential values against temperature facilitated not only identification of mass change processes but, with the differential data <u>always</u> created at the <u>same</u> (integer) temperatures whatever the sample, averaging procedures with, and comparisons between behaviours are simplified.

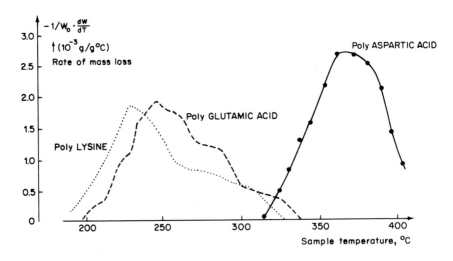

Figure 2. The Vacuum Thermogravimetry of Polyamino Acids: DTG Curves (5 mg, 4°C/min, 0.01 torr)

In Figure 2 is illustrated the transformed DTG curves from the observed data provided by the vacuum thermogravimetry of three polyamino acids. The differences in behaviour are clearly apparent.

A potential source of differences in the TG behaviour of proteins arises from interchainic side chain interactions specifically to identify the occurrence of potential actions resulting from the existence of salt bridges appropriate equimolar pairs of acidic and basic side chain amino acids and peptides were cocrystallised. The products from crystallisation were then also examined by vacuum thermogravimetry and compared with the observations from the separate components. The equimolar mixture of "acid" and "basic" components were more thermally stable than the components individually. There was observed in every case a weight loss at 160º-185ºC which was confirmed to be the result of water loss.

Observations with the examined amino acids and dipeptides indicated that in many cases the mass losses were dominated by sublimation or volatilisation. The effective thermal stability of peptides increased with increasing chain length.

With polyglycine, where two distinct structural forms are possible, markedly different and characteristic TG curves were observed for each, these are shown in Figure 3.

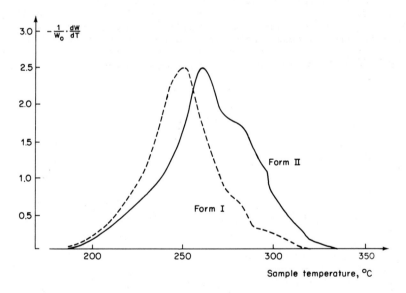

Figure 3. The Vacuum Thermogravimetry of Polyglycine: Effect of Structural Form. (5 mg, 4ºC/min, 0.01 torr)

The influence of side chains in dominating mass loss behaviour at temperatures up to 300ºC was confirmed by the TG observations with polypeptides supported by evolved gas analysis with qualitative mass spectrometric information.

The oligo-peptides and proteins show characteristic mass loss steps the nature of many of which were elucidated by correlation between the observations with the standard definitively modified substrates, further supported by evolved gas analysis.

SUMMARY

Thermogravimetric studies of amino acids, peptides and proteins are described. Programmed heating of rigorously dried substrates under continuous evacuation provided reproducible TG curves. The thermogravimetric data was assessed comparatively in the form of the temperature based rate of mass loss curves. Attempts are made to correlate features of these curves with the presence of specific chemical and physical features. On the basis of the current observations, combined with those previously reported for wool fibres, it is possible to suggest specific reactions induced in fibrous proteins and the temperatures at which these occur.

REFERENCES

1. J. S. Crighton, W. M. Findon and F. Happey, J. Applied Polymer Symposia, 1971, 18, 847.
2. M. Horio, T. Kondo, K. Sekimoto and M. Funatsu, Proceedings of 3rd International Wool Textile Research Conference, 1965, 2, 189.
3. W. C. Felix, M. A. McDowall and H. Eyring, Text. Res. J., 1963, 33, 465.
4. J. S. Crighton and F. Happey, "Symposium on Fibrous Proteins", Butterworths (Australia), (1967), p.409.
5. R. Hagege and J. Connet, Proceedings of 6th International Wool Textile Research Conference (Pretoria), 1980, 2, 221.
6. J. S. Crighton and P. N. Hole, Thermo Chimica Acta, 1978, 24, 327.
7. R. F. Schwenker and J. H. Dusenbury, Text. Res. J., 1960, 30, 800.
8. E. Menefee and G. Yee, Text. Res. J., 1965, 35, 801.
9. J. S. Crighton, F. Happey and J. T. Ball, "Conformation of Biopolymers", Academic Press, London, (1967), 2, 623.
10. J. S. Crighton and W. M. Findon, J. Therm. Analysis, 1977, 11, 305.
11. P. Alexander, R. F. Hudson and C. Earland "Wool: Its Chemistry and Physics", Chapman Hall, London, (1952).
12. J. S. Crighton and W. M. Findon, Proceedings of 2nd ESTA, Heyden, London, (1981).
13. J. S. Crighton and P. N. Hole, Proceedings of 5th International Wool Textile Research Conference (Aachen), 1975, 2, 499.

THERMODYNAMIC PROPERTIES OF THE POLYURO-
NATES-Ca^{2+} IONS INTERACTION IN AQUEOUS
SOLUTIONS

by

A.Cesàro, F.Delben and S.Paoletti
Laboratory of Macromolecular Chemistry
University of Trieste, 34127 Trieste,
Italy

INTRODUCTION

The aqueous solution properties of acidic polysaccharides have greatly attracted both biological and technological interest. Among the very large number of examples, widely varying for composition and structure, polyuronates occurring in the plant kingdom as alginate and pectate (Figure 1) have been extensively studied[1-3] Conformational and solution properties have been investigated also for the understanding of the mechanism of the gel formation/phase separation processes, which usually occur in the presence of "specific" counterions.

The general theme of the interaction between ions and polyelectrolytes, mostly referred to as ion binding, has also been the subject of many investigations. Mathematical framework has been provided by various theories for predicting thermodynamic and other solution properties of an aqueous system containing small ions interacting with a polyelectrolyte with known structural features[4-7]. Ion binding can be operationally defined[8] as:
a) non-localized binding of diffusable ions in the polymeric domain; b) localized, ion-pair, binding close to the polymeric charged groups; c) site binding of ions forming specific complex-like structures. The interaction of type c) can be studied, in principle, in comparison with the monomeric analogs; the interactions of both types a) and b) depend almost exclusively on the very polymeric nature of the substrate, as a consequence of the high electrostatic field generated by the charged macromolecule. The peculiar behavior of the polymeric chains also becomes evident when ion binding promotes a conformational transition. Such a transition, often cooperative but not esclusively, is effected by both the charge density screening (cases a and b) and the detailed geometry of the complex (case c)[9]. The knowledge of the energetics of these processes is of great importance for the understanding of the stability and the function of biological macromolecules.

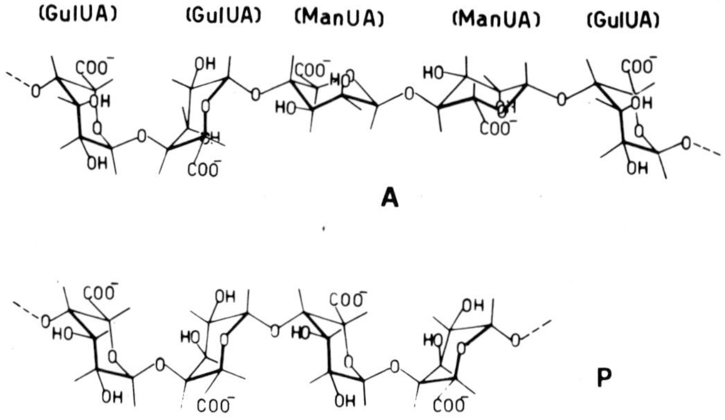

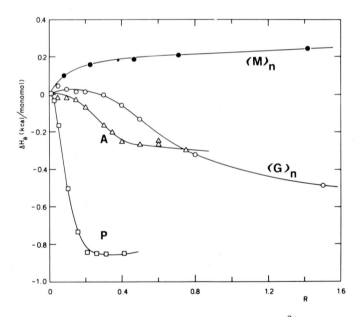

Figure 1. Schematic chain structures of Alginate (A) and Pectate (P). In a real chain of Alginate the monomers are largely distributed in omopolymeric block sequences of (Gul UA) and (Man UA), and in blocks of alternating sequence (Man UA-Gul UA), respectively. The composition of $(G)_n$ and $(M)_n$ largely coincides with pure (Gul UA) and (Man UA) blocks, respectively.

Figure 2. Enthalpy of mixing of polyuronates with Ca^{2+} ions in aqueous 0.05 M $NaClO_4$, as a function of the ratio, R, of moles of Ca^{2+} per monomol of polymer. Final polymer concentrations were 3.5×10^{-3} monomol/L.

In the case of ionic polysaccharides the thermodynamic approach, and especially the calorimetric one, has not been extensively exploited, despite of the large number of investigations on these systems. Although an exothermic enthalpy of binding of Ca^{2+} ions by guluronate-rich alginate was predicted more than ten years ago from the temperature dependence of the selectivity constants,[10] it is to be remarked that no direct determination of the enthalpy of binding has been reported since then. In a previous study[11] we have already shown that sodium pectate undergoes a conformational transition upon protonation, while no evidence of this kind has been provided for alginic acid or its oligomeric fragments, polyguluronic and polymannuronic acids[12] (Figure 1). We wish to report here some calorimetric data on the interaction of the above polyuronates with the divalent cation, Ca^{2+}. These results will be discussed together with other non-calorimetric data in order to evidence the occurrence of cooperative conformational transitions in these biopolymers, and to understand the molecular basis of their gel-forming ability.

RESULTS

The calorimetric results of mixing of polyuronates with Ca^{2+} ions (as calcium perchlorate) in aqueous 0.05 M $NaClO_4$ at 25° are reported in Figure 2. The results have been obtained with an LKB isothermal microcalorimeter 10700-2, batch type, following already established procedures. The experimental data, corrected for the heats of dilution of both polyuronate and $Ca(ClO_4)_2$, are plotted as the binding enthalpy change, ΔH_B (calories per monomol of polymer), versus R, the ratio of moles of Ca^{2+} per monomol of polymer. In the case of independent sites of interaction with a constant value of the enthalpy of interaction, the ΔH_B versus R plot is expected to be a Langmuir-like isotherm. This behavior is approached by the polymannuronate sample, $(M)_n$, while the trends of the ΔH_B curves for all the other polyuronates (alginate, pectate and polyguluronate, $(G)_n$) are significantly anomalous. The sigmoidal shapes shown in Figure 2 are taken as an evidence of a two-state mechanism in the interaction process. The exothermic values of the enthalpy of binding of Ca^{2+} ions at high R are in full agreement with the sign of the van't Hoff enthalpy values reported elsewhere[10].

The circular dichroism (CD) spectra of polyuronates in aqueous $NaClO_4$ both in the absence and in the presence of Ca^{2+} ions, are shown in Figure 3. The change of the molar ellipticity, ϑ, of the CD band associated with the $n \rightarrow \pi^*$ transition reflects the interaction of Ca^{2+} with the carboxylate groups[13,14]. The relative spectral differences, $(\vartheta_R / \vartheta_{R=0} - 1)$, are reported in Figure 4 as a

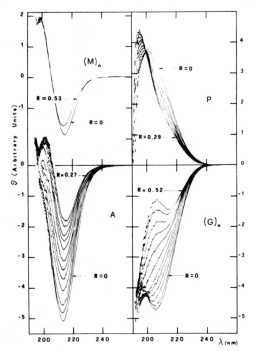

Figure 3. Effect of the increasing values of R on the CD spectra of polyuronates in aqueous 0.05 M $NaClO_4$ expressed as ϑ in arbitrary units. A: Alginate, polymer concentration, $C_p = 1 \cdot 10^{-2}$ M; $(G)_n$, $C_p = 3.5 \cdot 10^{-3}$ M; $(M)_n$, $C_p = 3.9 \cdot 10^{-3}$ M; P: Pectate, $C_p = 1 \cdot 10^{-2}$ M.

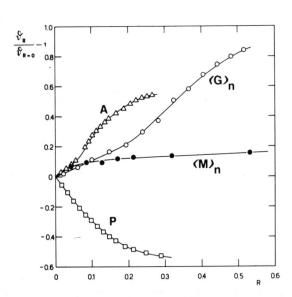

Figure 4. Relative CD spectral differences for various polyuronates as a function of R. Data taken from Figure 3 at λ_{max}

function of the ratio R, at a fixed wavelenght, λ. Although limited by the occurrence of gel formation at the highest value of R attained for each polymer, the trend observed in this plot is similar to that observed in the $\Delta H_B(R)$ plot.

DISCUSSION

One of the goals of our investigation was to discriminate, mainly by the use of calorimetry, among the different contributions to ion binding outlined in the Introduction, for the Ca^{2+}-polyuronate systems, and to determine : a) whether the polyuronate chain undergoes a conformational transition upon interaction with the cation; b) the enthalpy change associated with this transition. From the data presented in Figures 2 and 4 the mode of the interaction between Ca^{2+} and $(M)_n$ can be almost entirely interpreted as electrostatic (non-specific). Both the sign and the range of values of ΔH_B are qualitatively predicted on the basis of the polyelectrolyte theory developed by G.S.Manning[15]. Quantitative agreement could be achieved only knowing the free energy term of the counterion exchange process. On the other hand, the trend toward negative values of ΔH_B (on increasing R) shown by alginate, pectate, and $(G)_n$ is an evidence of an enthalpically favored (excess) process involving the polyuronate chain upon interaction with Ca^{2+}. The sigmoidal change in both the ΔH_B and the ($\theta_R/\theta_{R=0}$ -1) plots is interpreted in terms of a cooperative conformational transition of the polyguluronate and the polygalacturonate backbones (although the initial portion of the sigmoid is not clearly shown in the case of pectate). The difference of ΔH_B between alginate and $(G)_n$ can be ascribed to the different fraction of guluronate residues in the two polymers (about 45% and 95%, respectively). The enthalpy change associated with the full development of the process of binding can be tentatively evaluated to be -300, -600, -850 (cal/monomole), for alginate, $(G)_n$, and pectate, respectively. An additional contribution stemming from purely electrostatic interactions should be added to evaluate the full enthalpy change associated with the transition. Due to the inverse proportionality between the (electrostatic) ΔH_B and the linear charge density (ξ) of a polyelectrolyte[15], and since ξ should be about as twice as large for the Ca-$(G)_n$ and Ca-pectate systems than for $(M)_n$ (see later), this term should amount to about -100 cal/monomole.

Specific sites for the binding of Ca^{2+} ions by $(G)_n$ and pectate have been proposed by many authors[16-18]. They are supported by the low value of the activity coefficient of Ca^{2+} in the presence of $(G)_n$ as well as of pectate[19]. At variance with $(M)_n$, this interaction would result in the formation of interchain dimers[20], as the basic units for the well known phenomenon of gelation and formation of aggregates, popularized by D.A.Rees and co-workers as the "egg-bos" model[21]. Although not in disagreement with the above proposals, the cooperativity observed in the ΔH_B and ($\vartheta_R/\vartheta_{R=0}$ -1) curves suggests that the effective screening of the polyion charges by counterions (i.e. the counterion condensation, in terms of Manning's theory) may be indeed the switch for a conformational transition. This process gives rise to "nests" of the right size for further calcium ions[22]. If this view is correct, the specific "binding" site would be in fact a host/guest structure, which arises from a minimum in the conformational energy space of the polymer chain which has become accessible due to the combined effects of charge shielding and stereochemical fitting. Furthermore, it might not be a fortuitous coincidence that the values of the enthalpy of ion-induced transition already reported for different ionic polysaccharides (i-carrageenan, alginates, pectate) range from -500 to -1000 (cal/monomole of pyranose unit), and are independent of the nature of charged group on the polymer (OSO_3^-, COO^-), and of the valency and the type of counterion (H^+, Ca^{2+}, Cu^{2+}) (11,12,23).

MATERIALS

Alginate and pectate were commercial products, from Fluka and Sigma Chem.Co., respectively. Purification of the samples was carried out as already described[11]. Na$^+$ forms of $(G)_n$ and $(M)_n$ were kind gift of Dr.B.Larsen. Details on composition and properties of the samples are given in the following table.

Composition,% (^{13}C NMR)	Alginate	$(G)_n$	$(M)_n$	Pectate
Man UA	55	5	95	Gal UA[x] > 95
Gul UA	45	95	5	Neutral sugars < 5
Average MW (technique)	52,000 (viscosity)	4000 (GPC)	4000 (GPC)	18,000 (viscosity)
Average monomer length along the fiber axis, Å (X-Ray)		4.3	5.2	4.3

[x] Partially esterified to less than 8%.

ACKNOWLEDGEMENTS

The financial support from the Italian C.N.R. and from the University of Trieste is gratefully acknowledged.

REFERENCES

1. D.A.Brant,in "The Biochemistry of Plants", vol.3, J. Preiss Ed., Academic Press, New York, 180, Chapt.11.
2. D.A.Rees, Pure Appl.Chem., 1981, $\underline{53}$, 1.
3. O.Smidsrød,in "Proceedings of the 27th International Congress of Pure and Applied Chemistry", A.Varmavuori Ed., Pergamon Press, Oxford (1980), p.315.
4. R.M.Fuoss, A.Katchalsky, and S.Lifson, Proc.Natl.Acad. Sci.U.S., 1951, $\underline{37}$, 579.
5. T.Alfrey, P.W.Berg, and H.Morawetz, J.Polymer Sci., 1951, $\underline{7}$, 543.
6. F.Oosawa, "Polyelectrolytes", Marcel Dekker Inc., N.Y. (1971).
7. G.S.Manning, J.Chem.Phys., 1969, $\underline{51}$, 924.
8. G.S.Manning, Accounts Chem.Res., 1979, $\underline{12}$, 443.
9. G.S.Manning,in "Ions in Macromolecular and Biological Systems", Colston Papers No.29, D.H.Everett and B. Vincent Eds., Scientechnica, Bristol (1978), p.157.
10. A.Haug and O.Smidsrød, Acta Chem.Scand., 1970, $\underline{24}$,843.
11. A.Cesàro, A.Ciana, F.Delben, G.Manzini, S.Paoletti, Biopolymers, 1982, $\underline{21}$, 1.
12. A.Cesàro, S.Paoletti, F.Delben, V.Crescenzi, R.Rizzo, M.Dentini, Gazz.Chim.Ital., 1982, in press.
13. E.R.Morris, D.A.Rees, D.Thom, J.Boyd, Carbohydr.Res., 1978, $\underline{66}$, 145.
14. R.Kohn, T.Sticzay, Coll.Czech.Chem.Commun., 1977, $\underline{42}$, 2372.
15. G.E.Boyd, D.P.Wilson and G.S.Manning, J.Phys.Chem., 1976, $\underline{80}$, 808.
16. R.Kohn, I.Furda, A.Haug and O.Smidsrød, Acta Chem. Scand., 1968, $\underline{22}$, 3098.
17. E.R.Morris, D.A.Rees and D.Thom, Chem.Commun., 1973, 245.
18. R.Kohn, Pure Appl.Chem., 1975, $\underline{42}$, 371.
19. R.Kohn, B.Larsen, Acta Chem.Scand., 1972, $\underline{26}$, 2455.
20. P.Bucher, R.E.Cooper, A.Wassermann, J.Chem.Soc., 1961, 3974.
21. G.T.Grant, E.R.Morris, D.A.Rees, P.J.C.Smith, D.Thom, FEBS Letters, 1973, $\underline{32}$, 195.
22. O.Smidsrød, A.Haug, S.G.Whittington, Acta Chem.Scand., 1972, $\underline{26}$, 2563.
23. V.Crescenzi, C.Airoldi, M.Dentini, L.Pietrelli, R.Rizzo, Makromol. Chem. 1981, $\underline{182}$, 219.

ROLE OF "INTERMEDIATE WATER" FOR THE PHASE TRANSITION PHENOMENA IN SURFACTANT-WATER SYSTEMS AS REVEALED BY DSC

MICHIKO KODAMA, MIKA KUWABARA and SYÛZÔ SEKI

Department of Chemistry, Faculty of Science, Kwansei Gakuin
University, Uegahara, Nishinomiya-shi, 662 Japan

INTRODUCTION

In recent years, the binary systems of water and amphiphillic compounds such as lipids and biopolymers have been attracting the attention of many investigators. As is well known, the water content in vivo normally reaches more than 50-60 g% of a total weight and the amphiphillic compounds as the constituent of vivo is biologically activated for the first time by the presence of water. That is, an essential role of the water in vivo is pointed out.

On the other hand, surfactants having long hydrocarbon chains and polar head groups can be regarded, in a wide sence, as one of the amphiphillic compounds. Similarly to the lipid-water system, the surfactant-water system exibits the phase transition due to the order-disorder configurational change in their hydrocarbon chains and the transition temperature is expressed with the T_c curve in the phase diagram [1]. In this study, thermal analysis of two systems which consist of water and octadecyltrimethylammonium chloride ($1C_{18}$) having single hydrocarbon chain or dioctadecyldimethylammonium chloride ($2C_{18}$) having double chains were respectively performed by preparing about each 40 samples covering the water content from 0 to 95 g%. At a temperature above the T_c curve, $2C_{18}$-water system is allowed to exist as the bilayer-lamellar structure up to the highest water content, while $1C_{18}$-water system transforms from the lamellar to the hexagonal liquid crystals and finally to the isotropic micellar solution with the increase of water content [2,3]. This difference in the aggregation state of the surfactant molecules between the single and the double hydrocarbon chains was investigated from the viewpoint of roles of the new type of water, so-called "intermediate water" incorporated between the bilayers of the surfactant molecules [4]. Furthermore, the predominent roles of the "intermediate water" on the phase transition from the coagel to the gel were also pointed out.

EXPERIMENTAL

Surfactant, $2C_{18}$ was kindly supplied by Kao Soap Co. LTD., and was purified by recrystallization from acetone solution. $1C_{18}$ was obtained from Tokyo Kasei Co. LTD., and was recrystallized from acetone/methanol solution. The condition for complete dehydration was examined by the use of a Chan electrobalance Model 2000 and it was revealed that the complete dehydration was achieved only when the surfactants were dehydrated under high cacuum (10^{-4} Pa) and at a temperature above their transition temperature due to the hydrocarbon chain-melting. Each forty samples of $1C_{18}$-, and $2C_{18}$-water mixtures with different contents were prepared by adding a controlled amount of water to the completely dehydrated compound with the use of the microsyringe. The DSC measurements were performed with Mettler DSC TA 2000 in the temperature range from -20 to 140 °C, and with Daini Seikosha SSC/560U in the temperature down to -100 °C, both equipment by using the high-pressure-crucible. All the samples were preheated to about 100 °C in order to assure homogeneous mixing.

RESULT and DISCUSSION

Figure 1 shows the representative heating DSC curves among the forty samples of the $1C_{18}$-water mixture, in comparison with those of the $2C_{18}$-water mixture. Completely dehydrated $1C_{18}$ exhibits only a primary endothermic peak at 100.4 °C due to the partial melting of the hydrocarbon chain and with an increase in the water content, this peak is followed by successive appearences of new, second and third endothermic peaks at lower temperatures in this order. With further addition of water, the transition temperature of the final third peak reaches the limiting value of 15.5 °C corresponding to the so-called Krafft point of this system at the water content of about 50 g%.

The phase diagram of the $1C_{18}$-water system obtained from all the heating DSC curves is shown in Fig. 2 where the symbol T_c exhibits the variation of the transition temperature due to the chain-melting with the water content. It is revealed that the T_c curve of the $1C_{18}$-water system is composed of a total of three endothermic peaks and decreases stepwise to the limiting temperature at 15.5 °C, not showing that the primary endothermic peak itself decreases directly to that temperature. Similarly, $2C_{18}$-water system also shows the step-

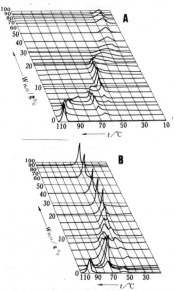

Figure 1. Typical DSC curves for [$1C_{18}$-water] system (A) and [$2C_{18}$-water] system (B).

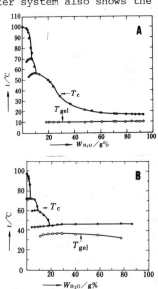

Figure 2. Phase diagrams of [$1C_{18}$-water] system (A) and [$2C_{18}$-water] system (B).

wise decreasing T_C curve, reflecting the composite thermal behaviors of a total four endothermic peaks as shown in Fig. 1. Thus, the T_C curve obtained here is apparently distinguished from the smooth T_C curves for the binary systems of water and lecithins or soaps reported by many other investigators [1,5].

As is well known, the so-called lamellar gel phase of the surfactant-water mixture comes into existance on cooling down to a temperature below the T_C curve where the hydrocarbon chain of the surfactant molecule adopts the ordered structure. Next, we shall proceed to the phase transition below the T_C curve, in connection with the ice-melting phenomenon. Each gel phases of $1C_{18}$-, and $2C_{18}$-water mixtures prepared by cooling is allowed to exist as a metastable, supercooled gel phase, down to -20 °C. The heating DSC curves staring from these supercooled gel systems (unannealed sample) exhibit only the endothermic peak due to the chain-melting, except for the endothermic peak due to the ice-melting, as shown in series I in Fig. 3. However, the supercooled gel phase is converted into a stable state, the so-called coagel phase, by suitable annealing treatment and the DSC curves starting from the coagel systems (annealed sample) after this treatment show a new endothermic peak (dotted area), successively followed by the endothermic peak due to the chain-melting, as shown in series II of Fig.3. Therefore, this new peak is corresponding to the phase transition of the coagel to the gel, to which we give the symbol T_{gel} described in the phase diagram of Fig. 2. The coagel-gel transition of the $1C_{18}$-water system is observed at a water content more than 17 g% up to the dilute aqueous solution of 95 g% and the peak area of this transition increases with the raise of the water content up to 50 g% (see Fig. 6). On the other hand, with the $2C_{18}$-water system, the coagel-gel transition appears above 11 g% water content and the peak area increases up to 18 g%, beyond which nearly the same area is observed (see Table 1).

When we pay an attention to the endothermic peaks due to the melting of ice in the temperature region around 0 °C shown in Fig. 3, a difference in their thermal behavior between the supercooled

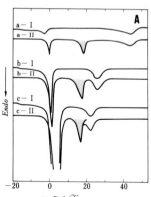

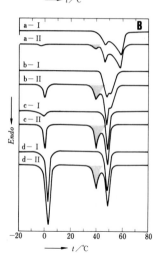

Figure 3. Comparison of DSC curves for unannealed (I) and annealed (II) samples in [$1C_{18}$-water] system (A) and [$2C_{18}$-water] system (B). Water content (W_{H_2O}/g%): a, 24.3; b, 44.46; c, 74.76 for (A) and a, 11.18; b, 17.40; c, 21.26; d, 39.80 for (B) (ref. 4).

gel (see series I in Fig. 3) and the coagel systems (see series II) will be noticed. With respect to the $1C_{18}$-water system, Fig. 4 shows the enlarged DSC curves around the ice-melting temperature at various water contents. In this figure, the $1C_{18}$-water supercooled gel system (series I) shows a broad endothermic peak starting around -15 °C and with futher addition of water, this peak is successively followed by an appearance of a sharp endothermic peak at 0 °C due to the melting of ice corresponding to the free water. This finding indicates the existance of the structural water in addition to the free water. While, the $1C_{18}$-water coagel system (series II) shows only a sharp endothermic peak at 0 °C and the area of this peak is revealed to be nearly equal to a total one of the broad endothermic peak plus the shap endothermic peak observed in the supercooled gel system at the corresponding water content, indicating that some of the free water in the coagel system is converted into the structural water which is hereafter refered to as the "intermediate water". It may be said that the phase transition of the coagel to the gel is accompanied by a change in the aggregation state of water molecule from the free water to the intermediate one.

With respect to the $2C_{18}$-water system, not only the coagel system but also the supercooled gel system exhibit only one sharp endothermic peak at 0 °C and the peak area in the former system becomes larger than that in the latter at the corresponding water content. This difference is just corresponding to the intermediate water. Thus it may be said that the intermediate water in the $2C_{18}$-water gel system does not crystallize on cooling down to -20 °C, while that in the $1C_{18}$-water gel system crystallizes rather easily.

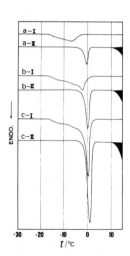

Figure 4. Enlarged DSC curves around the ice-melting temperature in [$1C_{18}$-water] system. Unannealed (I) and annealed (II) samples with the same water content are compared in pairs. Water content (W_{H_2O}/g%); a, 19.9; b, 24.3; c, 33.0.

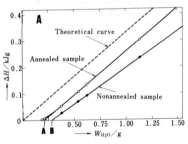

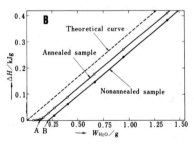

Figure 5. Relationship between ΔH due to the melting of ice and the amount of water added to 1 g of (A) $1C_{18}$ and (B) $2C_{18}$ (ref. 4).

In order to elucidate, quantitatively, a difference in the aggregation state of the water molecule between the coagel and the gel systems, the amounts of the water existing as the free water in both the coagel and the gel systems were examined from the enthalpy changes (ΔH) of the sharp endothermic peaks at 0 °C shown in series I and II of Fig. 3. The results obtained are shown in Fig. 5. A difference in ΔH between the theoretical curve and that of the coagel system corresponds to the amount of water which does not crystallize on cooling to -20 °C. This nonfreezable water, which is hereafter refered to as the "bound water", is strongly hydrated with the polar head groups of $1C_{18}$ and $2C_{18}$ molecules, respectively. The symbol A in Fig. 5 indicates that the bound water reaches the maximum value at the water contents of 17 and 11 g% for the $1C_{18}$-, and $2C_{18}$-water systems, respectively, both water contents corresponding to 1:4 in the molar ratio of the surfactant of the water. All the water added beyond these water contents is present as the "free water" phase coexisting with the coagel phase having the limiting amount of the "bound water", indicative of the existance of two kinds of water, i.e., free and bound water in the coagel system. Therefore the sharp endothermic peak at 0 °C shown in series II of fig. 3 is due to this free water. A difference in ΔH value between the curve for the coagel system and that for the supercooled gel system corresponds to the amount of the water which exists as the "intermediate water" in the latter system, whereas in the former system it exists as the "free water". With the $2C_{18}$-water system, the "intermediate water" reaches the maximum value at the water content of 18 g% (shown by the symbol B in Fig. 5) and all the water added beyond this content exists as the free water phase coexisting with the gel phase having the limiting amounts of the bound and intermediate waters. On the other hand, with the $1C_{18}$-water system, some of the water added more than 23 g% (shown by the symbol B in Fig. 5) exists as the intermediate water incorporated in the gel phase and the remaining as the free water coexisting with the gel phase, indicating that the amount of the intermediate water in the $1C_{18}$-water system increases with an increase in the water content, differently from the case of the $2C_{18}$-water system. It is obvious in Fig. 5 that in the $1C_{18}$-, and $2C_{18}$-water system, three kinds of water exist, i.e., free, intermediate and bound waters. Furthermore, it is interesting to note here that the free water coexisting with the coagel phase starts to appear at the same water content where the phase transition of the coagel to the gel shows itself for the first time (see series II in Fig. 3). This fact implies that the free water phase coexisting with the coagel phase is a requirement for the appearence of the coagel-gel transition. Thus, at this transition, some of the free water which was coexisting with the coagel phase is now incorporated between the bilayers of the $1C_{18}$ and $2C_{18}$ molecules and is converted into the intermediate water in the gel phase which causes loosening in the ordered packing of the polar head groups due to the electrostatic repulsion resulting from their ionization. However, there exists a difference in the allowed amount of the intermediate water between the $1C_{18}$-, and $2C_{18}$-water systems as mentioned above and this may be associated with a difference in the aggregation state of the surfactant molecule between the single and double hydrocarbon chains at the temperature above T_c curve, as will be discussed below.

Table 1 shows the entropy changes (ΔS) at the coagel-gel and the gel-liquid crystal transitions in the $2C_{18}$-water system at the water content more than 18 g% up to about 85 g%. As mentioned above, the phase transition from the coagel to the gel leads to a conversion to a less ordered structure of the polar head groups due to the incorporation of the intermediate water between the bilayers. Therefore, the degree of the disorder in the structure of the polar head groups is said to be determined exclusively by the amount of the intermediate water. This fact is quite consistent with the excellent constant values in the entropy data of the coagel-gel transition at the water content more than 18 g%, where the limiting amount of the intermediate water incorporated between the bilayers at this transition is achieved. On the other hand, the phase transition of the gel to the liquid crystal leads to the melting of the hydrocarbon chains anchored to the polar head groups. Therefore, the degree of the configurational disorder of the hydrocarbon chains is greatly governed by the degree of loosening in the lateral packing of the polar head groups. Accordingly, as indicated in Table 1, the entropy data concerning the gel-liquid crystal transition of the $2C_{18}$-water system also shows nearly the same value at the water content more than 18 g%, similarly to the coagel - gel transition. In conclusion, it may be said that the intermediate water plays an essential role not only in the coagel-gel transition, but also in the gel-liquid crystal one.

It has been well known that the $1C_{18}$-water system is accompanied by successive changes in the aggregation state of the $1C_{18}$ molecules, i.e., the lamellar (neat phase) to the hexagonal (middle phase) liquid crystals and finally to the micellar solution, with an increase in the water content at the temperature above T_C curve, while with the $2C_{18}$-water system, the lamellar liquid crystal is allowed to persist up to the dilute aqueous solution. Corresponding to these facts, $1C_{18}$-water system exhibits the characteristic entropy data shown in Fig. 6 where the entropy changes at both transitions can be divided into the three regions of A (below 17 g% water content), B (from 17 to 50 g%) and C (above 50 g%) according to the water content. In particular, the increase of both entropies starting around 17 g% water content in the region B seems to reflect the increment of the configurational disorders of the polar head groups and the hydrocarbon chains, respectively, due to the incorporation of the intermediate water between the bilayers and both disorders increase with an increase in the intermediate water (thus, an increase in the water content). Under these situations, the ionization of the polar head groups of the $1C_{18}$ molecule may proceed, so that the electrostatic repulsion force operating among them increases. This circumstance may require a change in the aggregation state of the $1C_{18}$ molecule in order to weaken its force. The fused state of the hydrocarbon chain of the $1C_{18}$ molecule above the T_C curve may cause this change by a rearrangement of the $1C_{18}$ molecule due to the

TABLE 1. ENTROPY CHANGES AT THE PHASE TRANSITIONS OF THE COAGEL-GEL AND THE GEL-LIQUID CRYSTAL IN [$2C_{18}$-WATER] SYSTEM.

$W_{H_2O}/g\%$	$\Delta S/J\ mol^{-1}\ K^{-1}$	
	Coagel→Gel	Gel→Liquid crystal
21.16	82.4	79.9
28.54	82.4	77.1
39.80	82.4	76.4
57.92	82.4	75.4
85.70	82.3	74.1

migrational motion. Accordingly, with an increase in the water content at the temperature above the T_c curve, there occures the successive conversions in the aggregation state from the planar (lamellar) to cylindrical (hexagonal) and finally to spherical (micellar) structures, that is, with an increase in the radius curvature of the interface formed by the polar head groups and the water in this order. Therefore, the entropy data concerning the transition of the gel-liquid crystal (or micellar solution) above 17 g% water content shown in Fig. 6 may be said to reflect the contribution of the orientational disorder due to the change in the aggregation state of the $1C_{18}$ molecule, in addition to that of the configurational disorder of its hydrocarbon chain mentioned above. The phase changes from the gel to the lamellar, hexagonal liquid crystals and the micellar solution seems to appear at the region A, B and C, respectively. In the region C, the micelle having the same degree of the configurational as well as the orientational disorder may be proposed. In the region B, the hexagonal liquid crystal, which is regarded as the intermediate phase between the lamellar liquid crystal and the micellar solution, may be accompanied by a gradual change from the planar interface to the spherical one with an increase in the water content. This may be closely related to the increase in the entropy change with an increase in the water content in the region B.

Finally, we should like to discuss the glass transition phenomenon of the supercooled gel phase found in the $2C_{18}$-water system. When the gel phase, which is allowed to be the supercooled state down to -20 °C, is successively cooled down to around -100 °C, the heating DSC curve for this sample exhibits the stepwise endothermic anomaly charactristic to the glass transition phenomenon (see Fig. 7). One component system of the completely dehydrated $2C_{18}$ no longer shows the glass transition and as is well known, the pure liquid water is not brought into its glassy state by the supercooling process [6]. Therefore, we propose here the new type of the glassy state, i.e., "glassy gel" named by us [7]. This glassy gel may be attributed to the co-operational interaction between the not-crystallized intermediate water and the partly ionized polar head groups. It may be added here that the glassy state of the gel phase in the $1C_{18}$-water system is not observed, reflecting that the intermediate water in this system is easily crystallized on cooling down to -20 °C as mentioned above.

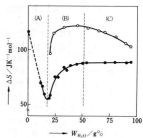

Figure 6. Entropy changes at the phase transitions of the coagel-gel (O) and the gel-liquid crystal (micellar solution) (●) in [$1C_{18}$-water] system.

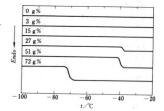

Figure 7. Glass transition phenomenon in [$2C_{18}$-water] system. Water content is shown on the left hand side of this figure.

REFERENCES

1. K. Fontell, *Mol. Cryst. Liq. Cryst.*, **63**, 59 (1981).
2. T. Kunitake, et al., *J. Am. Chem. Soc.*, **99**, 3860 (1977).
3. H. Kunieda, et al., *J. Phys. Chem.*, **82**, 1710 (1978).
4. M. Kodama, et al., *Thermochim. Acta*, **50**, 81 (1981).
5. D. Chapman, et al., *Chem. Phys. Lipids*, **1**, 445 (1967).
6. M. Sugisaki, et al., *Bull. Chem. Soc. Jpn.*, **41**, 2581 (1968).
7. M. Kodama, et al., *Mol. Cryst. Liq. Cryst.*, **64**, 277 (1981).

SPECIFIC HEAT ANOMALIES IN SOME ORGANIC COMPOUNDS[+]

Carlos A. Martín[*]

Instituto de Matemática, Astronomía y Física, Universidad Nacional de Córdoba, Laprida 854, 5000 Córdoba, Argentina

INTRODUCTION

Among the various response functions that may be used to study matter in the condense state, specific heat occupies a relevant place. It is, generally, a smooth temperature dependent quantity which may be well described by, for example, Debye's model [1]. However, in certain compounds, it may exhibit departures from that smooth behaviour, giving rise to the so called anomalies [2-5]. A careful analysis of these anomalies will provide various parameters [6] like, for example transition temperatures, hysteresis, heats evolved and critical point exponents [7-9]. It is the comparison of these parameters, which may be used to characterize a given anomaly, with those obtained from model calculations that will be of great help in order to propose a model for the anomaly (first, second or higher phase transitions [7-9], order-disorder, displacive [10], Schotky effect [11], etc.).

In this paper we report the anomalies in the specific heat in the following organic compounds: methylchloroform, acetonitrile, trichlorofuoromethane and 1,2,4,5-tetrachlorobenzene. The reported quantities are: a) the transition temperatures, T_c, detected cooling down and heating up the samples in the range 80-300 K. b) The enthalpy changes, ΔH, associated to each transition, and c) the classification according to the simplified scheme discussed in a companion paper [6]. The nomenclature used is that defined in that paper.

EXPERIMENTAL

The equipment used to gather the data is that described in Ref. [6]. The chemicals were supplied by Fluka: methylchloroform (91099), acetonitrile (00690), trichlorofluoromethane (91273) and 1,2,4,5-tetrachlorobenzene (87000). Methylchloroform, trichlorofluoromethane and 1,2,4,5-tetrachlorobenzene were used as received, while acetonitrile was additionally purified. Samples were prepared as indicated in [6].

In all cases reported in this paper, the time rate of change of the sample temperature, $\dot{T}_s$, has been below 5 K/h. More precise values for each individual case may be obtained from the curves shown in the Figs. 1-7. Also, in all these figures T_s and ΔT [6] are actually reported as thermocouple voltages. No scales are indicated in the drawings themselves, however the indications appearing in the figure captions must be understood as follows:

a) $V_1 \lesssim T_s \lesssim V_2$ indicate that the upper (lower) and lower (upper) edges of the figures correspond to the extreme sample thermocouple (Cu-CuNi) values V_1 and V_2. There should be no doubt as to what voltage assign to what edge since in the same figure caption it is indicated if we are dealing with a cooling down run or with a heating up one.

b) $\Delta T_{max} = V_3$ indicates the maximun value that ΔT may reach, and corresponds to the upper edge of the graph. The lower edge corresponds to $\Delta T = 0$ V.

c) $\Delta t = t_o$ indicates the time evolved from the left to the right edge of the figures.

d) The slight time shift between the two traces in necessary in order to allow proper functioning of the x-t chart recorder.

RESULTS AND DISCUSSION

Methylchloroform. It is known that this compound exhibits five phase transitions in the investigated temperature range [3]. Its transitions are identified as follows: liquid $\rightarrow$ solid Ia $\rightarrow$ solid Ib $\rightarrow$ solid II in a cooling down run. Heating up we obtain the sequence: solid II $\rightarrow$ solid Ib $\rightarrow$ liquid. These five transitions are shown in Figs. 1-4. These transitions have been reported by a number of workers [3, 12-14], therefore offering a good compound for checking and calibrating our equipment. Particularly, the transition II $\rightarrow$ Ib was used to calibrate the area under the ΔT vs t recorder traces in units of J, taking that area as corresponding to $\Delta H = (7,47 \pm 0.02)$ kJ/mol, which is obtained as an average of the values reported in Refs.[12-14]. The transition temperatures and enthalpy changes are given in Table 1. The following comments are in order:

1) The transition Ia $\rightarrow$ Ib [3] is of type E [6], therefore it is not possible to define a transition temperature as mentioned in Ref. [6]. Table 2 shows a few temperature pairs, the lower temperature, T_{sc}, is that at which the phase transition starts, while $T_{se} + \delta T$ is the temperature reached when the transition is finished. In our experiments $\dot{T}_s$ was about 2 K/h just before the transition.

2) No evidence was found of an anomaly in C_p at 206 K as reported in Refs. [3] and [12], and in agreement with Ref. [14].

3) As may be seen in Table 1, there is a small hysteresis in the transition Ib $\rightarrow$ II, in disagreement with the reported values in Ref. [3]. In general, large discrepancies are found for the transition temperatures with those reported in Ref. [3] in the cooling down runs. This may be due to the different cooling rates and/or sample impurities, since their cooling rate is far larger than ours and our sample is less pure than theirs (99.0% as compared to 99.9%).

4) Basing the calibration of the equipment in the enthalpy change corresponding to the II $\rightarrow$ Ib phase transition as

indicated above, good agreement is found in evaluating the ΔH corresponding to the transition Ib $\longrightarrow$ liquid.

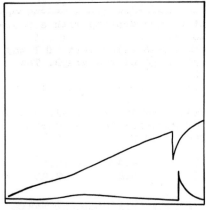

Figure 1. Methylchloroform. Transitions liquid $\longrightarrow$ Ia and Ia$\longrightarrow$Ib. $-1300 \mu V \geqslant T_s \geqslant -2300 \mu V$ $\Delta T_{max} = 1000 \mu V$, $\Delta t = 10$ h.

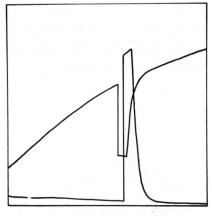

Figure 2. Methylchloroform. Transition Ib$\longrightarrow$II. $-1500 \mu V \geqslant T_s \geqslant -2500 \mu V$, $\Delta T_{max} = 500 \mu V$, $\Delta t = 20$ h.

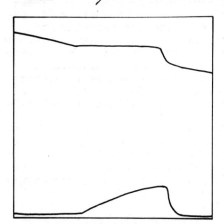

Figure 3. Methylchloroform. Transition II$\longrightarrow$Ib. $-900 \mu V \geqslant T_s \geqslant -1900 \mu V$, $\Delta T = 500 \mu V$, $\Delta t = 20$ h.

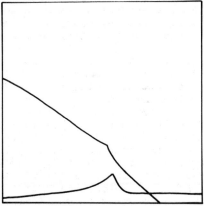

Figure 4. Methylchloroform. Transition Ib$\longrightarrow$liquid. $-900 \mu V \geqslant T_s \geqslant -1900 \mu V$, $\Delta T = 500 \mu V$, $\Delta t = 20$ h.

Acetonitrile. It is known that this compound exhibits two phase transitions in a heating up run [15]. They are solid $\beta \dashrightarrow$ solid α and solid $\alpha \dashrightarrow$ liquid. However, it was believed this compound solidified in phase β [16]. The existence of a total of four phase transitions was established, two during the cooling down: liquid $\dashrightarrow \alpha$ and $\alpha \dashrightarrow \beta$,

Table 1. Relevant results obtained during this research.

Compound	Transition	Type	T_c (K)	ΔH (kJ/mol)
Methyl-chloroform	liq→Ia	C	234.3	-1.30
	Ia→Ib	E	(*)	-1.53
	Ib→II	A	224.9	-8.00
	II→Ib	B	224.5	7.47
	Ib→liq	D	240.9	1.55
Acetonitrile	liq→α	A	229.9	-7.96
	α→β	E	(*)	-0.90
	β→α	B	218.0	0.80
	α→liq	D+B	228.7	6.67
Trichlorofluoromethane	liq→I	?	164.9	-10.10
	I→liq	B	165.4	7.90
1,2,4,5-tetrachlorobenzene	I→II	C	178.2	-0.028
	II→I	D	187.5	0.034

* See Table 2.

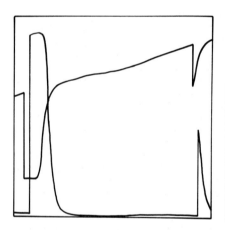

Figure 5. Acetonitrile. Transitions liquid→α and α→β. -1400 μV ⩾ T_s ⩾ -2400 μV, ΔT = 500 μV, Δt = 20 h.

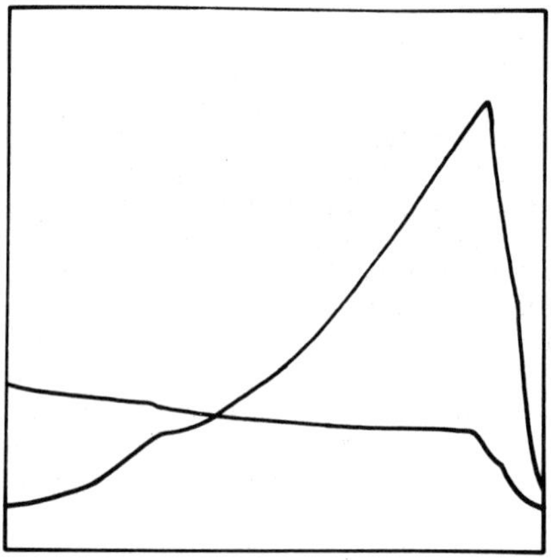

Figure 6. Acetonitrile. Transition α→liquid. -1400 μV ⩾ T_s ⩾ -2400 μV, ΔT = 150 μV, Δt = 20 h.

which are shown in Fig. 5, and two during the heating up: β→α and α→liquid. The β→α transition is similar to that shown in Fig. 3, while the α→liquid transition is shown in Fig. 6. Table 1 gives the various parameters associated with these four transitions. However, since the α→β transition is of type E no T_c may be define to characterise

Table 2. Data collected for transitions of type E.

Compound	Run	T_{sc} (K)	δT (K)
Methyl-chloroform	1	229.3	4.2
	2	227.6	4.1
	3	228.2	4.1
	4	226.2	3.9
Acetonitrile	1	209.7	6.1
	2	215.3	5.9
	3	212.0	5.9

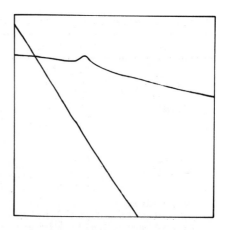

Figure 7. 1,2,4,5-tetrachlorobenzene. Transition II→I. $-2500\,\mu V \geq T_S \geq -3500\,\mu V$, $\Delta T = 100\,\mu V$, $\Delta t = 10$ h.

it [6]. The values measured in a few runs for T_{sc} and δT are given in Table 2. The α→liquid transition shows two relevant features, both of which may be seen in Fig. 6:

1) The fusion is not isothermal to a large extent, and only in the final part T_s approaches a constant; therefore indicating that C_p grows continuously up to a very large value in a small temperature range. Thus, this transition shows a mixed behaviour, beginning like one of type D and ending like one of type B. Unfortunately, due to our poor data acquisition system nothing may be said about the analytical behaviour of this increase in C_p as a function of T_s.

2) In all the runs we found that ΔT shows a fine structure which appears as a small peak in ΔT in the region where it is increasing and before reaching its maximun value. The explanation of this small peak must await further investigation.

Trichlorofluoromethane. Very little is known about this compound, and the main results are given in Table 1, showing that it exhibits only two transitions, liquid⇌solid I, correspondig to the solidification and to the fusion. Relevant features of these transitions are the following:

1) The shape of the T_s and ΔT curves during solidification was found to be non-reproducible. More precisely, it shows the supercooling at the beginning and the isothermal part at the end, and is the region in between the one that is non-reproducible. However, the isothermal part was taken as T_c since it was precisely reproduced in the various runs we did. Similarly, the area under the ΔT vs T curve is also perfectly reproducible.

2) The fusion transition takes place according to a type B transition, similar to that shown in Fig. 3.

1,2,4,5-tetrachlorobenzene. This compound exhibits a solid I⟶solid II phase transition in a cooling down run, and the reverse transition when heating up. The main features of these transitions are shown in Table 1. The transition II⟶I is shown in Fig. 7, and as may be seen the anomaly in the specific heat is very small and this is reflected in the barely noticeable change in T_S and in the slight change of ΔT, therefore producing a $\Delta S = 0.18$ J/(K mol) characteristic of a displacive phase transition [17]. Our results for ΔH and ΔS are similar to those found in analogous compound, Chloroamil [5], and in disagreement with those found by other authors [18]. However, eventhough the transitions seems to be second order, displacive and triggered by librational modes whose frequencies are drastically reduced when approaching T_c [18], it actually is not second order as is shown by the existence of quite a large hysteresis, therefore precluding this transition from being second order [8]. Therefore, the transition is first order, displacive. Incidentally, the ΔH appearing in Table 1 may be reduced by a factor of two, since it is by no means clear when the transition is finished, as is discussed in Ref. [6].

REFERENCES

[+] Partial finantial support to this research was granted by Consejo Nacional de Investigaciones Científicas y Técnicas and by Sub-Secretaría de Estado de Ciencia y Tecnología.
[*] Fellow of the Comsejo Nacional de Investigaciones Científicas y Técnicas.
1. C. Kittel, "Introduction to Solid State Physics", fourth edition, Wiley, New York, (1971).
2. R. G. S. Morfee, L. A. K. Staveley, S. T. Walters and D. L. Wigley, J. Phys. Chem. Solids, 1960, 13, 132.
3. L. Silver and R. Rudman, J. Phys. Chem., 1970, 74, 3134.
4. A. Inaba and H. Chihara, J. Chem. Thermodynamics, 1978, 10, 65.
5. H. Chihara and K. Masukane, J. Chem. Phys., 1973, 59, 5397.
6. C. A. Martín, companion paper presented at this Conference.
7. C. N. R. Rao and K. J. Rao, "Phase Transitions in Solids", McGraw-Hill, New York, (1978).
8. A. B. Pippard, "The Elements of Classical Thermodynamics", Cambridge, London, (1979).
9. H. E. Stanley, "Introduction to Phase Transitions and Critical Phenomena", Oxford, New York, (1971).
10. L. D. Lamdau and E. M. Lifshitz, "Statistical Physics", Addison-Wesley, Reading, (1970).
11. J. A. Sommers and E. F. Westrum, Jr, J. Chem. Thermody-

namics, 1976, $\underline{8}$, 1115.
12. T. R. Rubin, B. H. Lewedahl and D. M. Yost, J. Amer. Chem. Soc., 1944, $\underline{66}$, 279.
13. R. W. Crowe and C. P. Smyth, J. Amer. Chem Soc., 1950, $\underline{72}$, 4009.
14. R. J. L. Andon, J. F. Counsell, D. A. Lee and J. F. Martin J. Chem. Soc. Faraday Trans. I, 1973, $\underline{69}$, 1721.
15. W. E. Putnam, J. Chem. Phys., 1965, $\underline{42}$, 749.
16. A. Kaplan, Ph. D. Thesis, Universidad Nacional de Córdoba 1975.
17. G. P. O'leary and R. G. Wheeler, Phys. Rev. B, 1970, $\underline{1}$, 4409.
18. B. Pasquier and N. Le Calvé, J. Raman Spectrosc., 1977, $\underline{6}$, 155.

THERMOANALYTICAL INVESTIGATIONS ON RELEASE OF OXYGEN FROM A HEME-MODEL

O.Schneider, J.Moll*, M.Hanack and D.Krug*,
Institut für Organische Chemie,
*Institut für Anorganische Chemie,
Eberhard-Karls-Universität Tübingen,
Auf der Morgenstelle 18,
D-7400 Tübingen 1 / FRG.

INTRODUCTION

Oxygen plays an important role in all the life processes. In this respect the bonding of oxygen to heme is of special interest. Model substances like tetraphenylporphinatoiron(II) (TPPFe) or phthalocyaninatoiron(II) (PcFe) are well suited for the investigation of oxygen bonding and storage [1].

The identification of the oxygen-bound porphin derivatives and differentiation between bound oxygen and hydroxyl ions are often difficult [2].

In the case of PcFe, a few compounds of oxygenated PcFe have been investigated, but not completely identified. Some of these are:
{1}   a μ-oxo-dimer [3];
{2}   a dioxygen or peroxo-dimer with iron(II) or iron(III), respectivly [4];
{3}   a hydrated PcFe (= 2 PcFe + H_2O), where the protons should be bound to nitrogen-atoms of the phthalocyaninato-moiety [5]; and
{4}   a hydroxyl-ions containing derivative [6].

$$PcFe(III)-O-Fe(III)Pc \qquad PcFe-O_2-FePc$$
$$\{1\} \qquad\qquad \{2\}$$

$$H_2[\,PcFe(II)-O-Fe(II)Pc\,] \qquad PcFe\genfrac{}{}{0pt}{}{\overset{H}{\overset{|}{O}}}{\underset{\underset{H}{|}}{O}}FePc$$

$$\{3\} \qquad\qquad \{4\}$$

To determine their identities, the usually-employed elementary analysis for oxygen in these compounds is not applicable, because this method is strongly affected by traces of water and/or impurities like phthalimide [4, 7]. Moreover, due to the bulkiness of the phthalocyanine, elementary analyses cannot distinguish between bound oxygen, oxygen ions and hydroxyl ions. Other methods, e.g. spectroscopic studies, are equally unsatisfatory.

From previous reports it is known that oxygen-containing phthalocyanines {1-4} undergo chemical reactions with strongly coordinating ligands L, like pyridine or DMSO, to produce the appropriate PcFeL$_2$-adducts. Evolution of oxygen or water was not investigated nor were the thermal stabilities of the phthalocyaninato-iron-oxygen-compounds fully described [3 - 6].

In this paper we wish to report the thermal decomposition of PcFe-O-FePc (we call it "form D" because of its indeterminate structure) and of a new oxidized PcFe, form X, according to the redox-reaction

oxygenated-PcFe $\longrightarrow$ β-PcFe + [O$_2$]   (I)

To identify the evolved gases and to distinguish definitely between dioxygen and water, we carried out TG/DTG/DTA-, as well as EGA-measurements with helium, nitrogen and hydrogen as carrier gases [8].

The compound (PcFe)$_2$O, (form D), was synthesized according to literature methods [3].

If air is allowed to diffuse into a suspension of β-PcFe in DMF a new oxygen-containing product, form X, is produced. Its IR-spectrum is close to, but not identical with that of α-PcFe.

Compounds exhibiting similar IR-spectra are obtained from soluble PcFe(2-methylpyrazine)$_2$ by reaction with oxygen in chloroform.

Detailed information on these compounds including UV/VIS-, IR- and FIR-spectra is given elsewhere [8, 9].

TG/DTG/DTA IN NITROGEN

Simultaneously-performed TG/DTG/DTA on forms D and X show that an exothermic decomposition reaction has occured with weight losses in the range of 2 - 4 % (Table 1). The violet residue is β-PcFe (IR-spectrum) [10] as in all the other following decomposition experiments. The temperature must be kept below 500 °C. The weight loss for form D is greater than the calculated 1.4 % (reaction I).

EGA/DTA IN HELIUM

Thermal decomposition of forms D and X in helium-atmospheres leads to evolution of gaseous products in exactly the same temperature range. Less volatile products (deduced from the peak-shape), probably subliming organic impurities are also released. With a cold-trap (-50°C), we eliminated the less volatile products, but no significant reduction of the peak area was observed (as expected, if water was released from the oxygen-containing phthalocyanines). Therefore, structures {3} and {4} for forms D and X can be excluded, because evolution of H$_2$O + 1/2 O$_2$

e.g. for {4} should reduce the peak area at least by half if water is trapped at -50°C.

It was found that X produces roughly the same amount of gas per weight unit as D under the same measuring conditions.

EGA/DTA IN HYDROGEN

In the presence of phthalocyanine, oxygen reacts in a hydrogen atmosphere at temperatures >250°C to produce water [11].

With a cold trap (-50°C) between the oven and the EGA-detector, no signal due to water in hydrogen should be detected. Nevertheless a small peak was observed, because not all the oxygen reacted to produce water.

There are still other gases passing the cold trap at temperatures down to -100°C. To get more information about these gases, we carried out EGA-measurements using nitrogen as carrier gas.

EGA/DTA IN NITROGEN

Nitrogen as carrier gas is less frequently used for EGA because of its reduced sensitivity in comparison with hydrogen or helium. But its uniqueness for our problem is that oxygen, hydrogen or methane have higher thermal conductivities than nitrogen, whereas all other gases, which might be evolved in our systems, have thermal conductivities which are lower.

In figure 1 the very small negative signal at 400°C probably signifies the evolution of oxygen. This signal is followed by a sharp positive EGA-peak, 410 to 440°C, deriving from a gas with a poorer thermal conductivity than nitrogen. Simultaneously a sudden exothermic reaction is registered by DTA. Attempts to trap this gas, down to -120°C, failed. Liquid nitrogen, however, condensed the gas, which can be fully vaporized on sudden heating to -120°C.

Due to the trapping temperature, the thermal conductivity of the gas, and its strongly exothermic formation, we concluded, that carbon dioxide or perhaps carbon monoxide are evolved in addition to oxygen.

GC IN HELIUM

To get more information, we decomposed forms D and X in a closed system (modified Schlenk tube) in helium under reduced pressure. In the decomposition range of the compounds in addition to gas evolution a colourless coating was formed on the colder parts of the tube. This coating was identified by IR and TLC to consist mainly of phthalimide and phthalodinitrile.

The collected gas was analysed [12] on silica and molecular sieve columns (2 m, 60°C, He as carrier gas) and shown to contain about 86 % CO_2, 1.5 % CO, 1 % O_2 and 11.5 % N_2.

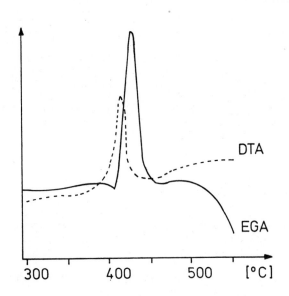

Figure 1: Thermal decomposition of form X.
DTA and EGA in nitrogen; heating rate 5 K/min.
DTA: ↑ exothermic; EGA: ↑ lower thermal conductivity.

CONCLUSIONS

We have proven that the oxygen-containing compounds (forms D and X) can be thermally decomposed at temperatures around 400°C, Table 1, to β-PcFe and intermediate oxygen. Unfortunately, a quantitative oxygen-analysis by EGA cannot be obtained, because the evolved oxygen reacts with PcFe to give a variety of gaseous and solid products. Formation of carbon dioxide from oxygen and PcFe at temperatures >200°C, noted also by another worker [13], was proved by means of oxygen-pulses in helium.

In hydrogen-atmosphere "oxygenated"-PcFe can be reduced to β-PcFe and water at temperatures >250°C, but carbon dioxide is still produced. At 200°C, no reduction occurs in hydrogen as also observed very recently by

Blomquist [14].

Therefore we adopt the idea [13] that PcFe at elevated temperatures strongly binds molecular oxygen which reacts spontanously with a carbon-delivering system, probably the Pc-moiety, to yield carbon dioxide, even in a pure hydrogen atmosphere.

That the parent PcFe can serve as a carbon-delivering system is understandable if we consider its thermal stability. In an inert gas atmosphere thermal destruction of PcFe started at 500°C (Figure 1) with an evolution of nitrogen and hydrogen (EGA in helium respectivly nitrogen) accompanied by weight loss (TG). Around 600°C two endothermic DTA signals show a severe decomposition reaction, and the residue gives only a diffuse IR-spectrum.

Quantitative oxygen analysis (by EGA) of the decomposition products of the oxygenated PcFe's fails, because of the complex nature of those products.

Combined EGA/GC would be helpful for this problem. Nevertheless, our thermoanalytical investigations as well as the chemical syntheses exclude {3} and {4} as structures for forms D and X. Our TG experiments exclude a composition like PcFe(O_2) with a calculated weight loss of 5.3 %. Moreover, we believe that PcFe(O_2) should have a lower decomposition temperature, than that found for form X [13].

On the basis of the improved thermal stability of form X and its slower reaction rate with pyridine, compared with form D [9], the description of derivatives with a PcFe : O ratio of 1 [4, 6] , and our knowledge that polymeric PcFe derivatives have higher thermal stabilities than their monomeric or dimeric analogues [15], we conclude that form X could be an oligomeric or polymeric oxo-bridged compound of structure [PcFeO]$_x$ [16].

Table 1: Thermoanalytical decomposition of β-PcFe and oxygenated PcFe compounds.

Compound	Method; atmosph.	Final temperature [°C]	Δm [%]	Ti Tm Tf [°C]
D	DTG (N$_2$)	480	2 - 4	369, 390, 4o6
D	EGA (N$_2$)	530	-	384, 394-, 424
D	EGA (He)	530	-	364, 394, 417
D	EGA (H$_2$)	530	-	368, 388, 398
D	DTA (N$_2$)	530	-	374, 390+, 402
X	DTG (N$_2$)	480	2.6-3	394, 411, 421
X	EGA (N$_2$)	560	-	409, 424-, 442 510+ beginning
X	EGA (He)	530	-	393, 416, 433
X	EGA (H$_2$)	530	-	366, 392, 429
PcFe	DTG (N$_2$)	650	8.5 Σ	589, 606, 625
PcFe	EGA (N$_2$)	620	-	501+ beginning
PcFe	EGA (He)	850	-	457 beginning 599, 625, ? 655, 755, ?
PcFe	DTA (He)	850	-	594, 599-, 604 ? , 611-, 617 739, 746+, 756

Heating rate 5 K/min.
Ti, Tm, Tf refers to the initial, maximum and final temperature of measured signal, respectively.
DTA: + indicates exothermic, - indicates endothermic signal.
EGA: + indicates better thermal conductivity than carrier gas and v.v.

REFERENCES

1. B.D. Berezin, "Coordination Compounds of Porphyrins and Phthalocyanines", J. Wiley & Sons, Chichester (1981).
2. T.G. Spiro (ed.), "Metal Ion Activation of Dioxygen", J. Wiley & Sons, New York (1980).
3. C. Ercolani, G. Rossi and F. Monacelli, Inorg.Chim. Acta, 1980, 44, L 215.
4. I. Collamati, Inorg.Chim.Acta, 1979, 35, L 303.
5. N.I. Budina, O.L. Kaliya, O.L. Lebedev, E.A. Lukyanets, G.N. Rodinova and T.M. Ivanova, Koord.Khim. (engl.transl.), 1976, 2, 720.
6. I. Collamati, Inorg.Nucl.Chem.Lett., 1981, 17, 69.
7. C. Ercolani, F. Monacelli and G. Rossi, Inorg.Chem., 1979, 18, 712.

8. J. Moll, Ph.D.Thesis, in preparation.
9. O. Schneider, Ph.D.Thesis, in preparation.
10. A.N. Sidorov and I.P. Kotlyar, Opt.i.Spektroskopiya (engl.transl.), 1961, 11, 92.
11. M. Calvin, E.G. Cockbain and M. Polanyi, Trans.Farad. Soc., 1936, 32, 1436.
12. We wish to thank Prof.Dr. Pauschmann for the GC-measurements.
13. H.H. Schmidt, Ph.D.Thesis, University of Hamburg, 1976. F. Steinbach and H.-J. Joswig, J.Catal., 1978, 55, 272.
14. J. Blomquist, L.C. Moberg, L.Y. Johansson and R. Larsson, J.Inorg.Nucl.Chem., 1981, 43, 2287.
15. O. Schneider and M. Hanack, Angew.Chem., 1982, 94, 68. O. Schneider and M. Hanack, manuscript in preparation.
16. While this work was in progress the structure of an oxo-bridged iron polymer with hemiporphyrazine as macrocyclic system was solved by single-crystal X-ray diffraction to be $[HpFeO]_x$ (A. Datz, W. Hiller, M. Hanack and J. Strähle, to be published). $[HpFeO]_x$ decomposes at exactly the same temperature like the compound X.

THERMOGRAVIMETRIC ANALYSIS OF POWDERED ACTIVATED CARBON IN BIOLOGICAL SYSTEMS

W.K. Bedford and T.R. Bridle

Wastewater Technology Centre, Environmental Protection Service
Environment Canada, Burlington, Ontario.

ABSTRACT

The quantification of powdered activated carbon (PAC) in biological sludges was achieved through the use of thermogravimetric analysis (TGA). The method comprised heating PAC, a municipal and industrial sludge and those same sludges spiked with various amounts of PAC under sequential inert/oxidizing atmospheres. A set of simultaneous equations describing weight loss of pure PAC and sludge under helium and oxygen atmospheres were developed. These equations were used to predict the PAC concentration in PAC/biomass (sludge) mixtures. Plots of percent PAC data versus percent weight loss data under oxygenic conditions for the sludges spiked with PAC yielded 99% correlation coefficient values.

This is a new application for TGA and with further research the development of a suitable methodology for laboratory PAC optimization studies and the control of full scale systems could be realized.

INTRODUCTION

Increasingly stringent effluent quality criteria are making demands on industrial and municipal wastewater treatment systems which can only be met by improving existing technologies. The addition of powdered activated carbon to the biomass of a biological plant is one such example of attempts to produce cleaner, safer effluents. The quantity of PAC which is added to a given system will be that amount which produces the desired results based on process, aesthetic and economic considerations. It is essential in any biological system to maintain a stringent biomass inventory, measured as volatile suspended solids. Theoretically, biomass concentration is determined by volatilization in air at 550°C, which, of course, also promotes oxidation of the activated carbon. To date, no acceptable methodology for the analysis of PAC in biological sludges has been derived. The inability to maintain an accurate biomass and activated carbon inventory in a process leads to a significant limitation in that the most cost-effective process conditions cannot be maintained. This is particularly true in cases where hydraulic overloads and subsequent PAC/biomass loss are common place. Numerous empirical differential volatilization procedures have been developed in order to conveniently estimate activated carbon concentrations [1], [2]. However, most of these techniques are based on assumptions, and are not consistently accurate over a broad range of activated carbon concentrations.

A technique for measuring PAC concentrations in PAC/biomass mixtures, using a Du Pont 1090 Thermal Analyzer is currently under development at the Wastewater Technology Centre (WTC). The approach taken was to utilize the ability of the thermogravimetric analyzer to generate direct weight loss data under both inert and oxidizing conditions. Development of the method was based on the premise that a series of simultaneous equations describing weight loss, under each condition, could be generated.

EXPERIMENTAL TECHNIQUE

Sludges from pilot and bench scale biological reactors operated at the WTC were used in this study. They comprised a municipal sewage sludge averaging 73.87% volatile suspended solids, and an industrial sludge from the treatment of coke plant wastewater, averaging 18.20% volatile suspended solids. Sludges were collected and spiked with various levels of PAC ranging from 7 to 75%. Samples for thermal analysis were prepared by air drying and grinding to -200 mesh. Thermograms were generated for pure PAC, biomass (mixed liquor) and PAC/biomass mixtures.

The 1090 Thermal Analyzer was programmed to operate under the following set of sequential conditions.

Method 1: Ambient temperature to 100°C at 10°C/minute
maintained at 100°C for 5 minutes,
Method 2: 100°C to 565°C at 20°C/minute
maintained at 565°C for 20 minutes,
Method 3: 565° to 575°C at 5°C/minute
maintained at 575°C for 10 minutes.

Methods 1 and 2 were purged with helium and Method 3 with oxygen at 50 cc/minute.

Method 1 achieves complete drying such that air dried sample weights are normalized. Methods 2 and 3 provided the appropriate temperature ramping and atmospheric conditions to generate the weight loss data under helium and oxygenic conditions. TGA operating conditions in Methods 2 and 3 were selected such that the temperatures and heating times fell within the range normally used for the determination of volatile suspended solids.

RESULTS

Thermograms for Hydrodarco C powdered activated carbon, municipal sewage sludge (ML-PP3) and industrial sludge (ML COKING) are shown in Figures 1 to 3. Inserted in the figures are tables of actual and average (for replicated runs) weight percent losses under helium and oxygen conditions. As expected, PAC exhibited very little weight loss under helium conditions. This weight loss is probably due to impurities in the PAC (this sample of PAC contained 25.87% ash). By contrast, the municipal and industrial sludges (mixed liquor) exhibited a major portion of their weight loss under helium conditions. This was attributed to decomposition of the low molecular weight acids, proteins and lipid materials present in bacterial samples.

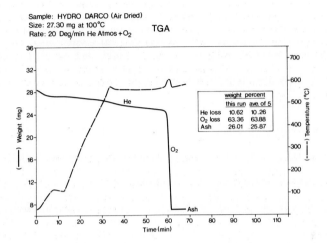

Figure 1. Hydrodarco (PAC) Thermogram

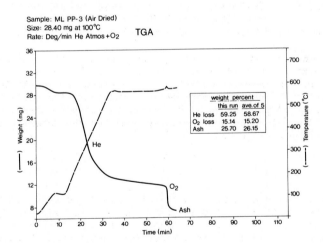

Figure 2. Municipal Sludge Thermogram

It can be hypothesised that provided no interactions between PAC and biomass occur, it would be possible to develop a set of simultaneous equations which describe weight loss of PAC/biomass mixtures under both inert and oxidizing atmospheres. Data from the pure PAC (Figure 1) and municipal sludge (Figure 2) data were used to generate the equations presented below:

$$W_1 = 0.1026 \text{ PAC} + 0.5867 \text{ MLSS} \tag{1}$$

$$W_2 = 0.6388 \text{ PAC} + 0.1520 \text{ MLSS} \tag{2}$$

where W_1 = weight loss under helium atmosphere, Method 2 (mg)

W_2 = weight loss under oxygen atmosphere, Method 3 (mg)

PAC = actual weight of PAC in sample (mg)

MLSS = actual weight of biological sludge in sample (mg)

The data from Figures 1 and 2 indicate that under a helium atmosphere PAC undergoes a 10.26% weight loss while the municipal sludge exhibits a 58.67% weight loss. Similarly, under oxygenic conditions PAC and municipal sludge showed weight losses of 63.88% and 15.20% respectively. These numbers are used in Equations 1 and 2 to describe total weight loss under helium and oxygen conditions.

The validity of these equations was tested using weight loss data generated from the spiked municipal sludge thermograms. The average weight loss (three runs) under helium and oxygen conditions was calculated and used in Equations 1 and 2 to predict PAC concentrations. A comparison of actual (spiked level) versus calculated PAC concentrations is shown in Table 1.

TABLE 1   MUNICIPAL SLUDGE DATA

| PAC Concentration (%) | | % Relative |
Spiked Level	Calculated	Difference
0	0	0
7.1	5.0	-29.5
14.2	14.4	+1.4
20.0	21.6	+8.0
26.2	28.1	+7.2
34.1	31.6	-7.3
38.7	35.5	-8.3
45.1	44.7	-0.8
50.2	51.2	+2.0
100.0	100.0	0

With the exception of the 7.1% PAC sample, very good estimates of the actual PAC concentrations were achieved with the relative difference ranging from -8.3% to +8.0%. The reason for poor estimation at low PAC concentrations is not immediately evident. However, it is

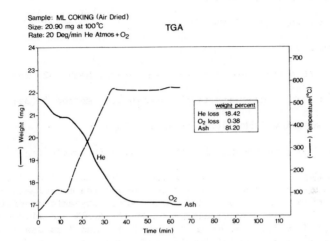

Figure 3. Industrial Sludge Thermogram

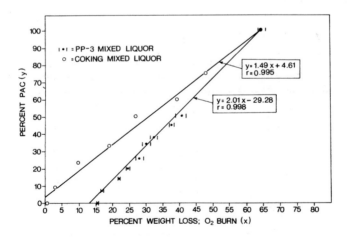

Figure 4. PAC/Biomass Standard Curves

speculated that sample nonhomogeneity may be a prominent factor.

From a plant operator's viewpoint, a desirable output would be a standard curve showing percent PAC versus, say, percent weight loss under oxygenic conditions. Such curves are plotted for the municipal (PP-3 mixed liquor) and industrial (Coking mixed liquor) sludge data obtained from the actual TGA runs.(Figure 4). It can be seen that excellent correlations were achieved (greater than 99%) for both sludges. The reproducibility for the municipal data, at each PAC concentration, is depicted by the range defined within the vertical bars on the figure.

The discrepancy between the two curves on Figure 4 indicates the sensitivity of the methodology to sludge volatility. For example, the municipal sludge at 73.87% volatile suspended solids exhibited a 15.20% weight loss under oxygenic conditions, whereas the industrial sludge at 18.80% volatile suspended solids showed only a 0.38% weight loss under oxygenic conditions.

SUMMARY

A preliminary methodology for the quantification of PAC in biological sludges has been presented. The method does, however, need verification using data generated from an operational PAC enhanced biological process. Furthermore, the method is sludge specific, and further research is required to define a relationship between sludge volatility and percent weight loss under oxygenic conditions. Additional studies are also required to identify the reasons for poor estimation at low PAC concentrations.

REFERENCES

1. T.R. Bridle, W.K. Bedford and B.E. Jank. Biological Nitrogen Control of Coke Plant Wastewaters. Progress in Water Technology, 1980, 12, p.667-680.
2. J.S. Lee, and W.K. Johnson. Carbon-Slurry Activated Sludge for Nitrification-Denitrification. Water Pollution Control Federation Journal, 1979, 51, p.111-126.

MICROCALORIMETRIC INVESTIGATIONS ON
MIXED MICROBIAL CULTURES

I. Lamprecht, B. Schaarschmidt
Institut für Biophysik, Freie Universität
D-1000 Berlin 33, Thielallee 63/67

INTRODUCTION

With the increasing versability and sensitivity of modern instruments attention of calorimetrists shifted more and more from the classical quantitative application to the modern analytical one. In this field of research one is not only interested in the amount of heat produced in a system but in the kinetics of thermogenesis, too. A typical example of these investigations are "fingerprint"-like power-time curves of growing microorganisms [1].
For this end it is necessary to have a well-defined medium and an isolated organism. The different peaks in the power-time curves are due to a consecutive decomposition of the initial material or to a production of metabolites which are used as energy or nutrient source in later steps by the culture. On the other hand, many microbiologically similarly acting substances as for instance several monosaccharides can be degraded simultaneously so that they exhibit only one common peak in the power-time curve.
When two or more different organisms are growing in a common culture on one medium the interrelations can become very complex. The resulting power-time curves are no simple superpositions of several heat profiles - so as spectra or as peaks in chromatograms can just be added together - for the metabolic products of one organism can serve as nutrient source for other organisms or can be poisoning to them. Therefore, characterization and identification of microorganisms by their fingerprint-like heat profiles [1] become nearly impossible in mixed cultures.
By slight changes in the outer parameters of the system such as oxygen tension, pH, temperature or substrate concentration the metabolic pathways can be altered considerably so that new heat profiles appear.
When microbes in a mixed culture live on the metabolic products of another species, then one of these products may be heat. If the system is large in diameter or otherwise adiabatically shielded against its surroundings the heat production leads to increasing temperatures of the culture. In the course of such an experiment different groups of microorganisms become active. At low tem-

peratures between 0 and 15 °C the psychrophilic microbes are predominant. They are of no importance in our investigations since starting temperature usually lies in the region of the mesophilic organisms between 20 and 40 °C. They are the prerequisits for all biological decompositions as they establish the growth conditions for the following thermotolerant and thermophilic organisms which develop between 40 and 60 °C or 50 and 90 °C, respectively. All these groups are ubiquitous in nature, as well in hay, straw and manure as in wool, cotton and palm-kernels - just to cite a few systems that exhibit self-heating or self-ignition. Normally in nature, thermal insulation is so bad that temperatures are established which are just slightly increased above the ambient ones. Only in compact material as in cotton, jute or wool barrels, in straw or hay bales or in large piles of compost or manure temperatures up to 90 °C are observed which are followed by chemical or physico-chemical processes till to self-ignition of the material.

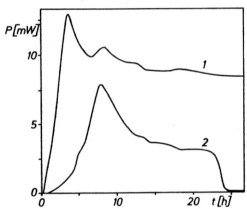

Fig. 1   Power-time curve of 50 ml of liquid cattle manure at 45 °C
1: in a vessel open to the atmosphere, 2: in a hermetically closed vessel.

Mixed cultures can be investigated calorimetrically in two different ways:
- adiabatically
- isothermally,

two ways that are found more or less profound in nature, too. The second one may be considered as a "cross-section" in temperature of the first one and provides information about one group of microorganisms active at the chosen temperature. Both methods are applied nowadays for the study of mixed cultures.

The first scientist working in this direction was supposingly Rubner [2] looking in a quasi-adiabatic system for lactic fermentation by non-defined microbes, among them fungi and yeasts. He observed only small increases in temperature - around 1 °C - so that only mesophilic organisms became active. With modern adiabatic set-ups temperatures up to 90 °C are obtained [3]. Often no calorimeters are used but temperatures in large bales, barrels or piles measured by thermocouples and heat leakages taken into account or just ignored [4,5].

Isothermal calorimetry asks for the microbial activity at a preset temperature and helps to compare the different groups of organisms in the system. For in adiabatic experiments thermophilic microbes may seem to be less active than mesophilic ones just by the fact that the readily accessible substrates were used up by mesophilic cells before. Often, these isothermal power-time curves are without any structure [6], sometimes they exhibit fingerprint-like shapes with high reproducibility [7].

As interesting as both ways may be one has to bear in mind that both are only approximations to the processes occuring in nature.

METHODS AND MATERIALS

Calorimeters. The isothermal experiments were performed in a batch microcalorimeter of the Calvet-type (SETARAM/Lyon-France) with 4 vessels of 100 ml content and a sensitivity of 52 μV/mW. The vessels could be closed hermetically so that no exchange of air with the surroundings was possible. The amount of substrate varied between 15 and 50 ml. Different temperatures were prechosen and kept constant throughout the experiment.

The adiabatic calorimeter, developed in the institute, consisted of a twin set-up of two Dewar flasks in an adiabatic shield and was described elsewhere in detail [8]. The 250 ml flasks were loaded with 25 to 100 ml of substrate so that ample space remained for air.

Substrates. Organic materials known for self-heating were used for these investigations: fresh and long-time stored cattle, pig and poultry manure, household litter of different origin and composition, compost from the garden, freshly cut gras. In connection with other experiments on ecologic heat production, material used by wood ants in constructing mounds was included in this research. The substrates were cut into smaller pieces - if necessary - to fit into the calorimetric containers. Loss of water during heating-up was taken into heat balance considerations by weighing the specimen.

No microbial determinations were as yet carried out. But it is known from other investigations that the microbial flora mainly consists of fungi, yeasts, bacteria and actinomycetes.

RESULTS

Figure 1 shows the power-time curve of liquid manure in an isothermal experiment at 45 °C. The differences appear because in (1) the calorimeter vessel was open to ambient atmosphere so that exchange of oxygen and carbon dioxid could take place, while in (2) the vessel was hermetically closed. There are no fingerprint-like structures. Modulations of the slope are artificially produced by outer parameters - mainly the oxygen supply - and are typical for the instrumental procedure, not for the biologic system. The later stationary phase of the open vessel can continue at this level for several days. Typical powers are in the order of 0.25 mW/g at 50 °C for half the vessel filled with liquid manure. The initial maximum amounts to nearly the double.

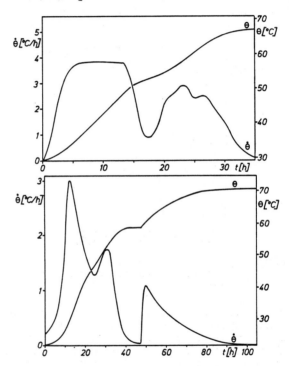

Fig.2 Temperature-time curves of two adiabatic systems. Top: 20 g freshly cut gras; bottom: 25 g pig manure. ⊙: temperature; ⊖: momentary increase in temperature.

Power-time curves of solid manure strongly depend on the degree of filling of the vessel. While output during

the stationary phase remains unchanged at approximately
0.50 mW/g manure, the maximum amounts to 2.2 mW/g for 15
g substrate, but to only 0.65 mW/g for 40 g. This is
plausible, as the first part is due to the readily accessible oxygen above the specimen.

Quite a different picture offer the temperature-time
curves or thermograms of adiabatic systems (Figs.2, 3).
The heat production as consequence of the biologic activity in the specimen leads to a steady increase in temperature, because heat is prevented from leaving the calorimeter. Temperature Θ as function of time or even
better the rate of temperature increase $d\Theta/dt = \dot\Theta$ which
is directly correlated with the biologic power via the
heat capacity of the system clearly exhibits distinct
phases of microbial activity: the mesophilic, thermotolerant and thermophilic ones. As was pointed out in the
INTRODUCTION the mesophilic period appears predominant
because of the consumption of readily digestable substrates.

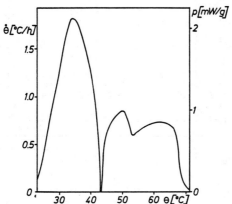

Fig. 3 Thermogram of 25 g household
litter in an adiabatic experiment.

Table 1 presents some selected figures from the
three phases. These values are only approximative as in
many experiments a clear distinction between the periods
- especially between the thermotolerant and thermophilic
one - is somewhat arbitrary. But the obtained mean values correspond well with data from literature.

Although very different materials were included in
the experiments the temperatures of maximum heat output
in the three phases fall into small intervals. Large
fluctuations are of course found in the maximum heat
productions themselves because microflora and digestable
compounds in specimens like pig manure and gras differ
considerably. Therefore, the full region of biologic

heat production is presented in Table 1 and the mean value only given for orientation.

The primary information drawn from adiabatic experiments are temperature-time curves. But often it is more suited to present the results as real thermograms as seen in Fig.3: rate of temperature increase or power as functions of temperature. In contrast to thermograms from DTA- or DSC-instruments, temperature here is not imposed by an external program, but a consequence of internal exothermic processes.

Table 1 Some characteristic data for mixed cultures of mesophilic, thermotolerant and thermophilic microbes.

		Mesophilic	Thermotol.	Thermophil.
Temper. of max. heat output	°C	36.0 ± 3.8	54.2 ± 2.6	64.0 ± 2.8
Upper limit of activity	°C	47.5 ± 3.2	59.1 ± 3.3	71.6 ± 4.4
Maximum heat production	$\frac{mW}{g}$	1.65 to 9.08	1.15 to 5.68	0.65 to 4.69
mean	"	4.67 ± 2.25	3.17 ± 1.73	2.14 ± 1.28
Percentage of total heat	%	50.1 ± 7.7	29.1 ± 5.5	19.7 ± 8.5

DISCUSSION

Many papers published in this field look for maximum temperatures or heat productions obtained by decomposing processes in natural material, but are not interested in the question of mixed cultures and the interrelations between the different microbial groups. Bryant [9] could show in the mesophilic range that degradation of sewage is carried out by a threefold consortium of interacting anaerobic microbes. While the first group hydrolyzes the original substrate by fermentation to shorter compounds, the two other groups use these compounds as energy and nutrient sources.

When microbial determinations are performed, bacteria fungi and actinomycetes are found. In self-heating hay mainly thermophilic moulds were observed, among them Absidia, Aspergillus and Mucor [5]. Glathe [10] isolated under corresponding conditions Aspergillus, Geotrichum, Mucor, Penecillium and Rhizopus.

Power-time curves or thermograms are not always as clearly structured as in Figs.2 and 3. Often, the interrelations of the microbes are so numerous or a special substrate so predominant, that only a steady and monotonous increase in temperature occurs.

Reproducibility of heat profiles is not very good, since a homogeneity of the biologic material is hardly to obtain without introducing artefacts. Size of larger particles (as leaves, branches etc.) change the physical consistence and therefore the diffusion characteristics of oxygen, so that the heat profiles change.

Table 1 presents the temperatures at which maximum heat production is observed within the three groups. They correspond to figures drawn from literature. In the decomposition of straw a first mesophilic peak occured at 40.0 °C, a second thermophilic one at 60.2 °C [11]. During destruction of hay maximum heat production was observed around 37 and 60 °C [10, 12].

As pointed out before thermogenesis can vary considerably for different organic material. The figures from Table 1 fall into the range published by other authors: 1.1 [5] and 1.4 mW/g [12] for hay; 1.6 to 1.8 mW/g in composting spruce bark [13]; 6.0 mW/g in the nest material of the red wood ant; 6.3 to 10.9 mW/g in straw [11]. These values appear really small as compared with the thermogenesis of microbes calculated per dry weight of cells. But if one takes into account the huge amount of biomass exposed to decomposition, the common interest in these energy sources become obvious.

The figures obtained with isothermal calorimetry are smaller but in the same order of magnitude as the above mentioned ones. Solid manure in our experiments developped 0.5 to 2.2 mW/g, barley straw 0.3 to 0.7 mW/g [7]. The heat production in the rumen content of sheeps was found to 3.0 mW/g [14].

Subdivision of the temperature profiles into microbial domains is consistent with other findings, too. Mesophiles contribute to the total energy budget with 53 [12], 55 [10] and 61 % [5] in the decomposition of hay, 56 % with straw [11], while thermophiles add 25 [5], 35 [12] or 40 % [10] in hay and 44 % in straw [11]. Figures for the thermotolerant portion range between 5 and 14 %.

The distribution between heat output of mesophilic and thermophilic organisms may change considerably in the naturally occuring processes, when a short mesophilic phase of heating-up is followed by a long lasting thermophilic phase at optimal but not maximum temperatures. In this case, heat leackage from the system protects it against unfavorable high temperatures which would inactivate the microbial populations. In an example of rotting litter mesophilic bacteria produced only 21 % of the total heat evolved by the system whereas from the initial temperature slope one would deduce 69 % [15]. Moreover, starting temperature can change the picture, too, as high initial values decrease the mesophilic period and lead directly to the thermotolerant one.

REFERENCES

1. R.D. Newell, in: "Biological Microcalorimetry (A.E. Beezer, Ed.), Academic Press, London, (1980)
2. M. Rubner, Arch.Hyg., 1906, 57, 244
3. H.M. Hussain, Z.allg.Mikrobiol., 1973, 13, 323
4. I.K. Walker, H.M. Williamson, J.appl.Chem., 1957, 7, 468
5. P.H. Gregory, M.E. Lacey, G.N. Festenstein, F.A. Skinner, J.gen.Microbiol., 1963, 33, 147
6. B. Schaarschmidt, I. Lamprecht, Experientia, 1976, 32, 1230
7. M.L.Fardeau, F. Plasse, J.-P. Belaich, Eur.J.Microbiol.Biotechnol., 1980, 10, 133
8. H. Bolouri, I. Lamprecht, in: "Angewandte chemische Thermodynamik und Thermoanalytik", Birkhäuser, Basel, (1979), 200
9. M.P. Bryant, in "Seminar on microbial energy conversion", Goltz, Göttingen, (1976), 107
10. H. Glathe, Zblt.Bakteriol., II.Abt., 1959, 113, 18
11. R.E. Carlyle, A.G. Norman, J.Bacteriol., 1941, 41, 699
12. H.P. Rothbaum, J.appl.Chem., 1963, 13, 291
13. G. Bagstam, Eur.J.appl.Microbiol.Biotechnol., 1979, 6, 279
14. W.W. Forrest, in "Les développements récents de la microcalorimétrie et de la thermogenèse", CNRS, Marseille, (1967), 405
15. F. Pöpel, Stuttgarter Ber.Siedlungswasserwirtsch., 1968, 41, 5

MICROCALORIMETRIC EXPERIMENTS ON SKIN
CELLS. INFLUENCE OF DRUGS ON HEAT
PRODUCTION

M.Pätel, B.Schaarschmidt, I.Lamprecht
Institut für Biophysik, Freie Universität
D-1000 Berlin 33, Thielallee 63/67

INTRODUCTION

Calorimetry of microorganisms is a well established method nowadays, and testing of bacteriocidic or bacteriostatic drugs by means of changes in power-time curves seems to be a helpful instrument in clinical work [1]. Until recently only few experiments were performed on human skin [2,3] or cell cultures of human tissue and the influence of medicaments upon them [4,5]. The aim of this paper is to show how calorimetry can be used for this end and to give some preliminary results of such an attempt.

Modern calorimetry with its high sensitivity comparable with that of classical Warburg technique or polarography offers a general approach to cell metabolism. For it is not bound to changes of gas pressure or of oxygen concentration only, but detects other, even unexpected heat effects as well. This can be of great importance in investigations with different drugs. Nevertheless, it is always interesting and helpful to use parallel test techniques to simplify the interpretation of the observed effects.

We concentrated our investigations on the drug anthralin, which is the most important and widely used pharmacon against several skin diseases, among them psoriasis, although its mode of action is largely unknown.

MATERIAL AND METHODS

Calorimeter. The experiments were performed with an isoperibolic batch calorimeter, type "BIOFLUX" from Thermanalyse/Grenoble-France. The vessels of this twin instrument had an active volume of 18 ml at an inner diameter of 1.6 cm. The working temperature in all experiments was 37 C, the sensitivity 62.4 μV/mW.

Medium. Cells were incubated in Dulbecco's modified Eagle medium which was supplemented by 15 % fetal calf serum, 25 mM HEPES, 100.000 units/liter penicillin, 100 mg/l streptomycin, and 250 μg/l amphotericin B to suppress microbial growth. The medium was adjusted to a pH of 7.4. When studying the influence of drugs on cell metabolism this medium was exchanged for an identical one containing the drug.

Drugs. The following drugs were used in these experiments: Anthralin (1,8-dihydroxy-9-anthron), anthraquinone and the anthralin dimer. Anthralin is known in therapy under the names Dithranol and Cignolin, too.

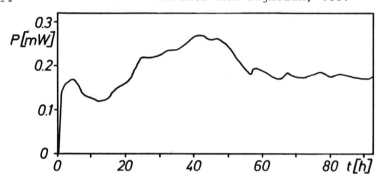

Fig.1 Power-time curve of the growth of a fibroblast culture. For details of the structure see text.

Cells. Human foreskin fibroblasts were precultivated in the medium over three passages to adapt them to the growth conditions and then stored in flasks until semiconfluence. In this state the cells were harvested by trypsinization in an HBSS-buffer. Finally they were seeded on charged plastic foils (Sterilin, 5 cm x 7 cm) in a tenter frame. 60 minutes later the cells were attached to the foil which could be then transferred to the calorimetric vessel in such a way that it covered the inner surface of the vessel. 16 ml of fresh medium were added. The reference vessel likewise contained 16 ml medium, but an untreated foil.

Cell viability was tested with the vital stain trypan blue. Only non viable cells take up this stain.

RESULTS

The growth of the fibroblast cultures is accompanied by a specific power-time curve. This heat profile is highly reproducible in different experiments and clearly structured. In principle, one can distinguish three different parts (Fig.1)
- a short initial phase which may be due to thermal disturbances, cellular adaptation processes or other surface reactions which are as yet unknown,
- a phase of cell growth by multiplication in which the metabolism of construction is predominant. In this phase one observes a heat output of 4.5 ± 0.6 μJ per newly formed cell,

- a phase of active, but not deviding cells which exhibit the metabolism of maintenance. Here, the rate of heat production amounts to 51.5 ± 5 pW per cell. Again this figure is in the order of magnitude which other authors found for similar cellular systems.

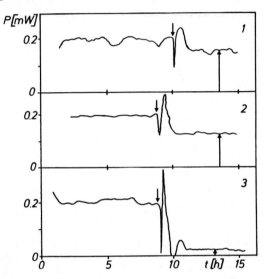

Fig.2 Power-time curves of the phases of maintenance and of the addition of anthralin (arrows) in the following concentrations:
1 - 1 μM anthralin in ethanol;
2 - 5 μM anthralin in ethanol;
3 - 10 μM anthralin in acetone.

When the culture reached confluence (i.e. the phase of metabolism of maintenance) the old medium was exchanged for a new one containing the wanted amount of the drug.

If concentrations of anthralin lower than 2 μM were offered to the cells no significant effect on the heat production of the cells could be observed. Increasing concentrations of drug, however, continuously decreased the power output of the system. The cultures took up new plateaus in the power-time curve which lasted for at least 20 h (Fig.2). Some "ondulations" were always present. For the moment it is not clear which are the reasons for these derivations.

A dose response curve of the anthralin action on heat output exhibits a two-phasic slope with a shoulder. Minimum or even no actions are observed below concentrations of 2 μM and a reduction of 50 % occured at 7 μM anthralin.

At 20 μM the rate of heat production nearly dropped to zero. Among the tested anthralin derivatives anthraquinone showed no influence on heat output in the concentration range normally applied for anthralin. In contrast, the anthralin dimer diminuished heat evolution at concentrations higher than 2 μM, being slightly more effective than anthralin itself. The dimer exhibited a 50 % decrease at a concentration of 4 μM. Both, monomer and dimer followed the same shape of dose-response curve.

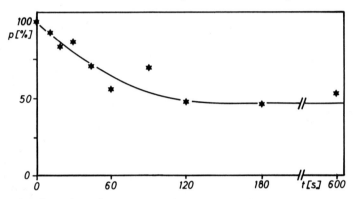

Fig.3 Kinetics of anthralin action on the rate of heat production. The abscissa represents the time t during which the culture was exposed to anthralin in a concentration of 10 μM.

In a few experiments cultures were exposed for limited times to a medium containing 10 μM anthralin. After subsequent washing and exchange of the medium the culture gave a significantly reduced heat evolution when the application time exceeded 10 s. The heat production was irreversibly inhibited and an application of 2 min already showed the same reduction as the permanent treatment (Fig.3). That means that anthralin uptake proceeded at a fairly high speed.

Anthralin caused lysis and cell death while the dimer and anthraquinone were inactive in this sense.

Parallel experiments on the respiration rate of fibroblasts showed that respiration is far more sensitive to anthralin than heat production or as survival in the sense of excluding trypan blue. 50 % reduction of respiration was obtained already with anthralin concentrations below 2 μM which was the threshold amount for the other effects (Fig.4). The same I_{50} value of respiration was found after addition of the dimer.

Figure 4 demonstrates the changes in the rates of heat production p and respiration r with the degree of inactivation I. For comparison respiration was transformed to power units, too, using the conversion factor of 21 kJ per liter O_2. The difference between p and r corresponds to the portion of fermentation of the cells to the heat production. The straight lines (p) and (r) were to be expected when the number of viable cells were decreased by a certain percentage. The differences between (p) and p or (r) and r exhibit the degree to which heat production or respiration are stronger influenced by anthralin than viability examined by trypan blue exclusion. From Fig.4 it becomes evident that anthralin treated human fibroblasts are still viable when heat production and respiration are already reduced.

So one may conclude from calorimetric experiments that human skin fibroblasts are more sensitive to the dimer than to anthralin. This is not correct for respiration measurements.

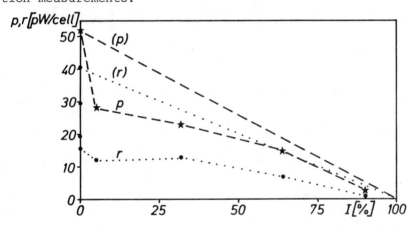

Fig.4 Dependence of specific heat production p and specific rate of respiration r (calculated per power units, too) as functions of the degree of inactivation I of cells by anthralin.

CONCLUSION

The cited experiments showed that microcalorimetry is a useful tool for monitoring cellular processes and the effects of therapeutic agents on human cell cultures. By means of these data, dose-response curves can be established for the different derivatives of anthralin. It seems to be possible to discriminate by calorimetric investigations between related and therapeutically similarly acting drugs. Calorimetry has approximately the same sensitivity as other methods but sometimes it gives additional and more detailed information than those.

Thus, it seems worthwhile to continue these experiments and to include further drugs into the program.

REFERENCES

1. A.E. Beezer (Ed.), "Biological Microcalorimetry", Academic Press, London, (1980)
2. L. Hellgren, K. Larsson, J. Vincent, Arch. Derm. Res. 1977, 258, 295
3. A. Anders, H. Schaefer, B. Schaarschmidt, I. Lamprecht, Arch. Derm. Res., 1979, 265, 173
4. M. Pätel, B. Schaarschmidt, U. Reichert, Brit. J. Dermat., 1981, 105, Suppl. 20, 60
5. B. Schaarschmidt, U. Reichert, Exp. Cell Res., 1981, 131, 480

TESTING THE EFFECT OF THREE CHEMICAL
SUBSTANCES FORMED IN LIVESTOCK FACILITIES
ON E. COLI BROTH CULTURES BY FLOW
MICROCALORIMETRY

J. Hartung
Institute for Animal Hygiene,
Hannover School of Veterinary Medicine
Bünteweg 17p
3000 Hannover 71
Federal Republic of Germany

INTRODUCTION

Biological microcalorimetry is used to study the effect of antibiotics, related drugs and water polluting compounds on microorganisms in medical microbiology, pharmacology and environmental analysis [1] to [6]. These reports suggest the possibility of an analytical procedure, based upon microcalorimetry for air polluting compounds too.

There are approximately 125 compounds identified in the air of pig fattening houses [7], mainly formed by aerobic and anaerobic digestion of indoor stored slurry [8]. There is only little knowledge regarding the role of the majority of these trace gases in the origin and progress of animal diseases in intense livestock production and there is a lack of suitable pre-testing methods in the laboratory scale to check the influence of these compounds on simple test organisms. Therefore it was tried to build up a microcalorimetric test method to screen the influence of some water-soluble chemical compounds on bacteria in broth culture.

This paper reports on flow microcalorimetric investigations on the action of the chemical compounds m-cresol, trimethylamine (TMA) and ethanol, appearing in the air of animal houses, on the heat changes in an Escherichia coli broth culture.

MATERIALS AND METHODS

Strain ATCC 11229 of Escherichia coli was used as test organism. Nutrient Broth No. 2 (Oxoid, Wesel, Germany), pH 7.5 was used as culture medium. M-cresol, trimethylamine (TMA), and ethanol were purchased from Merck, Darmstadt, Germany in analytical pure grade quality.

Heat evolution of the bacterial culture was measured by continuously pumping parts of the culture through a heat conduction flow microcalorimeter (type 2107-021,LKB, Bromma,Sweden) fitted with a flow-through cell. The calorimeter measured the rate of heat production Q=dQ/dt(W) of the suspension passing the tubular flow-through cell (volume 0.7ml). the p-t-curves were registered by a compensation recorder.

Outside the flow microcalorimeter was installed an incubation tank for preparing the broth culture and a mixing vessel for adding the test substances. The mixing vessel had a fixed volume of 10 ml and was equipped with a gastight injection port. The addition of the chemical test substances to the medium was affected by means of a syringe. Behind the mixing vessel a special device with ion sensitive electrodes was fitted to continuously control the oxygen and pH value of the medium. Incubation tank, mixing vessel, electrodes and flow microcalorimeter were thermostated at 37°C. Approximately 10 min after passing the mixing vessel the medium reached the flow through cell of the calorimeter. Figure 1 gives a diagram of the system.

The test substances were added to the medium approximately 1 h after having reached a steady state heat output. Starting with low test substance concentrations the amounts are increased until a reproducible decrease of the heat output level is recognized. This concentration is called the lowest necessary dose (LND). Increasing test substance concentrations decrease the heat output level. The dose at which the heat output level declines to zero is called the highest possible dose (HPD). Depending on the amount and nature of the test substance an initial endothermic or exothermic reaction is observed. It is followed by the decline of the heat output level caused by antimicrobial action of the test substance.

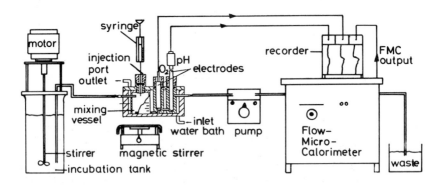

Fig. 1: Schematic Diagram of the Test System

RESULTS

Figure 2 shows the influence of different m-cresol concentrations on the heat output of the sterile nutrient broth (part A), of the grown through nutrient broth at continuous dilution of the test substance (part B), and of the grown through nutrient broth at constant m-cresol concentrations (part C). About ten minutes after the injection of the test substance an initial exothermic reaction is recorded. The height of this initial peak depends on the amount of the substance, the peak heightening with increasing amounts. In part A the heat output level is

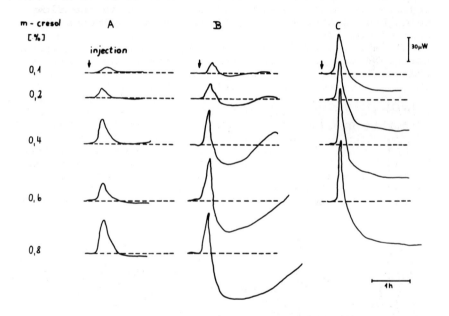

Fig. 2: Influence of Different m-cresol Concentrations on the Heat Output of the Sterile Nutrient Broth (A), of the Grown Through Broth at Continuous Dilution of the Test Substance (B), and of the Grown Through Broth at Constant Test Substance Concentration (C)

reached again within one hour. In parts B and C a decline below the heat output level is observed. This decline increases with increasing test substance amounts. In part B a distinct and reproducible decline of the heat output begins at 0.1% m-cresol and come to some µW. At 0.8% the highest possible decline is reached, showing a reading of 55 µW. Then the curve slowly increases, overshoots the heat output level and returns to it again. The procedure needs between 3 and 5 h, depending on the concentration used. In part C the curves fall and do not return to the baseline. The LND is already reached at about 0.05% and the HPD at about 0.6% m-cresol concentration in the medium. Below the LND-values of 0.1% resp. 0.05% there are also effects detectable but no sufficient reproducibility is given.

In figure 3 there are two curves indicating the development of the heat output in comparison to the germ content of the broth under the action of different m-cresol concentrations. The heat production is presented in relative numbers, 100 representing the uneffected heat flow level. The cell numbers are given in a logarithmic scale. Increasing m-cresol concentrations decrease the heat flow and the number of living cells monitored by plate count technique. A concentration of 0.1% m-cresol significantly reduces the heat flow. The germ number distinctly decreases with m-cresol concentrations of more than 0.4%.

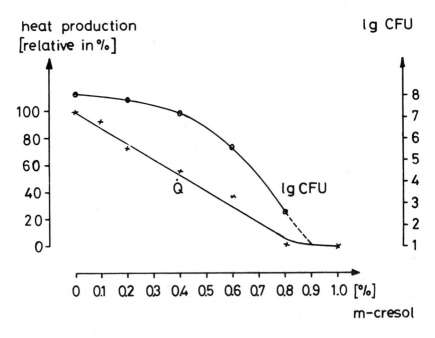

Fig. 3: Developement of Heat Production (%) and Germ Content (lg CFU) Under the Action of Different m-cresol Concentrations.
CFU = Colony Forming Unit

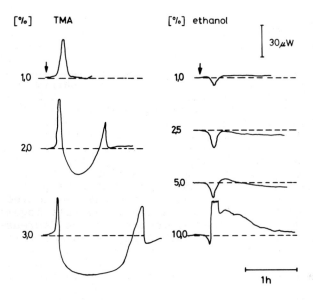

Fig. 4: Influence of Different Trimethylamine (TMA) and Ethanol Concentrations on the Heat Output of the Grown Through E.coli Broth at Continuous Dilution. The Arrows Indicate the Point of Injection.

Figure 4 gives the p-t-curves showing the effect of TMA and ethanol on E. coli in the test system at continuous dilution. For the compound TMA the LND is found between 1.0% and 2.0%, the HPD about 3.0%. Using the compound ethanol up to a concentration of 10% no distinct change in the heat output is observed.

During the experiments the oxygen electrode gave a constant reading of 0% in the grown through broth, the pH value ranged between 7.5 and 6.8.

DISCUSSION

The results show that the described flow microcalorimetric system would appear suitable for checking the influence of the chemical substances m-cresol, trimethylamine and ethanol on E. coli broth culture. M-cresol and TMA have a distinct effect on the heat production of the E.coli strain while ethanol does not affect the heat output of the culture in the concentrations used.

It is shown that the decrease of the heat output is directly dependent upon the nature and upon the concentration of the test substance. These findings correspond to results previously obtained with different antibiotics (4),(5).

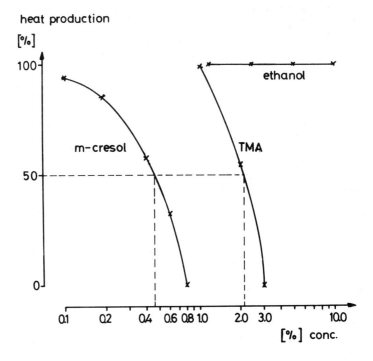

Fig. 5: Scheme for Comparing and Evaluating the Results Obtained fron the Three Test Substances in Regard to Concentration and Heat Production. The Dotted Lines Indicate the HD_{50}-value.

M-cresol shows the lowest reproducible dose effect at 0.05%. Dean and Rogers [9] used m-cresol in a concentration of 0.07% to slightly drop the mean generation time of a bacterial continuous culture. In our experiments the maximum heat flow decline is recorded under the action of 0.8% m-cresol, higher concentrations did not increase the heat flow decline. At this concentration only few living bacteria could be found in the broth. The high effectiveness of m-cresol on bacteria is to explain by the fact, that m-cresol acts on membranes and cell walls of microorganisms even in low concentrations [10].

When comparing the heat production and the germ content under the action of m-cresol it is noted that heat flow and germ content decrease when increasing the m-cresol concentration (fig.3). The diagram indicates that the decrease of the heat flow is earlier to observe and produces faster results than when using the viable plate count method.

In comparison to m-cresol TMA is needed in significant

higher concentrations to suppress the heat production of the microorganisms. Obviously TMA has no strong antimicrobial properties in liquid solution although it is toxic for laboratory rats in the ppm-range as shown by inhalation toxicity studies [11].

Using ethanol concentrations of up to 10% no appreciable differences in the heat flow could be observed. The most effective concentrations of ethanol against microorganisms are given about 60 to 70% [12]. Concentrations below 30% are largely uneffective [12]. Because of technical limitations not more than 1 ml resp. 10% could be injected in these experiments.

It is shown that there are clear differences in the capacity of the three test cubstances m-cresol, TMA and ethanol to influence the metabolism of the test organism E.coli. Figure 5 illustrates these differences in one scheme. By using this scheme the introduction of a "heat decrease 50 dose = HD_{50}" can be proposed that describes the test substance concentration in the medium at which only 50% of the maximum heat flow are left. Possibly a form like this can be the base for comparing and evaluating the results obtained from a greater number of different test substances.

REFERENCES

1. A.E. Beezer, R.D. Newell and H.J.V. Tyrrell, Analyt. Chem., 1977, 49, 34.
2. A.E. Beezer and B.Z. Chowdhry, "Biological Microcalorimetry" (A.E.Beezer, ed.), Academic Press, London, (1980), p.195.
3. K.A. Bettelheim and E.J. Shaw, "Biological Microcalorimetry" (A.E.Beezer, ed.), Academic Press, London, (1980), p.187.
4. P.A. Mardh, T. Ripa, K.E. Andersson and I. Wadsö, Antimicrobial Ag. Chemother., 1976, 10, 604.
5. E. Semenitz, J. antimicrob. Chemother., 1978, 4, 455.
6. B. Redl and F. Tiefenbrunner, European J. Appl. Microbiol. Biotechnol., 1981, 12, 234.
7. H.G. Hilliger and J. Hartung, "Reduction of odours in animal production" (A.A. Jongebreur, ed.), Elsevier, London, (1982), in press.
8. S.F. Spoelstra, Thesis, Wageningen-NL, 1978.
9. A.C.R. Dean and P.L. Rogers, Biochim. biophys. acta, 1967, 148, 774.
10. H. Weide and H. Aurich, "Allgemeine Microbiologie", Gustav Fischer Verlag, Jena, (1979), p.346.
11. F. Koch, G. Mehlhorn, R, Kliche and R. Lang, Math.-Nawi. Reihe, Leipzig, 1980, 29, 463.
12. W.B. Hugo, "Inhibition and Destruction of the Microbial Cell", Academic Press, London, (1971), p.78.

ANALYSIS OF THE THERMAL BEHAVIOUR
OF HUMAN FINGER NAIL CUTTINGS.

B.Marjit
Department of Anatomy
R.G.Kar Medical College, Calcutta.

D.Marjit
Physics Division
Forensic Science Laboratory
Government of West Bengal, Calcutta.

ABSTRACT

Epidermal cells of finger nails become clear and glossy due to the dissapearance of the nuclei in Stratum lucidum. The hard keratin of nail is a member of the albuminoid group of proteins and is very resistant to chemical changes. Because of the complexity of proteins no evaluation of thermoanalytical results is possible without prior anatomical investigations. However, the hard keratin in particular has a high content of sulphur and in general it contains amino acids in various sequences. These amino acids serve as useful model compound for tracing the decomposition mechanism of nails from their TG,DTG and DTA curves. The physical properties and the thermal behaviour of the refractile material of nails which consists of eleidin, a transformation product of keratohyalin were studied as a function of a temperature by subjecting the material to a controlled temperature programme with the help of a Derivatograph.

INTRODUCTION

Though much is known of the chemistry of nail however little use of this phase of nail examination has been made in for identity establishment ie. one species to another or of one individual to another. The chief chemical components of nail are protein, largly the stable and unreactive keratin of which the greater part of the nail is composed. Epidermal cells of human finger nail become clear and glossy due to the disappearance of the nuclei in Stratum lucidum. The hard keratin of nail is member of albuminoid group of proteins and is very resistant to chemical changes. Much is known of the structure of the keratins from the works of previous workers [1,2,3,4 and 5], though it is not certain that difference of significant nature exist between the keratin of different species or individuals in respect of their identity and age. Such attempt has recently been made by Bose and Marjit [6].

In this present work a programme has been undertaken to record the changes arises due to age of nail of human origin by the thermogramme produced by a Derivatograph (MOM,Budapest,Hungary). Different types of nail samples are collected for their comparison in respect of their identity, origin, sex and age. It is known that nails vary considerably in the stability of their structure and such variations may in time become the basis for signifacant differential studies. It is assumed that such variations are the results of differences in the keratin component of nail, though this is also largely surmise. It is sufficient to state that keratin is type of protein which occur in long chains with association or chemical combination of the reactive groups between chains so as to stabilize a structure of fibrils.

EXPERIMENTAL

As the processes taking part in the chemical reactions in the human nail cutting samples involve not only thermal effects detectable by the differential thermal analysis (DTA) but also usually quite substantial mass changes, an apparatus of the mark 'Derivatograph' was selected, which allows the simultaneous performance of DTA, thermogravimetry (TG) and derivative thermogravimetry (DTG).

Derivatograph Programme. The parameters of thermal analysis set on the Derivatograph were the following ones: sensitivities of TG, DTG and DTA were 100, 1/10 and 1/10 respectively. Medium sized platinum crucibles and Al_2O_3 as reference were used.

The samples were heated upto the measurements of 500C with a heating rate of 5C/minute and the entire experiment was performed in static air at 25C room temperature. Other informations about the experiment are given in Table 1.

Table 1

EXPERIMENTAL PARAMETERS ADOPTED FOR DIFFERENT NAIL SAMPLES

No	Samples Nail Cuttings	Age Years	Sex	Heating Rate			Starting Voltage
				Time min.	Temp Range	Rate C/min	
1.	M1	10	M	100	500C	5	35V
2.	F1	10	F	100	500C	5	35V
3.	M2	20	M	100	500C	5	35V
4.	F2	20	F	100	500C	5	35V
5.	M3	30	M	100	500C	5	35V
6.	F3	30	F	100	500C	5	35V
7.	M4	40	M	100	500C	5	35V
8.	F4	40	F	100	500C	5	35V

RESULTS AND DISCUSSIONS

A comparison of the results displayed in Tables 2 and 3 which show the shift of DTA and DTG peaks. The decomposition of nail cuttings occured tn three steps. The relative weight changes (TG) in the individual steps for the different nail cuttings were found to be dependent in respect of age and sex. The DTA peak temperatures were also showed significant difference between the samples. The area of the first exthermic DTA and DTG peaks in the case of aged nail cuttings are found to greater than that of unaged one.

Table 2

DIFFERENTIAL THERMAL PARAMETERS AND TEMPERATURE RANGE FOR DIFFERENT NAIL CUTTINGS.

No	Samples Nail Cuttings	DTA Peak Temperatures		Temperature Range $(T_i - T_f)$
		First Peak	Second Peak	
1.	M1	170C	230C	(90 - 170)
2.	F1	160C	220C	(80 - 160)
3.	M2	175C	255C	(99 - 180)
4.	F2	160C	235C	(90 - 170)
5.	M3	180C	260C	(100 - 190)
6.	F3	175C	245C	(95 - 180)
7.	M4	200C	288C	(120 - 200)
8.	F4	190C	280C	(100 - 190)

Table 3

TG and DTG INFORMATIONS OF HUMAN NAIL CUTTINGS IN AIR.

No	Samples Nail Cuttings	First Peak		Second Peak		Third Peak	
		Temp. C	Wt.loss %	Temp. C	Wt.loss %	Temp. C	Wt.Loss %
1.	M1	174	8.18	210	16.81	320	47.50
2.	F1	170	8.00	205	16.70	312	47.30
3.	M2	176	8.45	214	17.54	330	47.65
4.	F2	173	8.31	211	17.31	318	47.50
5.	M3	177	8.58	217	17.58	358	47.85
6.	F3	175	8.40	214	17.50	328	47.68
7.	M4	182	8.66	222	17.68	361	48.00
8.	F4	180	8.55	218	17.60	345	47.80

The heat transfer co-efficient (K) of the holders and the heat evolved by the reactions (dH) that took place through out the rise of furnace temperature of the aged and less aged nail cuttings may be shown as

$$dH = K \int_{t_1}^{t_2} dT \cdot dt.$$

Using the co-ordinates dT and t

$$S = \int_{t_1}^{t_2} dT \cdot dt.$$

dH = K·S., where S is the area of the peak on the DTA curve.

Tables 2 and 3 show that the nail cuttings undergo weight loss at temperature similar to those observed for human skin [7]. Weight losses occurring upto about 230C are presumably due to loss of water, which occurs in individual amino acid units by loss of one oxygen atom from the amide carboxyl group with two hydrogens from the amino acid chains.

It is known that the amino acid chains in keratin type of proteins are located not only by so called secondary valencies but also by other physical forces outside the chains. The nature and strength of these forces depend on the protein and on whether the properties of the protein changed or not. The differences between the curves for same age group with different sex and different age group with same sex are illustrated in Figures 1,2,3 and 4.

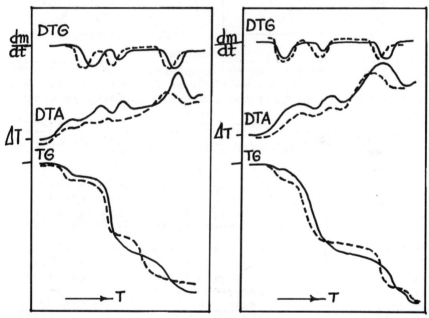

Figure 1

Simultanious DTG,DTA and TG Curves for human finger nail cuttings of ten years age group
Male (———) Female (----)

Figure 2

Simultanious DTG,DTA and TG Curves for human nail cuttings of twenty year age group
Male (———) Female (----)

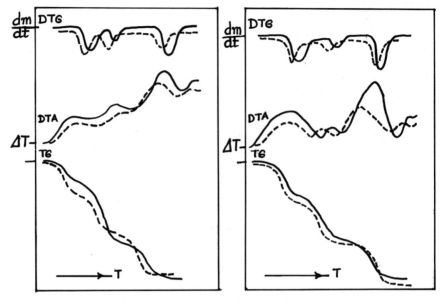

Figure 3
Simultanious DTG, DTA and TG Curves for human finger nail cuttings of thirty year age group.
Male (———) Female (———)

Figure 4
Simultanious DTG, DTA and TG Curves for human finger nail cuttings of forty year age group.
Male (———) Feamale (———)

The DTA curve for the nail sample for relatively higher aged persons show a broad endothermic peak after the exothermic peak at about 160C, whereas the curve for the nail sample from relative lower aged person yeild an exothermic peak at about 250C. From this work it is clear that nail samples do not show weight changes upto 100C although the denaturation commences in nail during gradual change of age at lower temperature.

It also appears from the thermal spectra and the Table 2 that the thermal parameters of the nail cuttings ie. the temperature range of the aged samples become higher than that of the nail samples of less age. The shift of temperature leads to the information that the change of the chemical constituents due to increase of age raised the DTA and DTG peak temperatures. Similar observation has also been made in case of investigation of the thermal behaviour of human hair[6].

ACKNOWLEDGEMENTS

The authors are grateful to the nail donars for the experiment.

REFERENCES

1. W.V.Mayer, Thesis, Stanford University, 1949, October.
2. J.L.Stoves, Proc. Roy. Soc., 1945, <u>62</u>, 132.
3. P.L.Kirk, S.Magagnose and D.Salisbury, J.Crim. Law & Criminal., 1949, <u>40</u>, 236.
4. C.H.Dabforth, Am. Med. Press., 1925.
5. B.Lastig, A.Kondritzer and D.Moore, Arch. Biochem., 1945, <u>8</u>, 57.
6. S.K.Basu and D.Marjit, 4th All India Forensic Sc. Conference., 1981, May.
7. B.Lorant, Differential Thermal Analysis, Vol-2, Academic Press, London., 1972, 500-504.

APPLICATIONS OF THERMAL METHODS IN THE
PHARMACEUTICAL INDUSTRY - PART 1

M.J. Hardy
Beecham Pharmaceuticals
Biosciences Research Centre
Yew Tree Bottom Road
Epsom, Surrey
England

INTRODUCTION

A discussion of the main applications of thermal methods in the pharmaceutical industry.

1. The determination of purity. The general method is outlined and data on precision is presented. This is compared to hplc data for both high purity (ca. 99.5%) and low purity (ca. 96%) compounds. Preliminary data on comparison of the standard calculation to that of Ramsland is presented.

2. Drug-excipient compatibility prediction. Compatibility is defined and the DSC/DTA method of prediction is outlined, giving examples. Mention is made of complex/eutectic formation in relation to dosage form compatibility.

3. Polymorphism. A brief introduction to polymorphism is followed by a suggested polymorphic scan procedure. This includes DSC, hot stage microscopy (HSM) and TGA runs. The importance of comparison of this data to other methods, such as IR and final confirmation by XRD is stressed. Finally DSC curves are given for a compound which has two fairly stable forms. The DSC curves are explained by HSM.

1. THE DETERMINATION OF PURITY.

Melting point determinations have long been used as a method of identification of organic compounds. These compounds could then be classified as "pure", if the melting point was sharp or "impure", if there was a broad melting range (see Kofler's method [1]).
More recently DSC methods have been established which are based on the well known Van't Hoff equation:

$$T_s = T_o - \frac{RT_o^2 X_2}{\Delta H_f} \cdot \frac{1}{F} \qquad \text{equation 1}$$

where T_s = sample temperature (K)
T_o = theoretical melting point of pure compound (K)
R = gas constant (1.987 cal mol^{-1}K^{-1})
X_2 = total mole fraction impurity
ΔH_f = heat of fusion of pure compound (cal mol^{-1})
F = fraction of sample melted at T_s

The equation will be valid under the following circumstances [2 - 4]:
1. The compound does not decompose at or near its melting point.
2. The impurities form a eutectic with the main component, i.e. the impurities are soluble in the main component in the liquid phase.
3. The impurities are not soluble in the main component in the solid phase (i.e. no solid solutions are formed).
4. The system is at constant pressure.
5. The heat of fusion is independent of temperature.
6. $LnX_1 \simeq -X_2$ This limits the determination to high purity samples (>95%) mole purity.
7. $T_sT_o \simeq T_o^2$. This is true, except at cryoscopic temperatures.
8. There are no other "thermal events" in the vicinity of the melting region, i.e. no volatile losses, polymorphic transitions etc.

The main source of error in this determination is in the measurement of area. Area measurement is first utilised in the calculation of the cell constant E. Table 1 shows the variation of the area generated on the melting 1.804 mg of indium. The areas are measured by planimeter, so this manual measurement will not be the most precise. The coefficient of variation obtained was 2.4%.

TABLE 1 - AREA GENERATED BY THE MELTING OF 1.804 mg OF INDIUM (MEASURED BY PLANIMETER)

Run	Area (in^2)	Cell 'Constant' E
1	4.88	0.8367
2	4.90	0.8333
3	5.04	0.8101
4	5.06	0.8069
5	4.90	0.8333
6	4.95	0.8249
7	5.13	0.7959
8	5.18	0.7882
9	4.99	0.8182
10	5.21	0.7837
Average	5.02	0.8131
SD	0.12	0.0193
CoV	2.4%	2.4%

Area measurement is also necessary for the sample and the imprecision can be seen in the calculation of the ΔH_f value. The sample batches have a low purity (in DSC measurement terms) of between 95.4% and 96.8%. The coefficient of variation in this case is 7.6%.

TABLE 2 - VARIATION OF ΔH_f FOR A LOW PURITY COMPOUND (AREAS MEASURED BY PLANIMETER)

Batch	Weight taken (mg)	Area (in^2)	ΔH_f (cal mole^{-1})
5001	1.610	4.25	4707
5002	1.781	5.19	5196
5003	1.836	6.08	5905
5004	1.998	5.47	4906
5005	1.653	4.83	5210
5006	1.673	4.58	4881
5007	2.141	5.81	4839
5008	2.236	7.03	5606
5009	2.257	6.61	5222
5010	1.994	5.31	4748
Average	-	-	5122
SD	-	-	390
CoV	-	-	7.6%

These errors will of course be reflected in the final purity calculations.

For high purity sample of 99.5% or better a variation of 2.4% on the impurity can be expected. This will imply a purity of 99.5 $\pm$ 0.12%. Table 3 shows some data on such a high purity sample. The impurities present in this compound are low, so although they are unknown, exact mole % figures will not vary much from the % w/w values. Also the molecular weight of this compound is 320.4 and the likely impurities will, in this case, have similar values. The results are certainly within the expected experimental variation.

TABLE 3 - COMPARISON OF HPLC AND DSC DATA FOR A HIGH PURITY COMPOUND (ca. 99.5%)

Batch	Purity by hplc % w/w		Assigned Purity* % w/w $\equiv$ mole %	Purity by DSC mole %	
P3/79	-	-	99.49	99.38	-0.11%
P4/77	99.8	+0.5%	99.31	99.28	-0.03%
	99.4	+0.1%			

*see text

The Hplc results for this compound are also recorded. Batch P3/79 was the assigned reference so there was no available main peak hplc data. However the second batch, which was analysed twice, does give results with the expected precision.

TABLE 4 - COMPARISON OF DSC AND HPLC DATA FOR A LOW PURITY COMPOUND
(ca. 96%)

Batch	Purity by Hplc % w/w	Assigned Purity % w/w = mole %*	Purity by DSC mole %	Detected % Hplc	Assigned % DSC
5001	95.8	96.0	96.9	−0.2	+0.9
5002	95.7	96.6	95.0	−0.9	−1.6
5003	94.5	95.0	93.7	−0.5	−1.3
5004	96.2	96.5	96.5	−0.3	0
5005	97.6	96.8	98.1	+0.8	+1.3
5006	96.0	95.7	96.0	+0.3	+0.3
5007	95.7	96.1	96.4	−0.4	+0.3
5008	96.4	96.0	96.5	+0.4	+0.5
5009	96.1	96.1	97.0	0	+0.9
5010	96.0	95.4	95.4	+0.6	0
		*see text		±0.7	±1.3

Low purity compounds present further problems particularly with respect to the assignment of the purity. The series of batches recorded in table 4 contained varying amounts of mainly water, with traces of isopropanol. The melting point of the compound was 116.5° ±1.0°C. The evidence suggested that these solvent impurities either had vapourised or did not form a eutectic with the sample. Therefore they would not be detected by DSC. The main organic impurities have very similar molecular weights to the main component (365.5) and therefore we felt justified in assigning the detectable mole % as equivalent to the % w/w for this compound (all the batches are synthesised by an identical method).

Once we have this assigned purity the Hplc and DSC data could be compared. The DSC precision of ±1.3% is surprisingly good for such a difficult sample and compares favourably with the Hplc precision of ±0.7%.

More recently Ramsland [5] has suggested an iterative multiple regression method for the analysis of stepwise data. This method is based on the equation:

$$T_s = T_o + \frac{RT_sT_o}{\Delta H_f} \ln\left[\frac{X_1^s}{X_1^1}\right] \qquad \text{equation 2}$$

where T_s = sample temperature (K)

T_o = theoretical melting point (K)

R = gas constant (1.987 cal mol^{-1}K^{-1})

X_1^s = mole fraction of main component in solid phase at T_s

X_1^1 = mole fraction of main component in liquid phase at T_s

ΔH_f = heat of fusion of pure main component (cal mole^{-1})

It can be seen that if we assume that:

A $X_2^s = 0$

B $\ln X_1^1 = -X_2^1$ (X_2^1 = mole fraction of impurity in the liquid phase at T_s)

C $T_s T_o = T_o^2$

then this equation degenerates to the well known Van't Hoff equation (equation 1).

This equation can be manipulated to a form that is solvable by an iterative multiple regression method

where $T_s = T_o + B_1 \dfrac{T_s}{(F-L)} + B_2 \cdot \dfrac{1}{F-L}$ equation 3

B_1 and B_2 are the multiple regression coefficients which solve for X_1

$$X_1 = \exp\left[\dfrac{\Delta H_f}{RT_o}\left(B_1 + \dfrac{B_2}{T_o}\right)\right]$$ equation 4

F is the fractional area and L an error function (cf. the linearisation technique of the usual Van't Hoff equation). This is initially set to zero and the equation solved (L is only truely zero when F = 1). The values obtained from this allow an estimation of L for each F value. The process is then repeated until a value of X_1 does not vary from a previous estimation by more than say 0.005%.

The equations will also give a partition coefficient $K = X_2^s / X_2^1$ so showing the degree to which any solid solution is formed.

"Because L is an error function it can only be neglected when P>>L i.e. at high values of F and X_1".

This statement by Ramsland is a justification of the equation's use, and is why the basic equation is limited to high molar purity samples (as L is ignored). Also by comparison of the results to those of the standard calculation will yield information as to the degree of solid solution formation and hence the validity of the basic equation.

TABLE 5 - COMPARISON OF THE STANDARD CALCULATION TO THAT OF RAMSLAND FOR DSC PURITY DATA

Batch	Assigned Purity mole %	Standard DSC Calculation mole %		'Ramsland' DSC Calculation mole %	
5001	96.0	96.9	+0.9	96.7	+0.7
5002	96.6	95.0	-1.6	95.1	-1.5
5003	95.0	93.7	-1.3	94.6	-0.4
5004	96.5	96.5	0	96.4	-0.1
5005	96.8	98.1	+1.3	98.1	+1.3
5006	95.7	96.0	+0.3	95.8	+0.1
5007	96.1	96.4	+0.3	96.2	+0.1
5008	96.0	96.5	+0.5	96.8	+0.8
5009	96.1	97.0	+0.9	97.1	+1.0
5010	95.4	95.4	0	95.1	-0.3
			+1.3%		+1.1%

Table 5 shows the batch to batch comparison of a low purity compound recalculated using the Ramsland method. Some improvement is noted even though the data was generated by the dynamic heating method. So the Ramsland method does seem to be worth more investigation.

2. DRUG-EXCIPIENT COMPATIBILITY PREDICTION BY DSC.

The Pharmaceutical Handbook [6] defines incompatibility in a medicine as "an interaction between two or more components to produce changes in the chemical, physical, microbiological or therapeutic properties of the preparation."

Chemical incompatibility will usually be due to redox, acid-base, hydrolysis or combination reactions; physical incompatibilities are changes in solubility, adsorption of a drug onto an excipient or formation of an eutectic; microbiological incompatibility is usually associated with the reduction in effectiveness of antimicrobial preservatives due to interaction with drug or excipient; and finally, therapeutic incompatibility is interaction of the drug with other drugs or foods taken concurrently, which effect changes in the therapeutic or toxic effect of the drug.

This discussion is limited to the prediction of physico-chemical incompatibility.

Once a drug is formulated into a dosage form, storage tests at various temperatures are initiated. These tests can last up to five years and provide the final data on, amongst other things, the compatibility of the drug with the chosen excipients. This compatibility is assessed by the comparison of "Main Peak Assay" results and "impurity profiles" of the active ingredient with time, using techniques such as HPLC, TLC etc.

This work is time consuming and expensive so it is most desirable to predict any possible incompatibility at an early stage. Jacobson and Reier [7] in 1969, and subsequently by other workers

[8 to 12] have reported the use of differential thermal analysis (DTA) for the prediction of specific drug-excipient physico-chemical incompatibility.

The DTA or DSC methods will record chemical reactions which either absorb or emit heat, changes of state, eutectic formation, etc., as a function of temperature or time. DSC curves of the pure drug substance and pure excipient are recorded. A simple additive superimposition of these curves is then compared to a DSC curve obtained from a well mixed 1:1 mixture of the drug and excipient.

If there are no differences between the theoretical and experimental mixture curves, thermal analysis will suggest that there is no interaction between drug and excipient, i.e. there is no physico-chemical incompatibility (see figure 1).

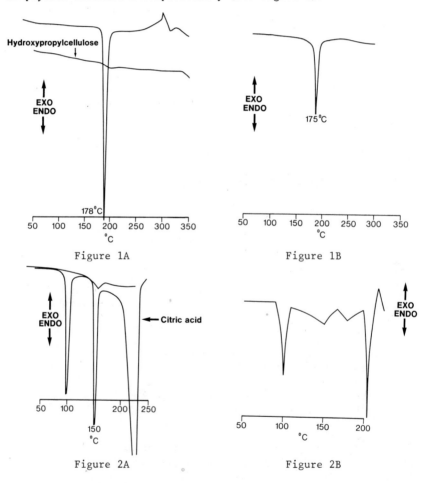

Figures 1 & 2 DSC curves of A/ Actives and excipients B/ 50:50 mixes of active and excipient.

However, problems of interpretation arise when there are differences in these curves (see figure 2). The drug and excipient may still be compatible, as the DSC technique necessarily requires elevated temperatures and these temperatures might induce reactions that do not occur at normal storage temperatures. (NB. The conventional storage tests at elevated temperatures, which use Arrhenius plots give similar problems). Also as the active/ excipient ratio is not usually 1:1 a series of runs which more closely match the chosen ratio might be necessary. Interaction will usually indicate either chemical reaction, adsorption or eutectic formation. This interaction might be advantageous (e.g. as a more desirable form of drug delivery system) or is an example of physico-chemical incompatibility.

These concepts of physical modification and complex formation have interested many workers [13 to 16]. Indeed at Beecham attempts have been made to design a drug delivery system based on eutectic formation. A development drug was required in a suppository formulation. It has an extremely low aqueous solubility and a low melting point (ca. 80°C. If a low melting eutectic (ca. 37°C) could be formed with a soluble excipient (e.g. a polyethylene-glycol), this would enhance solubility, dissolution and bioavailability.

DSC showed the "formulation" to melt in two phases at concentrations above 10% drug to 90% excipient, a drug/excipient phase and pure drug phase. The drug and excipient were clearly incompatible for the desired purpose of the formulation. This was confirmed by conventional melting and dissolution tests on the 'suppositories' (see figure 3).

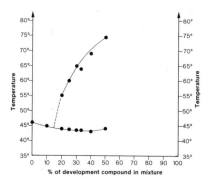

Figure 3. Partial phase diagram of an active/PEG 1500 mixture vs DSC peak maximum temperatures.

DSC can be used to screen excipient candidates. If any interaction is observed that excipient can be avoided in favour of one that shows no interaction. However, this interaction might indicate either complex or eutectic formation. This could be investigated (particularly if the drug has a low solubility) to find an alternative drug delivery system.

3. POLYMORPHISM.

Polymorphism is the ability of any element or compound to exist in more than one distinct crystalline species. This ability is of great importance in the pharmaceutical industry as each polymorph has different physical properties; density, hardness, crystal shape, solubilities, melting point, vapour pressure, optical and electrical properties.

Polymorphic systems are classified as either enantiotropic or monotropic. An enantiotropic system is defined when the free energies of the two polymorphs are equal at a temperature which is below their melting points at a given pressure (usually atmospheric). In this system a solid-solid transition temperature will be seen unless the transition occurs very slowly or is inhibited for some other reason. A monotropic system would be one in which this free energy equivalence is above their melting points so that no transition will usually be seen.

The formulation of a drug into its dosage form, must be related to the drugs physical properties. It is obvious therefore that a detailed study of the polymorphic forms is of the utmost importance.

Haleblian and McCrone [17], in their excellent review gave the following examples:

A/ Suspensions and Creams. Polymorphism can cause crystal growth which creates undesirable crystal sizes. This makes creams 'gritty' and may affect the bioavailability of creams and suspensions. Caking, causing non-uniform resuspension, can also be produced.

B/ Solutions. As each polymorph has different solubilities it is clear that the solubility data generated must consider this variation. The problem is always enhanced when low aqueous solubility and low dosage drugs are required to be formulated.

C/ Suppositories. Polymorphic changes here will affect the melting characteristics and hardness.

When a compound enters the development phase it is therefore most desirable to be able to answer the following questions [17]:

- i) How many polymorphic forms exist.
- ii) What are the relative stabilities of the polymorphs.
- iii) Is there a non-crystalline glass state and is it stable enough to consider as a dosage form.
- iv) Can any of the metastable forms be stabilised.
- v) What are the temperature stability ranges for each crystal form.
- vi) What are the solubilities of each form.
- vii) How can pure and stable crystals of each form be prepared.
- viii) Will the more soluble metastable form(s) survive processing micronising and/or tabletting).
- ix) Does the compound react with any chemical component to form a complex.

A study of this size is not usually practicable at an early stage of development although much of the data is generated by the time the drug has reached the market. At the earlier stages one is often limited to batch by batch comparison of data on DSC, microscopy, hot stage microscopy, m.p. and IR.

A probable 'polymorphic scan' could take the following form.

i) Preliminary DSC study ($-70°C$ to decomposition temperature). Identify any obvious transitions (melting, glass or solid/solid transitions).
ii) Preliminary hot stage microscopy (HSM) study ($-70°C$ to decomposition temperature). Compare to the DSC study.
iii) Preliminary TGA study (ambient to decomposition). Compare to both DSC and HSM studies for hydrates/solvates.
iv) Runs (DSC, HSM and TGA) on bulk and micro-recrystallisations from the melt. Compare to i), ii) and iii).
v) Runs (DSC, HSM and TGA) on recrystallisations from solvents. Compare to i), ii) and iii).
vi) NMR on any runs showing differences, to confirm structural equivalence.
vii) IR (ATR or nujol mull spectra). Further evidence for polymorphism.
viii) Since crystal XRD as final confirmation.

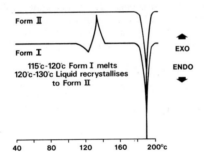

Figure 4. DSC curves of two polymorphs of a development compound.

Figure 4 shows the DSC curves of two forms of a development compound. Form II only gave a melting endotherm with $T_e = 182°$ to $184°C$. However form I gave a broad endotherm with peak maximum between $115°$ and $120°C$ immediately followed by an exotherm with its peak maximum at $120°$ to $130°C$ and a melting endotherm with $T_e = 182°$ to $184°C$. An HSM run on form I identified that form I slowly melted between $115°$ to $120°C$ then suddently recrystallised. This solid then melted between $182°$ and $184°C$. As form II is preferred for the formulation, batches can be quickly scanned by DSC, to ensure the metastable form I is not present.

Purity determinations, drug-excipient compatibility prediction and polymorphic form studies represent the most common applications of thermal methods to pharmaceutical research. Further, less common pharmaceutical applications will be discussed in part 2.

REFERENCES

1. L. Kofler and A. Kofler, "Thermomikromethoden zur Kennzeidung Organisher Stoffe und Stoffgemische", Verlag Chemie, Weinheim, (1954).
2. E. Marti, Thermochimica Acta, 1972, 5, 173-220.
3. E. Palermo and Jen Chiu, Thermochimica Acta, 1976, 14, 1-12.
4. P. Burroughs, Anal. Proc., 1980, 17, 231-234.
5. A. Ramsland, Anal. Chem., 1980, 52, 1474-1479.*
6. Pharmaceutical Handbook 19 ed. (1980), p.28.
7. H. Jacobson and G. Reier, J. Pharm. Sci., 1969, 58, 631-633.
8. H. Jacobson and I. Gibbs, J. Pharm. Sci., 1973, 62, 1543-1545.
9. M. Jacob et al., J. Pharm. Belg., 1979, 34, 96-98.
10. B. Muller, Acta Pharm. Tech., 1977, 23, 257-266.
11. K. Lee and J. Hersey, Comm. J. Pharm. Pharmac, 1977, 29, 515-516.
12. K. Lee and J. Hersey, Aust. J. Pharm. Sci. 1977, 6, 1-9.
13. W. Chiou and S. Riegelman, J. Pharm. Sci., 1971, 60, 1281-1301.
14. E. Shefter and K. Cheng, Int. J. of Pharmaceutics, 1980, 6, 179-182.
15. D. Grant et al., Int. J. of Pharmaceutics, 1980, 6, 109-116.
16. I. Abougola et al., 1st Int. Conf. on Pharm. Tech. (1977) p.142-155.
17. J. Haleblian and W. McCrone, J. Pharm. Sci., 1969, 58, 911-929.**
18. J. Clements, Proc. Anal. Div. Chem. Soc., 1976, 13, 21-25.

*Adapted with permission. Copyright (1980) American Chemical Society

**Reproduced with permission of the copyright owner.

APPLICATIONS OF THERMAL METHODS IN THE
PHARMACEUTICAL INDUSTRY - PART 2

M.J. Hardy
Beecham Pharmaceuticals
Biosciences Research Centre
Yew Tree Bottom Road
Epsom, Surrey
England

INTRODUCTION

A discussion of some less common applications of thermal methods in the pharmaceutical industry.

1. Stability prediction. The non-isothermal technique of Ozawa is discussed for DSC and an example given. The data generated was used to predict an half-life time and time to 10% decomposition. This can be used to assess storage times and safety of storage.

2. Eutectic point determination. The application of DSC for assessing correct freeze-drying conditions is discussed. Phase diagram construction is recommended for eutectic systems and an example is given relating to drug delivery design. The importance of observational techniques (e.g. hot stage microscopy) as confirmation, is mentioned.

3. Diastereoisomer ratio. A rapid DSC method for isomer ratio is discussed.

4. Hydration/solvation studies. The use of TGA for the study of bound or unbound solvent is presented.

5. Miscellaneous techniques. Other thermal techniques which might have some pharmaceutical applications are mentioned.

1. STABILITY.

The usual estimates of pharmaceutical stability - both for the bulk drug and the formulated product, are usually based on the conventional storage tests. These are in fact required by the various licensing authorities of each country before clinical trials can be staged. The tests involve storage of three batches of pure drug and each formulated product, at a range of temperatures up to 80°C. The results of these tests are usually interpreted using the Arrhenius equation [1,2]

$$K = Ze^{-E/RT}$$

where K is the specific rate constant at temperature T (K)
       Z is the Arrhenius frequency factor
       E is the activation energy
       R is the gas constant

The stored samples are periodically assayed for both purity and impurities by such techniques as hplc, gc, tlc, titration etc. to assess the degree to which any degradation has occurred. This enables $t_{0.1}$ or $t_{0.05}$ (times to 90% or 95% of initial assay) to be calculated. These times are most useful as the national registration bodies will usually require either a $\pm 10\%$ or $\pm 5\%$ variation on the assay.

It can be seen that this work is both very costly and time consuming so that some estimation of stability would be most desirable and of considerable help to the formulation pharmacists.

The variable heating rate method of Ozawa [3] forms the basis of our DSC stability work. A sample of the pure drug is heated at $10°C$ or $20°C\ min^{-1}$ from ambient temperature until decomposition has been observed. If this decomposition is endothermic, the compound is usually considered to be thermally stable. However, if an exothermic decomposition is obtained the ASTM method "Arrhenius kinetic constants for thermally unstable materials [4]" is then run. This will provide values for the activation energy (E) and the Arrhenius frequency factor (Z). The specific rate constant at various temperatures can then be calculated and hence the $t_{0.1}$ (time to 10% loss) and $t_{0.5}$ (half-life time) can be derived from the first order rate equation.

The validity of the results can be tested by isothermal ageing for the predicted half-life time at any specific temperature. The DSC curve of this 'aged' sample can then be compared to an 'unaged' sample; the area under the exotherm of the former should be half that of the latter. Alternatively the DSC predictions can be compared to pure drug storage data, if it is available.

TABLE 1 - STABILITY PREDICTION FOR ERGOCALCIFEROL

Temperature (°C)	Rate constant min^{-1}	Time to 10% decomposition $t_{0.1}$ (min)	Half-life time $t_{0.5}$ (min)
20	2.46×10^{-9}	4.27×10^{7}	2.82×10^{8}
50	2.03×10^{-7}	5.17×10^{5}	3.41×10^{6}
80	7.94×10^{-6}	1.32×10^{4}	8.73×10^{4}
100	6.59×10^{-5}	1.59×10^{3}	1.05×10^{4}
150	5.46×10^{-3}	1.92×10	1.27×10^{2}

Table 1 shows the data that was generated for a sample of ergocalciferol (vitamin D_2). The DSC ageing at 150° for 2 hours showed that the half-life prediction was within the expected experimental error; a loss of area of 46.6% was obtained where theory predicted 48.1% loss.

For particularly unstable compounds, the parameters determined above can be utilised in hazard evaluation [5 to 7] The accuracy of this data is most critical and it is advisable to check the values obtained by comparison to another technique such as accelerated rate calorimetry (ARC) [8]. The ARC technique is not limited to first order reactions so it has more general application in this safety evaluation area.

2. EUTECTIC POINT DETERMINATION.

Biological agents are often presented as reconstitutable formulations, which often require that freeze-drying is necessary during the final product preparation. This freeze-drying process depends for success on the conditions of the freeze-drying cycle as these will affect the physical appearance, chemical stability and the ease of reconstitution of the finished product [9]. These problems can be avoided by identification of eutectics, phase transitions and super cooling characteristics of the sample solution, allowing a suitable freeze-drying cycle to be designed.

DSC is an ideal technique for the identification of these transitions [9,10]. By the use of both heating and cooling cycles, varying β (the heating or cooling rate), the reversible and non-reversible transitions can be detected. Suitable isothermal annealing temperatures can be included if necessary, whilst results can be compared to trial freeze-drying runs. Initial work will usually be carried out on the various water/active and water/excipient binary mixtures in order to simplify the generated curves.

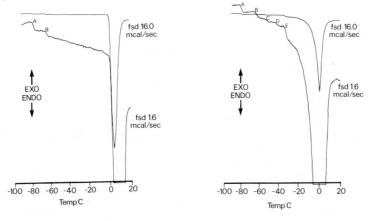

Figure 1. DSC curves of A/ water and B/ 20% w/v aqueous solution.

Figure 1 shows a controlled heating run for a frozen β-lactamase inhibitor solution and a similar run for water. The solid/solid transitions that occur in the water at -78°C and -63°C can also be seen in the sample solution (at -81°C and -65°C). The sample solution also shows three other possible solid/solid transitions; at -51°C, -43°C and -35°C. On comparison to freeze-drying runs, it was found that if the initial shelf temperature was -30°C to -40°C then the plug did not dry properly and melt back occurred). However, an initial temperature of -50° to -55°C ensured that this did not occur.

Whenever eutectic systems are considered, a phase diagram is often a most useful presentation for discussion. A series of DSC curves of different percentage amounts of a binary mixture can be run, plotting the percentage composition against the peak temperatures in the usual fashion.

The suppository formulation mentioned in part 1 of this discussion was easily visualised using such diagrams. Figure 2 shows the DSC peak onset temperatures plotted against the average molecular weight of five different polyethylene glycols (PEG's) which were mixed (50:50) with the development drug. Two peaks were obtained in each case equivalent to a drug/PEG eutectic and pure drug. As the PEG 1500 gave the drug/PEG eutectic with $T_e = 37.5°C$ (ideal for a suppository) the PEG 1500 was chosen for further work.

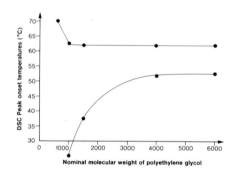

Figure 2. DSC peak onset temperature vs average molecular wt. of PEG.

Figure 3 in Part 1 of my discussion shows the partially generated phase diagram which shows that the formulation has to contain less than 15% drug for a single melting phase to be obtained. The temperatures are plotted as peak maximum as the conventional tests and hot stage microscopy showed this to be closer to the actual melting range of the suppository than the peak onset temperature. It is quite easy to see that the desired melting characteristics could not be obtained using this simple binary mixture.

3. DIASTEREOISOMER RATIO.

It is quite common that compounds of interest as drug candidates will have two or more assymetric carbon atoms and will therefore exist as diastereoisomers. These will have different physical properties and usually vastly different therapeutic activities. It is therefore important to have a quick and reliable method for quantitation of the isomer ratio. In our laboratories the ratio is generally determined by an hplc or gc method. However for occasional samples DSC provides a faster but equally reliable method. Some diastereoisomer pairs have been synthesised whose melting points differ by some $80^{\circ}C$. DSC curves can be run for various percentage mixtures whose ratios are known or can be measured by another technique (gc or hplc). The curves of unknown batches can then be visually compared with these known batches as a fast Q.C. test.

Obviously each isomer pair has to be separately assessed; considering melting point separation of the pure isomers, eutectic formation, required ratio (50:50 or 100:0), level of impurities, other interfering transitions, etc. For ratios in excess of 98:2 the purity calculation could be used to obtain more accurate values, although the analysis time would be much slower, unless fully automated data handling was available.

4. HYDRATION AND SOLVATION STUDIES.

TGA is an exceedingly useful tool for the determination of volatile solvents or water as a function of temperature. It will show if there is more than one type of water present (e.g. bound and unbound) and occasionally show mono-, di-, tri- etc. hydrate presence (e.g. calcium oxalate and calcium hydrogen phosphate). This type of work is not definitive and, as for polymorphic form studies by DSC, should always be confirmed by single crystal XRD whenever possible.

A technique such as NMR will initially show the presence of the solvent/water and some idea of the quantitative level. The TGA curve will then indicate if this is lost in one or more discrete amounts and hence imply unbound water or hydrate. It will also indicate if the compound appears stable at these 'water loss' temperatures. This will influence the choice of drying temperature (and time) for bulk drying, if the solvent/water is not desirable.

The water/solvent levels can be confirmed by gas chromatography (for water and solvents) or Karl Fischer titration (for water).

5. MISCELLANEOUS TECHNIQUES AND METHODS.

Other methods that might have some applications in the pharmaceutical industry are EGA, EGD, TMA and thermometric titration, although we have not had cause to utilize these techniques in our research environment.

This discussion (parts 1 and 2) presents the common applications of thermal methods in the pharmaceutical industry as used at Beecham Pharmaceuticals, Biosciences Research Centre. All the DSC curves were generated on a DuPont 990 DSC system.

REFERENCES

1. D.G. Pope, Drug and Cosmetic Industry, 1980, 127 (5), 54.
2. D.G. Pope, Drug and Cosmetic Industry, 1980, 127 (6), 48.
3. T. Ozawa, J. Thermal Anal. 1970, 2, 301-324.
4. Anon, ASTM Method E698-79.
5. A.A. Duswalt, Thermochim Acta, 1974, 8, 57.
6. L.C. Smith, Thermochim Acta, 1975, 13, 1.
7. D.R. Stull, AICh.E Monograph, Series 10, (1977), 73.
8. D.I. Townsend, Thermochim Acta, 1980, 37, 1.
9. R.M. Patel and A. Hurwitz, J. Pharm. Sci. 1972, 61 (11), 1806.
10. L. Gatlin and P.P. Deluca, J. Parent. Drug Assoc., 1980, 34 (5) 398.

APPLICATION OF THERMAL ANALYSIS TO ANALYTICAL
AND PROCESS RESEARCH IN THE PHARMACEUTICAL
INDUSTRY. James A. McCauley, Merck Sharp &
Dohme Research Laboratories, P. O. Box 2000,
Rahway, New Jersey 07065, U.S.A.

INTRODUCTION

Thermal analysis, particularly DTA, DSC, and TG, has been used throughout the multidisciplined pharmaceutical industry. Physical pharmacy, pharmacology, dosage formulation, membrane research, and biochemistry are areas in which thermal analysis has provided useful information and aided in the understanding of some basic biological processes. Thermal analysis and thermodynamics can also be applied to various types of problems encountered in analytical and process research in the pharmaceutical industry. This paper will discuss several such applications.

EXPERIMENTAL

The thermal analytical data discussed below were determined with the following instrumentation: Perkin-Elmer DSC-1B, DSC-2, and TGS-2 systems; DuPont 990 thermal analyzer, both DTA and TG systems. X-ray patterns, microscopic observations, thin layer chromatograms, and mass spectral analyses were obtained through established methods. Solubility measurements were performed at constant temperature with the saturated solutions analyzed by mass or ultraviolet absorption. All solubility data were obtained at least in duplicate.

RESULTS AND DISCUSSION

The characterization of bulk drug substances by melting point is a traditional analytical method used in the pharmaceutical industry. DTA and DSC can be used for such application. DSC is particularly useful not only for melting point characterization but also for purity determination. This aspect of DSC is well documented and will not be discussed further. DTA and DSC plus additional analytical techniques provide tools which are helpful in the resolution of problems such as solvation, decomposition, and polymorphism, which can arise during melting point characterization of bulk products.

Polymorphism occurs frequently in the pharmaceutical industry. DTA/DSC provide thermal and melting point data for the polymorphs. X-ray, IR, and microscopy indicate underlying structural differences. Solubility measurements at various temperatures under conditions where the polymorphs do not interconvert along with the thermal data can indicate the thermodynamic relationship between the polymorphs under investigation. For example, it has been established that indomethacin, an antiinflammatory agent, exists in at least two crystal forms designated Form I (161) and Form II (156). A third but rarer form [Form III (151)] is also known [1,2]. Solubility measurements at room temperature reveal that Form I (161) is less soluble than Form II (156), which in turn is less soluble than Form III (151). The combined thermal and solubility data indicate Form I (161) of indomethacin is the most thermodynamically stable form [3] and that the other two are monotropically related to Form I (161) and to each other with Form II (156) being more stable than Form III (151).

In the case of sulindac, another antiinflammatory agent, Form I (191) is more soluble at room temperature than Form II (186), indicating that the two forms are enantiotropically related. Extrapolation of solubility versus temperature exhibits a transition point at approximately 148°C. Unfortunately, no endothermal transition is observed with DTA or DSC when Form II (186) is heated from room temperature to its melting point. The kinetic process, in this case, of form conversion in the solid state is obviously too slow to be observed.

Phthalysulfathiazole, an antibacterial agent, is an example of a case where the use of a single thermal technique could possibly lead to erroneous conclusions. Figure 1 is a DTA curve for nonhydrated or nonsolvated phthalylsulfathiazole and appears to be a classic example of two monotropically related crystal forms. Melting of the less stable form is apparently observed at about 230°C followed by recrystallization to the more stable form with subsequent melting of the latter at about 275°C. Visual inspection during the DTA revealed discoloration and decomposition at the higher temperature. Thermogravimetry over the same temperature range shows approximately a 10% weight loss in the region of the first endotherm with a stable weight plateau extending from about 230° to the beginning of the second endotherm. A second weight loss of ill-defined nature corresponding to the higher endotherm and confirming the decomposition is also detected. The first endotherm and the first weight loss also indicate melting with decomposition and not polymorphic conversion. The product or products of the initial decomposition recrystallize from the melt and are thermally stable for an additional 20-30°C. Thin layer chromatography and mass spectral analysis of the material remaining after the first endotherm revealed that the major product of the decomposition is the phthalimide of sulfathiazole with sulfathiazole a minor product. The volatile products formed are water and phthalic anhydride.

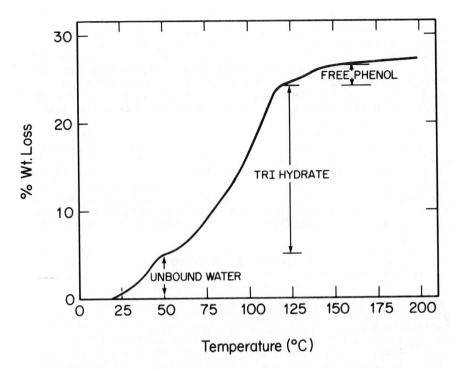

Figure 2. Thermogravimetric Curve for the Phenolate

Many of the compounds produced by the pharmaceutical industry contain chiral centers. Chiral centers are usually introduced by stereospecific synthesis or resolution of the corresponding racemate. The latter can be accomplished with a resolving agent or through spontaneous crystallization of the desired enantiomer. The choice of the resolution method depends to a great extent upon the nature of the racemate. If the racemate is a compound, a resolving agent is required. Spontaneous crystallization can be used only when the racemate is a mixture of enantiomers. The nature of the racemate can be ascertained through determining the binary melting point or phase diagram for the enantiomers. DTA and DSC are particularly suited for such determinations.

With the increased use of chiral catalysts which permit stereoselective but not necessarily stereospecific synthesis, it is possible to complete the resolution without a resolving agent even if the racemate is a compound. The extent and yield of the resolution are dependent on the enantiomorphic excess produced by the chiral catalysts and the ratio of the solubility of the desired enantiomer to the solubility of the racemic compound in the solvent chosen to complete the resolution. The yield (Y) of the resolution can be expressed as follows:

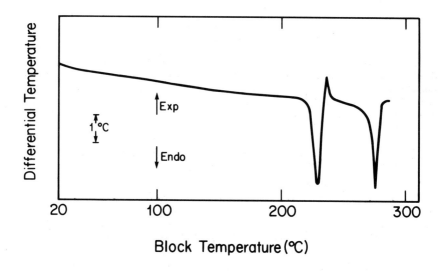

Figure 1. DTA Curve for Phthalylsulfathiazole

Thermogravimetry can be used in a more traditional way for the determination and quantification of volatile materials present in final products and process intermediates. Not only can water and other solvents be determined by thermogravimetry, but any volatile component which is separable from other volatiles can be determined. A particularly appropriate and concrete example occurred during the process development of diflunisal, an analgesic and antiinflammatory agent. A key process intermediate, a sodium phenolate, exists as a hydrate in the solid state. It was of interest to determine both the water content and the amount of free phenol. With a single thermogravimetric run, both water and the free phenol could be quantitatively estimated (Figure 2). The sodium phenolate is thermally unstable at elevated temperatures leading to some uncertainty in the free phenol content. A more precise method for the free phenol was developed. The free phenol was extracted from the phenolate with toluene. An aliquot of the toluene evaporated at 80°C under nitrogen in the TG apparatus. The extracted free phenol was then measured as the weight loss between 100 and 200°C. The sublimed material was collected and its identity verified by thin layer chromatography and mass spectral analysis. The method of standard additions (to the phenolate) was also used to quantitate the free phenol with the extractive procedure.

$$Y = \frac{(1-f_R)S_T - S_E}{(1-f_R)S_T} \qquad (1)$$

where f_R is the weight fraction of the racemic compound produced during the stereoselective synthesis, S_E is the solubility of the enantiomer expressed in weight units and S_T is the total weight of material to be resolved per unit weight of solvent (referred to as overall or system composition). In order for the resolution to be complete, the solubility of the racemic compound (S_R) must be at least equal to the product of the weight fraction of the racemate (f_R) and the system composition (S_T). Substitution of this relation into equation (1) gives

$$Y = \frac{1 - f_R(1+R)}{1 - f_R} \qquad (2)$$

where R is S_E/S_R. The solubility of the racemic compound for an ideal solution can be predicted from thermodynamics and is expressed as

$$\ln X_R = \frac{\Delta H_F^R}{R}\left[\frac{1}{T_R^0} - \frac{1}{T}\right] \qquad (3)$$

where X_R is the mole fraction of the racemate in solution, ΔH_F^R is the heat of fusion and T_R^0 is the melting point of the racemic compound, R is the gas constant and T is the absolute temperature. A corresponding expression can be written for the enantiomer. The combination of the thermodynamic expressions for the solubilities and equation (2) makes it possible to predict the yield of the resolution as a function of temperature. Figure 3 shows the yield temperature relationship for a chiral intermediate in the synthesis of an enzyme inhibitor. The heats of fusion (7.2 ± 0.3 and 5.3 ± 0.5 kcal/mole) and the melting points (177.5 and 136.8°C) for the racemic compound and enantiomer were determined by DSC. The chiral catalyst had a selectivity of 93.5% producing approximately 13% of the racemate.

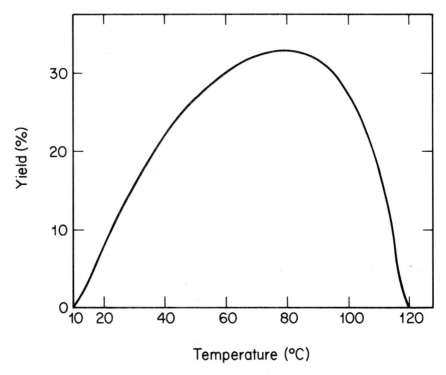

Figure 3. Yield from Enantiomorphic Excess Resolution as a Function of Temperature

REFERENCES

1. H. Yamamoto, Chem. Pharm. Bull. (Tokyo), 1968, <u>16</u>, 17.
2. L. Borka, Acta Pharm Suecica, 1974, <u>11</u>, 295.
3. A. Findlay, A. N. Campbell and N. O. Smith, "Phase Rule," Dover Publications, New York (1951) p. 35.

DYNAMIC PURITY MEASUREMENTS

P.D. Garn, B. Kawalec, J.J. Houser and
T.F. Habash
Department of Chemistry
The University of Akron
Akron, Ohio 44325

INTRODUCTION

Determination of purity by detection of phase boundaries and integration of heat absorption between boundaries requires detailed knowledge of the phase diagram for the system.

Because the knowledge of the true phase boundaries is seldom complete, it is customary to make a number of assumptions that have varying levels of approach to reality. The assumptions enable use of thermal methods, DTA or DSC, in purity determinations, taking advantage of the speed and simplicity of the technique and the high sensitivity of measurement. [1,2] The assumptions, to the extent that they are valid, provide a thermodynamic base that eliminates the need for a reference substance.

For pairs of materials that differ sufficiently in chemical properties and/or structure, the assumptions of ideality in the liquid phase and immiscibility in the solid phase are reasonable. This can be established for some systems; for example straight-chain hydrocarbons having differences in carbon number greater than ca. ten tend to be more or less soluble in one another. It is reasonable to infer that a great number of systems behave ideally and are therefore amenable to analysis by calorimetry or dynamic thermal methods.

On the other hand, pairs of materials that have some chemical similarity are suspect and those pairs that have substantial similarity can be expected to display some level of solid solution. This solid solubility is represented by a phase diagram such as that shown in Figure 1, in which there is sufficient mutual solubility that the entectic does not extend to either pure material.

Unfortunately, the impurities that are of greatest concern in any chemical processing, reaction or purification are materials that are precursors or homologs of the base material. It follows that there will be chemical similarity between the impurity and some part of the base material. Consequently, for most commercial syntheses, purity determinations are liable to error.

IMPROVEMENT IN METHODS

This research was undertaken primarily to improve purity determinations by establishing means of ascertaining whether or not a system may be treated by the *ideal liquid solution with zero solid solubility* methods and of correcting the error, particularly in DSC methods. Secondary goals were to ascertain whether or not the effects of two added impurities are additive and to generate data for different classes of compounds with the aim of providing correction

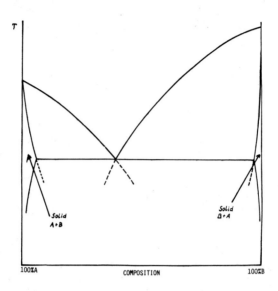

Figure 1. Phase diagram for a two component system showing the typical case of mutual solid solubility.

methods.

From Figure 1 it is clear that a material having an impurity level in the immiscibility region will form a solid solution upon reaching the temperature of the phase boundary between the two phases, then will begin to liquefy when the *solidus* is reached. It should be obvious too from the slope of this boundary that if the temperature is rising at a constant rate the amount of heat required to effect that liquefaction is initially very small and probably not easily recognizable as a clear deviation from the base line. This deviation grows slowly as the temperature increases and at some point higher in temperature (or later in *time*) is recognized as the start of the peak. This recognition may be subjective (by an operator) or objective by pre-set criteria (by a computer). The summation of the area under the curve would be unimportant if the areas was calculated from an extrapolation of the base line during the scan. Whereas the area is calculated from another baseline, viz., a baseline determined from the area of the tailing off portion of the peak spread uniformly from the beginning of melting to the peak temperature, the area between the scan and this corrected baseline in the interval between the true *solidus* temperature and the recognized inception of melting can become a significant fraction of the whole area and a predominant factor at low values of F, the fraction melted.

The experimental evidence of error due to "undetected melting" has led to empirical correction by adding an "x-factor" to the total and partial areas. This x-factor is evaluated by adding whatever correction provides the best straight line when T is plotted against 1/F. Because its value has no foundation in principle, the term can gloss over other features of the plot. A superior technique is the experimental determination of the *solidus* temperature.

The presence of a liquid can be detected readily by nuclear magnetic resonance. This possibility arises from the interaction of the applied field and certain nuclei, including hydrogen. In the absence of any perturbing local fields, as in a liquid, the resonance between the spin of the proton and the magnetic field is discrete, ca. 1 Hz in width. When the proton is in a fixed array of atoms, the many neighboring magnetic fields interact with the applied field to perturb the field experienced by the proton. The interaction is varying about each proton and for a large collection of protons there will be a virtual continuum of field strengths. Consequently the range of applied frequencies with which some of the protons will resonate is large, ca. 10^4 Hz. As a result, a very small fraction of liquid will appear as a peak above the wide band arising from the solid, as seen in Figure 2. The several points in Figure 2 can be extrapolated to their junction at the inception of melting.

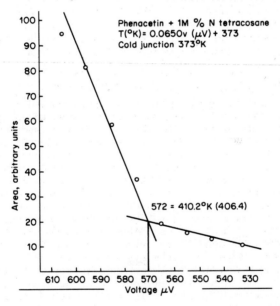

Figure 2. Area under nuclear magnetic resonance peaks versus signal of the temperature sensor for phenacetin contaminated with 1.0 mole percent tetracosane.

Some NMR spectrometers are capable of measuring the bands for solids with accuracy. In this case, the number of nuclei in the broad band compared to the number in the sharp band is a measure of the ratio of solid to liquid. Hence NMR can be used for purity measurements; this has been demonstrated by Herington and Lawrenson

[3]. NMR is not used ordinarily for purity measurements because it is time consuming, the sample temperature is less well known, and the instrumentation is more costly. Measurements have no superiority over DSC measurements *except* for detection of a liquid phase. This work makes use of that advantage to provide a correction to DSC data.

APPARATUS

A heating assembly was added to a Varian A-60 NMR to provide better control of the sample temperature than was possible with the temperature control supplied by the manufacturer. A nitrogen stream was heated at fixed voltages from a very well regulated power source; the sample was allowed to reach equilibrium, measured by a thermocouple in the sample; the NMR spectrum was taken; the input voltage was advanced and a new temperature equilibrium reached before the next measurement was taken. The power increments were small enough that the temperature intervals were suitably small, one to three degrees.

The purity curves were obtained on a Perkin-Elmer DSC-2 with associated interface and Tektronic Programmable Calculator. The purity calculations were done both by the program supplied with the instrument and by direct measuring on the strip chart with manual computation.

PROCEDURE

The purities of synthetic samples were determined according to standard methods, automatic data acquisition and calculation, manual measurements and calculation from a strip chart record using the standard procedure, and correction to manual data from NMR measurement of the solidus.

The samples were made from base materials carefully prepurified by zone-refining commercial materials in most cases, by repeated extraction for phenyl salicylate and by recrystallization for menadione. In each case the purification was carried out until no impurity was detectable by liquid chromatography and successive operations yielded no change in the DSC measurements.

Impurities were added to sizeable samples, e.g. 10g, to minimize error in weighing and homogenation. The melted and stirred samples were poured quickly on a chilled aluminum block to freeze the material quickly and eliminate segregation by diffusion. The preparation and measurements on the samples were carried out by the standard procedures.

Several samples were made up having two components to test whether or not the effect on the melting point is additive. The pairs of impurities were combinations of the impurities that were tested separately as single impurities.

The data and curves were evaluated not only by obtaining values for the purities of the specimens but also by comparisons of curves. That is, overlaps were prepared to enable detection of variations in form or irregularities that might vitiate the experiment.

Comparisons of repeat runs on the same specimen were made as well as the more common comparisons of successive specimens.

RESULTS AND DISCUSSION

As a test of the reproducibility of the quenched state, duplicate specimens of tin, n-hexatriacontane and phenacetin were melted and then quenched to about 50° below their melting points. The melting curves obtained, as in a purity run, of the first two were precisely superimposable, showing that the quenched state is reproducible. For phenacetin, the initial part of the melting peak was precisely reproducible but significant differences appeared as the melting proceeded, suggesting that some vaporization was taking place. The opened pans disclosed condensation on the top of the sample pan.

The metallic samples were run to evaluate the effect of thermal equilibrium within the specimen. The metals have thermal conductivities ca. three orders of magnitude greater than organics. The similarities of the T vs $1/F$ plots for metal and organics indicate that thermal equilibrium is not a problem; there must be other reasons for deviations of the ideal relationship. The possibilities include non-ideality of liquid and/or solid, partitioning in the solid, and—most important—solid solubility.

The measurements on phenacetin with benzamide, acetanilid and tetracosane were carried out to delineate the types of variations. The former two can be expected to show positive deviations from ideality whereas tetracosane should show almost no deviation because both molecules are neutral. Similarly, combinations of base material and impurities were carried out for menadione and phenyl salicylate.

The temperatures of the solidi in the organic systems determined by NMR, when used to recalculate the solidus, introduced substantial change in the calculated purity. These variations were in every case in the direction of better agreement of the measured value with the known composition of the sample.

REFERENCES

1. Thermal Analysis Newsletter No. 5, Perkin-Elmer Corporation, Norwalk, Connecticut, U.S.A.
2. Thermal Analysis Newsletter No. 6, Perkin-Elmer Corporation, Norwalk, Connecticut, U.S.A.
3. E.F.G. Herington and I.J. Lawrenson, Nuclear Magnetic Resonance Cryoscopy I. A. Method of Purity Control, J. Appl. Chem., 1969, Vol. 19, p. 337, 340-341.

ENTHALPY DETERMINATION BY DSC AND DTA
E. Marti, O. Heiber and A. Geoffroy
Central Function Research
Ciba-Geigy Ltd., CH-4002 Basel

The enthalpy determination by DSC and DTA instruments is important for the elucidation of a broad range of phenomena such as phase diagrams including purity determination, heat capacity and related phenomena, phase transition, chemical kinetics, vapor pressure and activity – Gibb's free energy function – among others. Furthermore a great many thermodynamic constants related to energy changes of pure substances or multi-component systems are accessible by extremely simple and reasonably accurate measurements. The measurements can be performed under a broad variety of different conditions with respect to sample weight, open and closed sample system, atmosphere, heating and cooling rate, or at selected constant temperatures. The caloric calibration is a focal point of the experimental performance of DSC and quantitative DTA instruments. The calibration is most easily carried out using a reference substance. Inorganic and organic substances as well as metals have been selected as reference materials, and the energy change of a first order transition is certainly most advantageous for purposes of calibration. The selection criteria of reference materials shall be discussed here taking the enthalpy of fusion of indium as an example because this substance is widely used for the calibration of DSC and DTA instruments. All the available original literature values of the enthalpy of fusion of indium shall be taken into consideration with the goal of establishing a most reliable mean value, and also for a statement of further necessities.

Some special values of the enthalpy of fusion for indium shall be outlined first. Andon et al. [1] of the "National Physical Laboratory" reported values obtained with three different samples of indium applying the method of adiabatic calorimetry. Andon et al. gave the following weighted values for the mean enthalpy of fusion:

$$\Delta H_f^{In} = 3259 \pm 2.5^* \text{ J mol}^{-1}$$

*The errors of the mean values and the error of a single measurement are given without additional comment in terms of confidence intervals on a 68 % level.

The melting point for indium was reported as

$$T_o^{In} = 429.79 \text{ K}.$$

All 24 measurements published by Andon et al. were submitted to the Grubbs test of outlying observations [2], [3] and no single value had to be discarded. Therefore all 24 values for the enthalpy of fusion measured with the three samples of indium were used in calculating the mean value:

$$\Delta H_f^{In} = 3258 \pm 3 \text{ J mol}^{-1}$$

with a standard error for a single measurement of 15 J mol^{-1}. The good agreement of the weighted and unweighted mean values are evidence for unbiased measurements.

Two further values of the enthalpy of fusion for indium of a purity better than 99.9995 % were determined in our laboratory following a indirect approach. The enthalpy of fusion for p-xylene Fluka puriss. was measured with a LKB 8700-1 Precision Calorimetry System by the dissolution of crystalline and liquid samples in toluene. The calibration was performed electrically. The enthalpy of fusion for p-xylene was determined as

$$\Delta H_f^{LKB} = 17016 \pm 42 \text{ J mol}^{-1}$$

a value which is in good agreement with the literature

$$\Delta H_f^{Lit} = 17058 \pm 80 \text{ J mol}^{-1}.$$

The enthalpy of fusion measured for p-xylene with the isoperibolic calorimeter was taken as the standard value. Indium and p-xylene were measured with a DSC-1B (Perkin-Elmer Corp.) and the enthalpy of fusion was calculated for indium as

$$\Delta H_f^{In} = 3238 \pm 39 \text{ J mol}^{-1}.$$

The same measurements with p-xylene and indium were performed with a DSC-2C of Perkin-Elmer Corp.. The result obtained is

$$\Delta H_f^{In} = 3332 \pm 17 \text{ J mol}^{-1}.$$

All primary literature values available to date are collected in Table 1.

The mean and the errors of the enthalpy of fusion for indium is calculated without any assignment of weights to the single values from the 17 experimental values listed in Table 1:

$$\Delta H_f^{In} = 3303 \pm 10 \text{ J mol}^{-1}$$

with a standard error for a single measurement of 42 J mol^{-1}. In other units still in use today the value is

$$\Delta H_f^{In} = 6.88 \pm 0.02 \text{ cal g}^{-1}.$$

DISCUSSION

The enthalpy of fusion for indium is widely applied in the calibration of DSC and quantitative DTA instruments. Many different literature values are reported and sometimes selected as "best" values in a range of 3132 to 3370 J mol^{-1}, respectively 6.52 to 7.02 cal g^{-1}. This range covers values with a relative scattering of 7.5 %. The DSC and quantitative DTA instruments, however, enable measurements of molar enthalpies in the range of a few hundred to a few ten thousand Joules down to 0.4 % relative error of mean and approximately 1.5 % relative error of a single measurement. The accuracy demanded of a reference material for these thermoanalytical methods should be very high compared to the instrumental error. Therefore a demand is put to reference materials to have a value of the enthalpy of fusion which is reproducible and accurate to or better than 0.1 %. The true value for the enthalpy of fusion in case of indium as a reference material regarding the above given limit should be known to within ±3 J mol^{-1}. The method with the best reproducibility cited here is that of adiabatic calorimetry, however, the two reported values differ by 25 J mol^{-1} or 0.8 % relatively.

The conclusions are clear:
- The true value of the enthalpy of fusion for indium is most likely within the range of 3280 to 3320 J mol^{-1}.
- Additional high precision measurements on indium are required to install indium as a real reference material.
- All the other organic and inorganic compounds as well as metals which shall be used as reference materials have to be compared with the accuracy of the enthalpy of fusion, the quality and stability typical for indium.

TABLE 1

Author	Year	ΔH_f^{In} J mol^{-1}	Method	Reference
Roth	1934	3260	Heating/cooling curves	5
Oelsen	1955	3260	Heating/cooling curves	6
Oelsen	1955	3268 ± 17	Heating/cooling curves	7
Predel	1964	3370	DTA	8
David	1964	3316	DTA calibrated with tin	9
Alpaut	1965	3340	DTA	10
Bros	1966	3269	Calvet calorimeter	11
Mechkovskii	1969	3360	DTA	12
Reznitskii	1970	3350	DTA	13
Malaspina	1971	3320	Calvet calorimeter	14
Widmann	1974	3290	DTA	15
Marti (1)	1974	3238 ± 39	DSC	4
Richardson	1975	3350 ± 30	DSC calibrated with alumina	16
Lowings	1978	3284 ± 12	DSC calibrated with tin and lead	17
Groenvold	1978	3283 ± 4*	Adiabatic calorimetry	18
Andon	1979	3258 ± 3	Adiabatic calorimetry	1
Marti (2)	1982	3332 ± 17	DSC	this work

* value calculated in this work

- High precision measurements demand not only one but several reference materials with properly selected melting points.
- The question arises if alternative calibration procedures exist.

REFERENCES

1. R. J. L. Andon, J. E. Connett and J. F. Martin, NPL Report Chem 101, July 1979.
2. F. E. Grubbs, Technometrics, 1969, 11 (1), 1.
3. F. E. Grubbs and G. Beck, 1972, 14 (4), 847.
4. E. E. Marti, "Analytical Calorimetry", proceedings, R. S. Porter and J. F. Johnson, Plenum Press, New York, 1974, Vol 3, p. 127.
5. W. A. Roth, I. Meyer and H. Zeumer, Z. anorg. allgem. Chem., 1934, 216, 303.
6. W. Oelsen, K. H. Rieskamp and O. Oelsen, Arch. Eisenhüttenw., 1955, 26, 253.
7. W. Oelsen, O. Oelsen and D. Thiel, Z. Metallkunde, 1955, 46, 555.
8. B. Predel, Z. Metallk., 1964, 55, 97.
9. D. J. David, Anal. Chem., 1964, 36, 2162.
10. O. Alpaut and Th. Heumann, Acta Met., 1965, 13, 543.
11. J. P. Bros, Bull. Soc. Chim. France, 1966, 8, 2582.
12. L. A. Mechkovskii and A. A. Vecher, Russ. J. Phys. Chem., 1969, 43, 751.
13. L. A. Reznitskii, V. A. Kholler and S. E. Filippova, Russ. J. Phys. Chem., 1970, 44, 299.
14. L. Malaspina, R. Gigli and V. Piacente, Rev. Int. Hautes Temp. Refract., 1971, 8, 211.
15. G. Widmann, Thermochim. Acta, 1975, 11, 331.
16. M. J. Richardson and N. G. Savill, Thermochim. Acta , 1978, 12, 221.
17. M. G. Lowings, K. G. McCurdy and L. G. Hepler, Thermochim. Acta, 1978, 23, 365.
18. F. G. Groenvold, J. Thermal Anal., 1978, 13, 419.

ACKNOWLEDGEMENT

The author wishes to thank the following persons for collaboration: Dr. M. Meyer and M. Szelagiewicz

OPTIMIZATION OF ACCURACY AND PRECISION
IN THE DYNAMIC PURITY METHOD

John E. Hunter, III and Roger L. Blaine,
E. I. Du Pont de Nemours & CO., (Inc.),
Analytical Instruments Division, Wilmington,
DE   19898   (U.S.A.)

INTRODUCTION

In the past few years, the determination of purity by differential scanning calorimetry (DSC) has become a widely accepted technique. The method is recognized by the U.S. Pharmacopeia and National Formulary, while ASTM Committee E37 on Thermal Measurements is presently preparing a standard test procedure for the method (Project TM-01-04A). The calorimetric method has the advantage of determining purity quickly (in less than one hour) and easily, while requiring neither the identification of the impurities nor the acquisition of standard samples for calibration. Furthermore, it requires only milligram specimen sizes.

It does, however, have several important limitations. The method is applicable only to: (1) highly pure samples (greater than 97 mole %); (2) materials which do not decompose during the melt; and (3) the detection of impurities which do not form a solid solution with the main component[1].

The calorimetric purity method is based on the van't Hoff equation:

$$T_s = T_o - \frac{RT_o^2 X}{\Delta H} \cdot \frac{1}{F}$$

Where: T_s = sample temperature (K)

T_o = melting point of the pure sample (K)

R = gas constant (8.314 J/mol K)

X = mole fraction impurity

ΔH = heat of fusion (J/mol)

F = fraction of total sample melted at T_s

To use the equation, a series of partial areas of the melting endotherm (with all partial areas including the onset of melting) are determined. These partial areas are divided by the total area of the melting endotherm to calculate a series of corresponding area fractions (F).

A figure created by plotting each of the reciprocal area fractions (1/F) versus its corresponding temperature (T_s) should (according to the van't Hoff equation) produce a straight line with an intercept temperature (T_o) and a slope of $-RT_o^2 X/\Delta H$. In practice, the 1/F versus T_s plot is observed to be concave upward. This non-linearity is attributed to undetected melting by some authors. To correct for this non-linearity, successive trial values of an empirical quantity x are added to each partial area and to the total area until, with the proper selection of x, a straight line plot of 1/F versus T_s is generated. The heat of fusion is calculated from the corrected total peak area. The impurity level is then calculated from the slope of the corrected 1/F versus T_s plot.

Many authors have given general guidelines for slow heating rates (0.5 to 1.0°C/min) and small specimen sizes (1 to 3 mg) which they have found necessary in order to obtain accurate results.[1-7] The present study undertook to make a systematic investigation of the effect and interaction of four experimental parameters on the accuracy and precision of calorimetric purity determinations: (1) specimen size, (2) heating rate, (3) impurity level, and (4) data acquisition rate.

The generally accepted technique for testing this method uses standards prepared by doping phenacetin with known levels of p-aminobenzoic acid (p-ABA)[2,3]. Such samples are under consideration by the National Bureau of Standards and by ASTM Committee E-37.02.01 as reference materials for the calorimetric purity method.

EXPERIMENTAL

A Du Pont 1090 Thermal Analyzer and 910 DSC Cell were used for this work. The Du Pont 1090 Dynamic Calorimetric Purity Program was used to conduct the purity analysis. The DSC Cell was calibrated taking mean values for the cell constant and onset slope from multiple runs of three pure indium specimens. The cell was purged with nitrogen at a rate of approximately 15 cm^3/min. Aluminum hermetic sample peans (with inverted lids to minimize dead volume and obtain better thermal contact) were used for most specimens; crimped aluminum pans were used for specimens larger than 7.5mg. Specimen and reference pans were chosen to have a weight difference of less than 0.5%, and care was taken to center the pans on the sample platforms in the DSC Cell. By plotting temperature versus time for heating the empty cell, it was found that the programmed heating rates used in this study (0.2 to 20.0°C/min) were accurate to within 1% over the temperature ranges employed.

The samples used were phenacetin doped with p-aminobenzoic acid to purity levels of 100, 99.3, 98, and 95 mole %. They were obtained from the National Bureau of Standards (courtesy of Dr. Charles M. Guttman). An additional sample of phenacetin doped with p-ABA to a 98.5 mole % purity level was obtained from a 1976 ASTM round robin test of the method.

In studying the dynamic calorimetric purity method, the experimental parameters were varied one by one, initially keeping the constant parameters in the area of their generally accepted literature values. Later, the constant parameters were held at levels beyond the generally accepted values in order to acquire information on the combined interaction of the various parameters. Both multiple specimens and multiple runs of individual specimens were employed in most of the investigations. The specimens apparently did not degrade even upon multiple remeltings, since the average calculated purity from the first two runs of two specimens consistently agreed with the average calculated purity from the eighth run, to well within the standard deviation.

RESULTS AND DISCUSSION

Specimen Size. The effect of specimen size was investigated with the sample of 98.5 mole % pure phenacetin, using the sampling interval of 1.0 s/data point at 0.5°C/min. The specimen size was varied from 0.50 to 14.54 mg. The mean purity value obtained at each specimen size was higher than 98.5 mole %. The mean purity value obtained increased (and thus the accuracy decreased) as the specimen size departed from 2 mg. By fitting a second order curve to three points near the optimum value (98.54 mole % at 2 mg) on the plot of calculated purity versus the logarithm of the specimen size (see Figure 1), the optimum specimen size, for greatest accuracy, was found to be 1.7 mg. Allowing for some deviation in specimen size around the optimum of 1.7 mg, an acceptable specimen size range of 1 to 3 mg was determined. Accuracy is believed to decrease with increasing specimen size because equilibrium is not attained in large specimens. With very small specimens, heat flow signal noise becomes significant, since it is very small, and thus the purity determination is again rendered less accurate.

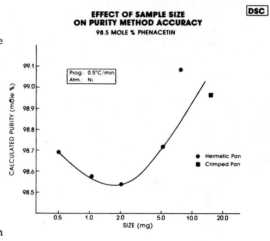

FIGURE 1

Heating Rate. The effect of heating rate was first studied with 1 mg specimens of the 98.5 mole % phenacetin sample. Heating rates were varied from 0.2 to 20.0°C/min. The sampling interval was varied inversely with the heating rate in order to maintain a constant number of data points per temperature interval. Wider temperature ranges were used as the heating rate was increased. The mean purity value obtained increased as the heating rate departed from the common 0.5°C/min. Accuracy is believed to decrease with increasing heating rate because of the non-attainment of equilibrium at high heating rates. Poor accuracy at 0.2°C/min may again be attributed to short-term heat flow noise making the determination of a baseline ambiguous.

A further investigation of the effect of the heating rate was made using 5 mg sample containing 98 mole % phenacetin. With this larger sample size, the most accurate purity value again occurred at 0.5°C/min. However, the heating rate was found to have a somewhat greater effect on the calculated purity than it did in the case of the smaller specimen size (see Figure 2). The effects of specimen size and heating rate thus appear to be partially additive.

An additional study was made of the effect of heating rate using 5 mg specimens of 95 mole % phenacetin. At this larger impurity level, the most accurate purity value occurred at 1.0°C/min. In this case, the calculated purity was even more sensitive to changes in the heating rate (see Figure 3). Thus, the deviations of specimen size, impurity level, and heating rate from their optimum values are shown to add to poor accuracy. This interaction is depicted graphically in Figures 4-7. The occurrence of the most accurate purity value at the somewhat higher heating rate of 1.0°C/min in this case is caused by the high impurity level. Low purity samples have very broad melting endotherms, and since the heat flow is then spread out over a wide temperature range, noise again becomes especially significant at very slow heating rates.

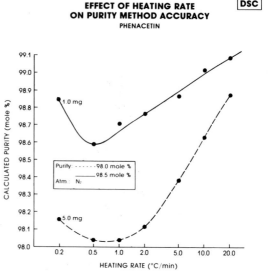

FIGURE 2

Impurity Level. The effect of impurity level was studied at various sets of conditions of specimen size (1 mg, 1.7 mg and 5 mg) and heating rate (0.5°C/min and 2.0°C/min). The data acquisition rate was maintained at a constant number of data points per temperature interval. The starting temperature had to be lowered as the impurity level increased since less pure samples melt over a broad range. The error in the purity determination was found to increase dramatically as the impurity level jumped from 2% to 5%. This increase became even larger as specimen size and heating rate departed from their optimum values (see Figures 5-7).

Data Acquisition Rate. The effect of data acquisition rate was studied using 1 mg specimens of the 98 mole % phenacetin sample and a heating rate of 0.5°C/min. The sampling interval was varied from 0.2 to 8.0 s/data point, which corresponds to 600 to 15 data points/°C at this heating rate.

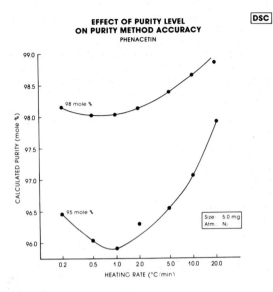

FIGURE 3

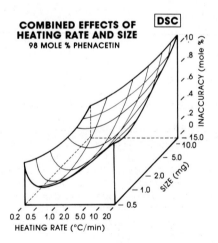

FIGURE 4

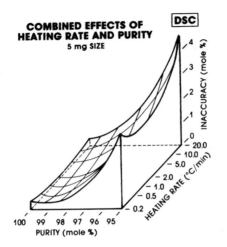

FIGURE 5

FIGURE 6 --

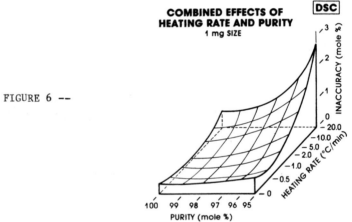

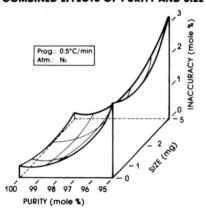

-- FIGURE 7

The accuracy of the purity determinations was found to be independent of the sampling interval (see Figure 8). This is believed to be due to the slow heating rate, which leads to the collection of a significant number of data points/°C even at the slowest data acquisition rate. The precision was found to decrease as the sampling interval departed from 1.0 s/data point (120 data points/°C), where the minimum standard deviation of 1.10 mole % occurred (see FIgure 9). The imprecision at the very short sampling intervals may be attributed to short-term noise making the determination of a baseline ambiguous. At the very large sampling intervals, the lower precision is believed to be caused by the limited number of data points on the melting curve leading to inaccurate integration and selection of partial areas. Thus, a large number of data points is beneficial to gain precision of the determination.

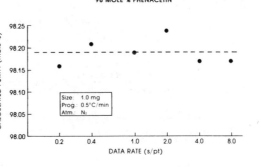

FIGURE 8

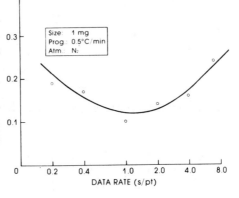

FIGURE 9

SUMMARY

In summary, the optimum conditions for dynamic calorimetric purity determinations are:

- data acquisition rate which gives approximately 120 data points/°C
- specimen size of 1.7 mg
- heating rate of 0.5°C/min
- impurity level of less than 2 mole %

Slight variations from these values may be permitted, and thus the general guidelines in the literature have been confirmed. The errors caused by deviations from these values are cumulative; thus, if any one of these parameters must deviate from its optimum value, it is of utmost importance to maintain the others at their optimum values in order to get the most accurate purity determination possible.

REFERENCES

1. Du Pont Company Thermal Analysis Application Brief #TA-80.
2. E. F. Palermo and J. Chiu, Thermochim. Acta, 14 (1976) 1.
3. E. E. Marti, Thermochim. Acta, 5 (1972) 173.
4. C. Plato and A. R. Glasgow, Anal. Chem., 41 (1969) 330.
5. H. Staub and W. Perron, Anal. Chem., 46 (1974) 128.
6. N. J. DeAngelis and G. J. Papariello, J. Pharm. Sci., 57 (1968) 1868.
7. P. Burroughs, Anal. Proc., 17 (1980) 231.

PREPARATION AND THERMAL PROPERTIES OF SOME L-PROLINE COMPLEXES OF NICKEL, COPPER AND ZINC

by

S. Z. Haider, K. M. A. Malik and M. M. Rahman
Department of Chemistry, University of Dacca, Dacca-2, Bangladesh

and

T. Wadsten
Arrhenius Laboratory, Department of Chemistry, University of Stockholm, S-106 91 Stockholm, Sweden

ABSTRACT

Some L-prolinate complexes of divalent nickel, copper and zinc have been prepared by reacting the appropriate metal salts with L-proline in aqueous medium. The thermogravimetric and differential scanning calorimetric analyses of these complexes as well as free L-proline are reported.

INTRODUCTION

Water hyacinth (Eichhornia Crassipes) grown in large quantities in many rural districts of South Asia and Africa and traditionally considered a nuisance, may be used for production of bio-gas and pollution control of waste water. Its antipolluting character may be attributed to its high capacity for absorbing various metal ions, such as Cr, Mn, Fe, Ni, Cu, Zn, Pb, Cd, Hg[1] and other chemical species. In order to explain the mechanism of absorption of the metal ions by water hyacinth from aqueous solutions, it was necessary to carry out some laboratory model experiments. It is known that the collagen in water hyacinth contains glycine, glutamine and proline along with other normal plant constituents. Therefore the synthesis of transition metal complexes of these amino acids was considered necessary. The glycine complexes have been, by and large, studied to a great extent and a varied literature exist in this field [2,3]. However, the transition metal complexes of proline and glutamine appear to have not been studied systematically. Only limited information is available on the synthesis, properties, structures, reactions and particularly their roles in the maintenance of ionic balance in plant nutrients, especially in water hyacinth. We have recently described the preparation and properties and X-ray structures of DL-prolinate complexes [4,5]. In the present paper we describe the preparation and thermal properties of some L-prolinate complexes of Ni, Cu and Zn.

EXPERIMENTAL

Preparation. The complexes were prepared by following the published procedure (6-8) with some minor modifications

1. **Ni (L-proline) I.** An aqueous solution of L-proline and an excess of freshly prepared nickel carbonate was heated on a steam bath with stirring for about three hours. The undissolved nickel carbonate was filtered off. The sky blue solution was concentrated to induce crystallization. The flaky shiny blue crystalline product was separated, washed with cold water and recrystallized from water.

Ni (L-proline) II. An aqueous solution containing freshly prepared nickel carbonate, L-proline and potassium carbonate in the molar ratio 2 : 10 : 5 was heated on a steam bath with stirring at about 90°C until it became syrupy. After reaching room temperature a blue product was obtained. 3 gms of the blue mass was refluxed in 60 ml methanol for 4 hours. The resulting suspension was filtered and the filtrate was evaporated to dryness under atmospheric condition. The product was redissolved in hot methanol and cooled to room temperature. The resulting product was then dried in a vacuum dessicator.

2. **Cu (L-proline) I.** An aqueous solution of L-proline with excess copper carbonate was heated on a steam bath with stirring and the resultant blue solution was filtered. The filtrate was condensed on the steam bath to induce crystallization. The deep blue crystals formed were separated, washed with cold water and were recrystallized from the same solvent by slow evaporation.

Cu (L-proline) II. This complex was also prepared in the same way using freshly prepared copper hydroxide instead of copper carbonate as one of the reactants.

3. **Zn (L-proline).** An aqueous solution of L-proline and an excess of freshly prepared zinc hydroxide were heated on a steam bath with stirring for about six hours. The resulting suspension was filtered hot and the filtrate was condensed on the steam bath to induce crystallization. A mixture of white crystals and transparent crystals was separated mechanically and each was washed with cold water. The products were air dried.

Thermal analysis. All samples presented in this paper have been analysed with Perkin-Elmer instruments. The thermogravimetric runs were performed in TGS-2 equipped with System 4 Programmer and the scanning results were obtained from DSC-2C connected to TADS. The amount of material used for thermogravimetry was 8-10 mgs and for DSC 3-5 mgs. For the sake of simplicity the resulting data from these techniques are put together according to ICTA recommendations. Some results are shown here, but the first derivative of the weight loss will be included in a future paper where the complete thermal stability will be discussed in terms of the crystal structures which are in progress. Certain calorimetric values are given in cal/g in the corresponding figures.

L-proline used in this study was obtained from Koch-Light Laboratories Ltd, England. All other chemicals were of standard reagent grade.

RESULTS AND DISCUSSION

L-proline (Fig. 1)

According to DSC observations, the onset for the melting is 232°C and the decomposition is in two separate steps, one at 238°C and the other at 270°C.

Ni (L-proline) (Fig. 2). The suggested molecular formula for this complex is $Ni(proline)_2(H_2O)_2$. The two water molecules are lost in the temperature range 140-200°C in two separate but close steps. This indicates that the water molecules may not be structurally equivalent. This step is followed by a very slow weight loss of 2%. The so formed material decomposes in air at 350°C and in argon at 400°C.

K-Ni (L-proline) (Fig. 3). This material loses 5.5% weight between 40 and 120°C without any further detectable changes in weight or heat. Within the same low temperature region, 11.5% is lost while heating in air, and further degradation starts at 250°C. According to the first derivative, one reaction is responsible for the first and two close effects are involved in the latter. On a very small amount of the material heated to 125°C for 30 minutes, the nickel content was found to be 11-12% by weight. This, together with the elemental analysis of the original sample, suggests a plausible formula as $KNi(proline)_3 \cdot 2-3H_2O$. More accurate formulation of the complex is not possible because of the very labile nature of one or more water molecules which are lost at or near the room temperature. This is in agreement with other findings[8].

Cu (L-proline) (I) (Fig. 4). This compound is assigned a molecular formula $Cu(proline)_2 \cdot 2-3 H_2O$. In air and argon, between 50-120°C, a weight loss of 13% is registered. This corresponds to 2.5 molecules of water (calc. 13.4%). In the narrow range, 120° to 135°C, there is an irreversible phase transition. Upon further heating in argon, nothing happens until 250°C where the final decomposition takes place. In air, however, there is a sharp weight loss of 7% between 180-190°C. This reaction must correspond to some oxidation process which is not yet fully understood.

The DSC experiments obviously would indicate the air reaction, but this is not evident from the figure. The reason for this is the closed construction of the standard aluminium sample holder in which an atmosphere of water is established which was liberated in an earlier stage (50-120°C).

Cu (L-proline) II (Fig. 5). The molecular formula and thermal properties are quite similar to Cu (proline) I. At temperatures above 220°C, however, some minor differences are observed, which at this stage of investigation are not yet interpreted.

Zn (L-proline) (Fig. 6). A distinct weight loss starts at room temperature and reaches equilibrium at 45°C. The shape of the thermograms indicates the presence of some very loosely bound water, possibly in the crystal lattice but not in the form of adsorbed water. The amount lost is 2.5%, which corresponds to one molecule of H_2O from a molecular unit of Zn (proline)$_2 \cdot H_2O$. The onset for melting is found to be 222°C which is immediately followed by decomposition, onset 230°C. The decomposition process involves several steps.

CONCLUDING REMARKS

It is found from this investigation that the thermal stability of the L-prolinate complexes of transition metals such as Ni, Cu and Zn are less stable than the corresponding DL-prolinato combinations. More detailed measurements are in progress.

REFERENCES

1. S. Z. Haider, K. M. A. Malik and M. M. Rahman, J. Bangladesh Acad. Sc., 1981, 5(2), 105.
2. R. A. Condrate and K. Nakamoto, J. Chem. Phys., 1965, 42, 2590.
3. K. Tomita, Bull. Chem. Soc. Japan, 1961, 34, 280.
4. S. Z. Haider, A. H. M. Ahmed, A. Habib, and K. J. Ahmed, J. Bangladesh Acad. Sc., 1979, 3(1&2), 81.
5. K. H. Ahmed, A. Habib, S. Z. Haider, K. M. A. Malik, and H. Hess, J. Bangladesh Acad. Sc., 1980, 4(1&2), 85.
6. A. Mathieson and H. K. Welsh, Acta Cryst., 1952, 5, 599.
7. C. Neuberg, H. Lusting and I. Mandl, Arch. Biochem. Biophys., 1950, 26, 77.
8. J. Hidaka and Y. Shimura, Bull. Chem. Soc. Japan, 1970, 43(9), 2990.

ACKNOWLEDGEMENT

Thanks are due to UNEP and the Commonwealth Science Council, London, for a financial grant to carry out the research work under the Project on Management of Water Hyacinth.

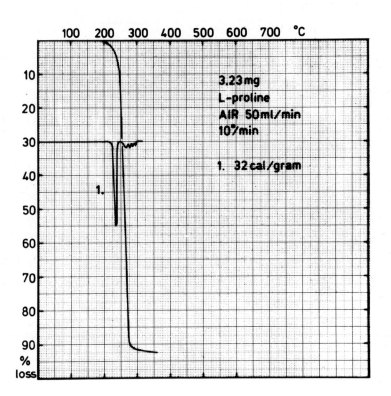

Figure 1. L-proline

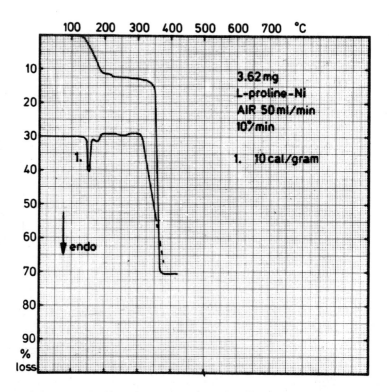

Figure 2. Ni(L-proline)

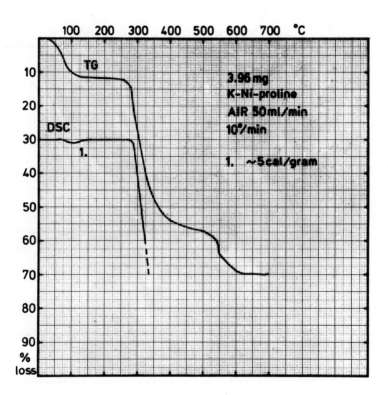

Figure 3. K-Ni(L-proline)

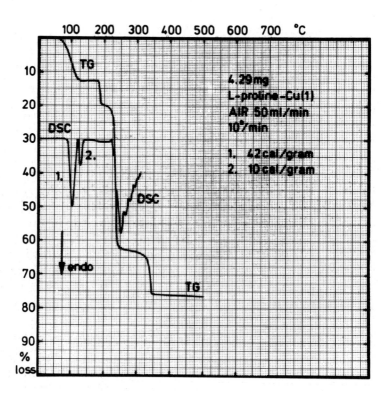

Figure 4. Cu(L-proline) (I)

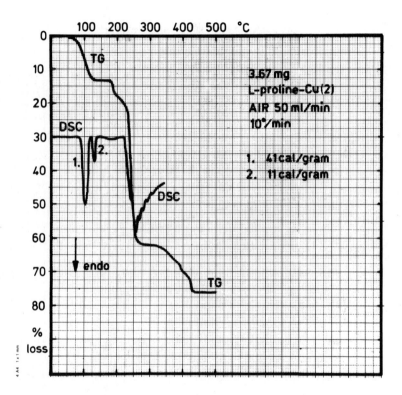

Figure 5. Cu(L-proline) (II)

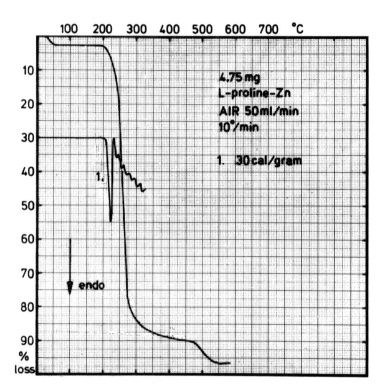

Figure 6. Zn(L-proline)

Section IV   Polymer Science

Plenary Lecture

RECENT DEVELOPMENTS IN THE AREA OF
THERMOTROPIC LIQUID CRYSTALLINE POLYMERS AND THEIR THERMAL ANALYSIS

Masatami Takeda
Department of Chemistry, Science University of Tokyo
Kagurazaka, Shinjuku-ku, Tokyo 162, Japan

1. INTRODUCTION

We are witnessing that thermotropic liquid crystalline polymers are being extensively developed by many researchers throughout the world. I will begin to discuss the early stage of this development.

Liquid crystal science has a long history which goes back about 100 years. However, since 1960 a rapid acceleration of research on liquid crystals has been seen. I would like to mention only the fact that the International Liquid Crystal Conferences had been started at Kent State University in 1963 by Glenn H. Brown, and successive meetings have been held almost every two years. The Eighth International Liquid Crystal Conference was held in Kyoto, Japan in 1980. The conference was attended by 562 participants representing 29 countries. This interest prompted some polymer scientists to try to develop thermotropic liquid crystal polymer systems with similar physical properties as found for the thermotropic liquid crystals. I can take up only a few examples of the research based on such an idea. More detailed information is available in other publications [1,2].

Amerik et al.[3] reported the polymerization of vinyloleate. This monomer shows the following general scheme of phase transitions based on thermal analysis and polarizing microscopy:

Cooling: Isotropic liquid $\xrightarrow{-18°C}$ liquid crystal $\xrightarrow{-45°C}$ crystal

Heating: Isotropic liquid $\xrightarrow{-18°C}$ liquid crystal $\xrightarrow{-32°C}$ crystal

They carried out the radiation induced polymerization of vinyl oleate in the liquid, the liquid crystalline, and the solid states. They found that the physical properties of the resulting polymer greatly depends on the phase state of the monomer. The polymer obtained in the isotropic state is an amorphous polymer, and the polymer formed in the liquid crystal or in the solid state is a crystalline polymer. They studied the polymerization of p-methacrylyloxybenzoic acid in the liquid crystalline state (dissolved in p-cetyloxybenzoic acid) and in the liquid state (solution in dimethylformamide) [4]. The polymerization of p-methacrylyloxybenzoic acid in the liquid crystalline state is characterized by a considerable increase in polymerization rate and polymer molecular weight (Mw = 6×10^5) over the liquid phase polymerization in DMFA (Mw = 5×10^4).

Paleos et al. [5] prepared the liquid crystalline monomers of N-(p-methoxy-o-hydroxy-benzylidene)-p-aminostyrene which have the following chemical structure, and the properties as shown in Table 1.

$$RO-\phi-CH=N-\phi-CH=CH_2$$

$$R=CH_3, C_{18}H_{37}O$$

They have studied the polymerization of N-(p-methoxy-o-hydroxybenzilidene)-p-aminostyrene in the nematic and isotropic phases. In addition, the polymerization in the nematic phase was conducted in a magnetic field. From the experimental results of rate constant of polymerization reaction and molecular weight of formed polymer, they have concluded that there is no significant difference in polymerization reaction between the isotropic and the liquid crystalline states. These results are not in accordance with those of Amerik et al. [4].

Blumstein et al. [6] have studied the polymerization of the p-methacryloxybenzoic acid in the mesophase solvent. Cetyl benzoic acid was used as the smectogen and heptylbenzoic acid used as the nematogen. The experimental results of this study lead to two interesting conclusions. One is that the isotactic triad decreases with increasing the conversion. Another is that the stereospecificity does not depend on the polymerization condition like originating from the nematic or smectic states.

I will now discuss the polymerization of cholesterylacrylate and cholesterylmethacrylate monomer and their polymerization. The transition temperatures of cholesterylacrylate and cholesterylmethacrylate are shown in Table 2. According to Table 2, these monomers have thermotropic liquid crystalline properties in a certain narrow temperature range, however, the transition temperatures changed with the experimental techniques.

Saeki et al. [12] carried out the thermal polymerization of cholesterylmethacrylate (ChMA) monomer in the cholesteric and isotropic states. The blue color, which is characteristic of the cholesteric mesophase of monomeric ChMA, disappeared during the progress of the polymerization. When the temperature of the reaction bath was lowered after standing with the ampule at 108°C for some time, the wall of the ampule became luminous again with a blue color over a limited temperature region below 108°C. The lowering of the region of temperature for the existence of the mesophase is of considerable interest. Careful observation of the temperature region of the color development found on polymerization revealed that it was reproducible for a given process of polymerization. We have failed to observe the same region of temperature for color development by blending of separate polymers made with monomer which had similar ratios of the isotropic temperature region. These results are summarized in Table 3. It is noted that the temperature region is lowered as the polymerization proceeds and the molecular weight of the polymer increases. We have no reasonable explanation for the observation described above, but we hope that it can be solved in the future.

The stereospecific structure of poly(ChMA) is not significantly different from the original poly(ChMA) obtained in the mesophase

Table 1. Properties of N-(p-alkoxybenzylidene)-p-aminostyrenes

p-Alkoxy group	Transition Temperatures °C		
	Solid mesophase	Mesophase liquid	Mesophase identity
MeO	111	124	nematic
$C_{18}H_{37}$	93.5	108	smectic

Table 2. Thermal Properties of the Monomers Cholesterylacrylate

	Solid (°C)	Cholesteric (°C)	Isotropic (°C)	ΔH* (cal/mol)	Ref.
Toth et al.		ca90	127	13.5	[7]
de Visser et al.	66, 65^a, 64^c	121^a, 122.5^b	125		[8]
Hardy et al.			125.8		[9]
Cholesterylmethacrylate					
de Visser et al.	64^c	[111.5]^b	114^b		[10]
Tanaka et al.	82∿86	105∿106** 104∿106***	111∿113**	10∿13	[11]
Saeki et al.	85	103	109	11.3	[12]

(a): determined by DTA, (b): microscopy, (c): X-ray,
*heat of fusion, **DSC, heating, ***DSC, cooling

Table 3. Conversions, molecular weights, and properties of poly(cholesteryl methacrylate)

	Polymerization Temp. (°C)	Polymerization time, min.	Conversion %	$\overline{M}_n$ ^{a}	Mesomorphic Temp. region °C
A	120	300	36	8800	—
B	140	360	57	50000	—
C	108	10	58	8000	109 ∿ 104
D	108	40	63	18500	108 ∿ 99
E	108	80	66	36000	98 ∿ 90
F	108	240	67	27000	63 ∿ 13
G	108	240	67	113000	below 29

^{a}Number-average molecular weight of the polymers was obtained by using a VPO (Mechrolab, Model-302) in benzene solution at 37°C, while that above 50000 was obtained by a GPC (Waters, ANA-PREP) in THF solution.

when obtained in the isotropic liquid phase. This agrees with the results of Blumstein et al. [6]. We did not examine thermotropic liquid crystalline properties of this polymer in this study.

Thermal analysis has been used to determine the glass transition temperature and the melting of the monomers and polymers. No one discussed these thermodynamic data at the early stage of development in this field.

2. MAIN CHAIN TYPE THERMOTROPIC LIQUID CRYSTALLINE POLYMERS

There are basically two approaches in polymer synthesis to obtain the mesomorphic properties in polymers. One type is to build a mesogenic or rigid group into the polymer backbone, and the other one is to attach mesogenic or rigid group to the main chain as a pendant group. The former polymer can be named "main chain" type polymer and the latter "side chain" type polymer. In this section I will discuss the development of the main chain type polymers. Since 1975 many researchers of main chain type polymers have appeared in the literature, I will discuss only a few examples among many.

Roviello et al. [13] have carried out the pioneering work of the main chain type polymer. They synthesized polyalkanoates of p,p'-dihydroxy-α,α'-dimethyl benzalazine which have the following formula:

$$[O-\phi-\underset{\underset{CH_3}{|}}{C}=N-N=\underset{\underset{CH_3}{|}}{C}-\phi-O-\underset{\underset{O}{||}}{C}-(CH_2)_n-\underset{\underset{O}{||}}{C}-]_x \quad n=6,8,10$$

The polymers will be noted as P8, P10, and P12 according to the number of carbon atoms in the aliphatic chain. Typical results are shown in Table. 4. The principal effect of lengthening the flexible spacer was to decrease T_i and T_m and to narrow the mesomorphic phase region. The entropy change (ΔS_i) decreases with length increase of the flexible spacer. However, the authors did not present a proper physical explanation for these observations. They concluded that all the examined samples show a first transition, assigned to the formation of an anisotropic liquid crystal, and a second transition assigned to the formation of an isotropic liquid. They showed for the first time the existence of a thermotropic liquid crystalline mesophase for a main chain type polymer using mainly DSC measurements. X-ray diffraction studies and DSC analysis suggested that the crystallization and orientation of the molecules in the polymer are quite different depending on the previous thermal history of the sample. The thermal history has a considerable influence on the phase transition or on the crystallinity in the polymer materials. A typical example is shown in Figure 1. [14]. The DSC-curve of a previously untreated sample shows four endothermic maxima T_1-T_4. The phase transition T_3 corresponds to the melting of the sample to give the mesophase, whereas the phase transition T_4 is that to the isotropic liquid phase, as determined by optical observation. X-ray analysis supports these assignments also.

The molar isotropization entropy (ΔS_i) is higher than that of the low molecular weight compound as shown in Table 5 [14]. The ratios of the entropy changes the nematic→isotropic (N→i) and crystal→isotropic (K→i) transitions were used to compare the degree of order in the phases involved. This ratio is typically 0.02-0.03 [15,16] for low molecular weight liquid crystals. However, the authors chose to calculate this ratio using Roviello's data, as shown in

Table 4. Thermodynamic data for the phase transition [13]

		$T_m(K)$	H_m^*	S_m^{**}	$T_i(K)$	H_i^*	S_i^{**}	A^a	B^b
	first heating	483	2.10						
	first cooling	453	1.57						
P12	second heating	476 483	0.71 1.32	1.49	514	1.90	3.79	38	0.71
	second cooling	454	1.55	3.41	489	1.80	3.68	35	0.52
	third heating	479	2.10	4.38	512	1.82	3.55	33	0.45
P10	first heating	476	1.90	3.99	529	2.33	4.40	53	0.52
P 8	first heating	511	2.53	4.95	568	2.63	4.63	57	0.48

*Kcal/mol
**cal/mol.K.
$^a A = \Delta T (T_i - T_m)$
$^b B = \Delta S_{n-i} / \Delta S_{k-i}$

Table 5. The isotropization entropy of low molecular compounds and its polymers. [13].

	ΔS(cal/mol.K)
$\{\phi\text{-C=N-N=C-}\phi\text{-O-C-(CH}_2)_{10}\text{-C-O}\}_n$ | | || || CH$_3$ CH$_3$ O O	3.70

H$_3$C-(CH$_2)_4$-C-O-ϕ-C=N-N=C-ϕ-O-C-(CH$_2)_4$-CH$_3$
        || | | ||
        O   CH$_3$ CH$_3$ O
	0.78

$\{\phi\text{-C=N-N=C-}\phi\text{-O-C-O-(CH}_2)_8\text{-O-C-O}\}_n$ | | || || CH$_3$ CH$_3$ O O	3.29

H$_3$C-(CH$_2)_3$-O-C-O-ϕ-C=N-N=C-ϕ-O-C-O-(CH$_2)_3$-CH$_3$
         || | | ||
         O   CH$_3$ CH$_3$ O
	0.80

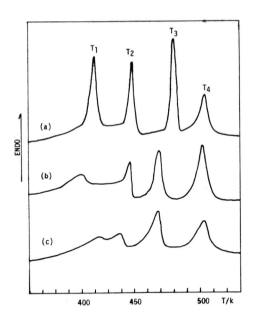

Figure 1. DSC traces [14]. (a): Sample annealed at T=446K for 20 h; (b): same sample, successive heating run; (c): untreated sample. T_1=410K, T_2=440K, T_3=440K, T_4=491K

Table 4. The $\Delta S_{N \to i}/\Delta S_{K \to i}$ ratio is between 0.4-0.7 which is a larger value than for the low molecular weight liquid crystals. It seems to me that there is more order in the nematic phase of the polymer relative to its solid phase than is the case for a low molecular weight nematic compound.

Roviello et al. [17] prepared copolymers of the formula $(-\phi-C(CH_3)=N-N=(CH_3)C-\phi-OOC(CH_2)_n COO-)_x$ (n=6 and 10). In this study they attempted to make a polymer system which can preserve the liquid crystalline structure (frozen nematic orientation) at room temperature by rapid quenching from the liquid crystalline state. They have stated that the mesophase structure is preserved at room temperature by quenching from the fluid liquid crystal phase. This study has raised two general questions: Can the frozen nematic and the smectic structure be realized in polymer system below the glass transition temperature, and what is the relation between this frozen structure and the crystallized structure of the polymer system? It seems to me that this is an interesting property of thermotropic liquid crystalline polymers when comparing them with low molecular weight liquid crystal compounds in which a nematic or smectic structure is changed to crystalline structure on the low temperature side.

Another interesting work is by Jackson et al. [18]. It gives stimulation to another area of the polymer liquid crystal science. They prepared a copolyester of poly(ethylene terephthalate) modified by p-acetoxybenzoic acid. The minimum melt viscosities at 275°C were obtained with polyesters containing 60-70% p-hydroxybenzoic acid. They have stated that the experimental results can be explained only from the formation of the liquid crystalline state, and the first thermotropic liquid crystal polymer is recognized. It is interesting that they do not refer to any thermoanalysis of these polymers, and recent attempts to characterize the thermotropic properties by DSC were not conclusive [15,19].

Griffin et al. [20] prepared polyesters having the following structure:

$$\{C-\phi-O(CH_2)_xO-\phi-C-O-\phi-O(CH_2)_yO-\phi-O\}_n$$
$$\parallel \qquad\qquad\qquad \parallel$$
$$O \qquad\qquad\qquad O$$

$x = 2\text{-}10, 12$
$y = 6, 8, 10$

These polyesters showed at leat two endotherms by DSC measurement. The highest transition was broad and corresponded to a transition from a mesomorphic phase to an isotropic fluid. These endotherms have been found to arise after annealing, and to be depended upon heating rate as I have already shown in Figure 1. As the heating rate is increased, the higher-temperature endotherm becomes smaller, and the lower-temperature endotherm becomes larger. The higher-temperature endotherm (the nematic→isotropic transition) was not noticeably affected by changing heating rates of 20°C/min or less. They also discussed the ratio of $\Delta S_{N \to i}/\Delta S_{K \to i}$. These ratios range from 0.05 to 0.23 with an average value of 0.13 for their polymers, which is close to the value of 0.11 obtained for the poly(ethyleneterephthalate) copolymer. [15].

In our research group, Iimura et al. [21] prepared polyurethanes by the reaction of 3,3',-dimethyl-4,4'biphenyl-diisocyanate (DBD) with α,ω-alkanediols and oligoethyleneglycol. Typical results of DSC measurements are shown in Figure 2. The DSC curves in Figure 2 have two peaks on heating and one peak on cooling. This tendency is just opposite to the usual monotropic low molecular weight liquid crystal materials. If these two peaks are related to the $T_{s \to m}$ and $T_{m \to i}$ transition temperatures, this behavior may be called the inverse monotropic liquid crystalline polymer. Figure 3 also shows the change of DSC curves with different scanning steps. Two peaks on heating and one peak on cooling are preserved until the 5th scanning step, however, a remarkable decrease of transition temperature is not explained as yet.

Thermal data analyzed by DSC are shown in Table 6. T_m and T_i increases with increasing length of the flexible part, and also depend on the molecular weight for the case of poly(DBD-HD).

DSC curves of DBD-oligoethylene glycol polymer are shown in Figure 4. The glass transition temperature is observed at 25.5, 60, and 90°C for di- tri- and tetra-ethylene glycol polymers, respectively. Higher transition temperatures observed at 190°C, of poly(DBD-DEG) become weak with successive scanning and disappear after the 4th scanning steps as shown in Figure 5. X-ray

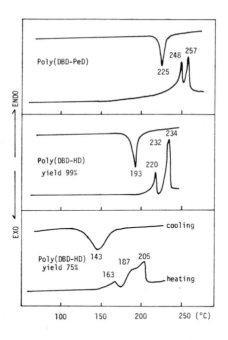

Figure 2. DSC traces of poly(DBD-alkanediol)s

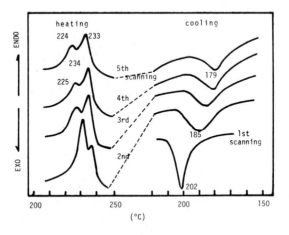

Figure 3. DSC traces of poly(DBD-OD) in different scanning steps: scanning rate: 10°C/min.

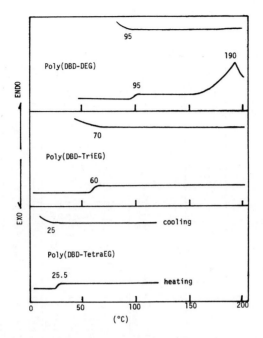

Figure 4. DSC traces of poly(DBD-oligoethylene glycol)s scanning rate: 10°C/min

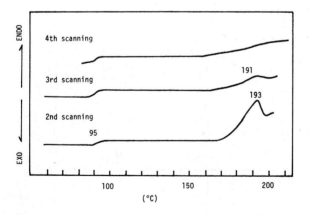

Figure 5. DSC traces of poly(DBD-DEG) on heating in different scanning steps. scanning rate: 10°C/min.

Table 6. Thermal data of poly(DBD-alkanediol)s and poly(DBD-oligoethylene glycol)s

Polymers	$T_m/°C$	ΔH_m kcal/mol	$T_i/°C$	ΔH_i kcal/mol	$T_g/°C$	$[\eta]$ dl.g^{-1}	$\overline{M_n}$	$\Delta S_{m-i}/\Delta S_{k-i}$
P5 Poly(DBD-PeD)	248	2.6	257	3.3		0.419		0.55
P6 Poly(DBD-HD)	163	0.2	205	1.8		0.148	4400	0.89
P6 Poly(DBD-HD)	220	1.0	234	1.7		0.434		0.63
P8 Poly(DBD-OD)	231	4.1	237	1.4				0.25
P10 Poly(DBD-DD)	202	0.3	221	2.0				0.87
P12 Poly(DBD-DoD)	174	0.1	186	1.7				0.94
P02 Poly(DBD-DEG)					95	0.285	15600	
P03 Poly(DBD-TriEG)					60	0.266	10300	
P04 Poly(DBD-TetraEG)					25.5	0.334	15700	

DBD: 3,3'-dimethyl-4,4'-biphenyldiyl diisocyanate.

Table 7 The combination of diols and diacid chlorides, and notation of the polymers

Diacid Chloride	Diol	Notation of polymer
ClOC-ϕ-N=N-ϕ-COCl ‖ O	HO-(CH$_2$CH$_2$O)$_n$H n=2,3,4	POEAOB-n n=2,3,4
ClOC-ϕ-N=N-ϕ-COCl	HO-(CH$_2$CH$_2$O)$_n$-H	POEAB-n

Table 8. Thermal properties of POEAOB and POEAB polymers

	ΔH_m (kcal/mol)	T_i (K)	ΔH_i (kcal/mol)	ΔS_i (cal/mol. K)	$\Delta S_{m-i}/\Delta S_{k-i}$
POEAOB-2	0.30	493	0.25	0.51	0.44
POEAOB-3	1.01	433	0.11	0.25	0.10
POEAOB-4	0.73	373	1.74	4.67	0.70
POEAB-2	2.84	461	0.65	1.41	0.19
POEAB-3	2.02	419	6.37	15.20	0.76
POEAB-4	3.76	370	2.99	8.08	0.44

diffractional patterns of these samples at room temperature which
have been cooled from the melt are shown in Figure 6.

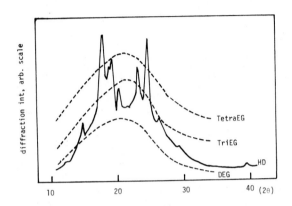

Figure 6. X-ray diffraction patterns obtained with poly(DBD-HD)
[full line], and poly(DBD-oligoethylene glycols) [dotted line]

Poly(DBD-alkanediol)s are particularly crystalline and Poly(DBD-oligoethylene glycol)s do not show thermotropic liquid crystalline polymer. It seems to me that both the flexible group of $(CH_2)_n$ and $(CH_2-CH_2-O)_n$ give quite different physical properties to the polymer system and adequate explanation is not given at present, however, these facts seem to be worthwhile to mention here.

Iimura et al. [22] have synthesized poly(oligoethylene glycol-p,p'-azoxybenzoate) and poly(oligoethylene glycol-p,p'-azobenzoate). In Table 7 data on the combinations of diols and diacidic chloride are summarized based in the preparation of the copolymers. DSC curves of these polymers are shown in Figure 7. In heating curves there are more than two peaks in three samples. In cooling curves, however, POEAB-3 and POEAB-4 polymers show two endothermic peaks, which can be assigned to $T_{i \to m}$ and $T_{m \to s}$ transitions, respectively. POEAOB-2 polymer shows clearly the existence of mesomorphic phase by microscopy. We have concluded that POEAOB-2,3,4 and POEAB-2 polymer are thermotropic liquid crystalline polymers and thermal properties of these polymers are shown in Table 8. The transition temperatures from the mesomorphic phase to the isotropic liquid of the POEAOB polymer is higher than that of the POEAB polymer. On the contrary, the transition temperature from the solid to the mesomorphic phase of POEAOB polymer is lower than that of the POEAB polymer. These facts indicated that the POEAOB polymers have a broader mesomorphic

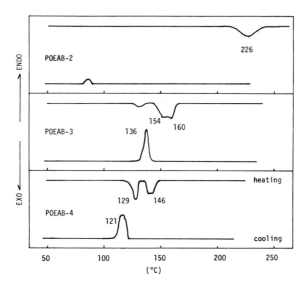

Figure 7. DSC traces of poly(oligoethylene glycol-p,p'-azobenzoate)s. The 2, 3 and 4 show di-, tri-and tetraethylene glycol, respectively. Scanning rate: 10°C/min.

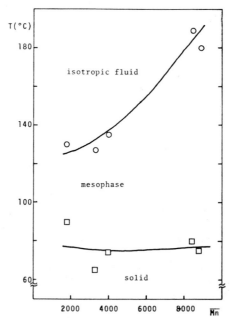

Figure 8. Molecular weight vs. transition temperature of POEAOB-3. (o): The transition temperature from mesophase to isotropic fluid; (□): The transition temperature from solid to mesophase

temperature range than the POEAB polymers. This tendency is the same as in the case of the corresponding low molecular weight, crystalline materials [23].

The effect of the average molecular weight of polymers on the transition temperature is studied in POEAOB-3 and shown in Figure 8. It is clear that $\overline{M}_n$ has more influence on $T_{m \to i}$ than on $T_{s \to m}$. This experimental results supports the observation that the high molecular weight compounds have a wider temperature region of mesophase existence.

Iimura et al [24] have synthesized also another type of azoxy- and azo-type polyesters by reacting of azophenol or azoxyphenol with dicarboxylic acid of different chain lengths. The transition temperatures from the mesomorphic phase to the isotropic fluid demonstrates in both polyesters an odd-even effect depending on the methylene units. A clear schlieren texture is observed for both polyesters.

Iimura et al. have also prepared the azoxy and azo type co-polyesters, as shown in Table 9 [25]. CoPOEAOB-2,4 shows the

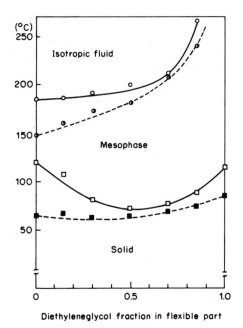

Figure 9. The transition temperature of CoPOEAOB-2,4 as a function of diethylene glycol composition. (□): $T_{s \to m}$ on heating; (o) $T_{m \to i}$ on heating; (●): $T_{i \to m}$ on cooling; (■): $T_{m \to s}$ on cooling.

Table 9. Azo- and Azoxy-type copolyesters

1. Polycoalkylazoxybenzoate), CoPAAOB-n,m

$$\{(CO-\phi-\underset{O}{N=N}-\phi-COO(CH_2)_n-O)\overline{\ }_x(CO-\phi-\underset{O}{N=N}-\phi-COO(CH_2)_m-O)\overline{\ }_y\}$$

2. Poly(azoxyphenolcoalkanoate), CoPAOPA-6,10

$$\{(O-\phi-\underset{O}{N=N}-\phi-OCO-(CH_2)_6-CO)\overline{\ }_x(O-\phi-\underset{O}{N=N}-\phi-OCO-(CH_2)_{10}-CO)\overline{\ }_y\}$$

3. Poly(azophenolcoalkanoate), CoPAPA-6,10

$$\{(O-\phi-N=N-\phi-OCO-(CH_2)_6-CO)\overline{\ }_x(O-\phi-N=N-\phi-OCO-(CH_2)_{10}-CO)\overline{\ }_y\}$$

4. Poly(azoxyphenolcooligoethylene glycol), CoPOEAOB-2,4

$$[CO-\phi-\underset{O}{N=N}-\phi-COO-(CH_2CH_2O)_2-O)\overline{\ }_x(CO-\phi-\underset{O}{N=N}-\phi-COO-(CH_2CH_2O)_4-O)\overline{\ }_y\}$$

n,m indicate the number of CH_2 units in the alkyl chains.

enantiotropic mesophase for the all copolymer compositions. The transition temperature from mesophase to isotropic fluid increases with decreasing diethylene glycol fraction. Figure 10 shows the

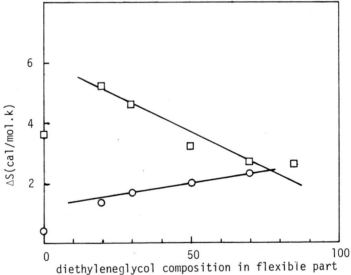

Figure 10. The transition entropy of CoPOEAOB-2,4 as a function of diethylene glycol composition. (□):$\Delta S_{s \to m}$; (○):$\Delta S_{m \to i}$.

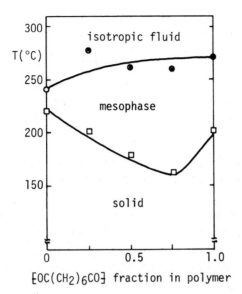

Figure 11. The transition temperature of CoPAOPA-6,10 as a function of ${OC(CH_2)_6CO}$ composition. (□): $T_{s \to m}$ on heating; (o): $T_{m \to i}$ on heating; (■): obtained by the microscopic observation.

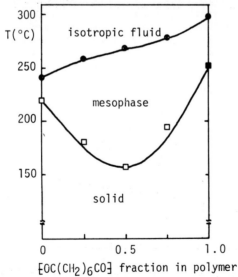

Figure 12. The transition temperature of CoPAOPA-6, 10 as a function of ${OC(CH_2)_6CO}$ composition. (□): $T_{s \to m}$ on heating; (●): $T_{m \to i}$ on heating: (■): obtained by the microscopic observation.

transition entropy dependence on the diethylene glycol composition in the flexible part. The transition entropy from the solid to the mesophase increases with increasing the flexible part in the co-polymer, while that from mesophase to isotropic fluid decreases with increasing the flexible part and the value of transition entropy ($\Delta S_{m \to i}$) is about 2 cal/mol.K.

The phase diagrams of CoPAOPA-6,10 and CoPAPA-6,10 are shown in Figures 11 and 12, respectively. The enantiotropic mesophase appears in both copolyesters, and the schlieren textures are found independent on the composition of flexible part. It is characteristic that the mesomorphic range becomes broader depending on the fraction of a flexible part. The transition entropy ($\Delta S_{m \to i}$) is 0.2-1.5 cal/mol.K in CoPAOPA and 0.1-0.5 cal/mol.K in CoPAPA. Even if these values are small compared to those in Roviello's work [14], it is concluded that the mesophase of these polymers is a nematic phase because of the observation by polarized microscopy and by X-ray diffraction.

Recently, Blumstein et al. [26] prepared the nematic and cholesteric thermotropic azoxy polyesters based on 4,4'-azoxy-2,2'-methylphenyl moiety with an asymmetric center in the polymer backbone. The thermal properties of these polyesters indicate an increase of isotropization entropy with the increase of average spacer length.

3. SIDE CHAIN TYPE THERMOTROPIC LIQUID CRYSTALLINE POLYMERS

The liquid crystalline polymers discussed in Chapter 1 were side chain type polymers. However, the thermotropic liquid crystalline properties were not clearly shown at that stage. Since many researches of thermotropic liquid crystalline polymers appeared in recent years, I can only take a few examples, among them the results of our research group.

The first example is the research of Strzelecki et al. [27]. They synthesized the following monomers:

A. $CH_2=CH-C-O-\phi-C=N-N-\phi-N=CH-\phi-O-C-CH=CH_2$
              $\|$ $\|$
              O O

B. $CH_2=CH-C-O-\phi-O-\phi-CH=N-\phi-R$
              $\|$
              O

   1B: $R = -CH_2-CH_2-CH_2-CH_3$
   2B: $R = -CN$

C. $CH_2=CH-CO-O-$cholesterol

In the case of homopolymerization of A. polymer they were able to show nematic and smectic texture by microscopic observation. Terpolymerization are carried out A, 2B and C monomer with weight ratio of A:2B:C are 3:5:0.09-0.3. Ternary polymer shows the cholesteric phase proven by texture observation.

Shibaev et al. [28] have synthesized the following series of the monomer:

$$CH_2=C-CH_3$$
$$|$$
$$C=O$$
$$|$$
$$HN-(CH_2)_n-COO-cholesterol$$

n=2,5,6,8,10 and 11

It is interesting to note that the same type of monomer has been synthesized by Kamogawa [29]. All monomers of ChMMA-n show a monotropic cholesteric phase. Polymerizations were carried out to make the homopolymers of this series of monomers and also copolymers of ChMA-n with n-alkylacrylate or n-alkylmethylmethacrylate. They discussed that the polymerizations of all ChMMA-n and ChMA from a melt yield polymers which show spontaneously optical anisotropy. The optical pattern is similar to the confocal texture of low molecular weight liquid crystals. They have studied the thermal properties of these polymers using thermoanalytical methods such as DTA and DSC and the thermomechanical method together with microscopy. The glass transition temperature (T_g) lies between 120-200°C, T_f (flow temperature observed by thermomechanical method) between 130-220°C and $T_{m \to i}$ between 180-220°C, depending on the alkyl chain length. T_g decreases with the alkyl chain length and also T_f and T_{m-i} decrease with increasing alkyl chain length. The value of $\Delta H_{m \to i}$ of PChMMA was 0.76 cal/g, which agrees well with the value of $\Delta H_{m \to i}$ for the corresponding low molecular weight liquid crystals. They have also observed that these polymers show the mesomorphic phase when cooled down to room temperature.

The same authors have also carried out a study of copolymerization of ChMMA-11 with butylacrylate (A4) or butylmethacrylate (MA-4) and showed that liquid crystalline phases resulted in a wide range of composition. The temperature dependence of the X-ray scattering of ChMMA-11 with A4 (37:63 weight ratio) also supports this conclusion. The intensity of the small angle scattering made a very important contribution by establishing the exsistence of thermotropic liquid crystalline polymers of the side chain type.

Finkelmann et al.[30] have made another important contribution to this kind of research. They have synthesized monomers of the following formula:

$$CH_3$$
$$|$$
$$H_2C=C-C-O-\phi-(CH_2)_n-C-O-cholesterol, \quad n=2,6,12$$
$$\qquad \parallel \qquad\qquad\qquad \parallel$$
$$\qquad O \qquad\qquad\qquad\quad O$$

The transition temperature from mesophase to isotropic fluid for the n=6 homopolymer is 182°C and ΔH_i is 1.82 cal/mol., and for the n=12 homopolymer it is 168°C and ΔH_i is 1.00 cal/mol. The homopolymers exhibit a smectic phase. Smectic polymer phases are obtained also by copolymerization of the monomers in which the spacer lengths are $n_1=2$, $n_2=6$, $n_2=12$. A similar result was obtained by Shibaev et al.[28]. They are successful in making a cholesteric polymer by copolymerization in which $n_1=2$, $n_2=12$ using monomers with 1:1 monomer ratio. This copolymer is the first enantiotropic cholesteric polymer. This copolymer has a T_i at 209°C and ΔH_i is 0.28 cal/mol. They have proposed [32] that inclusion of flexible spacers is necessary to have enantiotropic thermotropic crystalline polymers of the side chain type. They think that flexible spacers can decrease the steric effects of the main chain on the formation of orientation of the side chain mesogenic group in the liquid crystalline state. The flexible spacer can decouple the molecular motion of the side chain mesogenic group from the molecular motion of the main chain segment. Recently the concept of flexible spacer was also used in the field of main chain thermotropic polymers. However, the exact meaning of the flexible spacer in terms of the molecular motion of each segment of the polymer still needs elucidation.

Blumstein et al. [31] have prepared copolymers of cholesterol methacrylate (ChMA) with methylmethacrylate (MMA) and with butylmethacrylate (BMA). The film of the homopolymers was examined by X-ray diffraction. Sharp low-angle X-ray diffraction peaks at $2\Theta=2.5°$, corresponding to 35.5 Å of periodicity, and $2\Theta=15.5°$, corresponding to 6.3Å, were observed. Copolymers of ChMA with MMA and BMA as a function of concentration of ChMMA were also examined by the X-ray diffraction method, and similar low angle X-ray diffraction peaks were found, corresponding to the periodicity of 33.9 Å in the high concentration range of ChMA. They have also reported the T_g value of the copolymers of ChMA and BMA.

Recently our group has prepared the same type of polymers and performed cross-polarized microscopic observations as a function of temperature. The thin film cast from benzene solution is rather colorless at room temperature. With increasing temperature, the color development and the brightness are observed at certain temperature ranges above the T_g determined by Blumstein et al. [31]. The mesomorphic temperature ranges are shown in Figures 13a and 13b. The strong optical anisotropy observed in the ChMA rich copolymers is caused by a strong interaction between the cholesteric groups with some motional state of the homopolymer and copolymer chains. This evidence is correlated with the experimental results of the appearance of small angle X-ray diffraction described by Blumstein et al. [31] on the cooled sample. We have failed in the determination of T_g and T_i by the DSC method. We found only a large exothermic peak which is increased with increasing ChMA content. It seems to us that this appearance of the peak would be due to some rearrangement in the cholesterol groups in the copolymer system.

I will mention another recent important research of Finkelmann et al. [38,40,41]. They have synthesized the polymer of polysiloxane as the main chain with a mesogenic side chain, having flexible spacer. These polymers show nematic, smectic and cholesteric liquid crystalline phases depending on the mesogenic side group. These polymers show the lowest T_g which have ever been observed in liquid crystalline polymers.

Typical side chain type liquid crystalline polymers are summarized in Table 10. The enthalpy change for the transition from

Table 10. The transition enthalpy of side chain type polymers.

Polymers	Enthalpy(Cal/mol)	Ref.
$\{CH_2C(CH_3)\}_x$ $\|$ $O=C-O-(CH_2)_n-O-\phi-\phi-R$		33
n=2, R=OCH_3	0.67(n→i)	
n=6, R=OCH_3	1.71	
$\{CH_2C(CH_3)\}_x$ $\|$ $O=C-O-(CH_2)_n-C-O-cholesterol$ $\|\|$ O		31
n=2,6,12		
n=6	1.60(Sm→i)	
n=12	1.00(Sm→i)	
$\{CH_2C(CH_3)\}_x$ $\|$ $O=C-O-(CH_2)_n-O-\phi-COO-\phi-R$		34
n=2, R=OCH_3	0.55(n→i)	
n=2, R=OC_3H_7	2.20(Sm→i)	
n=6, R=OCH_3	0.50(n→i)	
$\{CH_2C(CH_3)\}_x$ $\|$ $O=C-NH-(CH_2)_n-C-O-\phi-O-C-\phi-R$ $\quad\quad\quad\quad\quad\quad \|\| \quad\quad\quad \|\|$ $\quad\quad\quad\quad\quad\quad O \quad\quad\quad\quad\quad O$		35
n=5, R=OC_6H_{13}	0.30	
n=11, R-OC_6H_{13}	2.00	
$\{CH_2C(CH_3)\}_x$ $\|$ $O=C-NH-(CH_2)_{11}-C-O-\phi-\phi-OC_6H_{13}$ $\quad\quad\quad\quad\quad\quad\quad\quad\quad \|\|$ $\quad\quad\quad\quad\quad\quad\quad\quad\quad O$	1.10	36
$\{CH_2C(CH_3)\}_x$ $\|$ $O=C-O-(CH_2)_n-O-\phi-CH=N-\phi-C_4H_{11}$		36
n=6	1.90	
n=11	2.90	
$\{CH_3-Si-O\}_x$ $\|$ $CH_2-CH_2-R,$		36
R=$CH_2-O-\phi-\phi-OCH_3$	5.50(k→i)	

Table 10. The transition enthaply of side chain type polymers (continued)

Polymers	Enthalpy (cal/mol)	Ref
$R=CH_2-O-\phi-\phi-OCH_3$	5.50 (k→i)	36
$R=CH_2-CH_2-O-\phi-\phi-OCH_3$	23.21 (k→i)	
$R=CH_2-O-\phi-COO-\phi-OCH_3$	0.53 (n→i)	
$R=CH_2-O-\phi-COO-\phi-OC_6H_{13}$	2.78 (Sm→i)	
$R=CH_2-O-\phi-COO-\phi-CN$	0.45 (Sm→i)	
$R=CH_2-COO$-cholesterol	0.65 (Sm→i)	

$\{P=N\}_x$ with R substituents

	Enthalpy	Ref
$R=CF_3-CH_2-O-$	0.80	39
$R=Cl-\phi-O-$	0.20	

the anisotropic phase to the isotropic fluid is about 0.3-2.9 cal/mol, and the smectic phase has higher values than the nematic phase as was found for low molecular weight liquid crystals.

4. SOME REMARKS AND CONCLUSION

Of the more recent activities of the thermotropic liquid crystalline polymer research, I would like to point out that the study of the effect of electric and magnetic fields on the thermotropic liquid crystalline state is increasing. Krigbaum et al. [42] have studied the flow instability of polymers in the nematic phase induced by an electric field. Using the poly[(ethylene terephthalate)-co-1,4-benzoate] containing 30 and 60 mol. % p-oxybenzoyl units. They were able to observe the formation of Williams domain patterns in 60 mol. % p-oxbenzoyl unit using dc voltage at higher temperature. But the formation time was the order of hours, as compared to tenths or hundredths of a second for low molecular weight liquid crystalline material such as p-azoxyanisol.

Volino et al. [43] have studied the magnetic effect on the nematic phase in polymers. They were able to show the PMR spectra at 120°C which indicates the orientation of mesogenic groups in the magnetic field and determined the order parameter of the nematic phase as 0.72-0.88.

Recently Uematsu et al. [44] have reported that the copoly(γ-methyl, hexyl-glutamate) shows a thermotropic liquid crystalline cholesteric phase above 150°C. Polyglutamate has been an important material in the field of lyotropic liquid crystal research, however, the thermotropic nature of this polymer was never observed before.

I have discussed the recent development of the thermotropic liquid crystalline polymers. The thermoanalytical methods have been often used to determine the phase transition of liquid crystalline

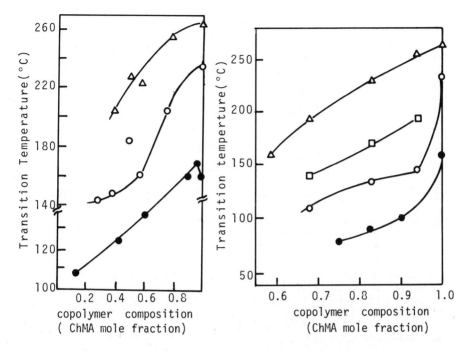

Figure 13. (a). Phase diagram of ChMA-MMA copolymer. (Δ): $T_{m \to i}$ on heating; (o): the temperature at which the polarized light is observed on heating; (●): T_g reported by Blumstein et al. [26].

(b). Phase diagram of ChMA-BMA copolymer. (Δ): $T_{m \to i}$ on heating; (o): the temperature at which the polarized light is observed on heating; (□): the temperature at which the strongest polarized light is observed; (●): T_g reported by Blumstein et al. [26].

polymers. However, comparing with the low molecular weight liquid crystal research, the results of DSC are some times difficult to interpret. The main reasons are that the phase transitions appear over a broader temperature range in the polymer system and that there is coexistence of the amorphous and the crystal phase in the polymer at low temperatures. One of the typical problems is the different results obtained from successive heating and cooling runs in DSC which I have shown in Section 2.

Since the research of thermotropic liquid crystalline polymers is a rather new field, one can expect that many unsolved problems in the thermoanalysis of these polymers will be solved in the near future with more advanced techniques of thermoanalysis and in co-operation with several other physico-chemical methods.

I am very grateful to Professor K. Iimura, Dr. N. Koide and Mr. K. Shoji and other coworkers in our Chemistry Department for their collaboration in the research in thermotropic liquid crystalline polymers for 10 years. I wish to express my appreciation to Dr. N. Koide for aid in the preparation of this manuscript. Acknowledgement is made to a Grant-in-aid for Scientific Research from the Ministry of Education.

REFERENCES

1. Liquid Crystalline Order in Polymers, Academic Press, 1978 edited by Alexandre Blumstein.
2. Handbook of Liquid Crystal, Verlag Chemie, 1980, edited by Hans Kelker and Rolf Hatz.
3. Y. B. Amerik and B. A. Krentsel, J. Polym. Sci., C16, 1385 (1967).
4. Y. B. Amerik, I. I. Konstantinov, and B. A. Krentsel, J. Polym. Sci., C23, 231 (1968).
5. C. M. Paleos and M. M. Labes, Mol. Cryst. Liq. Cryst., 11, 385 (1970).
6. A. Blumstein, N. Kitagawa, and R. Blumstein, Mol. Cryst. Liq. Cryst., 12, 215 (1971).
7. W. J. Toth and A. V. Tobolsky, J. Polym. Sci., B8, 289 (1970).
8. A. C. de Visser, K. de Groot, J. Feyen, and A. Bantjes, J. Polym. Sci., 9, 1893 (1971).
9. Gy. Hardy, F. Cser, A. Kallo, K. Nyitrai, G. Bodor, and M. Lengyel, Acta Chem. Acad. Sci. Hung., 65, 287, 301 (1970).
10. A. C. deVisser, K. de Grott, J. Feyen, and Bantjes, J. Polym. Sci., B 10, 851 (1972).
11. Y. Tanaka, S. Kabaya, Y. Shimura, A. Okada, Y. Kurihara, and Y. Sakakibara, J. Polym. Sci., B 10, 261 (1972).
12. H. Saeki, K. Iimura, and M. Takeda, Polym. J., 3, 414 (1972).
13. A. Roviello and A. Sirigu, J. Polym. Sci., B, 13, 455 (1975).
14. A. Roviello and A. Sirigu, Makromol. Chem., 181, 1799 (1980).
15. W. R. Krigbaum and F. Salaris, J. Polym. Sci. Polym. Phys. Ed., 16, 883 (1978).
16. E. M. Barrall, II and J. F. Johnson, Liquid Crystals and Plastics Crystals, G. W. Gray and P. A. Winson, Eds., Halsted., Ellis Horwood, Chichester, England, 1974, Vol. 2, p. 254.
17. A. Roviello and A. Sirigu, Europ. Polym. J., 15, 61 (1979).
18. W. J. Jackson, Jr., and H. F. Kuhfuss, J. Polym. Sci. Polym. Phys. Ed., 14, 2043 (1976).
19. H. J. Ladner and W. R. Krigbaum, J. Polym. Sci. Polym. Phys. Ed., 17, 1661 (1979).
20. A. C. Griffin and S. J. Havens, J. Polym. Sci. Polym. Phys. Ed., 19, 951 (1981).
21. K. Iimura, N. Koide, H. Tanabe, and M. Takeda, Makromol. Chem., 182, 2569 (1981).
22. K. Iimura, N. Koide, R. Ohta, and M. Takeda, Makromol. Chem., 182, 2563 (1981).
23. R. E. Rondeau, M. A. Berwick, R. N. Steppel, and M. P. Serve, J. Amer. Chem. Soc., 94, 1096 (1972).

24. K. Iimura, N. Koide, and R. Ohta, Rep. Prog. Polym. Phys. Japan, 24, 233 (1981).
25. K. Iimura, N. Koide, and R. Ohta, Rep. Prog. Polym. Phys. Japan, 24, 231 (1981).
26. A. Blumstein and S. Vilasagar, Mol. Cryst. Liq. Cryst., 72 (letters), 1, (1981).
27. L. Strzelecki and L. Liebert, Bull Soc. Chim. Fr., 597 (1973).
28. V. P. Shibaev, N. A. Plate, and Ya. S. Freidzon, J. Polym. Sci. Polym. Chem. Ed., 17, 1655 (1979).
29. H. Kamogawa, J. Polym. Sci., B 10, 7 (1972).
30. H. Finkelmann, H. Ringsdorf, W. Siol, and J. H. Wendorff, Makromol. Chem., 179, 829 (1978).
31. E. C. Hsu, S. B. Clough, and A. Blumstein, J. Polym. Sci., B, 545 (1977), (b) Y. Osada and A. Blumstein, J. Polym. Sci., B, 761 (1977).
32. H. Finkelmann, H. Ringsdorf, W. Siol, and J. H. Wendorff, Mesomorphic Order in Polymers and Polymerization in Liquid Crystalline Media, A. Blumstein Ed., (ACS symposium series; 74) American Chemical Society, 1978, p. 22.
33. H. Finkelmann, M. Happ, M. Portugal, and H. Ringsdorf, Makromol. Chem., 179, 2541 (1979).
34. H. Finkelmann, H. Ringsdorf, and J. H. Wendorf, Makromol. Chem., 179, 273 (1978).
35. V. P. Shibaev, V. M. Moiseenko, Ya. S. Freidzon, and N. A. Plate, Europ. Polym. J., 16, 277 (1980).
36. N. A. Plate and V. P. Shibaev, J. Polym. Sci. Polym. Symp., 67, 1 (1980).
37. V. P. Shibaev, Advances in Liquid Crystal Research and Applications, Ed. Lajos Bata, Pergamon Press, Oxford, 1980 p. 869.
38. H. Finkelmann and G. Rehage, Makromol. Chem. Rapid Commun., 1, 31 (1980).
39. H. R. Allcock, G. Y. Moore, and J. J. Meister, Macromolecules 9, 950 (1976).
40. H. Finkelmann and G. Rehage, Makromol. Chem. Rapid Commun. 1, 733 (1980).
41. H. Finkelmann, H. J. Kock, and G. Rehage, Makromol. Chem. Rapid Commun., 2, 317 (1981).
42. W. R. Krigbaum, H. J. Lader, and A. Ciferri, Macromolecules 13, 554 (1980).
43. F. Volino, A. F. Martins, R. B. Blumstein, and A. Blumstein, C. R. Acad. Sc. Paris 292, Serie II, 829 (1981).
44. J. Watanabe, Y. Fukuda, R. Gehani, and I. Uematsu, The Seventh Symposium on Liquid Crystals, Preprints 3V12, p 106 (1981), Okayama, Japan.

THE ATHAS DATA BANK OF HEAT CAPACITY OF
LINEAR MACROMOLECULES

Bernhard Wunderlich, S.-F. Lau and U. Gaur
Department of Chemistry
Rensselaer Polytechnic Institute
Troy, New York 12181

INTRODUCTION

Our effort to assess all data on heat capacity started in the 1960's. The initial summaries were published in 1970 [1]. The original data base was updated and computerized about five years ago [2]. The data bank is now incorporated within ATHAS, Advanced THermal AnalysiS, a laboratory for research and instruction. The data bank maintains a collection of more than 500 publications on heat capacities of polymers which include all the measurements ever reported. The publication list is updated every six months. From each publication, the following information is retrieved:

1. polymers studied
2. temperature range studied
3. thermal history and sample characterization
4. experimental technique used and claimed **uncertainty**
5. accuracy of representation of data.

Based upon the above information, the publication is critically assessed for quality and reliability of the heat capacity data. Judgments for evaluating the quality of heat capacity data are based upon the following:

a) characterization of the sample used for heat capacity measurements
b) experimental technique used and reported uncertainty
c) representation of data and
d) visual comparison of data with other investigations.

The principles followed in selecting acceptable data are, briefly, as follows:

a) Purity and morphological characterization of the samples are examined. Data on uncharacterized, commercial plastics are considered unreliable. The key characterization parameter useful for our analysis, besides molecular weight and thermal history, is the crystallinity of the sample. Crystallinity calculated from density measurement is preferred over its determination from the heat of fusion, X-ray and IR spectroscopy. Crystallinity determination from enthalpy and the heat capacity change at the glass transition are considered least reliable.

b) The experimental aspects are examined for each measurement. The reliability of the measuring technique, as determined from results on standard calibration materials, is considered important.

Below 200 K, adiabatic calorimetric results are in general more reliable than differential scanning calorimetric measurements. From 200 to 350 K, depending upon the sophistication of the instrument and averaging procedure used, the differential scanning calorimetry and adiabatic calorimetry data may be considered equally good. Above 350 K, in certain cases, due to metastability of the samples, differential scanning calorimetry results are often preferred over adiabatic calorimetry.

c) Heat capacity data represented only by small graphs which cannot be read accurately are not considered reliable. Tabulated data are considered more liable than graphical data. If raw heat capacity data points are reported by investigators, their data are curve fitted and the heat capacity function obtained is preferred over the tabulated and graphical values.

d) If any measurement shows obvious, significant deviations from other data sets of comparable samples, the data are questioned. Such discrepancies are usually several standard deviations and indicate systematic errors.

The acceptable heat capacity data are stored on magnetic tape in temperature intervals of 0.1 K up to 1 K, 0.2 K to 2 K, 1 K to 10 K, 2 K to 20 K, and 5 K to 30 K. At temperatures higher than 30 K, intervals of 10 K are used. If the data are not reported in the temperature intervals desired by us, the data are interpolated. The spline function technique is used to determine the interpolated heat capacity at desired temperatures. If unsmoothed data have been reported by the authors, the data are smoothed by curve-fitting prior to storage.

A Hewlett-Packard calculator (minicomputer) of type 9821A, equipped with magnetic tape deck and interfaced with thermal line printer HP 9866B and plotter HP 9862A is the data handling system. Computer programs have been developed to govern interpolation, curve fitting, data storage, tabulation and plotting of the heat capacity data. Programs have also been developed for assistance in comparison and critical assessment of the data and to further analyze the data to determine the best heat capacity values. Subroutines have been written to correlate the heat capacity data and the morphological properties to determine the heat capacity of completely amorphous and completely crystalline linear macromolecules. A detailed listing of these programs and the instructions for their operations are given in reference (2).

The data bank contains in over 800 tables all information on the heat capacity of polymers. These acceptable heat capacity data have been computer processed, to derive for the first time a comprehensive set of recommended data. At present, recommended data are available for 96 polymers. These recommended data are summarized in Table 1. They are being discussed in nine successive papers in the Journal of Physical and Chemical Reference Data (1981/83).

The recommended data are, whenever possible, extrapolated to the limiting macroconformation to determine the heat capacity of the crystalline and amorphous states. In cases where low temperature data are available, the thermodynamic functions; entropy and Gibbs energy have also been calculated.

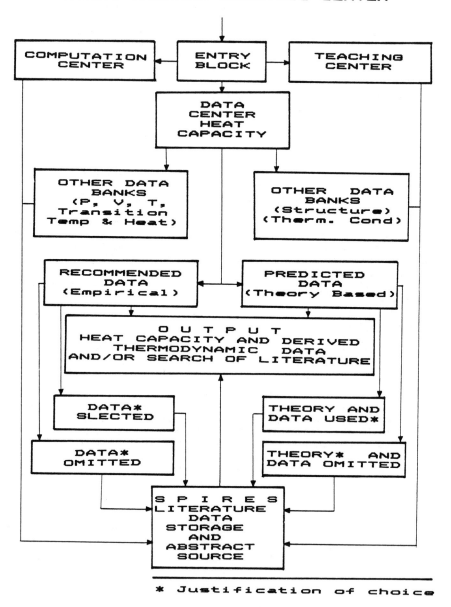

Figure 1. Block diagram of the proposed ATHAS Center for computation, Data and Teaching. Work has started on the computation center (3), the data center (2), the SPIRES block, and the computer assisted teaching modules.

Table 1.

Publication	No. of Polymers	Manuscript pages
1. Selenium	1	81
2. Polyethylene	1	91
3. Polypropylene	1	62
4. Polystyrene	1	52
5. Polyoxides	12	150
6. Acrylic Polymers	15	95
7. Other Carbon Backbone Polymers	24	170
8. Polyesters and Polyamides	10	120
9. Aromatic and Inorganic Polymers	31	103

The recommended heat capacity data are currently being analyzed in terms of chemical structure, structure of the polymers in the glassy, crystalline and molten states.

The further development of the data bank if support is possible will include extension to other parameters, the expansion from minicomputer base to a worldwide computer network (SPIRES). This expansion permits also the inclusion of computation and teaching modules. Preliminary work on all of these has been completed. Full scale development awaits funding. A summary of the plan is given in Figure 1.

ACKNOWLEDGEMENTS

This work has been supported by the Polymers Program of the National Science Foundation, Grant Number DMR 78-15279.

REFERENCES

1. B. Wunderlich and H. Baur, Adv. Polymer Sci., $\underline{7}$, 151 (1970).
2. U. Gaur, Ph.D. Thesis, Rensselaer Polytechnic Institute (1979).
3. Yu. V. Cheban, S. -F. Lau and B. Wunderlich, Colloid Polymer Sci., to be published 1982.

PRECISION AUTOMATED DIFFERENTIAL SCANNING CALORIMETRY OF POLYMERS

R. W. Connelly and J. M. O'Reilly
Research Laboratories
Eastman Kodak Company
Rochester, New York 14650

ABSTRACT

Automation of a D.S.C.II has been accomplished by interfacing with a HP-85 computer and through an HP 3497. Data collection and analysis have been facilitated by the flexibility of the HP-85 interfacing, basic programming and graphic display capabilities. Power temperature are monitored at 0.1°C over a range of heating rates (0.3 to 20°C/min). Energy calibration to an accuracy of 1% is obtained at 20°C/min and 2-4% at lower heating rates using indium and alumina standards. Several NBS standard polymers have been analyzed, polystyrene (NBS 705 and 706) and polyethylene (NBS 1475). Specific heat, glass temperature and enthalpy relaxation for polystyrene samples are recorded. Specific heat, enthalpy of melting, and melting behavior are presented for polyethylene. The application of precision automated DSC to enthalpy relaxation, melting of polymers and polymer reactions is discussed.

ON STRETCHING CALORIMETRY AND THERMOELASTICITY OF RUBBERS

G. W. H. Höhne, H. G. Kilian, P. Trögele
University Ulm
7900 Ulm (F.R.G.)

INTRODUCTION

A Stretching calorimeter is a very suitable instrument for measuring the change of thermodynamic potential functions of polymer systems during the stretching process, since both the heat and work exchanged can be measured simultaneously, thus allowing the calculation of the internal energy changes for instance. For that reason several such instruments has been constructed especially in polymer institutes (1, 2, 3, 4).

As the heat exchanged during deformation is very small, a high sensitive calorimeter is needed with relative large sample cells to take up the sample holder mechanics. Furthermore the traction rod is passing from the sensitive sample cell out of the calorimeter to the deformation device, thus forming a variable "heat-leak" which often affects the accuracy of the results and only allows measurements at room temperature.

We succeeded in constructing a stretching calorimeter with high sensitivity and a very good long-time baseline stability working in the temperature range between room temperature and temperatures > 100°C. With this calorimeter the thermoelastic behavior of several polymers was measured in the mode of simple elongation. The temperature dependence of results obtained on various rubbers shall be discussed here. The stress-strain behavior and the heat-flux during deformation can be described with the aid of a "Van-der-Waals-equation" of state (12, 13).

THE CALORIMETER

A Barberi-Calorimeter (5) (this is a modified Tian-Calvet-Calorimeter of heat-flux typ) was used in a version with two-sides open heat flux cell (Fig. 1). This allows constructing a stable stress-strain mechanics outside the calorimeter (Fig. 2). The heat-sensitive region covers the total sample and its traction rod clamp. Thus, the forces are not effecting the calorimeter itself. This may be pulled up for mounting the sample (Fig. 3). The whole calorimeter device is surrounded with a copper jacket. The jacket and the botten plate are kept at the constant measuring temperature by temperature controlled liquid pumped through a copper tube soldered on jacket and plate (Fig. 3). Further insulation to room temperature is done with help of rockwool and a plywood box (Fig. 2).

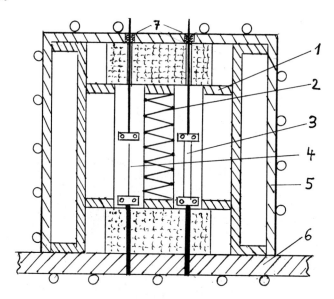

Figure 1: Stretching calorimeter. 1 = Heat flux calorimeter, 2 = Thermopile, 3 = Sample, 4 = Stretching dummy, 5 = Copper jacket, 6 = Stable botten plate, 7 = Copper brush.

The steel frame (Fig. 2) is calculated to guarantee a mechanical stability better than 0,1 mm for the maximum forces employed. The strain of the sample is done by a digital electronical controlled step-motor which moves a balance-cell with the sample traction rod hanging on it. The force effecting the balance cell is measured via a carrier frequency bridge.

To minimize the error of the heat-leak of the rod, this is made of stainless steel with low thermal conductivity and a minimum diameter. Thermal contacts to the thermostatised surroundings of the calorimeter is done via a copper brush with low friction in the top of the box. Never the less there is a rest heat leak variable with the strain. To decrease this effect we utilized differential heat flux measurements in a twin setup with an - apart from sample - identical stretching device in the reference cell (Fig. 3).

The most sensitive region of the calorimeter cell has a diameter of 9 mm and 40 mm length. The thermopile has a thermal sensitivity of 120 μV/mW with very low noise, thus allowing a resolution (thermal noise) of 0,2 μW at room temperature with a stability of the baseline of $\pm$ 0,4 μW per hour. At 100°C the thermal noise is 0,4 W and the stability about $\pm$ 0,6 W per hour. The calorimeter allows measurements of stretching heats down to 50 μJ between room temperature and 100°C with a resolution of $\pm$ 30 μJ. The time constant is 60 seconds.

Since every elastic material produces a "Thomson heat" during deformation (6):

$$dQ = l_o \cdot \alpha \cdot T \cdot df$$

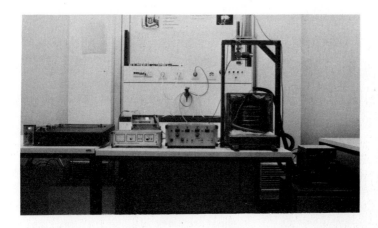

Figur 2: Stretching calorimeter device with carrier frequency bridge, recorder, amplifier, step motor control, calorimeter and thermostat (from left to right)

(Q = heat, l_o = length without force, α = linear thermal expansion coefficient, T = temperature in Kelvin, f = force)

The Thomson-heat of the traction rod, depending on force changes and length within the calorimeter cell, must always be subtracted from the measured heat. In our calorimeter this is usually found to be about 0,25 mJ/N)

EXPERIMENTAL RESULTS

Several Polymers have been investigated in the stretching calorimeter. The results on rubbers shall be presented here. Samples* of natural rubber (NR) and Styrene-Butadien-Copolymers (SBR) were stretched in the calorimeter up to 400 percent at roomtemperature, 60°C and 90°C. For low elongation degrees samples having the dimensions of 31,5 x 2,6 x 0,9 mm³ were used. Heat and work were measured in steps of 2 percent elongation. For the higher stretching degrees shorter samples (10 x 2,6 x 0,9 mm³) had to be used being elongated in steps of 25 % with a elongation rate of 4 and 12,5 percent respectively.

For higher forces the strain of the sample was corrected with regard to the elongation of the traction rod, just as the measured heat flux (resp. heat) with regard to the positiv "Thomson heat" of the rod depending on respective force change and sample dimension.

Some results are shown in Fig. 4 - 7. The measurements at roomtemperature are in good agreement with literature (7,8). Results at

*We are thankfull to Bayer AG Leverkusen for ceding the material to us.

Figure 3: Stretching calorimeter with calorimeter and its copper jacket pulled up. Sample is on right and dummy on left side.

higher temperatures are not known to us. The most interesting effect is the disappearance of the thermal interversion between 30° and 60° for natural rubber and betwen 60° and 90°C for SBR (Fig. 4). At the same time the singularity of the ratio heat to work observed at smallest elongation changes its sign (Fig. 7). The figured curves are averaged from several measurements, the reproducibility is about 2 percent the absolute error with respect to all corrections necessary may be 5 - 10 percent or ± 30 µJ for the small heats.

THEORY OF REAL RUBBERS (COMPUTED RESULTS)

The mechanics of rubbers has usually been described with an ideal network theory (9, 10, 11) using the strain-energy function

$$W = \frac{1}{2} N \cdot K \cdot T \left(\lambda^2 \frac{2}{\lambda} - 3 \right)$$

$$\frac{\partial W}{\partial l} = f = \frac{N \cdot K \cdot T}{l_o} (\lambda - \lambda^{-2})$$

(W = work, f = force, N = number of molecules, K = Boltzmann's constant, T = temperatur, $\lambda = l/l_o$, l = length)

Supposing the Gaussian chains having no volume no interaction and no conformational restrictions, the belonging equation of state is able to describe the mechanical behavior of rubbers at small elongations, only failing substantially for larger strains. Hence the thermoelastic behavior of rubbers during stretching cannot be described satisfactorily.

The analogy between an ideal network and an ideal gas has induced one of us to modify the equation of the ideal network in the same way done to formulate the Van-der-Waals-equation of gases, thus arriving at (12, 13)

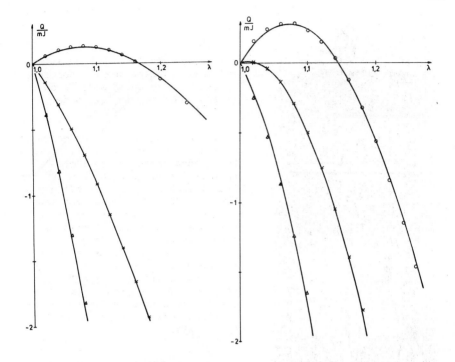

Figure 4: Heat of elongation for NR (left) and SBR (right) at various temperatures. Measuring points: O = 22°C, x = 61°C, ∆ = 90°C, ── = calculated.

	ideal equation:	Van-der-Waals-equation:

gas: $p \cdot V = N \cdot K \cdot T$ $\qquad p = N \cdot K \cdot T/(V-\tilde{b}) - \tilde{a}/V^2$

network: $f \cdot D^{-1} \cdot l_o = N \cdot K \cdot T$ $\qquad f = N \cdot K \cdot T/l_o/(D^{-1} - D_m^{-1}) - \bar{a} \cdot D^2$

As $\tilde{b}$ describes the minimum volume of the systems (volume of the molecules) and $\tilde{a}$ the interactions between the gas molecules, is D_m correlated to the maximum elongation of the finite chains and $\bar{a}$ to global interactions between the network. The Van-der-Waals-equation for real networks can be rewritten as

$$f = N \cdot K \cdot T/l_o \cdot D \left\{ D_m/(D_m - D) - a\, D \right\}$$

with $a = a/NKT$ and $D_m = \lambda_m - \lambda_m^{-2}$, $(\lambda_m = l_{max}/l_o)$.

To consider the correct temperature dependence one uses the factor $\langle r^2 \rangle / \langle r_o^2 \rangle$ (14) which takes in account the variation with temperature of the mean square length of chains in the network in the case of different potential energies for the rotational isomers. $\langle r^2 \rangle$ is the

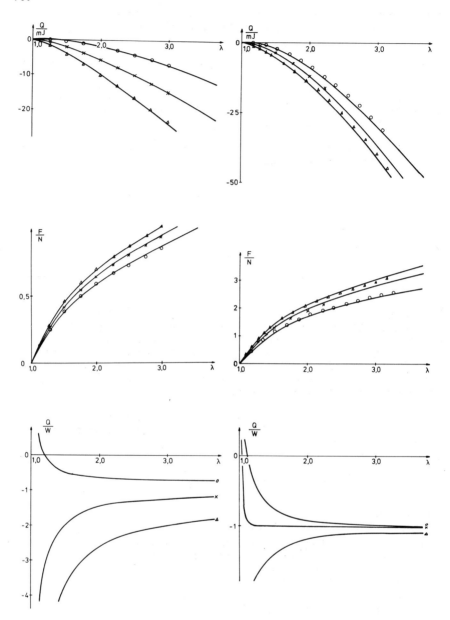

Figure 5-7: Heat, force and heat to work ratio of elongation for NR (left) and SBR (right) at various temperatures. Measuring points: O = 22°C, x = 61°C, △ = 90°C, — = calculated with the "Van-der-Waals-equation of state".

actual mean square length in the network, $\langle r_o^2 \rangle$ the mean square length of the corresponding free chain. Thus we get following equation:

$$f = N \cdot K \cdot T/l_o \cdot \langle r^2 \rangle / \langle r_o^2 \rangle \cdot D \cdot \{D_m/(D_m - D) - a \cdot D\} \quad \text{(A)}$$

<u>Temperature Dependence.</u> This equation of state should describe the stress strain behavior at various temperatures if we know the temperature dependence of all the parameters involved. For $\langle r^2 \rangle$ we find (15-17) that $d \ln \langle r^2 \rangle = 2/3 \cdot \beta \cdot dT$, with $\beta = d \ln V/dT$ as the thermal expansion coefficient in the unstrained state. Letting β be independent of temperature we get:

$$\langle r^2 \rangle = \langle r^2(T_o) \rangle \cdot \exp(2 \cdot \beta \Delta T/3) \quad \text{with } \Delta T = T - T_o$$

Likewise we get from $d \ln \langle r_o^2 \rangle = \mu \cdot dT$

$$\langle r_o^2 \rangle = \langle r_o^2(T_o) \rangle \exp(\mu \cdot \Delta T)$$

The reference temperature T_o should correspond to the state with $\langle r^2(T_o) \rangle / \langle r_o^2(T_o) \rangle = 1$.

The Van-der-Waals parameters a and D_m are not independent from one another. The empirical relation $a = a_o/\lambda m^2$ has been used (18). As λ and λ_m depends on l_o which is related to temperature via $l_o(T) = l_o(T_o) \cdot (1 + \alpha \cdot (T - T_o))$ both a and λ_m have a temperature dependence. Another possible understanding is, that the length of the statistical segments S_o in the chains related to λ_m via $y \cdot S_o = \lambda_m$ (y = number of segments) varies with temperature.(19)

<u>Calculation of the Stretching Heat.</u> Under defined constant pressure and temperature it follows from thermodynamics that the deformation heat is

$$\left. \frac{dQ}{dl} \right|_{p,T} = \left. \frac{\partial E}{\partial \ell} \right|_{T,p} + p \cdot \left. \frac{\partial V}{\partial \ell} \right|_{T,p} - f$$

From the Maxwell relation one gets:

$$\left. \frac{\partial E}{\partial \ell} \right|_{p,T} = f - T \cdot \left. \frac{\partial f}{\partial T} \right|_{p,T} - p \left. \frac{\partial V}{\partial \ell} \right|_{p,T}$$

(E = internal Enegie, V = Volume, f = force, p = pressure, T = temperature). Combination of both equations gives:

$$\left. \frac{dQ}{dl} \right|_{p,T} = - T \cdot \left. \frac{\partial f}{\partial T} \right|_{p,T}$$

Thus we are enabled to calculate the stretching heat from equation (A). We succeeded in describing quantitative the heat-curve at roomtemperature only. At higher temperatures there should contribute another heat (or entropy) change than that derived from the mechanical equation of state. Since the stress-strain behavior is described at all temperatures without further assumptions, the additional heats should be related to temperature induced changes of the internal state in the oriented systems. These effects may be related to "rotatoric freedoms" which must be considered being frozen at temperatures below T_R. Let the number of "rotators" approximately be equal to

$$n(T) = n_o \cdot \exp(-U_{rot}/R/(T-T_R))$$

(U_{rot} = activation energy, T_R = freezing temperature, R = gas constant)

The entropy changes with temperature are postulated to be defined by

$$Q_{rot} = T \cdot \Delta S_{rot} = -T \cdot \Delta S_o \cdot n_o \cdot \exp(-U_{rot}/R/(T-T_R))$$

Whereby the rotational nature of these freedoms may be expressed by

$$dW_{rot} = 0 \implies \frac{\partial U_{rot}}{\partial \lambda} = T \cdot \frac{\partial S_{rot}}{\partial \lambda}$$

such that no contribution to the strain-energy arises. The number of "rotators" activated is considered to depend on progressive stress, leading to the approximation:

$$Q(\lambda, T) = Q_{rot} \cdot (\lambda - 1)$$

With this additional term the heat exchange obtained can be calculated with the same set of parameters as used in the strain-stress representation and a rotator activation energy U_{rot} = 2500 J/mole. The results of these calculations are drawn in the fig. 4 and 5.

Summary. We like to emphazise the consequence of the above interpretations: real networks seem to behave like Van-der-Waals conformational gases in the global range, locally they show the behavior as an elastic fluid. At elevated temperatures a transformation of a new type of internal freedom "rotational in its nature" (dW = 0!) seems to occur as a typical effect in real networks.

REFERENCES

1. A. Engelter and F. H. Müller, Kolloid-Z., 1958, 157, 89
2. Yu. K. Godovsky, G. L. Slonimsky and V. F. Alekseev, Vysokomol. Soyed, 1969, A 11, 1181
3. Yu. A. Sergeyev, E. Z. Fainberg and N. V. Mikhailov, Vysokomol. Soyed, 1972, A 14, 1, 250
4. D. Miller and G. W. H. Höhne, Thermochimica Acta, 1980, 40, 137
5. P. Barberi, Brevet C. E. A. EN 7019.831 (The Calorimeter may be bought from P. Barberi, 6 Allée de Lavandieres, F 98320 Le Meneil St. Denis, France)
6. A. Eucken, Lehrbuch der Physik. Chemie, Band 2, S. 826 (1949) and other
7. Yu. K. Godovsky, Polymer, 1981, 22, 75
8. W. Dick and F. H. Müller, Kolloid-Z., 1960, 172, 1
9. W. Kuhn and F. Grün, 1942, 101, 248
10. P. J. Flory, Statistical Mechanics of Chain Molecules, Intersience, New York, (1969)
11. L. R. G. Treloar, The Physics of Rubber Elasticity, 3rd Ed., Claredon Press Oxford, (1975)
12. H.-G. Kilian, Polymer, 1981, 22, 209
13. H.-G. Kilian, Colloid u. Polymer Sci., 1981, 259, 1084
14. P. J. Flory, A. Ciferri and C. A. J. Hoeve, J. Polymer Sci., 1960, 45, 235
15. M. V. Volkenstein, Configurational Statistics of Polymeric Chains, Interscience, New York, (1963)
16. P. J. Flory, Trans. Faraday Soc., 1961, 57, 829
17. P. J. Flory, J. Am. Chem. Soc., 1956, 78, 5222
18. H.-G. Kilian, Polym. Bull., 1980, 3, 151
19. U. Eisele, B. Heise, H.-G. Kilian and M. Pietralla, Die Angewandte Makromolekulare Chemie, 1981, 100, 67

POLYMER ORIENTATION MEASURED FROM LINEAR EXPANSION BY THERMAL ANALYSIS

Li-Hui Wang, C.L. Choy and Roger S. Porter
Polymer Science and Engineering Department
Materials Research Laboratory
University of Massachusetts
Amherst, Massachusetts   01003

INTRODUCTION

The physical properties of a drawn polymer depend on the degree of molecular orientation. Therefore, the orientation functions of drawn polymers are important quantities and, as such, a variety of techniques [1-8] have been developed to determine these parameters. For amorphous polymers, however, most of these methods such as nuclear magnetic resonance [1,2], polarized Raman scattering [3], infrared dichroism [4-6] and wide angle x-ray scattering [7] involve detailed experimental work and theoretical analysis, and thus are not convenient for routine characterization. Several of these methods also are of limited precision.

For uniaxially-oriented polymers, one simple and commonly-used measure of orientation is the birefringence, Δn, which is related to the second moment of the orientation distribution, f (also known as the Herman-Stein orientation function), by

$$f = \Delta n / \Delta n_{max} \qquad (1)$$

where Δn_{max} is the birefringence of the perfectly oriented polymer. Unfortunately, Δn_{max} is known for only a few polymers, such as polystyrene [5,6,9] and so, in general, Δn can only serve as a relative measure of molecular orientation. Yet another simple method which can give absolute values of f is the thermal expansivity [10-12]. This is of special interest since the measurement of this quantity requires very little time and effort, and there are commercially available instruments which cover a wide temperature range. The property is important in use applications and can be determined with precision.

Although previous workers [10,11] have measured expansivities to evaluate the orientation function, their results have not been critically assessed by comparison with other techniques. Recently, we have determined [12] the orientation function for extruded polystyrene from both birefringence and thermal expansivity measurements. The results show that these two methods give equivalent values within an uncertainty of 15%. In the present study, we have extended this work to poly(methyl methacrylate) (PMMA) in order to determine whether thermal expansivity can be generally used to provide a quantitative measure of the molecular orientation in amorphous polymers and an intercomparison among them.

The thermal expansivities along ($\alpha_\parallel$) and perpendicular ($\alpha_\perp$) to the draw direction of PMMA with extrusion draw ratios, $\lambda=1-4$, have been measured between 150 and 298K. As λ was increased from 1 to 4, $\alpha_\parallel$ decreases 2-3 times, whereas $\alpha_\perp$ increases only 20-35%. The orientation function, f, calculated from thermal expansivity using the aggregate model is found to change linearly with birefringence, indicating that each property provide a sensitive measure of molecular orientation. For PMMA, however, only thermal expansivity can give an absolute f, with results at 150K in reasonable agreement with previous studies using other techniques. At higher temperature, i.e. above ambient, PMMA side group motions are excited, expanding volume, and calculations based on the aggregate model may not be valid in this study on the dependence of orientation function, as measured by expansion, on temperature of draw.

REFERENCES

1. M. Kashiwagi, M.J. Folkes and I.M. Ward, Polymer, 12, 697 (1971).
2. M. Kashiwagi and I.M. Ward, Polymer, 13, 145 (1972).
3. D.I. Bower, in "Structure and Properties of Oriented Polymers" (ed. I.M. Ward), Applied Science, London, 1975.
4. B.E. Read, in "Structure and Properties of Oriented Polymers" (ed. I.M. Ward), Applied Science, London, 1975.
5. M.F. Milagin, A.D. Gabarayeva and J.T. Shishkin, Polym. Sci. USSR, A12, 577 (1970).
6. B. Jasse and J.L. Koenig, J. Polym. Sci. (Phys. Ed.), 17, 799 (1979).
7. M. Pick, R. Lovell and A.H. Windle, Polymer, 21, 1017 (1980).
8. R.S. Stein and B.E. Read, Appl. Polym. Symp., 8, 255 (1969).
9. R.S. Stein, J. Appl. Phys., 32, 1280 (1961).
10. J. Hennig, Kunststoffe, 57, 385 (1967); Colloid Polym. Sci., 259, 80 (1981).
11. W. Retting, Colloid Polym. Sci., 257, 689 (1979); 259, 52 (1981).
12. L.H. Wang, C.L. Choy and R.S. Porter, J. Polym. Sci. (Phys. Ed.), accepted.

STRUCTURAL RELAXATION AND RECOVERY OF GLASSY POLYMERS: THEORY

T. S. Chow and W. M. Prest, Jr.
Webster Research Center
Xerox Corporation, W-114
Webster, New York 14580

Introduction

Phenomenological treatments[1-3] have been proposed to explain volumetric and enthalpy relaxation and recovery of amorphous polymers near the glass transition. There are two successful phenomenological models at present. Kovacs and co-workers[1] suggested that any state of a glass can be expressed by the values N ordering parameters in addition to the usual thermodynamic variables. Based on Narayanaswamy's[2] early work, DeBott, et al,[3] applied the Boltzmann superposition principle and the Williams-Watts distribution function to describe structural relaxation of materials. It is the purpose of this paper to show that DeBott's equation can be derived directly from the multiordering parameter equations of Kovacs. The numerical calculation of structural relaxation of quenched and annealed polymers involves very sensitive and complicated kinetic behavior in the glass transition zone. A self-consistent iteration method of calculation is applied to ensure the stability and convergence of the solution. The effects of experimental rate, time, temperature and structural parameters, including the nature of distribution function, on the enthalpy and specific heat of polymeric glass are discussed.

Theory

The kinetic aspect of glass transition phenomena can be described by a non-dimensional parameter δ measuring the volume or enthalpy departure of a system from equilibrium. In Kovacs, et al's, model, δ is partitioned into N finite parts and each individual δ_i corresponds to the fractional departure from equilibrium. The system is governed by N distinctive differential equations.[1]

$$\frac{d\delta_i}{dt} = -\frac{\delta_i}{\tau_i} - \Delta\alpha_i q \qquad (i = 1,\ldots\ldots,N) \qquad (1)$$

where q is the experimental heating (q>0) and quenching (q<0) rate, $\Delta\alpha_i$ is the ith contribution to the excess thermal expansion coefficient of the liquid with respect to the glass and τ_i is the ith relaxation time. Each τ_i depends on temperature (T) and structural (δ) and can be written:[1,2]

$$\tau_i(T, \delta) = \tau_{ir}\, a(T, \delta) \qquad (2)$$

where "a" is the shift factor and τ_{ir} is the ith relaxation time at a reference temperature (T_r) in equilibrium. When N is very large, the ratio of δ_i/δ becomes very small and the linearized solution of Eq. (1) is:

$$\delta_i(t) = \delta_{i0}\, e^{-u(t,t_0)/\tau_{ir}} - \Delta\alpha_i q \int_{t_0}^{t} e^{-u(t,t')/\tau_{ir}}\, dt' \qquad (3)$$

$$(i = 1, \ldots, N)$$

where the reduced time variable

$$u(t,t') = \int_{t'}^{t} \frac{dt''}{a[T(t''), \delta(t'')]} \quad (4)$$

It has been found more convenient to characterize the structural change by a fictive temperature (T_f).[2-3] Defining

$$\delta_{io}/\delta_o = \Delta\alpha_i/\Delta\alpha = g_i \quad (5)$$

we obtain

$$T_f(T) = T + (T_{f0}-T_0)\phi(T,T_0) - \int_{T_0}^{T} \phi(T,T') \, dT' \quad (6)$$

where T_0 and T_{f0} set the initial condition and

$$\phi(t,t') \equiv \phi(u) = \sum_{i=1}^{N} g_i \exp[-u(t,t')/\tau_{ir}] \quad (7)$$

The summation can be replaced with an integral for $N \to \infty$. Eq. (6) can be used to treat quenching (q<o) or heating (q>o) initially at non-equilibrium state with any relaxation function and shift factor.

The Williams-Watts function has been found to be a good emperical expression for a wide variety of the relaxation data.[3-4] We write:

$$\phi(T,T') = \exp(-\{\frac{1}{q}\int_{T'}^{T}\frac{dT''}{\tau[T'',T_f(T'')]}\}^\beta) \quad (8)$$

with the relaxation time given by[2-3]

$$\tau = A \exp\left[\frac{xE}{RT} + \frac{(1-x)E}{RT_f}\right] \equiv \tau_T a_\delta(T_f) \quad (9)$$

whre A and X ($o \leq X \leq 1$) are constants. E is the activation energy near equilibrium and R is the gas constant. When a system starts initially from equilibrium ($T_{f0} = T_0$), Eq. (6) combined with Eqs. (8-9) leads exactly to the equation derived by *DeBolt, et al.* Thus, the analysis reveals that the two seemingly different models[1-3] are actually based on the same fundamental equations.

In the case of isothermal relaxation (q = o), Eqs. (3-5) give

$$T_f(\tilde{t})-T_0 = (T_{f0}-T_0)\exp\{-\int_{t_0}^{\tilde{t}}\frac{dt'}{a_\delta[T_f(t')]}\}^\beta \quad (10)$$

where where the non-dimensional time $\tilde{t} = t/\tau_T$. Eq. (10) is used to determine the non-equilibrium recovery during isothermal annealing.

Calculation

The structural relaxation and recovery below the glass transition have complicated temperature-structure dependence of relaxation time, scanning rate, pronounced non-linearity and memory effects. There can be orders of magnitude change in relaxation times during quenching, annealing and heating which result in only a few degrees change in the fictive temperatures. The stability and convergence become a major concern in the numerical calculation and proper choice of numerical procedure is very important.

The Gaussian quadrature procedure[5] is better than any other method to evaluate a definite integral; it is accurate and requires a few computational steps. In this paper, all integrations are estimated by using double precision and the 32-point Gaussian quadrature formula. To solve the non-linear integral Eq. (6) numerically, we use a successive iteration method.[5] Let the right hand side of Eq. (6) be defined by $F(T, T_f)$. For a given T, the kth iteration of T_f is

$$\overline{T}_f{}^{(k)} = F[T, T_f{}^{(k-1)}] \tag{11}$$

To ensure the stability and convergence of solutions, the actual value at step k is calculated by means of the relation

$$T_f{}^{(k)} = (1 - \omega) T_f{}^{(k-1)} + \omega \overline{T}_f{}^{(k)} \tag{12}$$

where ω ($0 \leq \omega \leq 1$) may be called a convergence parameter and $T_f{}^{(k-1)}$ is the previous accepted value. Quantity $\omega = 1$ is usually chosen in the case of heating (q>o), however, a smaller value of ω ($\omega = 0.25$) is needed for calculating η during quenching (q<o) and annealing Eq. (10) far below the glass transition temperature.

The following values are adopted as the input for numerical calculation in the section.

$$\Theta = 0.75 \text{ K}^{-1}, \chi = 0.35, \beta = 0.35$$
$$T_r = 375°K \text{ with } \tau_r = 1 \text{ sec. and } q = 20°C/\text{min.}$$

A comparison of the successive iteration method, Eqs. (11-12), with *DeBolt, et al's* sequence method, where integrations are evaluated by the more accurate Gaussian quadrature procedure, are shown in Figure 1. When numerical integration is estimated by the rectangular rule instead of the Gaussian quadrature procedure, the original *DeBolt, et al's* sequence calculation may result in even larger errors for polymer glasses annealed far below the glass transition.[6]

Figure 2 is a plot of the calculated fictive temperature (T_f) and corresponding temperture derivative (dT_f/dT) versus temperature (T) of a system, characteristic of polymer glasses, quenched from equilibrium and heated from non-equilibrium at different experimental scanning rates. The asymmetry of quenching and heating resembles a second- and first-order phase transition, respectively. When a system, initially at non-equilibrium, is heated, Figure 3 suggested that the initial values of T_{fo} have a stronger effect on dT_f/dT which is equivalent to heating after annealing at constant temperatures to the indicated fictive temperatures (T_{fo}). Annealing materials below the glass transition raises and narrows the transition peak.

The Williams-Watts distribution function of relaxation times is plotted in Figure 4 where corresponding variations of the transition width and height and the shift of dT_f/dT peaks are also calculated. When the reference relaxation time τ_r is shifted two orders of magnitude toward the short-time spectrum, we see a softer and broader dT_f/dT as it transfers to lower temperatures. Since both q and τ_r are related, shorter reference relaxation times have a similar effect as slower heating rates.

Conclusions

Many kinetic features of glass transition phenomena have been explained satisfyingly in terms of two models by Kovacs and DeBolt and their colleagues. The former partitions the departure from equilibrium into finite parts to be determined by a set of distinctive equations. We have shown that the latter is a continuous representation of the former. The two seemingly different models are actually based on the same phenomenological equations and independent of the choice of distribution function of relaxation times.

To calculate structural relaxation of polymers during quenching, annealing and heating involves orders of magnitude change in relaxation times which result in only a few degree changes in the fictive temperature. In order to ensure the stability and convergence of the numerical calculation, the Gaussian quadrature procedure and a successive iteration method are suggested to obtain an accurate solution.

Calculations reveal that the height, width and shift of heat capacity in terms of dT_f/dT is very closely related to the distribution function of relaxation times. Using the Williams-Watts distribution function, we have confirmed the important aspects of structural dependence of non-equilibrium glasses and the experimental scanning rate. Annealing amorphous polymers below the glass transition raises and narrows the transition peak.

References

1. A. J. Kovacs, J. J. Aklonis, J. M. Hutchinson and A. R. Ramos, *J. Polym. Sci., Polym. Phys. Ed.*, **17**, 1097 (1979).
2. O. S. Narasawamy, *J. Am. Ceram. Soc.*, **54**, 491 (1971).
3. M. A. DeBolt, A. J. Easteal, P. B. Macedo and C. T. Moynihan, *J. Am. Ceram. Soc.*, **59**, 16 (1976).
4. C. P. Lindsey and G. D. Patterson, *J. Chem. Phys.*, **73**, 3348 (1980).
5. B. Carnahan, H. A. Luther and J. O. Wilkes, "Applied Numerical Methods," Wiley, New York, (1969).
6. I. M. Hodges and A. R. Berns, *Macromolecules*, **14**, 1599 (1981).

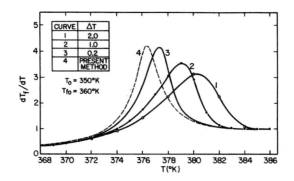

Figure 1: A comparison of the successive iteration (· · · · ·) and sequence (———) methods of calculation.

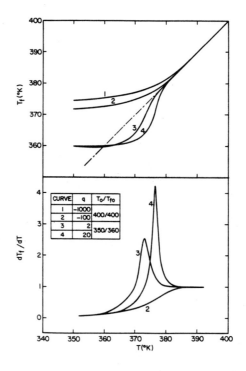

Figure 2: The calculated fictive temperature and corresponding temperature derivative of a system quenched and heated from the indicated initial equilibrium and non-equilibrium conditions and experimental rates.

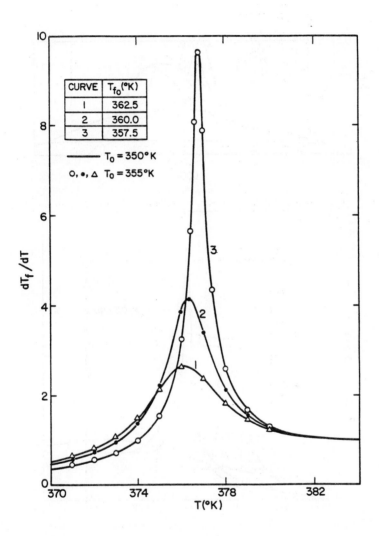

Figure 3: Effect of initial conditions on calculated dT_f/dT.

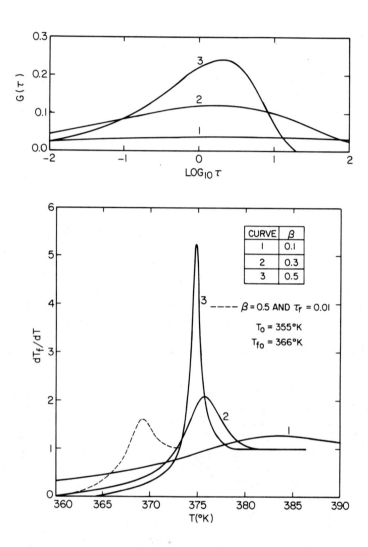

Figure 4: Effect of distribution function of relaxation times on calculated dT_f/dT.

AGING PROCESSES IN PARTIALLY COMPATIBLE POLYMER BLENDS

W. M. Prest, Jr. and F. J. Roberts, Jr.
Webster Research Center
Xerox Corporation, W-114
Webster, N.Y. 14580 U.S.A.

INTRODUCTION

Traditionally, two materials are considered to be compatible when their mixture forms a solution which has a single compositionally dependent glass transition temperature that is intermediate to the properties of either component (1). However, even in nominally "compatible" organic and inorganic alloys, this transition generally occurs over a considerably broader temperature range than in either of the components. In addition, the width of this transition is often a strong function of the composition of the mixture. For example, in the arsenic selenium alloys, the temperature width of the transition is a minimum near the stochometric As_2Se_3 composition (2). Similar effects are also observed in a variety of polymer blends including the classic example of a set of compatible polymers; the solid solutions formed by mixtures of polystyrene [PS] and poly (2,6 dimethyl 1,4 phenylene oxide) [PPO] (3-7). Furthermore, in partially miscible systems, such as blends formed from poly (*para*-chlorostyrene-co-*ortho*-chlorostyrene) and PPO, a significant broadening of this transition is observed at compositions immediately proceeding the miscible/immiscible phase boundary (8). As the result of these observations, the increase in the width of the glass transition region has been attributed microscopic inhomogenities in the alloy which result from fluctuations in the local concentration of the components (4-9).

This paper, which is a condensation of reference (10), proposes that a measure of these fluctuations can be obtained from the distribution of relaxation times that is implied by analyzing the results of isothermal aging experiments on polymer blends in terms of the phenomenological theories of the glass transiton (11). Selective annealing experiments are used to measure the composition and temperature dependence of the enthalpy recovery processes in blends of PS with PPO or poly(vinyl methyl ether) [PVME]. While the PPO/PS systems remain compatible throughout the compositional range, blends of PVME and PS can be selectively phase separated by annealing in the vicinity of the lower critical solution temperature [LCST] (12-

14). This technique is used to produce both compatible and selectively phase separated blends to test the concentration fluctuation hypothesis.

EXPERIMENTAL PROCEDURES

Poly (2,6 dimethyl 1,4 phenylene oxide)-polystyrene blends were prepared by freeze drying 2.5% w/w benzene solutions of reprecipitated PPO (General Electric Company), M_w = 37,000, M_n = 16,000, and anionic polymerized PS, M = 110,000 (Pressure Chemical Co.). PVME (PolySciences) was reprecipitated from toluene into boiling water and vacuum dried @ 333K for 24hrs. Fifty-fifty w/w blends of this PVME (M_w = 57,200, M_n = 28,700) and anionic PS, M = 37,000, were prepared by casting from a 5% toluene solution and drying at 373K.

Calorimetric measurements were made with a computerized Perkin-Elmer DSC-II differential scanning calorimeter based on a Tektronix 4051 graphics computer and a HP 3455A μDVM (7). Software programs were used to record, process and numerically analyze the data in terms of the relevant material parameters and to automatically control the time and temperature profiles imposed on the sample. The temperature calibration and time constants in scanning experiments were determined by measuring the fictive temperature of a thermally equilibrated glass which had a similiar thermal mass to the sample in question. The fictive temperature and the departure from equilibrium were calculated using the extrapolation of a least squares polynomial fit of the heat capacity of the liquid to define the enthalpy of the equilibrated glass.

Annealing studies were conducted automatically in the calorimeter. Each sample was quenched from the melt at 320K/min. and then reheated at 20K/min. The reference temperature, T_r, of each blend was defined as the temperature at the mid point of the subsequent recovery of the heat capacity from the glassy to the liquid state. The PPO/PS blends were heated to T_r +50K, cooled at 320K/min. to from 15 to 40K below T_r, isothermally annealed for from 1 to 1000 minutes and then quenched (320K/min.) to T_r - 50K. The heat capacity of the resulting samples were measured by heating at 20K/min. to T_r +50K.

The PVME/PS samples were equilibrated at 383K for 4 minutes and then cooled at 320K/min. to the annealing temperature ~ 15-40K below T_g. Measurements of the heat capacity were made at 20K/min. following isothermal annealing times of from 1 to 1000 minutes and quenching to 233K. The annealing experiments were subsequently repeated with an upper temperature of 433K and equilibration times of from 1 to 30 minutes. This thermal treatment in the vicinity of the LCST produced glasses with progressively broadened glass transition regions. A 383K, 5 minute equilibration treatment was interspersed between each annealing cycle in order to reform the compatible blend and insure the same starting point for each experiment.

Model calculations were based on the phenomenological theory developed by Kovacs, Aklonis, Hutchinson and Ramos (11). The numerical solutions of the coupled non-linear differential equations were obtained using the DGEAR subroutine of the IMSL software package which is an adaptation of a package designed by A. C. Hindmarsh (15) based on C. W. Gear's subroutine DIFSUB (16). All calculations were based on the backward differentiation method of Gear for the solution of stiff equations and run in double precision. Local truncation errors of less than 0.01% were maintained for each iteration and periodically evaluated at lower limits to insure the self-consistency of the calculations.

RESULTS

In both the PPO/PS and the PVME/PS blends, the widths of the glass transition regions (ie. change in Cp from the glass to the liquid state) of the unannealed samples are strongly dependent on the composition and method of preparation of the sample. In the PPO/PS blends the recovery to the liquid state occurs over a progressively wider temperature range as the PPO concentration increases. Surprisingly, the addition of even small amounts (<1.5%) of PS to PPO drastically increases the width of the glass transition region and produces a very disturbed glass. In the PVME/PS blends, the width of the transition is a maximum at an intermediate concentration which is a function of the molecular weight of the components. Equilibrating these blends in the vicinity of the LCST produces further increases the temperature domain over which the recovery occurs and eventually leads to the appearance of the individual transitions of each component.

The annealed blends exhibit similar trends in the apparent changes in the recovery processes. As the width of the transition is increased the recovery from the heat capacity of the glass occurs at progressively lower temperatures. In contrast to the response of the individual components, the aging induced changes in Cp occur at the low temperature end of the transition region. In other words, there is an increase in the apparent heat capacity of the glass rather than a shift in the temperature at which the heat capacity recovers to that of the liquid state. With time this low temperature shoulder in Cp intensifies, consolidates into a peak and occurs at progressively higher temperatures. This peak remains very broad and is observed in the transition region, rather than at higher temperatures, even after extended annealing times. Thus on aging, the recovery of the enthalpy of these blends occurs not only over a broad range of temperatures, but at surprisingly low temperatures compared to the aging characteristics of a simple glass.

The PVME/PS blends exhibit similar low temperature recovery processes. Even though the breadth of the transition of the compatible PVME/PS blend is significantly larger than that of even the broadest PPO/PS blend, the characteristic annealing peaks occur over a similar temperature range. However, in contrast to the behavior of the PPO/PS blends, the annealing characteristics of the PVME/PS blends are a function of the thermal history of the system. The recovery processes of the annealed partially compatible PVME/PS blends are spread over a much larger temperature domain. That is, the longer the blend is held in the vicinity of the LCST, the more disperse its recovery.

DISCUSSION

The sub-Tg recovery processes observed during the heating of isothermally annealed blends are analogous to those observed in stressed glasses (17) and in the glasses formed from PMMA and its copolymers (18). From the prospective of the phenomenological models (11), this type of unusual behavior implies the existence of processes which either: increase the enthalpy of the glass (ie. drive it further from equilibrium); decrease the average relaxation time of the system; or introduce a new set of rapidly relaxing (short relaxation time) elements. This latter process, which implies the existence of a broad distribution of relaxation mechanisms, is consistent with the nonuniform mixing model if one identifies the individual relaxation times with domains containing different concentrations of the blended components. That is, annealing at a given temperature should have the greatest affect on the domains with the lowest glass transition temperature (ie. shortest relaxation times). Conceptually this occurs because the effective glass transition temperatures of these regions are closer to the annealing temperature so that the associated recovery of the enthalpy occurs rapidly. The recovery of progressively associated with higher Tg (ie. longer relaxation time) regions becomes more important with additional annealing, giving rise to the increase in the magnitude and temperature of the apparent sub-Tg peak in Cp.

While the heating rate data are consistent with the mechanism of a broadened distribution of relaxation mechanisms, the analysis of the functional form of the isothermal recovery of the enthalpy of the PPO/PS blends leads to a different conclusion. The shape of the experimentally accessible portion of the isothermal response function (ie. the time dependence of the fictive temperature) is found to be independent of concentration. These curves, which describe the initial phases of the recovery to equilibrium, can be superimposed by only a shift in the average relaxation time of the system. In addition, apart from a similar change in this average relaxation time, the temperature dependence of the shift of these response functions is also independent of composition. These results indicate that the breadth of at least the short time (ie. experimentally accessible) portion of the underlying distribution of relaxation times is not a function of concentration and therefore, the data can be explained without the necessity of invoking localized fluctuations in the concentrations of the constituents. Instead, the broadening of the transition in the PPO/PS blends appears to be the result of a compositionally dependent change in both the average relaxation time and the structure of the system.

Aging processes similar to those of the PPO/PS blends are observed in the PVME/PS blends both before and following phase separation with the important distinction that the annealing characteristics are a function of the thermal history of the system. The compatible blends formed by slow cooling exhibit the low temperature recovery processes seen in the PPO/PS blends. However, the recovery processes observed in similar PVME/PS mixtures quenched from the lower critical solution temperature, are dispersed throughout the slightly broadened glass transition region. Furthermore, the shape of the isothermal response function of these PVME/PS blends is broadened by thermal treatments in the vicinity of the LCST. These observations strongly suggest that the increase in the width of the glass transition region of these partially compatible blends is the result of a broadening of the distribution of relaxation times caused by fluctuations in the concentrations of the components.

CONCLUSIONS

Annealing studies demonstrate that the broadening of the glass transition region in the compatible polymer blends formed from PPO and PS occurs because of differences in the structural properties of the glass rather than as the result of a change in the distribution of relaxation processes. In contrast to this, an increase in the width of this distribution is required to model the similar broadening which occurs as the result of the onset of the phase separation process in partially compatible blends of PVME and PS. Thus, in this example of a partially compatible polymer blend, the increase in breadth of the glass transition region can be interpreted as arising from fluctuations in the concentrations of the components which occurs during the microphase separation process.

REFERENCES

1- L. Bohn, *Rubber Chem. Technol.* 1968, **41**, 495
2- P. Chaudhari, P. Beardmore, M. B. Bever, *Phys. Chem. Glasses* 1966, **7**, 157
3- J. Stoelting, F. E. Karasz and W. J. MacKnight, *Polym. Eng. Sci.* 1970, **10**, 133
4- A.F. Yee, *Polym. Eng. and Sci.* 1977, **17**, 213
5- J.R. Fried, F.E. Karasz and W.J. MacKnight, *Macromolecules* 1978, **11**, 150
6- R.E. Wetton, W.J. MacKnight, J.R. Fried and F.E. Karasz, *Macromolecules* 1978, **11**, 158
7- W.M. Prest, Jr., D.J. Luca and F.J. Roberts, Jr., *Thermal Analysis in Polymer Characterization*, E. A. Turi, Ed., Heyden & Son, Inc., Philadelphia Pa., (1981), pp. 24-42

8- P. Alexandrovich, F. E. Karasz and W.J. Macknight, *Polymer* 1977, **18**, 1022
9- J. G. Phillips, *Physics Today* 1982, **35**, 27
10- W. M. Prest, Jr. and F. J. Roberts, Jr., to be published
11- A.J. Kovacs, J.J. Aklonis, J.M. Hutchinson and A.R. Ramos, *J. Polym. Sci., Polym. Phy. Ed.* 1979, **17**, 1097
12- M. Bank, J. Leffingwell and C. Thies, *J. Polym. Sci.: Part A-2* 1972, **10**, 1097
13- T. Nishi and T.K. Kwei, *Polymer* 1975, **16**, 85
14- D.D. Davis and T. K. Kwei, *J. Polym. Sci., Polym. Phy. Ed.* 1980, **18**, 2337
15- A.C. Hindmarsh, *Lawrence Livermore Lab. Report* UCID - 30001, Rev. 3, Dec 1974
16- C.W. Gear, *Numerical Initial Value Problems in Ordinary Differential Equations*, Prentice-Hall, N.J. 1971
17- W.M. Prest, Jr. and F.J. Roberts, Jr., *Ann. N.Y. Acad. Sci.* 1981, **371**, 67
18- W.M. Prest, Jr. and F.J. Roberts, Jr., *Bull. Am. Phys. Soc.* 1982, **27**, 392

ANALYSIS OF POLYMER BLENDS AND COPOLYMERS BY
COUPLED THERMOGRAVIMETRY AND AUTOMATIC TITRATION
Shirley G. Fischer and Jen Chiu, Polymer Products
Department, E. I. du Pont de Nemours & Company,
Experimental Station, Wilmington, DE 19898

INTRODUCTION

Thermogravimetry (TG) has been widely used for polymer stability studies. Frequently, compositional analysis of polymer systems can also be made on the basis of weight losses in various temperature ranges and under various atmospheres. However, definitive determination of the composition or structure and establishment of decomposition mechanism or kinetics rely on the use of ancillary techniques such as GC, IR, MS, etc. coupled with TG [1-7]. Another useful but often neglected technique for this purpose is to couple titrimetry to TG to selectively and quantitatively determine certain components in the TG effluence.

Some studies utilizing this concept have been reviewed by Paulik and Paulik [8]. Investigations of ammonia evolution in zeolites and transition metal amine chlorides [9] and adsorbed water analysis in ion exchange resins [10] are representative of more recent applications of this technique. However, it has not been widely exploited in the analysis of complex polymer systems.

Mey-Marom and Behar monitored the elution of HBr continuously as a function of temperature with a bromide ion selective electrode during the decomposition of a brominated polymer [11,12]. On-line titration was not utilized.

Some work on poly(vinyl chloride) (PVC) was reported by Carel [13], where both the effluent and residue were examined for total chloride content in order to determine the effectiveness of ingredients added to promote chloride retention in the residue. Effluent gases were trapped and titrated.

This work describes the analysis of several polymer blends and copolymers by monitoring a specific component evolved under precise temperature and flow conditions with an on-line automatic titrator. Precision and accuracy determinations will also be presented.

EXPERIMENTAL

A Du Pont 1090 Thermal Analyzer was used in conjunction with a Du Pont 951 Thermogravimetric Analyzer (Wilmington, DE) which was coupled to a Radiometer RTS 822 Recording Titration System (London Co., Cleveland, OH). A schematic diagram of this coupled apparatus is shown in Figure 1. The Du Pont 1090 and 951 are used in the conventional fashion to obtain weight loss as a function of temperature. However, a quartz sample boat was used instead of the original platinum pan in order to reduce catalytic reactions of some evolved products. The effluent gas from the TG is interfaced to the titration cell, where the sensing electrodes and autoburette are situated. The autoburette delivers titrant to the cell as required. This is sensed by the electrodes connected to the pH meter.

The volume of titrant added as a function of time or temperature is recorded by the servograph.

The core of this instrument, where interfacing occurs, is highlighted in Figure 2. The TG effluent delivery tip consists of a quartz tube with a female 12/5 spherical joint which is connected directly to the end of the TG quartz furnace tube. The neck of the delivery tip is bent at a right angle and tapered to a capillary of 1/8 inch diameter. After intensive studies on various sizes, this diameter has been found to provide optimum recovery of most eluted components. The vertical length is approximately 4-1/2 inches and was inserted directly into the sample inlet of the titration cell. A gas flow of 100 mL/minute is recommended.

Standard saturated calomel (Radiometer Model No. K4040) and glass (Radiometer Model No. G2040C) electrodes and double junction (Orion Research, Inc. Model 90-02-00) reference and chloride (Orion Research, Inc. Model 94-17A) ion selective electrodes were used.

RESULTS AND DISCUSSION

I. E/VAc Copolymer. Degradation of ethylene-vinyl acetate copolymer (E/VAc) is known to yield acetic acid quantitatively as shown in the TG scan in Figure 3(a). The first step in the degradation is due to liberation of acetic acid. The second major step observed is a result of the breakdown of the polymer backbone. The weight loss steps are well resolved and the first weight loss due to generation of acetic acid has been used to determine the VAc content [14]. However, the method would be in error if other volatiles, such as residual monomer or solvent should be released. In this case, coupled TG-titration provides precise analysis of the copolymer composition without ambiguity.

The titration curve obtained corresponding to the TG curve [Figure 3(a)] is found in Figure 3(b). The titrator is used in a pH stat mode, whereby an operator pre-set pH is maintained. As the liberation of acetic acid occurs, the resulting decrease in pH is compensated by the automatic addition of standard NaOH titrant. The total amount of titrant added is a quantitative indication of the amount of acetic acid liberated. The curve obtained [Figure 3(b)] shows the volume of titrant added as a function of temperature, which was easily converted from the original time scale output. Note that the titration curve shows only one step, that due to acid liberation, despite the two-step degradation that occurs as shown by TG.

The titration results obtained from a series of E/VAc copolymer samples, used in an ASTM round robin, with varying VAc content are plotted vs. those obtained by saponification in Figure 4. The solid line delineates ideal recovery, while the points indicate actual experimental data. Excellent agreement is observed.

A precision study was conducted utilizing one of the E/VAc samples. Results from this analysis are given on Table 1. A relative standard deviation of better than 1% is obtainable.

Table 1. PRECISION OF TG/TITRATION - DETERMINATION OF VINYL ACETATE

Sample No.	% VAc	
	TG	Titrimetrically
1	19.3	20.3
2	19.6	19.9
3	19.3	19.8
4	19.6	20.1
5	19.4	19.8
6	19.4	20.0
Average	19.4	20.0
σ	0.13	0.18
% RSD	0.67	0.90

II. **PVAc-PVAl Blends and Copolymers.** Analysis of poly(vinyl acetate) (PVAc)-poly(vinyl alcohol) (PVAl) blends or their copolymers is often difficult. A typical saponification method requires hours of refluxing. The IR method is restricted to narrow ranges of VAc content [15]. The TG method is not applicable because acetic acid and water moieties are liberated in the same temperature range (Figure 5). Here, TG coupled with titration is ideally suited to resolving this problem. One can selectively analyze for acetate content or hydroxyl content by monitoring either acetic acid or water evolution. Some results based on the acetic acid liberation are shown in Table 2 for several partially hydrolyzed PVAc samples and a blend of PVAc and PVAl. Generally, good recovery was observed.

Table 2. PVAc/PVAl SYSTEM

Sample	Mole % Unhydrolyzed	
	Nominal	TG/Titration
PVAc + PVAl	13.5	13.4±0.7
Part. Hyd. PVAl - A	12.5	13.7±0.4
B	11.7	11.8±0.3
C	12.0	13.9±0.8
D	12.4	12.3±0.4

III. **VCl/VAc Copolymers.** PVC produces hydrogen chloride (HCl) quantitatively upon thermal degradation by a mechanism very similar to that of PVAc [16] and, thus, can be effectively studied by TG coupled with titration. However, HCl evolution is very rapid and a slower heating rate (2°C/minute) and faster titration rate (20 mL/minute) are necessary to obtain complete recovery. Ordinarily, heating rates of 10-15°C/minute can be used.

With the present Radiometer System, ion selective electrodes (ISE) can be used as an endpoint detector for the titration. This provides a convenient method for analysis of vinyl chloride-vinyl acetate (VCl/VAc) copolymers since HCl liberated can be titrated by standard silver nitrate solution selectively in the presence of acetic acid. Figure 6(a&b) shows a typical TG and ISE scan obtained for a VCl/VAc copolymer sample. Some results utilizing the ISE method are given in Table 3. The agreement between experimental and expected recovery is good.

Table 3.   VCl/VAc SYSTEM

Sample	Mole % VCl	
	Nominal	TG/Titration - ISE
PVC	100	99.3 + 1.5
VCl/VAc	90	90.2 + 1.5
VCl/VAc	83	85.3 + 0.5
PVAc	0	0
PVCl + PVAc	92.6	90.3 + 1.3

CONCLUSIONS

The coupled TG-titration apparatus described here has been shown to be effective in polymer compositional studies utilizing both acidimetric and ion selective electrode methods. The other capabilities of the titrator, such as redox, complexometric, Karl Fischer titrations et. al., could also be utilized to selectively monitor a large variety of species in the effluence of the TG.

REFERENCES

1. J. Chiu, Anal. Chem., 40, 1516 (1968).
2. J. Chiu, Thermochimica Acta, 1, 231 (1970).
3. H. G. Wiedemann, "Thermal Analysis," Vol. 1, P. D. Garn and R. F. Schwenker, Jr., Ed., Academic, New York, (1969), p. 229.
4. D. E. Smith, Thermochimica Acta, 14, 370 (1976).
5. C. A. Cody, L. DiCarlo and B. K. Faulseit, Am. Lab, 13(1), 93 (1981).
6. J. Chiu and A. J. Beattie, Thermochimica Acta, 50, 49 (1981).
7. J. Chiu and A. J. Beattie, Thermochimica Acta, 40, 251 (1980).
8. J. Paulik and F. Paulik, "Wilson and Wilson's Comprehensive Analytical Chemistry," G. Svehla, Ed., Volume XII, W. W. Wendlandt, Ed., Part A, Elsevier Scientific Publishing Co., New York, (1981), Chapter 6, p. 35.
9. G. T. Kerr and A. W. Chester, Thermochimica Acta, 3, 113 (1971).
10. J. Kristók and J. Inczedy, J. Thermal Anal., 19, 51 (1980).
11. A. Mey-Marom and D. Behar, Thermochimica Acta, 30, 381 (1979).
12. A. Mey-Marom and D. Behar, J. Appl. Polym. Sci., 25, 691 (1980).
13. A. B. Carel, Proc. International Cable Symp., 27, 299 (1978).
14. J. Chiu, Appl. Polymer Symp., 2, 25 (1966).
15. B. B. Baker and C. E. Day, E. I. du Pont de Nemours & Company, Wilmington, Delaware; private communication.
16. N. Grassie, "Degradation and Stabilization of Polymers," G. Geuskens, Ed., John Wiley & Sons, New York, (1975) Chapter 1, p. 1.

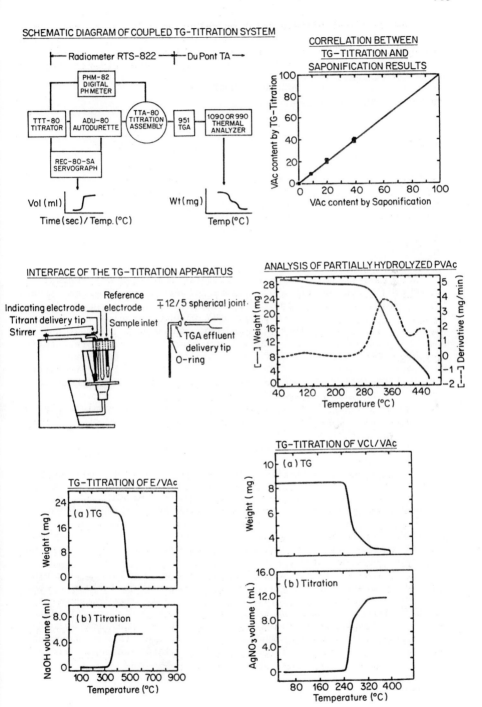

CREEP COMPLIANCE MEASUREMENTS BY THERMO-
MECHANICAL ANALYSIS - PARALLEL PLATE
RHEOMETRY (TMA-PPR)

R. B. PRIME
INTERNATIONAL BUSINESS MACHINES CORPORATION
SAN JOSE, CA 95193

INTRODUCTION

Two types of parallel plate rheometer experiments are possible: (a) when the sample fills the space between the plates and exudes from between them after the load is applied (constant sample radius, equal to radius of plates); and (b) when the radius of the test plates is larger than that of the sample throughout the test and the sample volume is constant. The Wallace plastimeter is representative of the first (a) and the Williams plastimeter of the second (b). Both methods require a flat-ended cylindrical sample. Viscosities from 10^1 to 10^8 Pa.s* can be measured with low shear rates of .0001 to 1 sec^{-1} being generated.

In this paper the method (b) TMA-PPR technique is employed in the measurement of creep compliance and apparent viscosity of a polystyrene reference material and an electrophotographic toner. In a separate study (1) it was shown that the time-temperature dependencies of both noncontact and hot-roll fixing of the toner to paper were identical to the time-temperature dependencies of the toner rheology measured by TMA-PPR.

Parallel plate rheometry is not a new technique. Its use for measuring "plasticity" of unvulcanized rubber stocks dates back to the 1920's. Theoretical solutions for the rate of approach of two parallel plates separated by a viscoelastic medium date back even farther (2-6). The technique has been applied in several polymer and nonpolymer applications.

Sone (7) applied the method (b) technique, and oscillating plate and cone-and-plate viscometry to the rheological properties of butter. Shama and Sherman (8) used parallel plate rheometry to study aging (staling) of cake and margarine. Dienes and Klemm (5) determined the viscosity-temperature behavior of polyethylene and vinyl chloride-acetate resin compounds from method (b) measurements. Dienes (9) further showed that parallel plate rheometry was also

---

*1 Pa.s = 10 poise

capable of measuring elastic and delayed elastic components of deformation. He applied the technique to polyethylene, vinyl chloride-acetate, and cellulose ester compounds. Longworth and Morawetz (10) measured melt viscosities of styrene homopolymers and copolymers with methacrylic acid over a wide range of shear rates, using the method (b) technique. Cessna and Jabloner (11) pioneered in the application of thermomechanical analysis (TMA) to parallel plate rheometry. Utilizing a DuPont Thermomechanical Analyzer and method (a) technique they investigated the early stages of curing of several polyaromatic thermosetting resins.

BACKGROUND THEORY

Parallel plate rheometry measures the approach of two parallel plates moving toward each other under the action of a force with a viscoelastic medium between them. The experiment measures creep compliance of the viscoelastic medium, $J(t)$, which is the time-dependent strain relative to a constant stress*

$$J(t) = \frac{strain(t)}{stress} = J_g + \Sigma J_i (1-e^{-t/\tau_i}) + t/\eta \qquad (1)$$

In noncross-linked thermoplastic materials, such as those investigated in this study, the deformation can be separated into an elastic component (J_g, the glassy compliance), a time-dependent elastic component (characterized by a distribution of relaxation times, τ_i), and a viscous or nonrecoverable component (t/η, where η is the viscosity).

As described earlier two types of experiments are possible: (a) when the sample fills the space between the plates and exudes from between them after the load is applied and (b) when the radius of the test plates is larger than that of the sample throughout the test and the sample volume is constant. Dienes and Klemm (5) describe the physics of both types of experiments. However, the solution they present is subject to the limiting condition that the separation of the plates h must be small compared to the sample radius R; Dienes and Klemm suggested that h should be less than $R/10$. Gent (6) modified the basic equation to avoid this limitation. The creep compliance, in terms of Gent's modified equation and method (b) measurements becomes

$$J(t) = \frac{3V}{F} \left\{ \left(\frac{1}{h} - \frac{1}{h_o} \right) + \frac{V}{8\pi} \left(\frac{1}{h^4} - \frac{1}{h_o^4} \right) \right\} \qquad (2)$$

where h_o is the initial sample height (= plate separation at $t = 0$), h is the sample height at time t after the load is applied, V is the sample volume (constant), and F is the applied force. Gent demonstrated the applicability of Equation 2 over the entire deformation range with experimental measurements (Method b) of a coal-tar pitch sample of $h_o/R_o = 3.9$. Independent measurements of viscosity were in

---
* In the parallel plate method, stress is only approximately constant, decreasing through the test. In this study the stress decrease ranged from 0.1% at 48°C to 56% at 130°C.

excellent agreement with results from the parallel plate method. Note that the solution given by Dienes and Klemm (5) does not include the term $(1/h - 1/h_o)$.

In this study measurements were made according to method (b), and the TMA-PPR data converted to creep compliance by means of Equation 2. At high temperatures and/or long times a predominantly viscous region could be identified, where the slope of the creep compliance curve, $\Delta \log J(t)/ \Delta \log t = 1$. Under these conditions $J(t) \cong t/\eta$, which provides for the measurement of viscosity. The mean shear rate associated with the viscosity measurements was calculated from the following relationship given by Longworth and Morawetz (10)

$$\bar{\gamma} = \frac{2\pi^{\frac{1}{2}} h_o^{5/2} F}{3V^{3/2} \eta} \tag{3}$$

where $\bar{\gamma}$ is the mean velocity gradient in the direction parallel to the plates. The mean shear stress was calculated as $\bar{\tau} = \eta \bar{\gamma}$.

EXPERIMENTAL

1. <u>Materials.</u> Materials used in this study were a narrow molecular weight polystyrene and an electrophotographic toner. The polystyrene was Waters Associates Standard No. 25710: $\overline{M}_n = 49,000$, $\overline{M}_w = 51,000$. The tonner was IBM 790 toner, for use in the IBM 3800 Printing Subsystem. It is a blend of thermoplastic resins and carbon black. By DSC the toner exhibits a T_g between 65-70°C, and a secondary transition between 90-100°C.

2. <u>Test Specimens.</u> The test specimens were right circular cylinders, which were molded at 100°C using the simple mold depicted in Figure 1. Samples of radius 2.55 ±.03 mm were prepared. Sample heights varied between 1.3 and 4.4 mm, with taller samples generally used at lower temperatures. Sample weights and volumes were measured to insure the absence of air bubbles. It was observed that tempera-

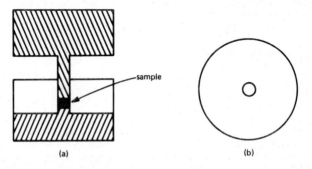

Figure 1. Schematic for three-section mold used to prepare TMA-PPR samples. (a) cross-sectional view. (b) top view of center section. Sample diameter was 5 mm for this work; 2 mm diameter samples for higher pressure studies were prepared with a similar mold.

ture gradients in the sample could be kept to $<1^{\circ}C$ providing sample thickness was <4mm at temperatures between 55 and $112^{\circ}C$, at a flow rate at ~35 cc/m.

3. <u>Instrumental</u>. Measurements were made using a Perkin-Elmer TMS-1 Thermomechanical Analysis System. A special probe was made with an enlarged flat circular foot (~4mm in radius); see Figure 2. A special device was constructed for applying the load to the sample. The device is attached to the TMA analyzer and greatly facilitates the adding and removal of weights to the TMA sample loading pan. Brass washers were epoxied to the weights so that they could be lowered onto the loading pan by means of the hook-chain-dial mechanism of the loading device. Temperature calibration was accomplished by step-heating of naphthalene and benzoic acid pellets.

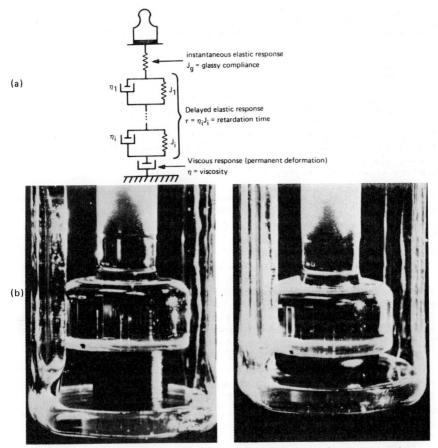

Figure 2. Creep compliance measurements. a) Spring and dashpot model representing elastic, time-dependent elastic (viscoelastic) and viscous components of the deformation. b) Illustration of TMA technique at beginning and end of measurement.

4. <u>Test Procedure and Data Analysis</u>. An outline of the TMA-PPR test procedure is given below. An N_2 purge of 35 cc/m was used. For the studies reported here, a force of 4.91×10^4 dynes (50.1g load) was applied to the samples, producing an initial pressure of $2.40 \pm .04 \times 10^4$ Pa ($3.48 \pm .06$ psi):

a. Place test specimen between parallel plates of the TMA, as illustrated in Figure 2. Approximately zero load should be on the sample.* Place thermocouple close to sample and at approximately one-half the average sample height.

b. Raise furnace assembly and heat to a temperature somewhat below the test temperature. Add the weight to the sample loading tray and remove after a short time. The time and temperature should be just sufficient to allow the sample to "mold" to the TMA-PPR faces without causing significant change in sample height. Allow sample to recover viscoelastically (2-15 hours). Heat to the test temperature. When the temperature is stable ($\sim$10m), lower the weight onto the sample loading pan and record sample height versus time. A fast recorder chart speed is recommended for the early portions of the experiment.

c. A computer program was used to convert the TMA-PPR data to creep compliance via Equation 2. Input into the computer was the probe displacement-time data, the instrument sensitivity, RT sample height, the change in sample height on heating to temperature, sample diameter, sample mass, and load on the sample. Output included plots of h vs t, $J(t)$ vs t, and pressure versus time. If a viscous region of the creep compliance curve could be identified, the program would also compute the slope ($\Delta \log J(t)/ \Delta \log t$), the apparent viscosity, mean shear rate, and mean shear stress.

<u>RESULTS</u>

1. <u>General</u>. Figure 2 illustrates the TMA-PPR measurement. Figure 3 shows the h versus time data for a toner sample at 103.5°C; Figure 4 shows the same data converted to creep compliance. At these conditions the toner exhibits viscoelastic response over approximately the first 400 seconds, beyond which the response is purely viscous. Apparent viscosity, mean shear rate, and mean shear stress were calculated for the viscous region and are presented in Table 1. Since parallel plate rheometry is a constant force measurement, the pressure decreases throughout the experiment. The pressure-time profile corresponding to the data of Figures 3 and 4 is shown in Figure 5.

2. <u>Polystyrene</u>. The TMA-PPR method was tested for accuracy by measuring the narrow molecular weight polystyrene at two temperatures, and comparing the results against accepted data (12). As shown in Figure 6, the comparison is excellent. Agreement between log viscosity is $\sim 1\%$. It is concluded that the TMA-PPR method is capable of generating quantitative rheological data.

3. <u>Toner</u>. Creep compliance of IBM 790 toner was measured over a wide range of temperatures and pressures. Data at an initial pres-

---

* A set of washers of differing weights, which could be added to the sample loading tray, was found to be useful in obtaining zero load.

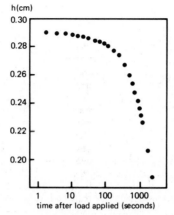

Figure 3. Plate separation = sample height data for toner at 103.5°C. h_0 = 0.293 cm at temperature.

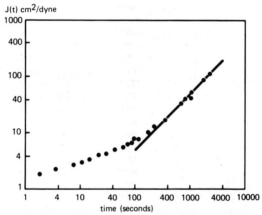

Figure 4. Creep compliance of toner at 103.5°C, from Equation 2 and data of Figure 3. Region of constant slope = 1 identifies viscous portion.

sure of 2.4×10^4 Pa (3.5 psi) and over a temperature range of 48-130°C are shown in Figure 7. At 48° an estimate of the glassy compliance, $J_g \cong 1 \times 10^{-10}$ cm^2/dyne, is obtained; the Young's modulus is the reciprocal of J_g. At longer times viscoelastic behavior is observed. Between 62 and 100°C the deformation is predominantly viscoelastic, while at 130° it is mostly viscous.

Time-temperature superposition (13, 14) was used to generate a master curve by shifting the creep compliance curves along the time axis until they overlapped. Note that the creep compliance data have been normalized for temperature. The extent of the shifting is the shift factor a_T, where a_T is a ratio of corresponding times for the same behavior to occur at different temperatures. The shift factor is arbitrarily chosen to have a value of unity at some reference temperature. Figure 8 is the master curve obtained by shifting the

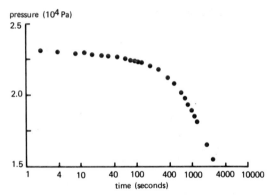

Figure 5. Pressure profile through TMA-PPR experiment. Toner at 103.5°C, $P_o = 2.37 \times 10^4$ Pa.

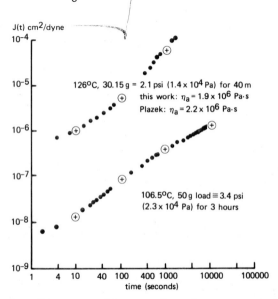

Figure 6. Creep compliance of polystyrene reference material. ● 50,000 MW, this work (TMA-PPR) ⊕ 46,900 MW (Reference 12).

creep compliance data of Figure 7; note that the elastic and viscous regions are separated by ~14 decades of time.

Figure 9 is the shift factor-temperature curve, obtained from several TMA-PPR and capillary rheometer measurements. The significance of this curve is that it relates time and temperature. For electrophotographic toners this is important both at low temperatures where caking (or blocking) is a concern, and at high temperatures where fixing occurs. As an example, from Figure 9 it would be expected that the same amount of flow that occurs in the nip of a hot roll fuser in 10 msec at 140°C would require 20 msec at 130°C. An

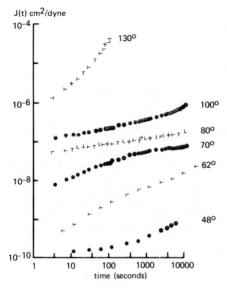

Figure 7. Creep compliance curves for IBM 790 toner. Initial pressure = 2.4×10^4 Pa.

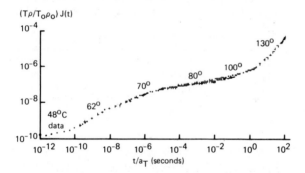

Figure 8. Creep compliance master curve for toner, from data of Figure 7, plotted against reduced time t/a_T, for $T_o = 130°C$.

interesting aspect of the shift factor curve is the deviation from the standard WLF [14] behavior of unfilled, monocomponent polymers (dashed line) at ~ 90°C. This is attributed to the multi-component nature of toner, where one of the components functions as a solid diluent.

Toner viscosity data are presented in Table 1. The non-Newtonian nature of toner is exemplified by the dependence of viscosity on shear rate (111°C data). Note, by comparison with Eq. 1 and from the definition of the shift factor as a ratio of times that, when the deformation is predominantly viscous, the shift factor measures the temperature dependence of viscosity.

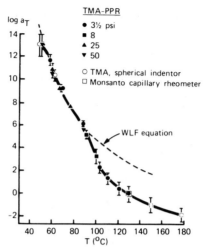

Figure 9. Shift factor - temperature curve for IBM 790 toner. a_T = 1@130°C (arbitrary).

Table 1 Viscosity data for IBM 790 toner from TMA-PPR data. η_a = apparent viscosity, $\bar{\dot{\gamma}}$ = mean shear rate, $\bar{\tau}$ = mean shear stress, slope = $\Delta \log J(t) / \Delta \log t$.

T (°C)	h_0 @ T (CM)	SLOPE	η_a (Pa.S)	$\bar{\dot{\gamma}}$(SEC^{-1})	$\bar{\tau}$(Pa)
103.5	0.293	1.04	2.85x10^7	0.00067	1.91x10^4
110.8	0.229	0.99	1.51x10^6	0.0099	1.52x10^4
111.0	0.134	0.98	1.80x10^6	0.0050	0.91x10^4
121.0	0.203	1.02	2.33x10^5	0.059	1.38x10^4
130.0	0.155	0.98	7.24x10^4	0.139	1.00x10^4

Based on the TMA-PPR rheological data of IBM 790 toner, the temperature dependencies for both noncontact and hot-roll fixing were shown to be identical to the temperature dependence of the toner viscosity [1]. For example, when judged by a scratch test, the pressure (p), time (t), temperature (T) dependence of hot-roll fixing was found to be pt^2/a_T. This correspondence between the fixing of toner images and fundamental flow properties of the toner shows that flow of toner dominates the fixing process.

DISCUSSION AND CONCLUSIONS

By comparison with reference data for polystyrene, it was shown that the TMA-PPR technique can yield quantitative rheological data. The utility of the technique was demonstrated by generating the complete creep compliance master curve for an electrophotographic toner, and by showing that the time-temperature shift factors (a_T) from TMA-PPR measurements were identical to those for noncontact and hot-roll fixing of the toner to paper. Rigid temperature control, precise measurement of sample thickness with time, and modern data collection and analysis capabilities makes thermomechanical analysis an ideal method for parallel plate rheometry.

ACKNOWLEDGEMENTS

It is a pleasure to acknowledge Dr. Thor Smith of IBM Research for his guidance and many helpful discussions. Mel Astrahan, an exceptional summer student, deserves credit for many of the creep compliance measurements and for the computer program.

REFERENCES

1. R. B. Prime, Photo. Sci. Eng., 1982, in press.

2. M. J. Stefan, Akad. Wiss. Wien Math., - Naturwiss. Kl. Abt. IIa, 1874, 69, 713.

3. A. Healy, I.R.I. Trans., 1926, 1, 334.

4. J. R. Scott, I.R.I. Trans., 1931, 7, 169.

5. G. J. Dienes and H. F. Klemm, J. Appl. Phys., 1946, 17, 45.

6. A. N. Gent, Br. J. Appl, Phys., 1960, 11, 85.

7. T. Sone, J. Phys. Soc. Japan, 1961, 16, 961.

8. F. Shama and P. Sherman, Soc. Chem. Ind. SCI Monograph #27, 1967, 77.

9. G. J. Dienes, J. Colloid Sci., 1947, 2, 131.

10. R. Longworth and H. Morawetz, J. Poly. Sci., 1958, 29, 307.

11. L. C. Cessna, Jr. and H. Jabloner, J. Elastomers Plast., 1974, 6, 103.

12. D. J. Plazek, J. Phys. Chem., 1965, 69, 3480.

13. J. D. Ferry, "Viscoelastic Properties of Polymers," (2nd Edition), John Wiley, New York, (1970), Chapter 2C.

14. J. D. Ferry, R. F. Landel, and M. L. Williams, J. Appl. Phys., 1955, 26, 359.

RECENT DEVELOPMENTS IN CHARACTERIZATION OF
POLYMERS BY DYNAMIC MECHANICAL ANALYSIS

by: Philip S. Gill and R. L. Blaine,
E. I. Du Pont de Nemours & Co., Inc.
Wilmington, Delaware   19898

INTRODUCTION

Recent developments have provided significant improvements in the capability and ease of operation of the 982 Du Pont Dynamic Mechanical Analyzer (DMA) used in conjunction with the Du Pont 1090 Thermal Analysis System. These improvements in terms of mechanical design, temperature control and data manipulation result in greater accuracy of the measured viscoelastic variables, more precise temperature measurement and control, faster sample throughput and ease of sample and instrument set-up. The ability to handle a broader range of sample geometries has also extended the versatility of the DMA in being able to characterize both very stiff composite samples and very soft elastomer materials.

This paper will provide examples to illustrate these new techniques and capabilities and also provide details of the specifics of these added features.

EXPERIMENTAL

Data were generated using the Du Pont 982 DMA in conjunction with the 1090 Thermal Analyzer for control and data analysis. Advanced DMA software was used with automatic system calibration and correction factor calculations.

The Du Pont 982 Dynamic Mechanical Analyzer operates on the mechanical principle of forced, resonant vibratory motion with fixed, selected amplitude. Two parallel, balanced sample support arms, free to oscillate around low-hysteresis flexure pivots, are the heart of this unique system.

The sample is rigidly clamped in place between the two arms, making the arms and sample part of a combined resonant system. The position of one arm and pivot is fixed, while the other arm and pivot are movable by means of a precision mechanical slide to accomodate a wide variety of sample lengths. Once the desired arm spacing is established for the sample, the arm and pivot are locked in place to maintain the two arms in parallel. The system is displaced and set into oscillation by a driver at an amplitude selected by the operator. As the system oscillates, the sample is subjected to the type of deformation similar to that shown in Figure 1. The amplitude is measured by a linear variable differential transformer (LVDT), the core of which is attached to the fixed-pivot arm. The output of the LVDT contains both frequency and amplitude information needed to calculate the viscoelastic properties measured by the 982 DMA.

Normally, a system so displaced would oscillate at the system resonant frequency, with constantly decreasing amplitude due to loss of energy (damping) within the sample. The electronics of the 982 are designed to compensate for this loss of energy in the samples. The amplitude signal from the LVDT is fed into a circuit which in turn provides an output signal to the electromechanical driver. This supplies additional energy to the driven arm forcing the system to oscillate continuously at constant amplitude.

The frequency of oscillation is directly related to the modulus of the sample under investigation, while the energy needed to maintain constant amplitude oscillation is a measure of damping within the sample.

Quantitative frequency and damping signals are transmitted to the 1090 Thermal Analyzer where they are digitized and stored in disk memory along with sample temperature and time signals. Quantitative calculation of the viscoelastic parameters such as tensile modulus, shear modulus, loss modulus and tan δ is performed using the data analysis software available with the 1090 Thermal Analyzer. Calculated results are subsequently plotted or tabulated on the printer/plotter output.

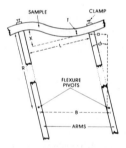

Figure 1 **982 DMA SAMPLE DEFORMATION**

RESULTS & DISCUSSION

A summary of the various kinds of end-use applications for Dynamic Mechanical Analysis techniques is shown in Table 1. Examples illustrating some of these applications are given here:

Engineering Design Data. Figure 2 shows modulus data calculated from the 982 DMA compared to that obtained by independenet determinations for a broad spectrum of material moduli. A six decade range in modulus was covered from steel at 186GPa to neoprene at 2.06MPa. Excellent correlation was shown between the two measurement techniques. Precision data from repeated experiments showed better than ±5% coefficient of variation for modulus values and ±0.005 for tan δ

precision for the same series of materials.

- Curing of Thermosets
- Polymer Blend Compatibility
- Correlation of Impact Stability with Damping
- Observation of Plasticizer Effects
- Measurement of Subtle Transitions
- Sound and Vibration Dissipation Correlation with Damping
- Engineering Design - Mechanical Data
- Characterization of Stiff Composites and Metals
- Analysis of Soft Elastomers
- Characterization of Supported Systems (Coatings, Prepregs, Adhesives)

Table 1: **DMA APPLICATIONS**

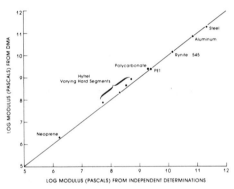

Figure 2 **COMPARATIVE MODULUS**

Detection of Subtle Transitions. Figure 3 shows the DMA output for low density branched and high density linear polyethylene. The three transitons shown -- alpha, beta and gamma at $80°C$, $-3°C$ and $-115°C$ --- are readily detected by the DMA technique but are very difficult to observe by traditional thermal analysis techniques. Each of these transitions corresponds to specific molecular motions which have significance in terms of structure/property relationships:

- Alpha-transition is associated with crystalline relaxations occuring below the melting point of polyethylene.

- Beta-transition is due to motion of the amorphous regions side chains or branches from the main polymer backbone. The intensity of the Beta-transition varies with the degree of branching.

- Gamma-transition is due to crankshaft rotation of short methylene main chain segments and can influence low temperature impact stability of polyethylene.

Lower modulus values are also apparent for the low density polyethylene compared to the high density material. Subtle distinstions in the temperatures for modulus changes can also be observed.

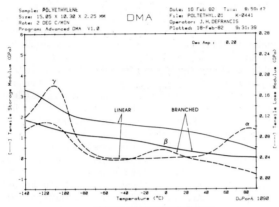

Figure 3: **LINEAR vs BRANCHED POLYETHYLENE**

Characterization of Stiff Composites. Figure 4 shows the 982 DMA modulus determinations for a series of composite laminates with varying ply orientations and compares them with independent bending modulus calculations made by other techniques. The designation 0/45/90 refers to the orientation of the fibers in the ply with respect to the longitudinal direction of the sample. Each data point represents a different stacking arrangement of the oriented fiber plys. The results followed the predicted trend: as the zero-degree ply was moved toward the outer surface of the laminate, a progressive increase in modulus is observed. The numbers represent the angle of orientation of the ply relative to the bending axis and each set is a repeat unit as designated by //.

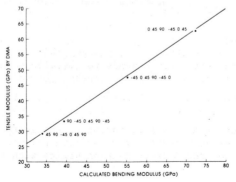

Figure 4 **COMPOSITE LAMINATE — PLY ORIENTATION EFFECT**

Thermoset Curing Behavior. The viscoelastic changes which take place as a result of crosslinking reactions in a thermosetting polymer can be followed by DMA with respect to both time and temperature. The glass transition temperature, the onset of gelation and vitrification, as well as the modulus and damping of the final cured material, can all be measured by the DMA technique.

An example is shown in Figure 5. In this experiment, a partially cured graphite-reinforced epoxy prepreg is held horizontally in the instrument and subjected to a simulated cure cycle using the linked method capability of the 1090 Programmer. The material was slowly heated to an isothermal pre-cure temperature of 107°C and held for a predetermined time before being subjected to a final post-cure heat treatment at 170°C to achieve optimum crosslinking. The viscoelastic changes occurring during the treatment are interpreted from both the elastic response (resonance frequency) and the damping response. The initial glass transition (Tg), where the uncured resin first softened, is observed at 25°C as both a damping peak and a frequency decline. This temperature gives an indication of the extent of precure. During the isothermal step, damping and frequency show an increase due to progression of crosslinking reactions. During final cure, a damping peak and a frequency plateau are observed, indicating vitrification as cure completion in the resin. Such information from DMA experiments can be used for optimization of cure cycles, as well as for investigation of variations in raw material performance.

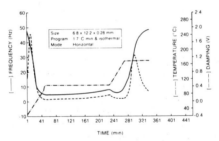

Figure 5 **GRAPHITE — EPOXY PREPREG CURE CYCLE**

PHENOMENOLOGICAL BEHAVIOR OF POLYMERS
IN STRESS-FIELD DIFFUSION
Benoy B. Chowdhury
M&T Chemicals Inc.
Rahway, N.J. 07065

INTRODUCTION

Among the use-related properties of polymer materials, manifestation of moisture vapor permeability through electronic device encapsulants has received considerable attention [1-5] due to its deleterious effects on system reliability. Tests in simulated environments consisting of temperature, pressure and voltage stress show that macroscopic behavior of polymers in the diffusion process is related to both the external stress-fields and those operative in the polymers from internal sources. Polymers vary widely in their thermal properties as a result of structural differences and exhibit dielectric relaxation modes that are very different from one another. Thus, stress-field diffusion affords a study of the applicability of constitutive relations to the phenomenological behavior of the polymers. Steady-state diffusion through polymer network is dependent on volume anisotropy [6] and its understanding requires a number of assumptions and some intuitive modifications of the classical laws of diffusion.

THEORETICAL
   Thermal Stress.

Use of phenomenological equations of irreversible thermodynamics permit addition of a stress term,
$$\sigma = (D_{AB}Q^*C_A/RT^2)(dT/dx)$$
to Fick's First Law of diffusion:
$$\underline{J} = -D_{AB}(\delta C_A/\delta x)\underline{i} + (\delta C_A/\delta y)\underline{j} + (\delta C_A/\delta z)\underline{k} + (Q^*C_A/RT^2)(dT/dx)$$
where $\underline{J}$ is the mass flux expressed as a vector sum of diffusion gradients in the x, y and z directions, multiplied by the diffusion coefficient D_{AB}; Q^* is the heat of transport.

The mobility, (D_{AB}/RT), of the vapor molecules is the result of an effective force, $\psi = (Q^*/T)(dT/dx)$ exerted by the temperature gradient on each molecule of vapor in a potential field $V(x,y,z)$. Therefore, $\psi = -\nabla V$ where ∇ is the gradient operator. The mean diffusion velocity, $v = B\psi = -(D_{AB}C/kT)\nabla V$; here B = mobility, C = vapor concentration and k = Boltzmann constant. The diffusion velocity thus calculated from experimental results in a gradient field can be used in a constitutive relation proposed by Aifantis [7]:

$$\rho v = -D\nabla \rho + M^*\nabla \sigma \quad \ldots\ldots\ldots\ldots (1)$$

where ρ is the mass of permeated vapor, v is its velocity, D is the diffusion constant and σ is the trace of the stress tensor. In a temperature field, the phenomenological constant M^* stands for the coefficient of transport and can be evaluated by using a heat conduction type of equation:

$$M^* = (\Delta H/C_p\rho) \quad \ldots\ldots\ldots\ldots\ldots (2)$$

Cp is the specific heat of vapor, 100°C and ΔH is the enthalpy change during diffusion. ΔH at half thickness of the polymer, where the uptake of vapor and heat flow are symmetrical, is equal to

$$1/2 \, \rho L_v \, (dR/dt) \qquad \qquad (3)$$

L_v = latent heat of vaporization and $R = B(t/L^2)^{1/2}$; i.e., mass of vapor taken up per unit mass of the polymer. B can be evaluated from experimental results:

$$B = 4C_0 (D/\pi)^{1/2}.$$

C_0 is the equilibrium regain at infinite time t_∞, D is the average diffusion coefficient for the concentration range 0 to C_0 of vapor, t denotes time and L is the material thickness.

The contribution of thermal energy to the kinetic energy of diffusion is difficult to assign, since mechanical energy of fluid eddies also contribute to the total energy. A rationale is developed here for the change in thermodynamic internal energy function U, leading to a variation of the mean energy <E> of the polymer into its components of work energy and the uptake of thermal energy <dq>:

$$dU = TdS - pdV + \Sigma \mu_i dn_i$$

at constant temperature and pressure. S stands for entropy, V denotes volume and μ refers to chemical potential. Then,

$$d\langle E\rangle = \Sigma P_i dE_i n_i + \Sigma \, E_i dP_i n_i \qquad \ldots (4)$$

where P_i is the probability of the polymer being in a quantum state whose momentary eigenvalue is E_i. The individual diffusion steps, n_i, are independent of each other and the vapor/solid system is assumed to be in thermal equilibrium at every step of the process. Hence, the first term in equation (4) corresponds to the mean work energy being expended on the system and the second term corresponds to the thermal energy uptake by the system. Under these conditions, the variation of <dq> with temperature can be found experimentally by plotting total permeability values against stepwise increase in temperature, and the magnitude of the required activation energy for the diffusion process, calculated from the linear portion of an Arrhenius plot, provides a measure of the relative diffusivity of the polymer.

Pressure Field.

Compressive stress of external pressure on polymer pore volume decreases diffusivity. Data obtained in concomitant temperature/pressure field can be treated by a graphical technique to obtain permeability values at zero pressure gradient. An equation has been specifically devised for this purpose:

$$\frac{p}{V_t(p-p_0)} = \frac{p}{p_0} \cdot \frac{(C-1)}{C} + \frac{1}{C}$$

where p = head pressure, p_0 = post-diffusion pressure, V_t = permeability value, C = total vapor concentration on the entering side of the polymer. The slope of a linear plot of the left-hand side of the equation against p_0/p is $x = (C-1)/C$ and its intercept on the ordinate $y = (1/C)$. Permeability value at zero pressure gradient is

$$V_t' = (1/x+y)$$

Voltage Stress.

The theory of voltage-induced polarization phenomenon in polymer electrets, whereby dipolar alignment causes a charge accumulation which can be measured by a thermal discharge technique, is discussed elsewhere [8]. A direct relationship between the level of polarization and vapor permeability has been found in this study.

Interactive Forces.

In gas-solid diffusion, the diffusive force, ψ, depends upon the degree of elasticity of the solid through its deformation gradient which determines its stress component [7]. Green and Adkins [9] have shown that the constitutive equations for the diffusive force depends on moments rather than velocities. Adopting a constitutive assumption that the mechanical response of the polymer is indifferent to the presence of the vapor at equilibrium, the energy involved in the solid deformation gradient is the only consideration for the diffusive force in the exchange of momentum between the solid and the vapor. For this reason, creep compliance is studied here from stress relaxation effects in a deformation mode over a temperature range. Time effects are extrapolated from a change in temperature scale. From the theory of such effects [10], a dependence of relaxation transition on the time scale of the experiment relative to some basic parameter of the polymer is expected. This parameter (τ) falls in the middle of the visco-elastic range and corresponds to the discontinuity in the uniform thermal expansion at T_g which changes to a sudden onset of expansion of free volume. It therefore represents a diffusion boundary on the time scale.

EXPERIMENTAL

The apparatus and methods used for measurement of diffusion coefficient and permeability are essentially those described by Vanderkooi and Riddell [5]. Polymer materials used can be grouped into three classifications. Class I materials are generally epoxy, Class II are generally silicon polyimides and Class III are generally urethanes. For proprietary reasons the actual structures are not given here.

RESULTS AND DISCUSSION

Data given in Table 1 show stress buildup effect on permeability for successive measurements on the same sample at constant temperature. Care was taken to ensure complete removal of moisture between measurements.

Table 1. PERMEABILITY VALUES OF SUCCESSIVE MEASUREMENTS (ml.mil/100 in^2.24 hrs.atm)

Material Class	No.	Day 1	Day 2	Day 3
II	1.	5.33x10^5	5.86x10^5	6.19x10^5
	2.	5.47x10^5	5.86x10^5	6.24x10^5

Thermal activation energy calculated from Arrhenius plots confirm that lower diffusivity results from higher activation energy requirement. For example, Class III material, which is more permeable than Class II material, has a typical value of 34.5 KJoules whereas Class II material has a typical value of 38 KJoules. Conversely, on a comparative basis, lower ΔH values have been found to correspond to lower diffusivity. Additionally, non-linear diffusion in polymers, attributed to structural inhomogeneity [11], also has its origin in the non-uniformity of enthalpy change due to the same cause. Based on a contrived molecular approach, whereby the single molecule parameter for vapor diffusivity, γ (= $[Ax22415] \div (V_t/\Delta x)N$; A is the polymer surface area available for diffusion, V_t is total permeated volume, Δx is material thickness and N is Avogadro number) is plotted against total permeability, linear relationships are obtained as can be seen in Figures 1 and 2 for Class I and Class II materials. The scatter in data for Class I material is also found when calculated vapor/solid ratio is plotted against γ, as shown in Figure 3. Representative ΔH values, calculated for different thicknesses of these two classes of materials, show more uniformity for Class II than for Class I material. This data is given in Table 2.

Table 2. COMPARISON OF ΔH VALUES
(CALCULATED BY USE OF EQN.3)

ΔH
(gm.cals)

Class I	Class II
3.95×10^{-7}	3.08×10^{-6}
3.02×10^{-7}	3.06×10^{-6}
2.88×10^{-7}	3.04×10^{-6}
1.83×10^{-7}	3.11×10^{-6}
2.43×10^{-7}	3.11×10^{-6}
3.40×10^{-7}	3.06×10^{-6}

Figures 4 and 5 are included here to illustrate separation of pressure effects from two- and three-stress component permeability values. The values of V_t' for three different samples of Class II material, free of pressure gradient but containing thermal and residual voltage stress are:

Sample 1	Sample 2	Sample 3
(ml.mil/100 in^2.24 hrs.atm)		
3.24×10^5	3.33×10^5	3.29×10^5

and 2.5×10^5 for thermal stress alone for a Class III material.

Voltage stress increases permeability considerably as can be seen from the values in Table 3.

Table 3. VOLTAGE-FIELD STRESS EFFECTS ON PERMEABILITY

Material Class	No.	After Voltage Breakdown	After Polarization
		(ml.mil/100 in^2.24 hrs.atm)	
II	1.	4.39×10^5	8.79×10^6
	2.	4.60×10^5	8.96×10^6
	3.	4.93×10^5	9.13×10^6

The value prior to voltage stress was around 5×10^4.

A direct correlation is found between permeability and voltage-induced polarization, as measured by discharge current:

Table 4. CORRELATION BETWEEN LEVEL OF POLARIZATION AND PERMEABILITY

Total Charge (μcoulombs)	Permeability Value after Polarization (ml.mil/100 in^2.24 hrs.atm)
9.81	8.79×10^6
10.92	1.02×10^7
34.28	2.95×10^7
49.19	5.18×10^7
72.08	∞

For the above study, polymer structures were suitably modified to induce different levels of polarizability.

Dynamic mechanical stress-relaxation characteristics in Figures 6, 7 and 8 for the three classes of materials show the magnitude of τ values varying in the following manner:

Class I > Class II > Class III

which is in reverse order of their permeability in keeping with the reasoning given earlier.

Finally, calculated molecular diffusion velocity $v = V_t \times D'/kT$, where D' = diffusion coefficient x γ, for different thicknesses of Class I material, was used to evaluate equation (1). Values in Table 5 show that the product of the mass of permeated vapor and its molecular diffusion velocity follow the variation of the applicable stress factor in each case.

Table 5. EVALUATED STRESS FACTORS FROM EXPERIMENTAL DATA

ρv (gm.cm/sec)	D (cm^2/sec)	$\nabla \rho$ (gm)	M^* (gm.cm/sec)	$\nabla \sigma$ (dimensionless)
15.80×10^{-13}	3.2×10^{-7}	1.74×10^{-11}	3.95×10^{-7}	4.00×10^{-6}
9.59×10^{-13}	2.2×10^{-7}	1.33×10^{-11}	3.02×10^{-7}	3.18×10^{-6}
8.85×10^{-13}	2.2×10^{-7}	1.27×10^{-11}	2.88×10^{-7}	3.07×10^{-6}
2.85×10^{-13}	1.5×10^{-7}	0.81×10^{-11}	1.83×10^{-7}	1.41×10^{-6}
5.15×10^{-13}	1.9×10^{-7}	1.07×10^{-11}	2.43×10^{-7}	2.12×10^{-6}
13.60×10^{-13}	3.1×10^{-7}	1.50×10^{-11}	3.40×10^{-7}	4.00×10^{-6}

CONCLUSIONS

The results of this study show that external as well as evolved thermal stress, pressure gradient and voltage fields are effective parameters that can account for force fields acting on vapor diffusion in polymers. Furthermore, experimentally determined parameters such as diffusion velocity and heat of transport related to the constitutive properties, can be used in a constitutive equation specifically formulated for stress-field diffusion, to evaluate the magnitude of stress factors that are operative in polymers from internal sources.

REFERENCES

1. S.M. Lee, J.J. Licari and A. Valles, "Properties of Plastic Materials and How They Relate to Device Failure Mechanism", Proceedings Reliability Physics Symp., 1965, 8, 1.
2. W.L. Hunter and R.J. Zettek, "A Study on the Entry of Water into Plastic-Encapsulated Semi-conductors", SRDL Process Technology Laboratory Technical Report No. 48, July 1978, Motorola Corporation, Phoenix, Arizona.
3. R.G. Hadge, M.N. Riddell and L. O'Toole, "Quick Vapor Permeability Test for Molded Plastics", SPE Journal, 1972, 28, 12.
4. N. Vanderkooi and M.N. Riddell, "Dynamic Permeability Method for Epoxy Encapsulation Resins", Proceedings Reliability Physics Symp., 1976, 14, 219.
5. N. Vanderkooi and M.N. Riddell, "Dynamic Permeability Test Method Checks Encapsulants Quickly, Accurately", Materials Engineering, March 1977, p. 58.
6. P.G. Shewmon, "Diffusion in Solids", McGraw-Hill Book Co., New York, (1963).
7. E.C. Aifantis, Ph.D. Dissertation, University of Minnesota, 1975.
8. J. Van Turnhout, "Thermally Stimulated Discharge of Polymer Electrets", Polymer Journal, 1971, 2, 173.
9. A.E. Green and J.E. Adkins, "A Contribution to the Theory of Non-Linear Diffusion", Arch. Rational Mech. Anal., 1964, 15, 235.

10. I.M. Ward, "Mechanical Properties of Solid Polymers", Wiley-Interscience, London, (1971), p. 81.
11. J. Crank and G.S. Park, "Diffusion in Polymers", Academic Press, New York, (1968), pp. 141-162.

ACKNOWLEDGMENT

The author wishes to thank Ivor L. Simmons of M&T Chemicals for his support of this work.

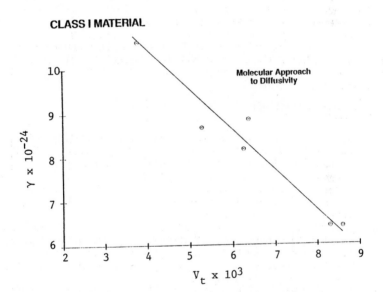

Figure 1. Molecular Approach to Diffusivity -- Class I Material

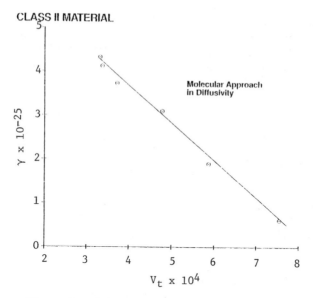

Figure 2. Molecular Approach to Diffusivity -- Class II Material

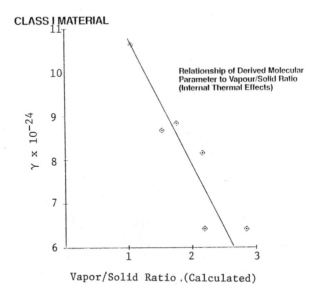

Figure 3. Relationship of Derived Molecular Parameter to Vapor/Solid Ratio

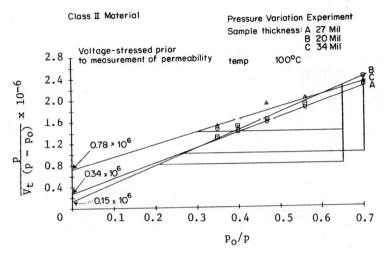

Figure 4. Pressure Variation Experiment -- Class Material

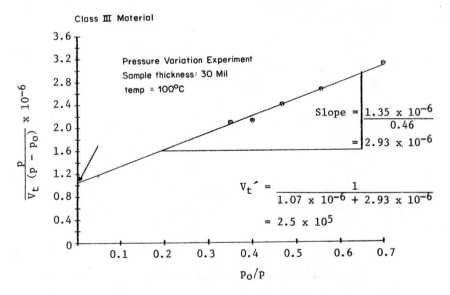

Figure 5. Pressure Variation Experiment -- Class III Material

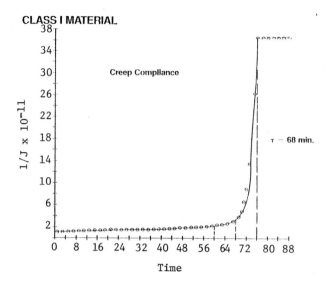

Figure 6. Creep Compliance -- Class I Material

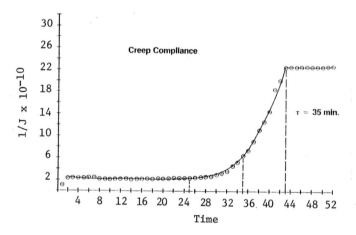

Figure 7. Creep Compliance -- Class II Material

CLASS III MATERIAL

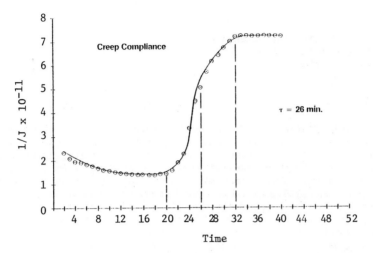

Figure 8. Creep Compliance -- Class III Material

THE ROLE OF ZINC STEARATE CRYSTALLIZATION IN THE
REINFORCEMENT OF SULFONATED EPDM

R. A. Weiss
Institute of Materials Science
University of Connecticut
Storrs, CT 06268

INTRODUCTION

The modification of hydrocarbon polymers by the introduction of small amounts of chemically bonded ionic groups has received considerable attention over the past decade. These materials, ionomers, have been the subject of two relatively recent monographs [1,2], an American Chemical Society symposium [3], and two Gordon Research Conferences in 1979 and 1981. An important characteristic of ionomers is that the intermolecular interactions that take place between the ionic groups give rise to properties similar to those of a crosslinked polymer. The "ionic-crosslink" is not, however, permanent, and at sufficiently high temperatures or in the presence of suitable solvents ionomers exhibit viscous flow or can be dissolved, characteristics of non-crosslinked polymers.

One specific ionomer system that has been described in recent years is sulfonated ethylene-propylene-diene terpolymer, S-EPDM. [4-6]. These materials which contain small amounts of metal sulfonate groups exhibit a strongly associated network at room temperature, yet at elevated temperatures the ionic associations are sufficiently relaxed to allow for melt processability in conventional plastics processing machinery, eg., extrusion and injection molding.

The viscosity of S-EPDM at normal processing temperatures is too high for this material to be considered in any practical commercial melt process. [6] It is possible, however, to achieve significant improvements in melt flow by the addition of various polar additives that interact preferentially with the ionic associations. [7,8] While these additives, "ionic-domain plasticizers", are effective in lowering the melt viscosity, they also, in general, have a detrimental effect on the mechanical properties of the ionomer.

One particular ionic-domain plasticizer that not only improves the melt flow of S-EPDM, but also simultaneously improves the mechanical properties of the resultant system is zinc stearate (ZnST). [9] The improvement of the melt flow at elevated temperatures is due to the preferential solvation of the ionic association by the ZnST resulting in a much weaker crosslink network than in the absence of the polar plasticizer. The improvment of the tensile strength of S-EPDM containing ZnST is believed to result from the formation of a separate crystalline ZnST phase that interacts with the sulfonate groups of S-EPDM. It is noteworthy that the behavior of ZnST in S-EPDM is different from its behavior in unsulfonated EPDM. In EPDM ZnST macroscopically separates--that is, it exudes from the polymer. At concentrations of greater than 2%(wt) in EPDM

ZnST exhibits its normal crystalline melting point as measured by DSC indicating the formation of undiluted crystals. On the other hand, ZnST is stable in S-EPDM, though it does exist as a separate crystalline microphase as evidenced by the presence of a melting endotherm by DSC. The melting point is depressed below concentrations of 30%(wt) ZnST which suggests a dilution phenomenon. [9] In addition, the mechanical properties exhibit a strong time-dependence after being cooled from the melt [10], and it is this time-dependent mechanical behavior that is the subject of the present investigation. In this paper, we will describe some thermal analytical studies of the crystallization of ZnST in S-EPDM and attempt to relate this to mechanical property results determined thermomechanically and by conventional static testing.

EXPERIMENTAL

The materials studied were supplied by Dr. Ilan Duvdevani of Exxon Research and Engineering. The S-EPDM, TP-303, contains about 30 meq. zinc-sulfonate per 100g polymer; its synthesis is described elsewhere [6]. Zinc stearate was added to the polymer on a heated two-roll mill.

Thermal measurements were made with a Perkin Elmer differential scanning calorimeter, DSC-2, equipped with a mechanical cooling accessory. Samples of approximately 10mg were crimped inside aluminum pans and all measurements were made in a dry nitrogen atmosphere. For the aging studies, a cooling rate of 40K/min was used in cooling from the melt to room temperature. Heating scans were made at 10 and 20K/min.

Thermomechanical measurements were made with a DuPont TMA, and tensile measurements were made at room temperature with an Instron Universal Testing Machine using compression molded samples. These results are not given here, but will be presented in the paper at ICTA.

RESULTS AND DISCUSSION

The effect of ZnST concentration on the polymer glass transition temperature is shown in Table I. The small but definite increase in Tg with increasing ZnST concentration is opposite the normal effect that plasticizers have on Tg. These data suggest a strong interaction betwen the polymer and the ZnST phases, an interaction most likely occurring between the metal salts of the sulfonate groups and the stearate. Thus the presence of the stearate in the solid state acts to restrict the mobility of the polymer backbone, an antiplasticizer effect, while above the crystalline melting transition of the stearate the effect is normal plasticization.

Upon cooling the samples from the melt, 420K, to room temperature a crystallization peak(s) was observed in the samples containing the higher concentrations of ZnST, but no crystallization was observed in samples containing less than 16% ZnST.

Melting of the zinc stearate was observed, however, in all the samples, albeit at temperatures depressed from the melting temperature of crystalline ZnST, indicating that some crystallization takes place while the sample ages at room temperature. Enthalpy of crystallization date, given in Table II, demonstrate that only a small fraction of the zinc stearate can be accounted for in the crystallization endotherm observed above room temperature by a dynamic experiment.

If the S-EPDM/ZnST samples are cooled below room temperature a second crystallization is observed; for example, in a sample containing 33.3% ZnST this is observed near 280K, c.f. figure 1. The large supercooling suggests extremely small crystallities of ZnST are forming. This is consistent with the fact that films of these materials are relatively clear indicating that the ZnST crystallites are sufficiently small so as not to scatter visible light.

Isothermal crystallization thermograms for a sample containing 23.1% ZnST are shown in figure 2. The heating scans run after holding at the isothermal conditions noted for 10 minutes are shown in figure 3. The feature of greatest interest in figure 3 is the appearance of an endotherm about 20° above the isothermal crystallization temperature in addition to the endotherm corresponding to the melting point of pure ZnST. The two melting endotherms in these scans suggest two different sizes of ZnST crystals and help account for the ZnST not observed by the data in Table II. At this time, the areas under the different melting peaks have not been calculated, so neither the relative amounts of the different crystals, nor whether all the ZnST is present in the crystalline form is known.

As stated earlier, the mechanical properties of these materials exhibit a strong time dependency. This has been heretofore believed to be due to the time necessary to develop the ionic associations, or clusters. Further, it is believed that the improvement of the mechanical properties of these materials is related to the presence of microscopic ZnST crystallites. It is suggested here that the time-dependent mechanical properties may in fact be a result of the time needed for the ZnST to crystallize and anneal. Thus, we propose that the crosslink structure in these polymers changes from physically associated pendent ionic groups in the case of the neat S-EPDM to phase separated ZnST crystals interacting with the polymer sulfonate groups through coulombic interactions. Data will be presented at the meeting which at least qualitatively support the hypothesis.

We suggest, therefore, that the thermal behavior of these polymers should also exhibit time-dependent effects. This is confirmed by figures 4-6 which are heating thermograms of samples containing 9.1, 16.7, and 33.3% ZnST, respectively, as a function of aging time at room temeprature ($\sim$298K) after cooling from 420 at 40K/min. The crystallinity of the samples increases with increasing annealing time and the low temperature melting peak moves to higher temperatures indicating a more perfect or larger crystal. The major physical aging effect appears to occur in the low temperature endotherm which results from the large supercooling crystallization.

CONCLUSION

While any conclusions based on these initial results must be considered speculative, these data suggest that the time-dependent mechanical properties may be due to the time-dependent morphological changes due to crystallization. In particular, the low temperature melting endotherm which corresponds to small crystals appears to be most affected by physical aging. We are currently running the appropriate mechanical properties-physical aging experiments, and we are determining the relative amounts of crystalline material in a more quantitative manner. These results will be presented at ICTA.

REFERENCES

1. L. Holliday, Ed., "Ionic Polymers", Applied Science Publishers, London, (1975).
2. A. Eisenberg and M. King, "Ion-Containing Polymers", Academic Press, New York, (1977).
3. A. Eisenberg, Ed. "Ions in Polymers", Advances in Chemistry Series, 187, Amer. Chem. Soc., Washington, D. C. (1980).
4. N. H. Canter, U. S. Patent 3,642,728, 1972.
5. C. P. O'Farrell and G. E. Sernick, U. S. Patent 3,836,511, 1974.
6. H. S. Makowski, R. D. Lundberg, L. Westerman and J. Bock, in "Ions in Polymers", see reference 3 above, p. 3.
7. R. D. Lundberg, H. S. Makowski and L. Westerman, in "Ions in Polymers", see reference 3 above, p. 67.
8. H. S. Makowski and R. D. Lundberg, in "Ions in Polymers", see reference 3 above, p. 37.
9. H. S. Makowski, P. K. Agarwal, R. A. Weiss and R. D. Lundberg, Polym. Preprints, 20(2), 281 (1979).
10. Personal observation of the author and from conversations with Dr. I. Duvdevani, Exxon Research and Engineering Co.

TABLE I. EFFECT OF ZnST ON T_g

wt % ZnST	T_g(K)
0	218
1.0	217
2.9	217
4.8	216
9.1	218
13.0	220
16.7	221
23.1	223
28.6	223
33.3	224
37.5	225

TABLE II. CRYSTALLIZATION DATA FROM 420-300K, COOLING AT 20K/MIN.

Sample wt % ZnST	T_c (K)	ΔH_c Cal/g	%xyl ZnST*
100	367/360	35.8	100
37.5	356	12.8	36
33.3	352	11.1	31
28.6	344	8.3	23
23.1	334	5.5	15
16.7	307	0.4	1

* based on ΔH_c of 100% ZnST

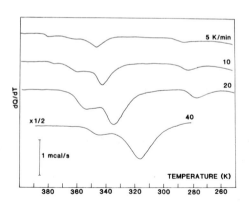

Figure 1. Cooling thermograms at various rates for S-EPDM containing 33.3% ZnST.

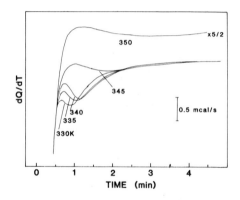

Figure 2. Isothermal crystallization at various temperatures for S-EPDM containing 23.1% ZnST.

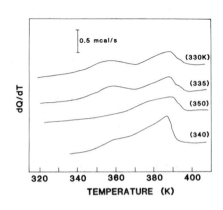

Figure 3. Heating thermograms at 20K/min for S-EPDM containing 23.1% ZnST after isothermal crystallization for 10 minutes at noted temperatures.

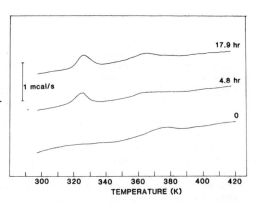

Figure 4. Heating thermograms of S-EPDM containing 9.1% ZnST after annealing at room temperature for various times.

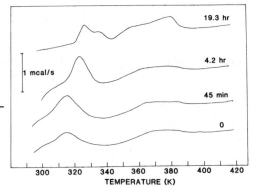

Figure 5. Heating thermograms of S-EPDM containing 16.7% ZnST after annealing at room temperature for various times.

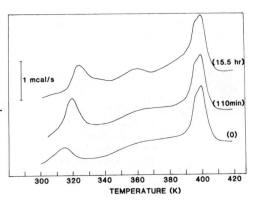

Figure 6. Heating thermograms of S-EPDM containing 33.3% ZnST after annealing at room temperatures for various times.

MULTIPLE MELTING IN NYLON 1010
by
Fu Shu-ren, The Chem. Research Institute of Kwangchow, Academia Sinica, WuShan Canton, China
Chen Tao-yung, Department of Chemistry, The Chinese University of Hong Kong, Hong Kong

Introduction

Semicrystalline polymers, obtained by slow cooling from their melts, or rapid cooling from their melts, but annealled at definite temperatures, show double melting peaks in their DSC curves. There are two explanations for this phenomena. Bell and Murayama[1] attributed the high temperature peak to the melting of folded chain crystals and the low temperature one to the melting of extended chain crystals. Roberts[2], Holdsworth and Tunner-Jones[3] and Ikada[4], however, had renounced the above viewpoint, and chaimed that the low temperature peak is due to the melting of the crystallites originally formed when the material is cooled from its melt, while the high temperature peak is due to the melting of the crystallites formed from recrystallization of the material melted at the low temperatures, during the scanning process. These two controversial viewpoints existed concurrently for many years and became the main subject for investigation in this field.

Double melting behavior on several semi-crystalline polymers such as Nylon 66[1], PET [2)(3)(5)(6)isopolystyrene[7], polyethylene[8] and polypropylene[9)(10)(11)has been extensively studied in the past ten years. To the author's knowledge, no investigation on Nylon 1010 has been done. This paper is by means of studing the multiple melting behavior of Nylon 1010 to demonstrate stuctural reorganization point of view of Roberts and Holdsworth.

Experimental

Starting material was Nylon 1010 bristles. It was heated to $230^{\circ}C$ in a Shimadzu DSC-20 instrument, kept at this temperature for 10 minutes and then allowed to crystallize at a cooling rate of $5^{\circ}C/min$. This erases the thermal history of starting material. Weight of sample used was 8 mg. All the experiments were carried out under a N_2 atmosphere.

Melting behavior of samples was examined using a scanning rate of $10^{\circ}C/min$, unless otherwise indicated. (Double peaks were resolved in a manner as shown in Fig 1.) Peak area was determined by a cut-and-weigh method.

Results and Discussion

(1) Melting behavior after partial scan

Nylon 1010 sample was heated to melting in the DSC instrument and then allow to crystallize at a cooling rate of $5^{\circ}C/min$. The sample was, then, scanned at a heating rate of $10^{\circ}C/min$. The result is shown in Fig 2(a). Two melting endothermic peaks (I and II) were observed. The sample was then allowed to cool and rescanned to a temperature just beyond that of peak II as shown in Fig 2(b). The sample was taken out and quenched into ice-water. Upon rescanning of the quenched sample, only one peak (H) was observed as shown in Fig 2(c).The maximum temperature of the peak

H is the same as that of peak I, while the area of the peak H equals the sum of the areas of peaks I and II, as shown in Table I. This implies that crystallites (HM, IM) symbolized by peaks H and I are much the same in degree of perfection. Crystallites (IIM) symbolized by peak II become very imperfect or amorphous after melting and quenching; they continuously melt and recrystallize to high melting crystals during the rescanning process as suggested by Holdsworth and Turner-Jones[3]. Therefore. peak II did not appear and the amount of high melting crystals increased. On the contrary, if peaks I and II were originated from the melting of two morphologically different crystallites, one would expect the areas of peak I and H be equal[7].

(2) Melting behavior after partial scan and isothermal annealing

The same sample used in Experiment I was scanned to $192^{\circ}C$ (Just beyond peak II) and then kept isothermally at this temperature for 3.5 minutes. A broad exothermic peak (Rec) was observed. Continued scan showed the final endothermic peak H. The whole process is shown in Fig 3. The temperatures and areas of the peaks are shown in Table II. It is seen that the area of peak Rec is approximately equal to that of peak II. This again proves that crystallites symbolized by peak II converted to crystallites symbolized by peak H; this conversion took place by melting and recrystallization[2]. If peaks I and II were originated from the melting of two morphologically different crystallites, one would expect no exothermic peak and the areas of peaks I and H be equal.

(3) Effect of heating rate on melting behavior

The same sample was scanned at various heating rates. Results are shown in Fig 4 and Table 3. It is seen that as heating rate increased, both the temperature and area of peak I decreased while those of peak II increased; the sum of areas of peaks I and II, however, remained approximately constant.

These results may also be explained on the basis of melting and recrystallization during scanning process. High heating rate supressed recrystallization, decreased the area of peak I; low heating rate facilitated recrystallization, increased the area of peak I[8]. As heating rate increase, recrystallization supressed, degree of perfection of crystallites IM decreased, so, the peak temperature of peak I decreased. The peak temperature of peak II remained fairly constant at lower heating rates because it is the melting temperature of the original crystallites. However, it increased a few degrees at higher heating. This could be attributed to thermal lag[8]. If peaks I and II were originated from two morphologically different crystallites, then heating rate should not affect their relative positions.

One may also observe that at lower heating rates, the valley between the two endothermic peaks I and II lies below the dotted line and appears as an exothermic peak as shown in Curve C of Fig 4. This exothermic peak is a proof of recrystallization[7].

(4) Effect of Cooling rate on melting behavior

In order to understand more clearly the nature of recrystallization process during DSC scanning, the effect of cooling rate on melting behavior was studied, since cooling rate determines the condition of crystallization from the melt. Nylon 1010 samples were melted completely in DSC and allowed to crystallize at different cooling rates. Different samples, thus, obtained were scanned all using a heating rate of $10°C/min$. Results are shown in Fig 5. It is seen that both the temperature and area of peak II decreased with increased cooling rate, where-as the area of peak I increased, while its temperature was essentially uneffected. The quenched sample showed only one endothermic peak. All the samples obrained with a cooling rate from $1°C/min.$ to $40°C/min.$ show exothermic peaks between peaks I and II. The ratio of the area of the exothermic peak to that of peak II above the dotted line increased with increased cooling rate. If this ratio were taken as an expression for the driving force of recrystallization, it is apparent that the driving force increased with increased cooling rate. This is understandable because at higher cooling rate the crystallites formed are less perfect; their melting point is lower than that of more perfect crystallites. The melt could be visualized as a supercooled liquid of perfect crystallites, and has a tendency to recrystallize. The higher the cooling rate, the higher is the degree of supercooling and the greater is the amount of recrystallization[12]. It is also apparent that crystallites formed at a higher cooling rate has lower melting point and also smaller heat of fusion. This explains the decreasing of area and temperature of peak II with increased cooling rate. Thus, in the range of cooling rate between $1°C/min.$ to $40°C/min.$ scanning reveals the process of melting and recrystallization. At very high cooling rate, the crystallites formed are very imperfect or amorphous. Under the same scan rate, the process of melting and recrystallization become burried in the overall melting curve[5]. As a result, only the final endothermic peak is observed, which characterizes the melting of those crystallites formed by continuous melting and recrystallization. At a very slow cooling rate, crystallites formed from the melt are highly perfect. Their melting point become close to that of the crystallites formed by recrystallization. Therefore peak II and peak I almost coincided and no exothermic peak, characteristic of recrystallization, could be seen.

(5) Effect of annealing temperature on melting behavior

Several samples were prepared by cooling their melts at a rate of $5°C/min.$, anmealed at respective different temperatures for 1.5 hr. and then quenched into ice-water. These samples were rescanned with the same heating rate of $10°C/min.$ Results are shown in Fig 6. A small endothermic peak at a temperature higher than its annealing temperature was observed on each melting curve. The peak temperature was higher and its area was larger for samples annealed at higher temperatures.

When the annealing temperature was close to that of peak II, the temperature, height and area of peak II increased while the area of peak I decreased. When the sample was annealed at the peak II temperature, the temperature, height and area of peak II increased further and the area of peak I decreased further. The two peaks I and II practically coincided, and the former was barely visible. When the sample was annealed at a temperature between peak II and peak I, only a single sharp peak was observed.

The above observations may be explained as following. When the sample is annealed at a certain temperature, only recrystallization process takes place (3)(9). Crystallites, whose degree of perfection corresponding to the annealing temperature, is stabilized. These crystallites undergo melting and recrystallization in subsequent scanning, produce crystallites of higher perfection. Therefore a melting endothermic peak at a temperature higher than its annealing temperature will be seen(9). As the perfection of stabilized crystallites increases with annealing temperature, so is the size of the endothermic peak(13). When the annealing temperature is close to that of peak II, IIM crystallites become more perfect than those before annealing; therefore, their melting point and heat of fusion increases. Since IIM crystallites have become more perfect, its melt, as a supercooled liquid of crystallites IM, has less degree of supercooling, i.e. the driving force for recrystallization to crystallites IM becomes smaller. Less crystallites IM are formed upon rescanning and the area of peak I decreases. When the sample is annealed at the peak II temperature, IIM crystallites acquire further perfection, thus, have very samll tendency to recrystallize. So, both area and temperature of peak II increase further while peak I becomes barely visible. When the sample is annealed at a temperature between that of peak II and peak I, then quenched into ice-water, there will not be IIM crystallites in the quenched sample because crystals IIM become very imperfect or amorphous. Upon rescanning, these very imperfect ctystallites or amorphous material undergo the process of partial melting and recrystallization. Finally, it forms more perfect high melting crystallites which are much the same as IM crystallites in degree of perfection. Therefore, the melting curve shows only a single sharp peak H, which has the same peak temperature as that of peak I.

(6) Multiple melting behavior after annaling treatment at successively lower temperatures

A Nylom 1010 sample was heated to melt. It was then cooled slowly at a rate of $5^{\circ}C/min.$ to a temperature below that of peak II and annealed at that temperature for 1.5 hr. The sample was taken out and quenched into ice-water. The quenched sample was reheated to a second temperature below that of the first annealing temperature and annealed for another 1.5 hrs. and again quenched imto ice-water. It was again reheated to a third temperature below that of the second annealing temperature and annealed for another 1.5 hrs. and quenched into ice-water. The final sample was scanned at heating rate of $10^{\circ}C/min.$ Three additional peaks III IV,

V, were observed as shown in Fig 7. The temperatures of these peaks are all a few degrees higher than their respective annealing temperatures and their heights and areas increased with increased annealing temperature. This result may be interpreted as follows.

Nylon 1010 crystallized from the melt contains an assembly of crystallites with different degrees of perfection, depends strongly on the previous thermal history. Crystallites of different degrees of perfection, of course, have different melting points. The higher the degree of perfection, the higher is the melting point. When a sample is annealed at a certain temperature, crystallites whose degree of perfection corresponds to that temperature is stabilized. Upon scanning, the melting curve will show an endothermic peak due to the melting of the crystallites stabilized at the annealing temperature. Since, during scanning, the crystallites melt partially and recrystallized into more perfect crystallites, therefore the melting endothermic peak occurs at a temperature somewhat higher than the annealing temperature. If annealing treatment were carried out at several temperatures from higher to lower, successively, several types of crystallites with different degrees of perfection will be stabilized. Upon scanning they will melt at temperatures somewhat higher than their respective stabilizing temperatures. Therefore several endothermic peaks are observed. The fact that the areas of these additional peaks decreased with decreased annealing temperature could be attributed to the same reason as mentioned in previous paraph. However, another reason may also play a part in this situation. When the sample was annealed at the first time, there were large amount of imperfect crystallites that could be stabilized into crystallites corresponding to the annealing temperature. In the second annealing, the amount of imperfect crystallites were less than that in the first annealing because some crystallites had already be stabilized. Therefore the amount of crystallites stabilized in the second annealing were also less than that stabilized in the first annealing, and the heat of fusion of the former should be less than that of the latter. Heat of fusion of the crystallites stabilized in the third annealing should be less further.

Conclusion

It may be concluded from above experimental results that the DSC peaks of Nylon 1010 do not reflect directly its actual state at room temperature before scanning. The primary and secondary crystallites in the sample undergo a series of continuous partial melting and recrystallization in the scanning process. Peak II may be attributed to the melting of primary crystallites formed from the melt during slow cooling. Peak I may be attributed to the melting of crystallites recrystallized more perfectly after the partial melting of primary crystallites. Peaks III, IV and V may be attributed to the melting of secondary crystallites stabilized at their respective annealing temperatures.

References:

1. J.P. Bell and T. Murayama, J. Polym. Sci. (A-2) 1969, 7, 1059.
2. R.C. Roberts, J. Polym. Sci. (B), 1970, 8, 381.
3. P.J. Holdsworth and jurner-Jones, Polymer, 1971, 12, 195
4. M. Ikeda, Kobunshi Kagaku, 1968, 25, 87.
5. H.J. Berndt and Adelgund Bossmann, Polymer, 1976, 17, 241.
6. S. Fakirv et all, Polymer, 1977, 18, 1121.
7. P.J. Lemstra et all, J. Polym. Sci. (A-2), 1972, 10, 823.
8. D.P. Pope, H.H. Wills, J. Polym. Sci. Phys. Ed., 1976, 14, 811.
9. W. Ken Busfield and chris S. Blake, Polymer, 1980, 12, 35.
10. M. Lengyel et all, Acta Chimica Academiae Scientiarium Hungaricae, Tomus, 1977, 94(4), 309.
11. Xi Xue-ying and Fu Shu-ren, (Chemistry Letter), The Chem. Research Institute of Kwangchow, Academia Sinica, China, 1981, 1, 11.
12. G.E. Sweet and J.P. Bell, J. Polym. Sci. (A-2), 1972, 10, 1273.
13. Zachmann, H.G. and Stuart, H.A. Maktomol. Chem. 1960, 41, 131.

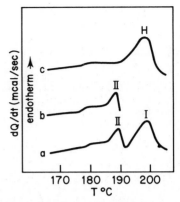

Fig 2. Effect of partial scanning on melting behavior. (a) continuous scan; (b) sample scanned just beyond the peak II; (c) rescan of sample b after quenched to ice-water.

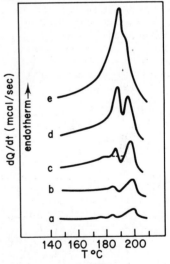

Fig 4. Effect of heating rate on melting behavior. (a) 2.5°C/min; (b) 5°C/min; (c) 10°C/min; (d) 20°C/min (e) 40°C/min (The area under the curves are not directly comparable since different chart speeds were used)

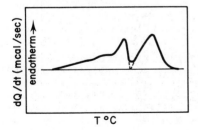

Fig 1. Method of endothermic peak resolution

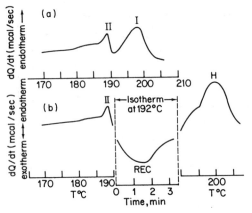

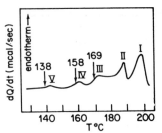

(g) 40°C/min; (h) The molten sample was quickly cooled to room temperature; (i) The molten sample was quenched to ice-water.

Fig 3. Effect of partial scanning then keeping isothermally on melting behavior. (a) continuously scanned at 10°C/min to 210°C; (b) scanned at 10°C/min to 192°C, kept 3.5 min at 192°C, scanning continued at 10°C/min to 210°C

Fig 7. Effect of annealing a sample at succesively lower temperatures, indicated by arrows

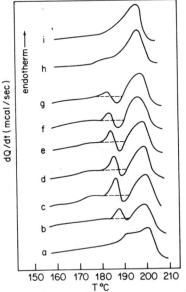

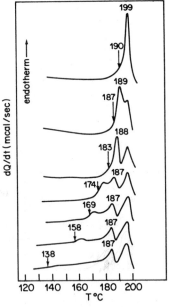

Fig 5. Effect of cooling rate on melting behavior. the sample was rescanned at 10°C/min after each different cooling rate. (a) 0.5; (b) 1; (c) 2.5; (d) 5; (e) 10; (f) 20;

Fig 6. Effect of annealing temperature on melting behavior. The sample was scanned to various annealing temperature (indicated by allows) and kept 1.5hr and quenched to ice-water and then rescanned at 10°C/min

Table 1

scanned mode	continuous scan		rescanning after partial scan
peak area (cm^2)	A_{II}	A_I	A_H
	3.00	2.51	5.40
peak temperature (°C)	T_{IIM}	T_{IM}	T_{HM}
	187	199	199

Table 2

scanned mode	continuous scan		keeping isothermally	continuing scan after keeping isothermally
peak area (cm^2)	A_I	A_I	A_{REC}	A_H
	3.00	2.51	3.06	5.26
peak temperature (°C)	T_{IIM}	T_{IM}	T_{REC}	T_{HM}
	187	199	192	199

Table 3

heating rate (°C/min)	peak II		peak I		peak II + peak I
	T_{IIM}(°C)	A_{II}(cm^2)	T_{IM}(°C)	A_I(cm^2)	$A_{II}+A_I$(cm^2)
2.5	186	—	201	—	—
5	186	2.10	200	3.03	5.13
10	187	3.00	199	2.51	5.51
20	189	3.57	197	1.97	5.54
40	191	5.23	195	0.33	5.56

A STUDY OF ISOTHERMAL CRYSTALLIZATION AND LAMELLAR THICKENING IN POLY(ETHYLENE-TEREPHTHALATE) USING DIFFERENTIAL SCANNING CALORIMETRY.

Y.S. Yadav, P.C. Jain and V.S. Nanda
Department of Physics and Astrophysics,
University of Delhi, Delhi-110007,
India.

A systematic study of isothermal crystallization from melt in poly(ethylene-terephthalate) has been carried out using DSC. Phenomenon of double melting endotherms has been observed. At $T_c <$ 500K, this phenomenon is attributed to recrystallization during melting. At $T_c > 510K$, the lower temperature peak is due to incomplete crystallization. Results show that the process of isothermal lamellar thickening involves simultaneously the molecular fractionation as well as annealing.

INTRODUCTION

Poly(ethylene-terephthalate) (PET) samples crystallized isothermally from melt often show double melting endotherms. Appearance of such endotherms is intimately connected with the crystallization conditions and the heating rate employed in DTA or DSC. Conflicting theories about the origin of this phenonomenon have been put forward (1-6). It has been attributed to different crystal morphologies as well as partial melting and recrystallization. However, no clear picture has yet emerged. Although extensive work on isothermal crystallization of PET has been carried out but most of it is confined to crystallization kinetics. The isothermal lamellar thickening and molecular fractionation during crystallization have not been investigated. To provide a better understanding of the isothermal crystallization process and the origin of multiple melting endotherms, a systematic study of this process from melt has been carried out. The present paper deals with the results of this investigation.

EXPERIMENTAL DETAILS

In the present study a commercial grade sample of PET has been employed. It had an average molecular weight (M_η) of 37,200.

In every case 5-7 mg of the sample was first heated upto 550K in the sample can of Perkin-Elmer DSC model DSC1B and kept at this temperature for about 30 min. This procedure was adopted to wipe out any previous morphological history of the sample. The sample was then rapidly cooled to the crystallization temperature (T_c) and kept at this temperature for a desired length of time. The crystallization was terminated by cooling the sample rapidly (> 32K/min.) to the room temperature. The DSC curves for melting of the isothermally crystallized samples were recorded at a scan rate of 8K/min. However, to avoid any degradation of the polymer, crystallization or melting was always carried out in dry nitrogen atmosphere. To investigate the effect of heating rate, samples crystallized under identical conditions were heated at different rates.

RESULTS AND DISCUSSION

Some typical DSC melting curves of PET samples crystallized isothermally under different conditions are shown in Figures 1 and 2. A general feature of these is the appearance of two melting peaks. The results obtained can be described in two sets, one for crystallization temperatures such that 480K < T_c < 500K and another for T_c 500K.

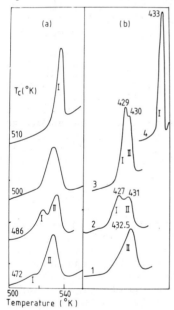

Fig. 1. DSC melting:(a) Effect of T_c and (b) effect of heating rate, curves 1,2,3 and 4 are for heating rates 4,8,16 and 32K/min.

At T_c < 500K, the rate of crystallization is so rapid that no partial crystallization studies are possible. Therefore, at these temperatures samples were crystallized for long enough times such that by keeping them for longer times at these temperatures no further increase in the extent of crystallization was observed. The set of curves shown in Figure 1 represent the melting behaviour of such samples crystallized at different temperatures. In the increasing order of peak temperatures, the two observed melting peaks have been labelled as I and II. At a constant heating rate, the area under peak II is found to decrease with increasing T_C (Fig. 1a). Further, for samples crystallized at a given temperature, on increasing the heating rate the area under peak I is found to increase at the expense of area under the peak II. This effect is demonstrated by the DSC curves shown in Figure 1(b). From this result it can be concluded that at $T_C < 500K$, the peak I is due to the melting of the crystals formed during isothermal crystallization while peak II is a consequence of re-crystallization or reorganization during heating. With increasing T_C the thermal stability of the crystals formed increases resulting in a decrease in the probability for recrystallization to take place. Therefore, at a constant heating rate the area under peak II is expected to decrease with increasing T_C which is in agreement with our observations. At $T_C \cong 500K$, only one peak is observed indicating formation of thermally stable crystals which do not undergo any crystallization during heating.

Figure 2 shows some typical DSC melting curves of samples crystallized partially at $T_C > 500K$. The two sets of curves (Figs. 2c and 2d) shown in this figure are for crystallization temperatures 520K and 525K, respectively. The different curves in each set are of samples crystallized at a given T_C for different times. As in Fig. 1, here also the two peaks have been labelled as I and II but in a different manner, peak I usually occurs at a higher temperature than the peak II. The onset time of peak I is found to increase with T_C. On heating at a constant rate the samples crystallized at a given T_C but for different times, the area under peak I increases with increasing crystallization time at the expense of the area under peak II.

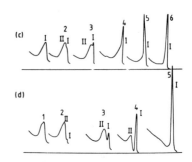

Fig. 2. DSC melting curves showing the effect of crystallization time: (c) $T_C = 520K$, curves 1-6 are for 20, 30, 50, 100, 180 and 600 min. (d) $T_C = 525K$, curves 1-5 are for 1, 60, 120, 240 and 900 min.

On varying heating rate of the samples crystallized under identical conditions, the relative areas under the two peaks do not show any variation. However, with heating rate the two peak temperatures show a variation. With increasing heating rate peak II temperature decreases while the peak I temperature increases showing the normal phenomenon of super heating. Based on these observations it can be concluded that for $T_c > 500K$, the double melting endotherm phenomenon is not due any recrystallization during heating but the two peaks represent melting of two different fractions. Peak I is due to the melting of the isothermally crystallized fraction while peak II is due to the melting of the fraction which remained uncrystallized at the termination of isothermal crystallization but got crystallized on rapid cooling. Since the later fraction crystallizes at a lower temperature, it undergoes recrystallization during heating. Therefore, at lower temperatures ($T_c < 515K$) the resolution between the two peaks is poor. With increasing T_c, the peak I moves to higher temperatures and the separation between the two peaks increases.

The time dependence of the two peak temperatures at different T_c is shown in Figure 3. It is observed that, at a given T_c, with increasing crystallization time the peak I temperature increases while that corresponding to peak II decreases. Similar behaviour has been observed in polyethylene (7-8). The temperature dependence of peak I can be understood interms of isothermal lamellar thickening whereas that of peak II is a consequence of molecular fractionation. Unlike polyethylene, the time dependence of peak I temperature in PET is suggestive of thickening of lamellae predominantly due to annealing of the initially formed crystals. In PET the initially formed crystallites are small which grow bigger in size on keeping at the crystallization temperature. This view is supported by the results of WAXD(8). At any given T_c, with increasing time more and more material gets crystallized and therefore, in a polydisperse sample the average molecular weight of the fraction which remains uncrystallized would decrease. Since the melting temperature is a

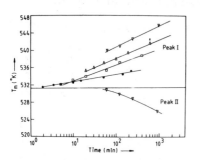

Fig. 3. Time dependence of melting temperature, the points ●, □, △ and ▽ are for T_c = 512, 515, 520 and 525K.

function of average molecular weight, the peak II temperature is expected to decrease with increasing time.

On the basis of present study it can be concluded that with increasing T_c, the PET crystals formed are thermally more stable. The isothermal lamellar thickening in PET is dominantly due to annealing of the initially formed crystals. There is some evidence for molecular fractionation.

REFERENCES

1. J.P. Bell and T. Mureyama, J. Polym. Sci., A-2, 1969, 7, 1059.
2. P.C. Roberts, Polymer, 1969, 10, 117.
3. R.C. Roberts, J. Polym. Sci., Pt.-B, 1970, 8, 381.
4. D.L. Nealy, T.G. Davis and C.J. Kibler, J. Polym. Sci. A-2, 1970, 8, 2141.
5. P.J. Holdsworth and A. Turner-Jones, Polymer, 1971, 12, 195.
6. G.R. Sweet and J.P. Bell, J. Polym. Sci. A-2, 1972, 10, 1273.
7. J. Dlugosz, G.V. Fraser, D. Grubb, A Keller, J.A. Odell and P.I. Goggin, Polymer, 1976, 17, 471.
8. Y.S. Yadav, Ph.D. Thesis (University of Delhi, 1981).

CRYSTALLINITY AS A SELECTION CRITERION FOR ENGINEERING PROPERTIES OF HIGH DENSITY POLYETHYLENE

D. Mark Hoffman
University of California
P.O. 808
Livermore, Ca 94550

ABSTRACT

As the result of a design change by Lawrence Livermore National Laboratory (LLNL) engineers, the polymer group was asked to select and characterize a high density polyethylene (HDPE) on the basis of four design requirements. These were: (1) high purity, (2) high density, (3) low thermal expansion, and (4) good mechanical properties. Since polyethylene is semicrystalline, the degree of crystallinity of the polymer should be directly proportional to the density and mechanical properties and inversely proportional to thermal expansion. The degree of crystallinity and crystallization kinetics are easily measured by differential scanning calorimetry and these results were used to characterize about 25 commercial high density polyethylenes. The degree of crystallinity and melting point of these polyethylenes were used to select seven polymers for further study. The overall crystallization kinetics of these seven polyethylenes with ~75% crystallinity were analyzed and compared with theoretical predictions based on the Avrami equation and results from the polarizing microscope.

---

*Work performed under the auspices of the U.S. Department of Energy by Lawrence Livermore National Laboratory under contract No. W-7405-Eng-48.

THE EFFECT OF CARRIER GAS ON RATES OF CRYSTAL-
LIZATION OF ISOTACTIC POLYPROPYLENE OBTAINED
BY DIFFERENTIAL SCANNING CALORIMETRY

by

Jocelyn M. Willis, G. R. Brown and
L. E. St-Pierre
Department of Chemistry
McGill University
Montreal, Quebec H3A 2K6

INTRODUCTION

Numerous comparative studies have been made of the utility of the use of the differential scanning calorimeter (DSC) as compared to more conventional techniques, e.g. dilatometry, for the study of the kinetics of overall (bulk) crystallization of polymers. One such study [1], for example, suggests that differential scanning calorimetry is not as sensitive a technique as dilatometry, particularly for slow crystallization rates. Such kinetic studies as the present that use the DSC in the isothermal mode are generally deemed more precise than dynamic, i.e., temperature programmed DSC studies because the baseline can be determined with less ambiguity. However, there now exist many improved methods of determining the baseline for dynamic DSC scans [2-6].

The temperature calibration of the DSC is of paramount importance in kinetic studies, especially when rate constants are derived. Previous studies using organic polymers and various standards indicate that there are thermal lags in the sample which become more pronounced as the heating rates are increased and are dependent on the sample size [4,7-11]. Explanations for this thermal lag have been offered and corrections by means of calculations and improved calorimeter designs have been proposed [12-15].

The present study reports the effect of the carrier gas on kinetic parameters for the crystallization of isotactic polypropylene as determined with a Perkin-Elmer DSC in the isothermal mode. An apparent increase in the rate of crystallization was obtained when the carrier gas was changed from nitrogen to helium.

EXPERIMENTAL

Materials: The isotactic polypropylene (iPP), supplied by Aldrich Chemicals, was used as received and had an equilibrium melting temperature of 440 K as determined by DSC. The carrier gases, nitrogen and helium, were prepurified specialty gases supplied by Linde.

Procedure: Studies of the kinetics of crystallization were made using a Perkin-Elmer DSC, either model 1B or 2C, in the isothermal mode. The DSC was calibrated for temperature, to within ± 0.2 K, prior to each series of scans using benzoic acid (reagent, ACS, Matheson, Coleman and Bell, m.p. 395 K), naphthalene (certified thermometric standard, Fisher Scientific, m.p. 353 K) and tin (Baker analyzed reagent, m.p. 505 K). The melting point was taken to be the first inflection from the baseline. The DSC-1B was also calibrated differentially, using benzoic acid, prior to each series of scans. New calibrations were made whenever the carrier gas was changed.

Accurately weighed samples of iPP, weighing ~10 mg, were sealed into standard Al pans, supplied by Perkin-Elmer, with Al lids. References were prepared by compressing finely divided silica (Aerosil Ox50 by Degussa, Germany) into Al pans so as to contain the same mass as the polymer samples.

The flow rate of carrier gas was measured using a soap bubble flow meter attached to the gas outlet of the DSC. Each DSC scan was comprised of five steps using the same polymer sample for each series. These steps are:

i. The gas flow rate was set to the desired value, generally 0.30 ± 0.02 ml/s.
ii. The polymer was melted at 460 K for 10 min.
iii. The DSC was manually set to the desired crystallization temperature. At the same time the recorder was started.
iv. The zero time of the crystallization was set when the average temperature pilot light was illuminated. (That the illumination of this pilot light corresponded to the equilibration to the temperature set on the DSC was verified in separate experiments by monitoring the EMF readings on a thermocouple placed within the sample pan inside the calorimeter.)
v. The crystallization of the polymer was recorded as the change in signal with time.

Data Analysis: All the DSC thermograms obtained for the crystallization of iPP were of the type illustrated in Fig. 1. The weight fraction of crystallized material at any time can be evaluated by taking the ratio of the total heat evolved up to that time to the total heat released as represented by the equation

$$X_t = \int_0^t (dH/dt)\,dt / \int_0^\infty (dH/dt)\,dt$$

Evaluation of the integrals was done by the "cut and weigh" method.

The definition of the correct baseline is of critical importance to the determination of valid areas. With the DSC-1B the baselines were obtained at each crystallization temperature by scanning a sample pan containing silica versus a reference pan also containing an identical weight of silica following the five steps outlined above. With the DSC-2C baselines were obtained by scanning the sample pan containing iPP versus the reference pan containing silica, following the five steps outlined above, at 410 K, i.e., at a temperature where the iPP does not undergo crystallization.

RESULTS AND DISCUSSION

Crystallization data were obtained at temperatures ranging from 378 to 400 K under nitrogen and under helium using the DSC-2C. The plot of $(1 - X_t)$ as a function of time represents the decrease in the weight fraction of uncrystallized iPP with time. Figure 2 illustrates the variation in rate with temperature and exemplifies the effect of the carrier gas. At a given temperature the difference in rate due to the change in carrier gas was largest at lower crystallization temperatures, i.e., the largest undercooling. Figure 3 is a typical Avrami plot of the data and illustrates the reproducibility obtained with the DSC-1B. Similar excellent reproducibility was also obtained with the DSC-2C.

The nonlinearity of the Avrami plots was generally most noticeable at the longer crystallization times and was very sensitive to the choice of baseline. In general a more pronounced curvature occurred when the baseline was chosen so as to include contributions from secondary crystallization occurring after the primary crystallization has neared completion because of impingement of spherulites.

The average value of the Avrami exponent n for the crystallization process under nitrogen was 3.08 ± 0.09 and that for helium was 3.0 ± 0.3, in excellent agreement with values obtained previously [16]. These Avrami exponents were obtained by a linear regression of the central, linear parts of the plot. Since it is expected that the same crystallization process occurs irrespective of the carrier gas it is not surprising that the value of n is not substantially different for the two gases.

To compare the rates of crystallization under each gas rate constants were calculated using an average Avrami exponent of 3.0 in the equation

$$z = (\ln 2)/t_{\frac{1}{2}}^n$$

where $t_{\frac{1}{2}}$ is the time required for crystallization to reach 50% completion. Figure 4 illustrates the dependence of the rate constants on the temperature for both gases using data obtained with the DSC-2C.

The change in the rate of crystallization that resulted from a change in the carrier gas was quite unexpected. Since the Avrami n is unaffected by the change in the carrier gas it seems unlikely that this change in rate arises from a change in the nature of the crystallization process, especially since the iPP samples were contained in sealed DSC pans.

As a further test experiments were made using a sample pan that had holes pierced into the lid. DSC scans were obtained with samles under nitrogen and helium and the results were compared with those for completely closed sample pans. This would indicate whether there were any interactions between the iPP and the carrier gas during crystallization. The times corresponding to the maximum peak height, t_{max}, were measured and indicated a negligible between the open and closed sample pans, the largest being 2%, well within experimental error. This is further evidence that the carrier gases do no effect the kinetics of the crystallization directly.

To verify this conclusion the rates of spherulite growth were measured directly at 395 K by photomicroscopy. The radial growth rates obtained were 0.107 ± 0.005 and 0.106 ± 0.003 cm/min under nitrogen and under helium, respectively. Apparently the gas has no effect on the rate of spherulite growth.

Since the effect of the carrier gas on the rate of crystallization measured with the DSC is, apparently, due to some temperature effect of the instrument experiments were made to obtain a better understanding of this behavior. First of all, the sensitivity of the DSC was evaluated by comparing the areas of the melting endotherms of tin for each gas at a flow rate of 0.30 ± 0.01 ml/s.

TABLE 1

THE EFFECT OF CARRIER GAS ON PEAK AREAS FOR THE
CRYSTALLIZATION OF ISOTACTIC POLYPROPYLENE

Crystallization Temperature (K)	Peak Areas (cm^2) Nitrogen	Helium
396	74 ± 4	67 ± 4
392	70 ± 3	64 ± 3
388	79 ± 3	–
382	68 ± 3	63 ± 3
380	68 ± 2	61 ± 2

Within experimental error the sensitivity was unaffected by the change in carrier gas. These results indicate that, inspite of apparent temperature discrepancies, quantitative or at least identical heat transfer is obtained for both carrier gases.

Measurements of the crystallinity of iPP when crystallization has reached completion gives further evidence for a temperature effect in the DSC. Table 1 shows that the total area under the crystallization curve is greater for iPP samples crystallized in nitrogen than for those crystallized in helium at a given temperature setting on the DSC. This is in keeping with the general observation that as the supercooling decreases the chains have more time to rearrange to a stable conformation before crystallizing. At the same time amorphous polymer is forced out from within the spherulite giving rise to higher crystallinity.

As a further test of the behavior of the DSC the effect of changing the flow rate of the carrier gas was studied. For the crystallization of iPP at 390 K the t_{max} values varied minimally when the flow rate of the gas, either nitrogen or helium, was changed between 0.11 and 0.44 ml/s. Hence, it would seem that convection processes are not the cause of the observed effect of carrier gas on the kinetics of crystallization.

Figure 5 illustrates the effect of sample size on the measured half-time for crystallization of iPP at 384 and 396 K under nitrogen using the DSC-2C. The half-time appears to increase as the sample size decreases indicating a decrease in the rate of crystallization. These results suggest that heat transport within the sample is also of importance although perhaps not as significant as heat transfer from the sample pan to the DSC holder.

CONCLUSION

The rates of crystallization of iPP, as obtained isothermally with the DSC, are clearly dependent on the carrier gas. This effect seems to be related to the thermal conductivity of the gas. It is of interest that the rate constants obtained with the two gases (Fig. 4) can be made to correspond by shifting the nitrogen data up in temperature by ~4 K. This seems to be an unreasonably large temperature discrepancy but it must be kept in mind that, in the temperature range considered in this study, the rate increases exponentially with temperature. It is of interest that a similar shift in crystallization isotherms to higher temperatures has been obtained when iPP film was wrapped in Al foil [17]. A dependence on heat transfer has also been reported in a study of the kinetics of crystallization of polymers by light depolarization [18,19].

It is apparent that considerable care must be exercised in the use of the DSC in the study of the crystallization of polymers. It seems that the carrier gas has an important role in the transfer of heat from the sample to the pan holder, at least in Perkin-Elmer type DSC's.

REFERENCES
1. A. Booth and J.N. Hay, Polymer, 1969, 10, 95.
2. W.P. Brennan, B. Miller and J.C. Whitwell, I & EC Fundamentals, 1969, 8, 314.
3. C.M. Guttman and J.H. Flynn, Anal. Chem.,1973, 45, 408.
4. J.H. Flynn, Thermochem. Acta, 1974, 8, 69.
5. J.P. Dumas, J. Phys. D, 1978, 11, 1.
6. T.C. Ehlert, Thermochem. Acta, 1977, 21, 111.
7. S. Strella and P.F. Erhardt, J. Appl. Poly. Sci. 1969, 13, 1373.
8. M.J. O'Neill, Anal. Chem. 1966, 38, 1331.
9. S. Strella, J. Appl. Poly. Sci. 1963, 7, 569.
10. N. Peppas, J. Appl. Poly. Sci. 1976, 20, 1715.
11. J.M. Barton, Thermochem. Acta, 1977, 20, 249.
12. H.M. Heuvel and K.C.J.B. Lind, Anal. Chem. 1970, 42, 1045.
13. M.J. Richardson and N.G. Savill, Thermochem. Acta, 1975, 12, 213.
14. R.T. Marano, Thermochem. Acta, 1978, 26, 27.
15. S.L. Randzio and S. Sunner, Anal. Chem. 1978, 50, 704.
16. G. Turturro, Ph.D. Thesis, McGill University, Montreal, 1981.
17. Yu.K. Godovskii and G.L. Slonimskii, Poly. Sci. USSR, 1966, 8, 441
18. J.H. Magill, Nature, 1961, 191, 1092.
19. J.H. Magill, Polymer, 1961, 2, 921.

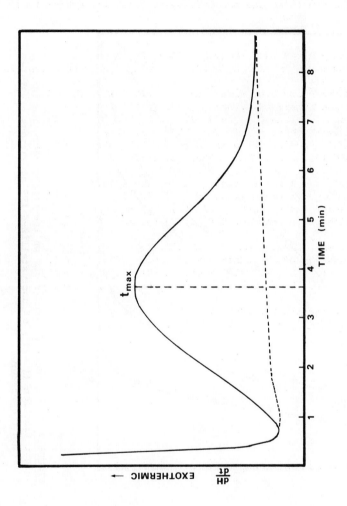

FIGURE 1 A typical DSC thermogram for the isothermal crystallization of iPP

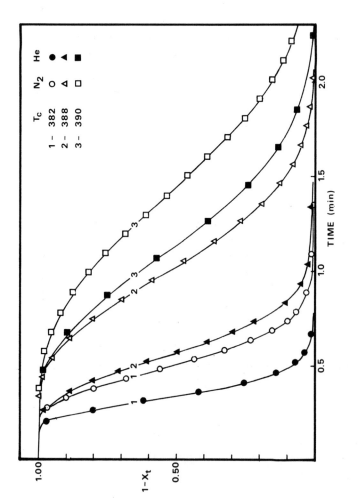

FIGURE 2 The effect of carrier gas on the rate of decrease in weight fraction $(1 - X_t)$ of uncrystallized iPP at various crystallization temperatures, t_{T_c}.

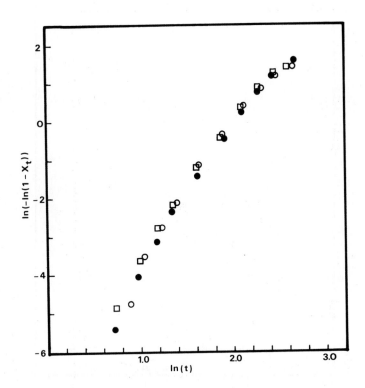

FIGURE 3 A typical Avrami plot of three separate DSC scans of the crystallization of iPP at 395 K under nitrogen.

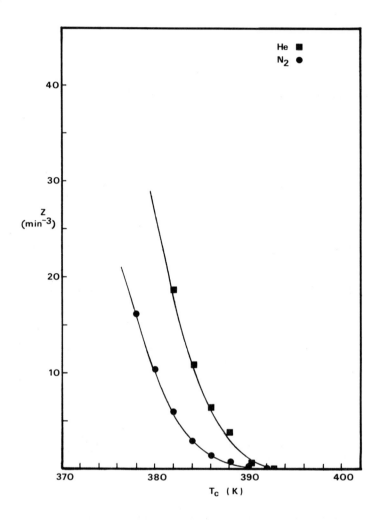

FIGURE 4 The effect of carrier gas on the temperature variation of the rate constants, z, for the crystallization of iPP.

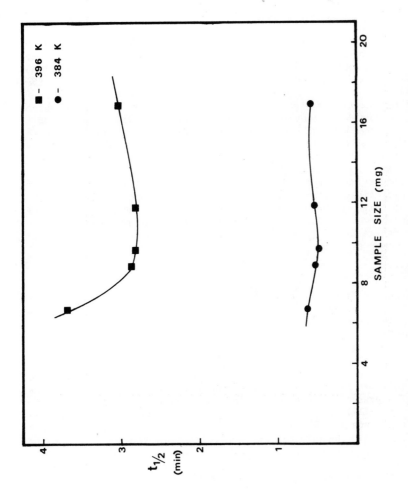

FIGURE 5 The effect of sample size on the half-time of crystallization of iPP under nitrogen.

ANALYSIS OF MICROSTRUCTURE AND DOMAIN FORMATION
IN SULFONATED ETHYLENE-PROPYLENE-ETHYLIDENE
NORBORNENE IONOMERS
J. J. Maurer
Exxon Research and Engineering Company
P. O. Box 45
Linden, New Jesey 07036

ABSTRACT

Exploratory Differential Scanning Calorimetry (DSC) studies revealed a broad endotherm which encompassed the transition or relaxation regions observed in Thermomechanical Analysis (TMA) studies of sulfonated ethylene-propylene-ethylidene norbornene (S-EPDM) ionomers. This paper presents a detailed DSC examination of S-EPDM to determine the procedural and polymer variables which influence the DSC endotherm in these ionomers. The results indicate that the DSC endotherm is due to ionomer associations which vary with S-EPDM sulfonate content. In these molded samples, there is an apparent induction period, following which the endotherm slowly increases in intensity over a period of several months. After longer storage periods, there is a major change in endotherm characteristics, suggesting a corresponding change in the nature of the associated ionomer species. The paper concludes with a brief description of the influence of heating rate and reference systems on the detection of fine structure in the morphology of these complex ionomer systems.

INTRODUCTION

Sulfonated EPDM ionomers (S-EPDM) have substantially different viscoelastic properties compared to the unsulfonated ethylene-propylene-5-ethylidene-2-norbornene (EPDM) base polymer due to association and/or interaction of the ionic groups [1-3]. A detailed Thermomechanical Analysis (TMA) study revealed complex behavior in S-EPDM and exploratory Differential Scanning Calorimetry (DSC) studies revealed a broad endotherm which encompassed the transition or relaxation regions observed in the TMA thermograms [4]. The present investigation is concerned with a detailed examination of S-EPDM to determine the procedural and polymer variables which influence the DSC endotherm in these ionomers, to relate this information to the transition and relaxation characteristics observed in TMA studies of these polymers and to examine the utility of DSC and TMA for elucidating the nature of the ionomer associations and microphase-separated domains in these systems.

EXPERIMENTAL

Polymer Preparation and Properties. Ethylene-propylene-5-ethylidene-2-norbornene copolymer (EPDM) was sulfonated and neutralized with magnesium acetate by the methods described by Makowski et al [5]. EPDM composition (wt.%) was: ethylene: 43; propylene: 54; 5-ethylidene-2-norbornene: 3. EPDM inherent viscosities (dl/g; decalin at 135°C); Sample A: 0.8; Sample B: 1.0; S-EPDM sulfur contents (wt%); Sample A: 0.69; Sample B: 0.39.

Sample Preparation. Compression molded pads (8-10 mil thick) were prepared at 218°C in a PHI (Model OL-433-1-H-C) hydraulic press as previously described [4]. The pressure was then released and the samples were cooled in place (~2.5 hrs) under gradually decreasing temperature and pressure. Cast films were prepared from 2.0 wt.% S-EPDM solutions in 90 wt. percent heptane and 10 wt.% hexanol. The solution was added in several increments to a Teflon\-coated aluminum foil tray. Solvent was evaporated in a nitrogen purged system at 90°C. The resulting film was dried at 90°C in a vacuum oven for 24 hours.

TMA Procedure. A DuPont 990 controller was used in conjunction with the 941 Thermal Mechanical Analyzer operated in the penetrometer mode. The following conditions were employed: load: 5 gm; heating rate: 5°C/min., temperature range: - 150 to 340°C.

DSC Procedures. A DuPont DSC cell was used in conjunction with the DuPont 990 controller. Except where indicated, (see Figure Captions), operating conditions were as follows: reference: empty aluminum pan; atmosphere: nitrogen; heating rate: 10°C/min. Sample size: 14± 1 mg. Encapsulated samples were prepared via the DuPont Encapsulating Press.

RESULTS AND DISCUSSION

The influence of ionic group association and/or interaction on viscoelastic properties of S-EPDM is illustrated in Figure 1 which shows T_g, the glass transition of the base polymer EPDM at -61°, the rubbery plateau extending from ca. 0 to 240°C., and the onset of the viscous flow region near 240°. TMA studies of a variety of S-EPDM ionomers revealed a major, time-dependent transition or relaxation region (near ~100°C) in the rubbery plateau region in molded samples and an additional, minor transition region in the viscous flow region of these polymers. These features were attributed to associations of the ionic groups, possibly into microphase-separated domains. An initial comparison of the DSC and TMA thermograms for one of these S-EPDM ionomers, Figure 2, indicated a broad DSC endotherm which encompassed the TMA-detected "transition" near 100°C. The minimum in this endotherm occurred at about the end of the 100°C transition region as shown in Figure 2.

DSC Characteristics of Aged Samples. A primary purpose of this study was to employ DSC in a study of aged, molded S-EPDM samples and relate the results to similar TMA data. A DSC study of this type is summarized in Figure 3 which reveals the following interesting features:
  (A) Three hours aging: little or no evidence of an endotherm.
  (B) Six days aging: strong endotherm has developed; minimum ca 150°C.
  (C) Thirty-three days aging: endotherm area increased; minimum remains near 150°.

(D) Seven months aging: major change in endotherm characteristics: weak transition region ca 75°C; the major endotherm of curves B and C is not evident; a weaker, more diffuse endotherm is noted above 150°C.

These data suggest that the observed endotherms are due to ionomer associations which form slowly in molded S-EPDM samples. In addition, curve D suggests that the type and/or nature of these species changes during long-term storage at ambient conditions. In turn, this suggests the utility of DSC in studies to relate the type and intensity of ionomer associations to viscoelastic and mechanical properties of S-EPDM ionomer systems. Experiments are planned to relate these DSC observations to similar TMA experiments. Another interesting feature of these data is that the T_g of the polymer remained relatively constant (-59°± 2°C). This value is the same as that of unsulfonated EPDM and thus is consistent with the concept of microphase-separated, ionic domains or associated species [6].

Encapsulation of Samples Immediately After Molding. Two questions which arise regarding interpretation of the DSC endotherms in these S-EPDM ionomers are: (a) is the endotherm related to mechanical relaxation effects in the molded samples and (b) does water pick-up during ambient aging of molded samples contribute to the observed changes in endotherm characteristics. In order to examine the potential significance of these factors, a series of molded samples were encapsulated in the standard DSC pans immediately after molding. The samples were then aged in this condition in order to minimize water pick up and relaxation effects. Comparison of the fresh sample (five hours after molding) with a sample which had been aged for 21 days (Figure 4) shows thermogram characteristics similar to those of the previous aging study (Figure 3). Thus, the DSC endotherm does not appear to result from a gradual pick-up of water in samples exposed to the atmosphere. A similar evaluation of sample encapsulation immediately after molding was conducted using a fractionated polymer which had originally been in the form of a cast film. As shown in Figure 5, the influence of aging on the DSC endotherm characteristics was similar to that of the whole polymer (Figure 3 and 4). This experiment also indicates that the endotherm is not very sensitive to molecular weight and/or compositional uniformity. Further, it is not likely to be due to byproducts of the sulfonation or neutralization reactions, since most of these would have been removed from the fractionated sample.

Comparison of TMA and DSC Characteristics. In order to compare the thermomechanical and calorimetric properties of molded S-EPDM, a sample which had been aged for 45 days at ambient conditions was evaluated by the TMA penetrometer procedures previously reported [4]. As shown in Figure 6, the TMA thermogram reveals a well-defined transition region in the same general temperature interval as that in which the DSC endotherms occur (see for example Figure 3, Curves B and C). Thus, the two techniques confirm the existence of the effect, and we note that there is a mechanical property change in the same temperature region as the endotherm in the calorimetric experiment.

Influence of Physical Form. As another means of assessing the influence of molding conditions and sample history on the DSC endotherm, several types of samples were compared. One, which was not molded was a free-flowing polymer crumb (the "as received" form of the polymer), a second was a molded sample prepared from a film cast from a 90 heptane/10 hexanol solution and a third was a molded pad prepared from polymer crumb. The composition of these samples is shown in the Figure captions and in the EXPERIMENTAL section. As shown in Figure 7, similar, well-defined endotherms were observed in all of these samples. Thus the endotherm does not appear to be due to residual effects from the molding process. Second, the original form of the polymer (film vs. crumb) does not appear to influence development of the species responsible for the endotherm in the molded sample. Third, it is suggested that water does not play a large role in the formation of the endotherm-related species. For example, the high surface area crumb was exposed to ambient conditions for much longer periods than the molded sample. Both exhibit a strong endotherm. However, the aggregates and/or microphase-separated regions may be less well developed in the crumb sample since the endotherm minimum occurs at lower temperature than in the molded samples.

Influence of Heating Rate and Reference Type. These DSC studies, in combination with the previous TMA studies, indicate the complex nature of the transition and/or relaxation processes in S-EPDM ionomers. Further evidence of this complexity was uncovered in two additional studies. One of these was directed toward optimizing experimental variables. This study revealed an apparent unusual heating rate effect as shown in Figure 8. As the heating rate was increased from 2 to 5°C/minute, the apparent endotherm area increased, but there was very little change in the location of the endotherm minimum or termination temperatures. The endotherm does appear to start at a lower temperature at the higher heating rate, but this is difficult to establish due to the baseline slope. However, as shown in Figure 8, substantial changes were evident when the heating rate was increased to 20°C/minute. In this case, the endotherm appears to have been shifted to a substantially higher temperature region. In addition, there is an apparent sharp transition in the ~180°C region. The nature and significance of this observation is currently under study.

Another interesting study relates to the evaluation of S-EPDM as a reference in DSC studies. The objective here was to seek to enhance the endotherm features which might be related to ionomer aggregates or microphase-separated domains. It has been shown that the ionomer interactions or associations which appear to be involved in the DSC endotherms observed in molded S-EPDM form slowly over a period of ~ one week. If one assumes that samples which have been "aged" for much shorter periods than this would have little of the features responsible for the DSC endotherm in the aged samples, then the use of these "fresh" samples as reference materials in the DSC experiment should, in the ideal case, lead to a thermogram whose features are primarily due to ionomer associations, aggregates or microphase-separated regions.

An evaluation of this approach is shown in Figure 9. The thermograms in Curves A and B were obtained using a two-day old, molded sample as a reference. DSC heating rates were 5°C/min and 10°C/min respectively. Several features of these thermograms are noteworthy: (a) the low intensity of the endotherms (note the high sensitivity scale) which would be consistent with the removal of most of the thermogram intensity due to cancellation of the other polymer effects; (b) the endotherms appear to have several features; (c) comparison of Curves B and A suggest that there may be two regions of the endotherm which vary in intensity as the heating rate is changed. These and other observations lead to the question of whether the nature of the associated ionic species changes due to the thermal history applied to the polymer <u>during</u> the DSC experiment. The thermogram shown in Curve C of Figure 9 was obtained in a DSC experiment in which the polymer reference was thermally treated <u>in a separate DSC run,</u> immediately prior to its use as a reference, to minimize the presence of time dependent, ionomer associations. The endotherm area in this system is much larger than that of Curve B which was obtained at the same heating rate and sensitivity. Thus, it appears that the use of fresh, unaged polymer samples as reference materials may enable enhanced sensitivity and resolution regarding thermogram features related to associated ionomer species. These exploratory studies also suggest that modification and/or extension of the DSC and TMA techniques may enable enhanced detection and resolution of the detailed nature of S-EPDM association and of the "morphology" of these systems. Further studies in this area are in progress.

REFERENCES

1. D. Rahrig and W. J. MacKnight, Polym. Prepr. Am. Chem. Soc. Div. Polym. Chem., <u>19</u> (2), 314 (1978).
2. R. M. Neumann, W. J. MacKnight and R. D. Lundberg, Polym. Prepr. Am. Chem. Soc. Div. Polym. Chem. <u>19</u> (2), 298 (1978).
3. R. A. Weiss, Proceedings of The 10th North American Thermal Analysis Society Conference, October 26-29, 1980, Boston, Mass., p. 45.
4. J. J. Maurer and G. D. Harvey, Proceedings of The 11th North American Thermal Analysis Society Conference, October 18-21, 1981, New Orleans, La, p. 53.
5. H. S. Makowski, R. D. Lundberg, L. Westerman and J. Bock, Polym. Prepr., Am. Chem. Soc., Div. of Polym. Chem., <u>19</u> (2), 292 (1978).
6. W. J. MacKnight and T. R. Earnest, Jr., Journal of Polymer Science: Macromolecular Reviews, <u>16</u> 41 (1981)

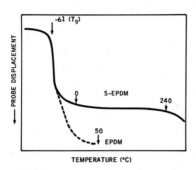

Figure 1. Influence of Sulfonation on EPDM Flow Characteristics.
(A) Unsulfonated EPDM; Inherent viscosity (decalin, 135°C): 1.0:
(B) S-EPDM (Sample B).

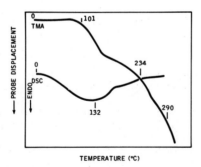

Figure 2. Comparison of DSC and TMA for S-EPDM (Sample A).
(A) DSC: dT/dt: 10°C/min.
(B) TMA: dT/dt: 5°C/min.

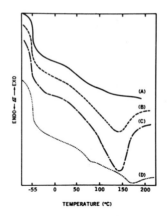

Figure 3. Influence of Aging on DSC Endotherm of Molded S-EPM (Sample A). The molded pads aged at ambient conditions. Samples were encapsulated immediately before the DSC experiment.
    (A) Aged 3 hours   (13.1 mg)
    (B) Aged 6 days   (13.4 mg)
    (C) Aged 33 days   (13.2 mg)
    (D) Aged 7 months   (15.0 mg)

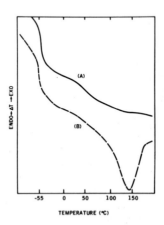

Figure 4. Endotherm characteristics of aged S-EPDM which had been encapsulated immediately after molding (Sample A).
    (A) Aged 5 hours   (14.0 mg)
    (B) Aged 21 days   (13.1 mg)

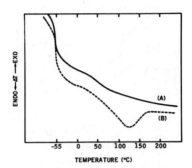

Figure 5. Influence of aging on DSC endotherm of S-EPDM fraction (from Sample B) which had been encapsulated immediately after molding.
(A) Aged 1 hour and 45 min.   (13.8 mg)
(B) Aged 21 days   (14.0 mg)

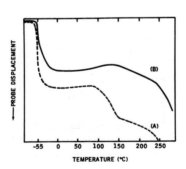

Figure 6. Thermomechanical Analysis of aged, molded S-EPDM (Sample A). (Molded sample aged at ambient conditions for 45 days; not encapsulated. 5 gm load; 20 mil pad; heating rate: 5°C/min; displacement scale: 10 mils/inch; Penetrometer mode).
(A) Aged 45 days; initial TMA run.
(B) Following (A), the sample was cooled to room temperature and rerun in place. Probe repositioned on the sample.

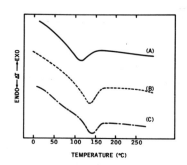

Figure 7. Comparison of DSC endotherm characteristics for different physical forms of S-EPDM (Sample B). Encapsulated samples.
(A) Crumb (as received); (11.1 mg).
(B) Molded pad (aged 41 days); (12.8 mg).
(C) Cast film (from 90 heptane/10 hexanol); aged at ambient conditions for one month; (16.6 mg).

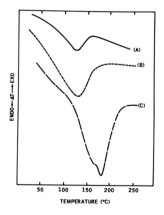

Figure 8. Influence of heating rate on DSC endotherm characteristics (Sample A). (Molded pad aged at ambient conditions. Encapsulated on the day the DSC run was made).
(A) Aged 77 days; (12.8 mg); heating rate = 2°C/min.
  ΔT scale = 0.1°C/inch.
(B) Aged 78 days; (13.0 mg); heating rate = 5°C/min.
  ΔT scale = 0.1°C/inch.
(C) Aged 80 days; (12.0 mg); heating rate = 20°C/min.
  ΔT scale = 0.2°C/inch.

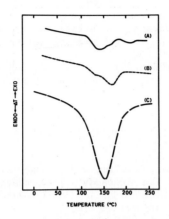

Figure 9. DSC evaluation of S-EPDM (Sample A) using thermally treated S-EPDM (Sample A) as a reference. (Molded samples were aged at ambient conditions. Both sample and reference were encapsulated just prior to molding).
(A) Sample A: Aged two months (13.8 mg).
Reference: Sample A; aged two days after molding; (13.9 mg); heating rate: 5°C/min. ΔT scale: 0.1°C/inch.
(B) Sample A: Aged two months; (13.8).
Reference: Sample A; aged two days after molding (13.6 mg); heating rate: 10°C/min. ΔT scale: 0.1°C/inch.
(C) Sample A: Aged two months (13.9 mg).
Reference: Sample A aged at ambient conditions for two days after molding. It was then heated to 280°C at 5°C/min. in the DSC unit and then cooled to room temperature. This polymer was then used as the reference for curve (C) (13.9 mg); heating rate: 10°C/min; ΔT scale: 0.1°C/inch.

RAPID QUANTITATIVE CHARACTERIZATION OF ELASTOMERS BY DYNAMIC LOAD THERMOMECHANICAL ANALYSIS

Rudolf Riesen, Mettler Instrumente AG
CH-8606 Greifensee, Switzerland

Waldemar Bartels, Isotech AG, Hegmattenstr. 20
CH-8404 Winterthur, Switzerland

INTRODUCTION

The increasing demand for technically high developed elastomers (e.g. rubber, sealing compounds) calls for rapid and reliable characterization methods for research, development, production and application. In a broad temperature range, the physical and chemical properties can easily be determined by the thermoanalytical methods DSC, TG and TMA. In this field, of special interest is the mechanical behavior of the elastomers as a function of the temperature. In the past, numerous standard procedures using dedicated test systems have been developed to quantify properties like the expansion coefficient, the elastic modulus, the dimensional stability, Youngs's modulus in flexure, the flow behavior. However, these tests are often time-consuming, expensive and not suited for routine analysis [1].

Recently, the applicability of DSC, TG and TMA for the characterization of elastomers has been reviewed [2,3]. Thermomechanical analysis (TMA) has been used for several years to deliver important information (e.g. exparsion coefficient, transition temperature, thermal dimensional stability and history effects) on elastomers [4]. At the present time, a unique thermomechanical analysis system with dynamic load (Mettler TMA40) allows rapid determination of the mentioned properties, in particular the elastic behavior. Elastomers now may be characterized either by qualitative comparison of dynamic load TMA curves or by monitoring numerical results (e.g. glass transition and decomposition temperatures in additon to the mechanical data).

Dynamic load TMA (DLTMA) is herein defined as the technique in which the deformation of a substance under changing load is measured as a function of the temperature whilst the substance is subjected to a controlled temperature program (dynamic or isothermal). As from now, the changing load will be a cyclic alternating load between two values. Hence, this technique provides information as a function of temperature and force at the same time.

This powerful tool for R&D as well as for quality control will be discussed for different types of joint and glass packings to be used for building construction. Up to now, these products have been characterized by standard tests [5] (e.g. recovery at room temperature) which often is not sufficient for a practical judgment. Some results of the DLTMA and standard methods are compared.

EXPERIMENTAL

As yet described, a Mettler TA3000 system with a TMA40 measuring cell has been used. The static or dynamic load is automatically applied on the TMA probe. The alternating load has a cycle time of 12 seconds.

There has been analysed four types of commercial sealing compounds to be used for building construction: polysulfides, polyacrylates, polyurethanes and a silicon rubber (see Table 1). Following the instructions of the manufacturers, the samples have been cured between two blocks as it is indicated in the standard tests for the determination of the recovery [6,7]. Cut discs of 6mm diameter and 2mm thickness were used for TMA measurements.

The quartz probe was 3mm in diameter with a flat end.

Table 1:   Types of elastomer used (commercial joint sealants for building construction, Note: the names are registered ^{R}).

1. Polysulfides

1.1 Paltox Thiokol
1.2 Eurolastic TK41
1.3 Heinoxan 3000 1.
1.4 Heinoxan 3000 2.
1.5 Thiokol Standard

2. Polyacrylates

2.1 Paltox Acryl
2.3 Terostat AC 40
2.2 Terostat AC 20
2.4 Compactal ACU

3. Polyurethanes

3.1 Sikaflex 1a
3.2 PALTOX 2000

4. Silicon rubber

4.1 Karosil E

RESULTS AND DISCUSSION

The difference between the information content of a DLTMA and a TMA curve is given in Fig. 1. The DLTMA curve shows clearly the region of the glass transition (Tg = -50°C), the increasing rubber elasticity up to 150°C, the plastic deformation under load (beginning at Td = 130°C) and the viscous flow above Tb = 200°C. Further characteristic temperatures are indicated in Fig. 1, e.g. Ta, the beginning of the "blow up" and the three temperatures T_1, T_2, T_3, at which the elastic moduli have been determined (see Table 2).

The four different classes of elastomers show very different shapes of DLTMA curves (Fig. 2). Besides a relaxation (?) effect at -50°C the silicon rubber behaves ideally rubber-elastic over the whole temperature range. Polysulfides and polyurethanes show at higher temperatures (150°C) viscous flow. Polyacrylates will soften above 50°C indicating a pseudoplastic behavior (see also Fig. 4). In all cases the length change due to the load change (change of the applied stress) is reciprocal to the E modulus; hence, the DLTMA also provides a rubber hardness.

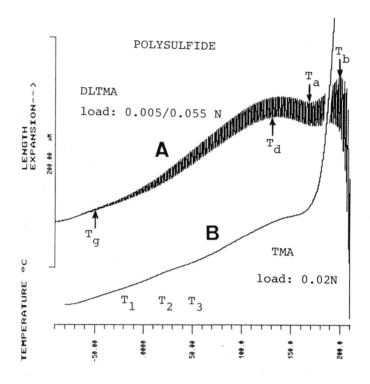

Fig. 1: Dynamic Load TMA (A) and TMA (B) curves of a Polysulfide (1.1). Heating rate 10 K/min. from -90° to 210°C
Sample thickness 2.62 mm; alternating load of 0.005 and 0.055 N

DLTMA measurements are not only suited to distinguish different classes of elastomers but also allow quality control, i.e. detection of irregularities in the properties of one substance as it is shown in Fig. 3. The poor crosslinking of the polysulfide (sample 1.3 compared to 1.4) leads to a higher weakness and to a creep flow at low temperatures giving poor recovery after extension tests

The elastic behavior may be expressed numerically in terms of the Young's or E modulus which is defined as the ratio of the stress applied on a sample to the deformation caused by this stress. In Table 2 some values are listed as a function of the temperature for the same samples shown in fig. 2. The moduli measured by DTMA are compared to the moduli resulting from the tension test [5]. In general the moduli correspond to each other in the trend and magnitude. The existing differences are caused by the different measuring procedures: In DLTMA the samples are strained rapidly to a deformation of not more than 3 % whilst the standard test extends the samples more than 25 % in a slow extension rate of 5 mm/min. Hence, in the second procedure (standard tests) the samples may deform pseudoplastically giving lower moduli. But this is compensated by the large extension (25 %) giving high E moduli due to a non-linear relationship between stress and deformation (compare samples no. 4.1 and 2.3).

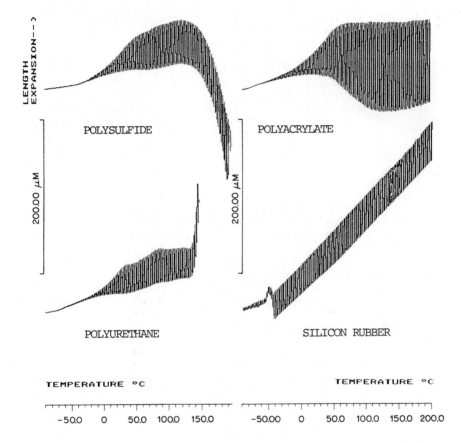

Fig. 2: Typical DLTMA curves of four different elastomers (sealing compounds) in the temperature range of -90° and 200°C.
Heating rate 10 K/min. Alternating load of 0.00 and 0.04 N.
Sample thickness: polysulfide (1.5) 2.77 mm, polyacrylate (2.3) 2.59 mm, polyurethane (3.1) 2.30 mm, silicon rubber (4.1) 2.66 mm

A pseudoplastic deformation may be detected easily in the manner shown in Fig. 4. The deformation of the polyacrylate (pseudoplastic) changes with time, hence giving varying E moduli. The recovery measured with the standard test [6] shows the same great difference (95 % for 4.1 to 60 % for 2.3) as it may be detected in Fig. 4.

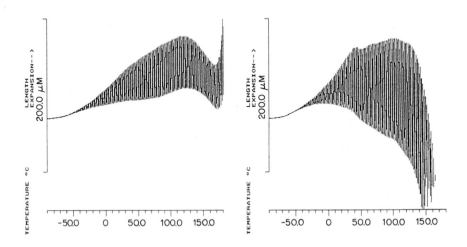

Fig. 3: DLTMA curve of a good (on the left, no. 1.4) and poorly vulcanized (no. 1.3) polysulfide product, where the plastic flow begins at a lower temperature due to a light crosslinking. Heating rate 10 K/min, Temperature range -90 to 180°C. Sample thickness 2.44 mm (no. 1.4) and 2.60 mm (no. 1.3)

Table 2: E moduli at three temperatures (-20, 20 and 50° C) for four different samples (1.5, 2.3, 3.1 and 4.1) measured by DLTMA (curves in fig. 2) and by the standard extension test at elongation of 25 % [5].

		E-Modulus [N/cm2]			
		-20°C	+20°C	+50°C	
polysulfide	1.5	746	54	34	DLTMA
		340	88	52	stand.test
polyacrylate	2.3	210	50	24	DLTMA
		260	56	40	stand.test
polyurethane	3.1	373	57	45	DLTMA
		360	136	80	stand.test
silicon rubber	4.1	82	72	64	DLTMA
		200	192	180	stand.test

All the samples analyzed were freshly prepared and therefore accessible to both methods, DLTMA and standard test, because the latter require standardized sample preparation. Naturally aged samples taken from a joint of a building, only may be measured by thermal analysis.

In the same way, other elastomers - e.g. rubber, softened polymers, films and threads - may be rapidly characterized and controlled by one single TMA instrument using the DLTMA mode.

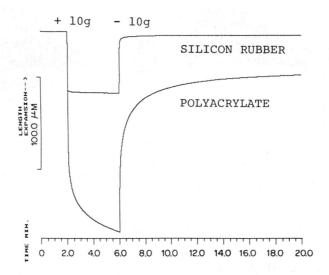

Fig. 4: Isothermal TMA curves of a partially plastic (polyacrylate 2.3) and a highly elastic sample (silicon rubber 4.1). 2 minutes after the start, the static load of approximate 0 N has been increased by 10 g (0.1 N) and reduced again after 6 min. Due to the partially plastic deformation, the polyacrylate does not recover the original size. Deformation after 6 min.: silicon rubber 2.8 %, polyacrylate 8.3 %.

REFERENCES

1. D.W. Brazier, G.H. Nickel, Thermochimica Acta, 26 (1978) 399
2. D.W. Brazier, Thermal Analysis of Polymeric Materials"
   Polymerteknisk orientering 1.80; Polymerteknisk Selskab,
   University of Denmark 1980
3. A.K. Sircar, Proceedings of the 11th North American Thermal Analysis Society Conference, Vol. II, paper No.8
4. John J. Maurer, "Thermal Methods in Polymer Analysis", 1977
   Eastern Analytical Symposium Series, The Franklin Institute
   Press, Philadelphia, p. 129
5. W. Bartels, Material und Technik, $\underline{2}$(1974) No. 2 and 3;
   EMPA, Dübendorf, Switzerland
6. DIN 52455; Testing of materials for joint and glass packings to be used for building construction
7. DIN 52458 Testing of sealing compounds, determination of the recovery.

THERMAL CHARACTERIZATION OF AN ETHYLENE-VINYL ALCOHOL COPOLYMER

W. R. R. Park
Dow Chemical Company
Midland, MI 48640

INTRODUCTION

Ethylene-vinyl alcohol copolymers have found increasing usage in co-extruded barrier constructions in recent years. They are made by complete or partial hydrolysis of ethylene-vinyl acetate copolymers and have extremely good gas barrier properties, particularly when dry.

One of the best of these barrier copolymers, EVAL EC-F, manufactured by Kuraray in Japan, was chosen for study. This polymer has a 33:67 mole ratio of ethylene:vinyl alcohol. All of the acetate groups have been hydrolyzed.

EXPERIMENTAL AND DISCUSSION

The characterization was carried out on DuPont thermal equipment using DSC, TGA and both 990 and 1090 programmers. It is well known that EVAL EC-F is very hygroscopic. Samples exposed to 100% R.H. for 65 and 240 hours absorbed 2.3 and 5.0% by weight of moisture. Larger pellets of the same material absorbed 1.6 and 3.2%. An immersed sample, after 260 hours, showed a weight gain of 9.7%. This latter figure is near the equilibrium moisture content. Further exposure gave insignificant weight gains. These values of moisture pickup were determined by running weight loss under nitrogen in the TGA for 20 minutes at 190°C. The polymer pellets showed no continuing weight loss at this temperature and usually lost all their absorbed water in less than ten minutes.

Table I summarizes the desorption curves for pellets which had initially equilibrated at 5.0% by weight of moisture content. Again, the TGA was used under isothermal conditions to gather the weight loss versus time data. Kuraray recommends that this material contain less than 0.3% moisture for successful extrusion. Table I, then, allows one to choose appropriate drying conditions for EVAL resin which has been humidified.

Table I
ISOTHERMAL TGA WEIGHT LOSS OF EVAL EC-F* AFTER 3 HOURS UNDER DRY N_2 FLOW

	Percent
30°C	0.25
40°C	0.45
65°C	1.4
80°C	2.3
100°C	4.3
121°C	4.8

*Equilibrated at 5.0% water content

A less precise measure of moisture content of EVAL resin can be obtained from the location of its major DSC melting peak. When scanned at 20°C/min, 5 mg samples containing 0.0, 0.2, 2.6 and 5.0% moisture gave T_{melt} values of 190, 188, 183 and 171°C. All gave a T_{melt} of 190°C after an initial scan up to 220°C which was hot enough to cause complete dehydration without any degradation.

Next, the thermal decomposition behavior was examined by TGA. A sample was first held at 210°C. Its temperature was then increased to 290°C in 20°C steps after varying periods of isothermal heating. Table II shows that no significant decomposition occurs up to 230°C but that at 250°C, weight loss (decomposition) has become appreciable.

Table II
THERMAL DECOMPOSITION OF EVAL EC-F ABOVE ITS MELTING POINT

Time (Min.)	@	Temperature °C	Cumulative % Weight Loss Under N_2
27		210	0.3
20		230	0.4
90		250	3.4
40		270	4.9
20		290	7.5
Total 197			

Common plastics extrusion melt temperatures measured by insertion of a thermocouple into the melt stream at the nosepiece exit, will usually range from 225°C to 300°C and the average residence time inside the extruder will be two to three minutes. However, slow moving fractions may spend ten minutes or more at these elevated temperatures because forwarding in a single screw extruder is not by plug flow but rather involves a distribution of residence times. Thus, a finite percentage of a sample can be expected to spend quite a long time at elevated temperatures. It is the continuing decomposition of this fraction which is likely to cause the most serious degradation problems. Also, while the average melt temperature may be near 250°C, the instantaneous temperature reached by small volumes of resin under very high shear (e.g. between the extruder screw flight tip and the barrel) may be much higher. If one wishes, then, to simulate extrusion heat history it seems reasonable to include some conditions well above the likely average melt temperature.

To this end, a series of EVAL EC-F samples were aged, in open DSC pans, under nitrogen, for periods of five minutes at 220, 260, 300 and 340°C. This produced weight losses of 0.3, 0.5, 2.3 and 11.7%, respectively. A repetition of these exposures in air led to 0.0, 0.0, 2.5 and 18.4% weight loss, respectively.

When these samples were subsequently scanned by DSC it was found that their T_{melt} and Tg values had been progressively depressed.

Additionally, three co-extruded film samples of EVAL EC-F were scanned by DSC for T_{melt} and Tg. Table III summarizes the results.

Table III
EFFECTS OF HEAT AND EXTRUSION ON PHYSICAL PROPERTIES

Sample	Wt. mg	Heat History	Melt* °C	Tg** °C
EVAL EC-F	7.3	5 min @ 220°C in N_2	188.7	69.4
EVAL EC-F	6.5	5 min @ 260°C in N_2	187.0	68.4
EVAL EC-F	6.2	5 min @ 300°C in N_2	183.5	63.0
EVAL EC-F	5.6	5 min @ 340°C in N_2	167.6	53.5
EVAL EC-F	4.5	5 min @ 260°C in air	185.7	69.5
EVAL EC-F	4.9	5 min @ 300°C in air	184.0	67.3
EVAL EC-F	4.3	5 min @ 340°C in air	163.0	52.0
EVAL EC-F	4.0***	Ext. melt temp. 224°C	187.5	67.8
EVAL EC-F	2.5***	Ext. melt temp. 232°C	183.5	66.4
EVAL EC-F	3.4***	Ext. melt temp. 260°C	173.5	56.0

* T_{melt} defined as the peak of the melting endotherm taken at 20°/min
** Tg defined as the peak of the derivative of the DSC curve taken at 20°/min
*** Co-extruded between plies of polypropylene without a "glue" layer so that the EVAL ply can be easily separated

Figure 3 shows another behavior of extruded EVAL which needs to be considered when using it as one ply in a co-extruded barrier construction. This is the low melting peak which is present in the "FIRST MELTING" scan. It accounts for about 12% of the total heat of fusion of the film sample. At present, there is no information on how this peak, which is believed due to rapid melt quenching during co-extrusion, affects barrier properties. Barrier properties are usually determined on an annealled sample which will have a scan like that of the "SECOND MELTING" in Figure 3.

Note also that the "SECOND MELTING" scan shows a significantly lower heat of fusion than in the "FIRST MELTING" scan. This may be an indication that there was stress-induced crystallization during the co-extrusion process where a relatively thick melted extrudate was accelerated away from a die and onto a quench cooling roll. This also will have some undefined effect on barrier properties.

Finally, Figure 4 shows how rapidly virgin EVAL EC-F crystallizes from the melt on cooling. The crystallization exotherm peaks at about 20°C below T_{melt} on the cooling scan. Under the same conditions, polypropylene shows a 45°C difference and, thus, crystallizes much slower. EVAL EC-F appears to crystallize from the melt about as fast as polyethylene. Note also that the heat of fusion of virgin EVAL is lower than that of co-extruded film in Figure 3. It appears that the initial degradation, associated with melt extrusion, allows the development of a higher level of crystallization than in the virgin resin. This is hypothesized to be due to the degradation of a small, very high molecular weight fraction, to lower molecular weight values. The absence of this high molecular weight fraction increases the molecular mobility in the melt and this is reflected in the development of a higher level of crystallization (higher heat of fusion) on cooling.

CONCLUSIONS

1. Processing conditions have significant effects on decomposition, molecular weight degradation, quench endotherms, degree of crystallization, moisture content and, presumably, barrier properties. Thus, the need to evaluat barrier on the finished product rather than on an optimized lab sample, is highlighted.

2. An unexpectedly high degree of dependency of degradation on melt shearing was demonstrated.

3. Thermogravimetry can be used to precisely define both hygroscopic behavior and drying requirements of EVAL resins.

4. Thermal analysis can provide much of the evaluation data needed on a new resin. It can do so quicker and at much less cost than by extrusion evaluations and can also direct the more expensive extrusion evaluations to the most critical areas.

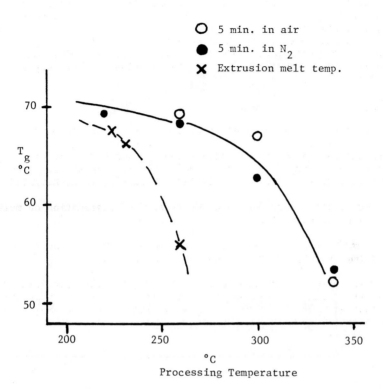

Figure 1: Tg vs $T_{process}$ for EVAL EC-F

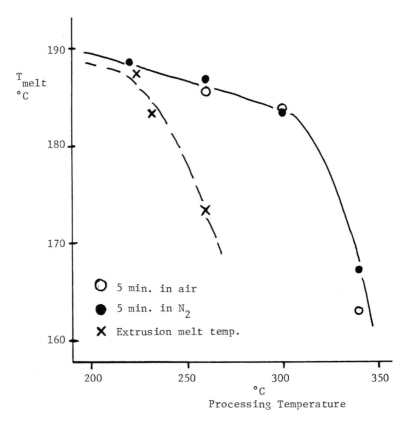

Figure 2: T_{melt} vs $T_{process}$ for EVAL EC-F

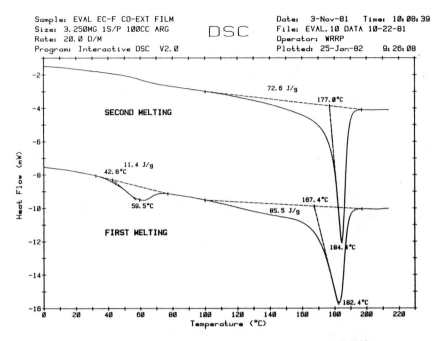

Figure 3: DSC Scans of Co-extruded EVAL EC-F Film

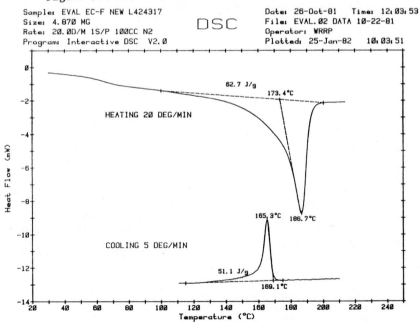

Figure 4: DSC Scans of Virgin EVAL EC-F Resin

POLYURETHANE/POLYMETHYLMETHACRYLATE SIMU-
LTANEOUS INTERPENETRATING NETWORKS (SINs)
- SYNTHESIS AND THERMAL BEHAVIOUR

S.K.SRIVASTAVA, A.B.MATHUR and G.N.MATHUR
DEPARTMENT OF PLASTICS TECHNOLOGY
HARCOURT BUTLER TECHNOLOGICAL INSTITUTE
KANPUR-208 002 (INDIA).

INTRODUCTION

Interpenetrating polymeric networks (IPNs) are being used as pressure sensitive adhesives, nose and vibration damping materials and electrical insulators etc. These network polymers are unique type of polyblends having physically dispersed components (no primary bonds in between them), woven together. Varying degree of phase mixing of the components may take place in these IPNs leading to improved or unique properties in comparison to just physical mixing of such components.

Simultaneous interpenetrating networks (SINs) are the type of IPNs in which the components are synthesized by polymerization of monomers and/or prepolymers simultaneously [1-5] using different polymerization reactions i.e. condensation and addition. Techniques of synthesis of SINs are significant as it leads to the formation of network polyblends of noncompatible polymers. Few workers [6,7] have studied the thermal behaviour of SINs having components of different molecular structures.

In the present investigation various SIN samples of polyether based polyurethane - polymethylmethacrylate have been synthesized using trimethylol propane (TMP) and divinylbenzene (DVB) as crosslinking (curing)agents for polyurethane and polymethylmethacrylate respectively. The SINs were characterized by infra-red spectroscopy. The comparative thermal behaviour of these SIN samples and their individual components has been evaluated by thermogravimetry (TG) and differential scanning calorimetry (DSC).

EXPERIMENTAL

Polyurethane based SIN samples were prepared by using the chemicals, given in table 1. SINs of different composition were synthesized, using prepolymer technique by varying the parameters i.e. proportion of the components, hard and soft segment ratio, chain length of soft segment and concentration of TMP in polyurethane components (table 2).

Polyether based isocyanate terminated prepolymers were synthesized by reacting poly(oxypropylene-b-oxyethylene) glycol (POPG) with excess of TDI at 60°C for 1 hr which was then mixed with MMA, AIBN and DVB (4%, v/v) mixture (previously polymerized to about 20%) and

TABLE-1 : DETAILS OF THE CHEMICALS USED

S. No.	Name	Description - Chemical name	Description - Properties	Source
1.	Pluronic L-31	Poly(oxypropylene-b-oxyethylene) glycol(POPG)	$\overline{M}n$,1000; functionality 2; Sp.gravity 1.01	BASF Wyandotte (N.J.USA)
2.	Pluronic L-61	-do-	$\overline{M}n$,2000; functionality 2; Sp. gravity 1.02	-do-
3.	Pluronic L-81	-do-	$\overline{M}n$,2750; functionality 2; Sp. gravity 1.01	-do-
4.	TDI	Toluene di-isocyanate	80:20 mixture of 1,4 and 1,6 isomers, mol.wt.174 Sp. gravity 1.22	Bayer AG (West Germany)
5.	BD	1,4 Butandiol	Mol.wt.90, Sp. gravity 1.02	Kotch Light Lab.Ltd. (England)
6.	TMP	Trimethylol propane	Mol.wt.134, Sp. gravity 1.23	Fluka Switzerland
7.	MMA	Methylmethacrylate	Mol.wt.100.12 Sp.gravity 0.94	BDH, (England)
8.	DVB	Divinyl benzene	Mol.wt. 130	Polychem Ltd.Bombay (India)
9.	AIBN	α,α'Azobis-(isobutyronitrile)	Mol.wt. 164	Fluka (Swiss)

BD, TMP solution (chain extender and crosslinker of polyurethane prepolymer). The viscous solution was then cured at 70°C for 20 hr in between teflon coated glass plates. Pure crosslinked polyurethane (CPU)(sample B,G and I) and polymethylmethacrylate (CPMMA)(Sample A) were also synthesized to compare the thermal behaviour of SINs with the pure homopolymeric components. The NCO/OH ratio was kept unity for the synthesis of polyurethanes.

Infra-red spectroscopy of the samples was done by Perkin-Elmer 599-B infra-red spectrophotometer. Dynamic and derivative thermogravimetric analysis (TG and DTG) of the samples was done by Stanton Redcroft TG 750 thermobalance in continuous nitrogen flow (10 ml/min), at the heating rate of 10°C/min using 6 mg sample for each run. Dynamic differential scanning calorimetry was done by Du Pont-990. The behaviour of these samples was seen upto 500°C in continuous nitrogen flow (50 ml/min) at

TABLE-2: COMPOSITION OF SINs, THEIR DESIGNATION AND THERMOGRAVIMETRIC (TG) DATA i.e. VALUE OF t_i (INITIAL TEMPERATURE OF WEIGHT LOSS) t_{max} (TEMPERATURE AT MAXIMUM RATE OF WEIGHT LOSS) AND ACTIVATION ENERGY 'E'.

S. No.	Components for crosslinked polyurethane (CPU) synthesis					SINs Composition		Designation of SINs	TG data of SINs		
	Soft segment, POPG, mole	Hard segment				CPU wt. %	CPMMA wt. %		t_i °C	t_{max} °C	E kJ mole^{-1}
		TDI mole	BD mole	TMP mole							
1.	–	–	–	–		–	100	A	215	392	109
2.	L-31 (1.0)	3.3	2.0	0.2		100	–	B	258	366	90
3.	,,	3.3	2.0	0.2		90	10	C	262	370	95
4.	,,	3.3	2.0	0.2		80	20	D	260	382	101
5.	,,	3.3	2.0	0.2		70	30	E	261	388	101
6.	,,	3.3	2.0	0.2		60	40	F	263	390	109
7.	,,	4.45	3.0	0.3		100	–	G	248	–	88
8.	,,	4.45	3.0	0.3		80	20	H	263	–	87
9.	,,	5.56	4.0	0.4		100	–	I	243	–	88
10.	,,	5.56	4.0	0.4		80	20	J	257	–	86
11.	,,	3.3	1.7	0.4		80	20	K	240	–	107
12.	,,	3.3	1.4	0.6		80	20	L	243	415	97
13.	L-61 (1.0)	3.3	2.0	0.2		80	20	M	237	406	129
14.	L-81 (1.0)	3.3	2.0	0.2		80	20	N	229	400	168

the heating rate 10°C/min.

The activation energy (E) was calculated by Coats and Redfern [8] method using TG thermograms to get a further insight into the thermal degradation behaviour of the SINs.

RESULTS AND DISCUSSION

Infra-red spectroscopy: On comparing the infrared spectra of the SINs and pure polymers it has been found that the positions and widths of all the bands of the SINs are same as that of their individual components. This indicates that in SINs no primary bond formation has taken place between the two individual networks.

Thermogravimetry: Thermogravimetric data are shown in figure 1 and table 2 and 3. The initial temperature of weight loss (t_i) of SINs (C-F, H and J) and individual components (A,B,G and I) clearly shows higher

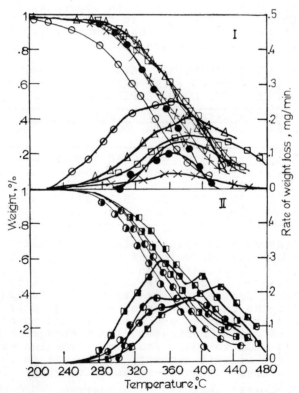

Figure 1 : Thermograms (TG) and derivative thermograms (DTG) of SINs (I)-○-A; ●-B; ✕-C; □-D; ▽-E; △-F; (II)-◐-G; ◼-H; ●-I; ◼-J. (Abbreviation A-J, as in table -2)

stability of SINs than pure individual components. Temperature where rate of weight loss is maximum, (t_{max}), of SINs has been found to increase with the increase in CPMMA content from 0 - 40 wt.% and the values of activation energy also increases with the increase in CPMMA content showing the increase in thermal stability of SINs with CPMMA concentration. This may be due to the radical scavanging of polyurethane radicals by unzipped monomers of PMMA [6,7] and thus increasing the weight retention, t_{max}, t_i and E values. The t_i values of CPU (sample B,G and I) has been found to decrease with the increase in hard segment concentration from 41.49 to 55.79 weight percent while in their SINs (sample D,H and J) no substantial change in the t_i has been observed. Derivative thermograms show bimodal nature of pure CPU and their SINs having 49.64 and 55.79% hard segments. The (DTG curve) also indicates that intermediate stable molecular species are formed in degradation of CPU having higher concentration of hard segments. The value of activation energy of SINs also decreases from 101 to 86 kJ mole^{-1} with increase in hard segment (TDI) concentration from 3.3 to 5.56 moles in CPU.

TABLE-3: THERMOGRAVIMETRIC DATA OF SINs SHOWING THE EFFECT OF BD : TMP RATIO AND $\bar{M}n$ OF 'POPG' ON THE WEIGHT LOSS TEMPERATURE.

SINs Designation	Temperature of wt loss, °C								
	10%	20%	30%	40%	50%	60%	70%	80%	90%
L	292	319	335	351	366	380.5	397	416	456
K	306	331	347	363	379	392	407	422	442
D	300	325	350	370	386	400	416	428	450
M	298	319	353	347	361	376	388	402	423
N	298	319	330	342	351	360	368	379	395

The increase in weight retention of SINs (i.e. above 30% weight loss) and value of activation energy of decreasing the TMP concentration in polyurethane component (sample L,K and D) shows the higher stability of SINs having lower degree of crosslinking (Table 2,3). On increasing the chain length of soft segment in polyurethane, the value of t_i and weight retention of SINs (sample D,M and N) decreases. The increase in activation energy from 101 to 168 kJ/mole with the increase in chain length of soft segment, may be due to the high energy required for chain scission of ether linkage.

<u>Differential scanning calorimetry:</u> DSC thermograms of SINs having 10, 20, 30 and 40 wt.% CPMMA (sample C-F) are shown in fig.2. On comparing these thermograms upto 125°C, it is clear that the broad endotherm due to second order transition shifts from 20°C to 102°C with increase

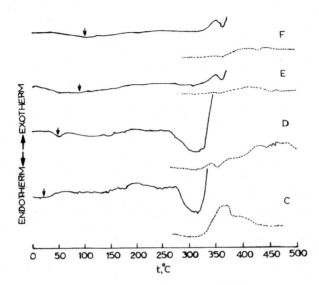

Figure-2: DSC thermograms of SINs
(Abbreviations - as in Table2)

in CPMMA content showing the increase in rigidity due to interpenetration of CPMMA networks in CPU. The shift of this transition region also indicates the homogeneous phase mixing of two non-compatible networks during SINs synthesis. The area of second broad endotherm at about 300°C has been found to decrease with the increase in CPMMA content. The decrease in the area of the exotherms at about 325°C with increase in CPMMA content may be due to the decrease in degree of chain scission in SINs.

On the basis of these results it can be concluded that the stability of SINs increases with the increase in CPMMA content as well as the second order transition shifts towards higher temperature. The concentration of soft and hard segment in CPU component also affects the stability of SINs.

REFERENCES

1. D.Klempner, H.L.Frisch and K.C. Frisch, J.Polym. Sci., 1972, A-2,8, 921.
2. K.C. Frisch, D.Klempner and S.Migdal, J.Polym.Sci., 1974, 12, 885.
3. H.L. Frisch, K.C. Frisch and D. Klempner, J.Polym. Engg.Sci., 1974, 14(9), 466.
4. S.C. Kim, D. Klempner, K.C. Frisch, W. Radigan and H.L. Frisch, J.Macromolecules, 1976, 9(2), 258.
5. V.V. Shilov, Yu.S.Lipatov, L.V. Karabanov and L.M. Sergeeva, J.Polym.Sci.,Polym.Chem. Ed. 1979, 17(10) 3083.

6. S.C. Kim, D. Klempner, K.C. Frisch, H. Ghiradela and H.L. Frisch, J.Polym.Engg.Sci., 1975, 15, 339.
7. S.C. Kim, D. Klempner, and K.C. Frisch, J.Appl. Polym.Sci., 1977, 21, 1289.
8. A.W. Coats and J.P. Redfern, Nature, 1964, 201, 64.

SYNTHESIS AND CHARACTERIZATION OF POLY-IMIDES BY THERMO ANALYTICAL TECHNIQUES

A.B. MATHUR, S.K. SRIVASTAVA, P.K. SINGH and G.N. MATHUR
DEPARTMENT OF PLASTICS TECHNOLOGY
HARCOURT BUTLER TECHNOLOGICAL INSTITUTE
KANPUR-208 002 (INDIA)

INTRODUCTION

Aromatic polyimides are one of the most stable and environmentally resistant heterochain polymers. A variety of materials based on polyimides and modified polyimides are available for special uses e.g. in aviation and space applications, where besides extensive weight and space savings a highly stable material is required. These polymeric materials are mainly being used as film components, wire enamel, adhesive and reinforced composites. This new class of thermally stable polymers withstand higher operational temperatures than do most organic polymeric materials in use today.

Polyimides are traditionally prepared by the reaction of a diamine and dianhydride in a polar solvent and a stable polyamic acid intermediate formed by this reaction (stage 1) is converted by thermal treatment to an insoluble polyimide (stage 2).

$$O\begin{matrix}CO\\ \diagdown \\ CO\end{matrix}R\begin{matrix}CO\\ \diagup \\ CO\end{matrix}O + H_2N - R' - NH_2$$

Dianhydride $\downarrow$ Diamine

$$\left[\begin{matrix}HOOC\\ \diagdown \\ HNOC\end{matrix}R\begin{matrix}CO - NH - R\\ \diagup \\ COOH\end{matrix}\right]_n \quad (1)$$

Polyamic acid

$\downarrow$ heat

$$\left[N\begin{matrix}OC\\ \diagdown \\ OC\end{matrix}R\begin{matrix}CO\\ \diagup \\ CO\end{matrix}N - R'\right]_n + 2n\ H_2O \quad (2)$$

Polyimide (where R is aromatic and R' is aromatic or aliphatic components).

Several workers [1-10] are engaged in the development of new types of polyimides with better properties for the last few years and still a lot of work is required to further improve the stability and physicomechanical properties to make these polymers of more wider applicability.

In the present investigation, the samples of polyimide have been synthesized by using dianhydrides and diamines of different molecular structures. These samples have been characterized by spectroscopic and elemental analysis method. Thermal evaluation of these samples have been carried out by thermogravimetry (TG) differential scanning calorimetry (DSC) and differential thermal analysis (DTA) and on the basis of these results a correlation of molecular structure with thermal properties has been discussed.

EXPERIMENTAL

Synthesis of polyimides: Polyimide samples were prepared by a two step method [1,2].

Materials : Pyromellitic dianhydride (PMDA) of Fluka AG as white powder and 3,4, 3',4' benzophenone tetracarboxylic dianhydride (BTDA) of Gulf Oil Co. as yellow powder, were purified by sublimation, under reduced pressure (1 mm of Hg) at about $200^{\circ}C$. It was than recrystallized before use. Diamines of Fluka AG, SD's and Upjon Co. were used for polymerization with dianhydrides (Table 1). N,N' dimethylformamide (DMF) of SD's was dried and distilled before use.

Polymerization : Polyimide samples were prepared by a two step method using purified dianhydride and diamines in DMF solution. Equimolar amount of diamine and dianhydride (i.e. 4×10^{-3} mole) were taken for polymerization. A solution of diamine in 5 ml of DMF was prepared in a 500 ml three necked flask equipped with a thermometer, nitrogen gas inlet condensor connected to a drying tube and a magnetic stirrer. After the dissolution of diamine, dianhydride solution (in 5 ml DMF) was added gradually over a period of 30 min at $0^{\circ}C$. The solution was stirred vigorously for 30 min and then the temperature was raised to $40^{\circ}C$ in 1 hr. This solution was held for 3 hr. The temperature was then raised slowly to $90^{\circ}C$ and in this duration the water content was removed by creating low pressure in the flask. The solution was then heated at $135^{\circ}C$ for more than 20 hr. Polymer was precipitated by pouring it into excessive amount of water. It was then filtered, washed with methanol and dried at 60°C for several hours. The percentage yield of the polymers is given in table 1.

Characterization and thermal evaluation: The molecular structure of polyimide samples was studied using a Perkin-Elmer 599 B infra-red spectrophotometer. A suspension of the polyimide powder in Nujol was used for infra-red spectroscopy. Characterization of the samples was also done by elemental analysis using Thomas CH-Analyzer 35 and Column N - analyzer 29.

TABLE-2: RESULTS OF POLYIMIDES SYNTHESIS AND THEIR DESIGNATION

S. No.	Diamine	Yield, %	Colour of the polymer	Designation of polymer
Pyromellatic-Dianhydride (PMDA) with				
1.	4,4' diamino diphenylmethane	81	Brown	A
2.	3,3' dichloro-benzidine	80.5	Light brown	B
3.	3,3' dichloro, 4,4' diamino diphenylmethane	83	Light saffron	C
4.	p-phenylene diamine	79	Dark violet	D
3,4, 3',4' benzophenone tetracarboxylic dianhydride (BTDA) with				
5.	4,4' diamino diphenylmethane	82	Yellowish brown	E
6.	3,3' dichloro benzidine	84	Yellow	F
7.	3,3' dichloro-4,4' diamino diphenylmethane (MOCA)	85	Light yellow	G

The thermal stability of polyimide samples was evaluated by thermogravimetry (TG), differential scanning calorimetry (DSC) and differential thermal analysis (DTA). Dynamic thermogravimetric traces (TG) were determined by using a Stanton Redcroft TG 750 thermobalance at a constant heating rate of $10^\circ C/min$, in a constant flow of nitrogen (10 ml/min). Dynamic differential scanning calorimetry of the samples was done by using a Du Pont 990 system with a heating rate of $10^\circ C/min$ in a constant flow of nitrogen (50 mil/min). Differential thermal analyzer (DTA), was used to further evaluate the thermal behaviour of polyimide samples in the presence of air at the heating rate of $5^\circ C/min$.

RESULTS AND DISCUSSION

The results of polymerization and the colors of the polyimide samples are given in table 1. The IR spectra of polyimides (A - G) shows the presence of bands at 1780, 1720, 1380 and 720 cm^{-1} and keto groupnband at 1670 - 1675 cm^{-1} (fig.1). The spectrum of Nujol (indicated by dotted lines) has been used to study the comparative evaluation of absorption bands of polyimide

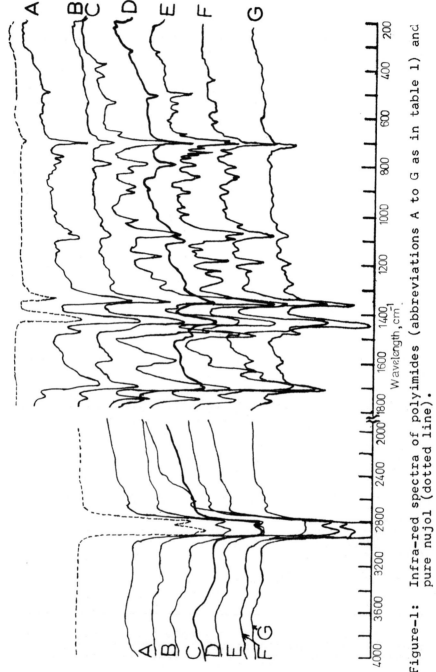

Figure-1: Infra-red spectra of polyimides (abbreviations A to G as in table 1) and pure nujol (dotted line).

samples. The molecular structural composition of polyimide samples has also been confirmed by elemental analysis.

The TG thermograms of polyimides (A-C and E-G) are given in fig.2. It has been found that the maximum weight loss of about 65%, takes place in polypyromellitimides (i.e. PMDA based polyimides, A to C) and after that a plateau is observed. The initial temperature of weight loss has been found in the range of 325 to 400°C. The polyimide 'D' has been found to be the most thermally stable as compared to other polyimide samples and only about 45% weight loss has been observed upto 775°C and after that no quantitative weight loss occurs. This stability may be due to the higher aromatic content in polyimides [1,2].

The BTDA based polyimides (E-G) have been found to have lower thermal resistance than polypyromellitimides, as clear from the weight loss behaviour and comparative maximum weight loss. The values of the activation energy for thermal degradation of these polyimides have been calculated from TG thermograms using Coats and Redfern's method [11]. The values of activation energy have been found in the range of 17 to 30 kcal/mole. The thermal stability of the polyimides, based on the diamine components (using PMDA as second component), is in the order-p-phenylene diamine > 3,3' dichlorobenzidine > 4,4' diamino diphenylmethane > 3,3' dichloro, 4,4' diphenyl methane, while the stability (comparing the initial weight loss) of polyimides, based on diamine components (using BTDA as second component) has been found in the order - 3,3' dichlorobenzidine > 3,3' dichloro 4,4' diamino diphenyl methane > 4,4' diamino diphenylmethane. The DSC curve indicates that all the polyimides show endotherm below 390°C and the exotherm have been found to start in the temperature range of 395 - 470°C. The DTA curve also shows broad exotherm starting in the temperature range of 380 to 450°C. Second order transition temperatures, as observed from thermograms are given in table 2.

TABLE 2 + VALUES OF THE SECOND ORDER TRANSITION OF POLYIMIDES

Polyimides (Abbreviations as in Table 1)	A	B	C	E	F	G
Value, °C	> 235	> 286	282	>260	330	> 220

Hence from the above mentioned data it may be seen that on increasing the aromatic content the thermal stability of the polyimide also increases. The substi-

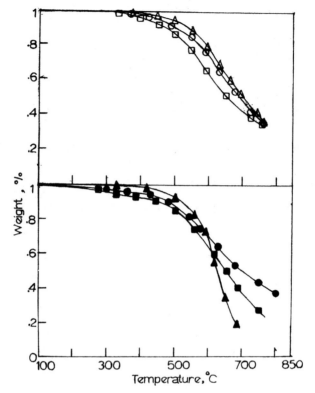

Figure-2: Thermograms (TG) and derivative thermograms of polyimides ; □—A, △—B, ○—C, ■—E, ▲—F, ●—G. (Abbreviations as in Table 1).

tuted chlorine atoms in the diamine also affects their thermal stability.

ACKNOWLEDGEMENT

The author (ABM) is grateful to the Council of Scientific and Industrial Research, New Delhi for the financial assistance.

REFERENCES

1. C.E. Sroog, J.Polym.Sci.Macromol.Rev., 1976, 11, 161
2. W.W. Wright, Developments in Polymer Degradation -3, (N. Grassie Ed.), Applied Science Pub.(England) 1981, P-1.
3. C.W. Tsimpris and K.G. Mayhan, Thermochimica Acta, 1971, 3, 125.

4. L.C. Scala and W.M. Hickam, J.Appl.Polym.Sci., 1965, 3, 245.
5. T.H. Johnston and C.A. Gaulin, Polymers in Space Research (C.L.Segal, M.Shen and F.N. Kelley, Eds.) Marcel Dekker Inc. New York, 1970, P-123.
6. J.K. Gillham, K.D. Hallock and S.J. Stadnicki, J. Appl.Polym.Sci., 1972, 16, 2593.
7. P.S. Carleton, W.J. Farrissey, Jr. and J.S.Rose, J.Appl.Polym.Sci., 1972, 16, 2983.
8. G.M. Bower and L.W. Frost, J.Polym.Sci., 1963, A-1, 3135.
9. C.E. Sroog, A.L. Endrey, S.V. Abramo, C.E. Berr, W.M. Edwards and K.L. Olivier J.Polym.Sci., 1965, A-1, 1373.
10. Y. Imai and K. Kojima, J.Polym.Sci., 1972, A-1, 10, 2091.
11. A.W. Coats and J.P. Redfern, Nature 1964, 201, 68.

THE INVESTIGATION OF THE COMPATIBILITY AND STABILITY OF
POLY (CHLORO-CO-FLUOROSTYRENE) AND PPO BLENDS BY DSC
AND TG ANALYSIS

Radivoje Vuković, Vjera Kurešević, [x]Frank E.Karasz and
[x]William J.MacKnight

Research and Development Institute, INA, Zagreb, Yugoslavia
[x]Department of Polymer Science and Engineering, University
of Massachusetts, 01003 Amherst, MA, USA

INTRODUCTION

In previous articles we described the preparation of some fluoro and chloro substituted styrene copolymers and their behaviour in the blends with poly(2,6-dimethyl-1,4-phenylene oxide) (PPO)[1-7]. Generally, it was found that compatibility in these systems depends on the copolymer compositions as well as on the position of the substituted halogen in benzene ring. Homopolymers of para-chlorostyrene (PpClS), ortho-chlorostyrene (PoClS), para-fluorostyrene (PpFS) and ortho-fluorostyrene are found to be incompatible with PPO independent on the blend compositions [2,5]. Copolymers of pClS and oClS with styrene P(S-pClS), P(S-oClS), containing less than 76% of pClS and 72% of oClS are miscible with PPO [2,3]. At the same time corresponding fluorostyrene-styrene copolymers exibit different behaviour in the blends with PPO. It was found that copolymers with less then 56% of pFS and 91% of oFS respectively are compatible with PPO [4,5]. Copolymers of oClS and pClS and oFS-pFS containing from 26 to 64% of pClS and 10 to 40% pFS are compatible with PPO [6]. Furthermore, compatible blends show different behaviour upon annealing at elevated temperatures.

In continuation of our work, we have copolymerized oFS with pClS and oClS and investigated the miscibility of prepared copolymers with PPO, especially behaviour of compatible blend at elevated temperatures up to 325°C. The investigations were performed by means of differential scanning calorimetry and thermogravimetric analysis.

EXPERIMENTAL

Samples. Copolymers of oFS with oClS and pClS of various compositions were synthesized by conventional free radical polymerization, yielding a high molecular weight product of atactic structure [4,5]. Copolymer compositions were determined by quantitative infrared analysis in chloroform solutions on a Perkin Elmer model 580 b infrared spectrometer using selected absorbance peaks. Molecular weights were obtained from light scattering and osmometric measurements. The weight average molecular weight was about 220000, and number average molecular weight about 100000.

PPO (General El.Co.), Mw 35000 Mn 17000 (GPC) in THF at 25°C). PPO was purified by precipitation of a toluene solution with methanol and was dried at 80°C in vacuum.

Blends. The copolymers were blended with PPO by coprecipitation from dilute toluene solution into a large quantity of methanol. The resulting powders were dried in vacuum at 80°C for 48 hrs. Films of the blends which were used in DSC experiments were obtained by compression moulding at temperature of 210°C [4].

Mesurements. A Perkin Elmer DSC 2 was used to study the glass transition temperature of the blends and the blend components. The Tg was taken as the temperature at half height of the transition of the entire step change observed.

Thermogravimetric analysis were carried out on a Perkin Elmer TGS 2 at a heating rate of 10 C.min^{-1} in nitrogen.

RESULTS AND DISCUSSION

Glass transition temperatures of copolymers were found to be dependent on the copolymer compositions and changed from 104°C to 130°C in the P(oFS-co-pFS) by changing copolymer composition from 13 to 77% of pClS in the copolymer. P(oFS-co-oClS) exibit similar behaviour. Tg's changed from 100 to 125°C for the oClS content in the copolymers of 14 to 80%.

As mentioned in the introduction, the compatibility studies were performed by annealing the samples at temperatures up to 325°C. Such high temperature raise the question of whether polymer degradation is occuring under the conditions of experiment. For that reason thermogravimetric analysis experiments were conducted for some copolymer samplers. The results of weights loss are shown in the Table 1. As it can be seen all samples lost negligable amounts of weight up to 325°C. Similar results were obtained for the copolymer-PPO blends. Although the TG technique detects only weight loss, it is usually good first approximation for testing the thermal stability of polymers.

Compatibility of Copolymer-PPO Blends. All copolymer-PPO blends containing 50/50 weight %, were studied by means of calorimetry

Table 1. POLY(CHLOROSTYRENE-CO-FLUOROSTYRENE) STABILITY
RESULTS OF THERMOGRAVIMETRIC ANALYSIS

Sample[x]	Weight loss (%) Temperature (°C)			
	300	325	350	375
P(oFS - 0.77 pClS)	0.3	0.8	1.3	2.8
P(oFS - 0.74 pClS)	0.3	0.7	1.1	2.5
P(oFS - 0.66 pClS)	0.2	0.7	1.2	2.6
P(oFS - 0.58 pClS)	0.3	0.8	1.3	2.7
P(oFS - 0.45 pClS)	0.2	0.8	1.2	2.5
P(oFS - 0.37 pClS)	0.2	0.6	1.2	2.2
P(oFS - 0.30 pClS)	0.3	0.7	1.2	2.2
P(oFS - 0.18 pClS)	0.3	0.8	1.1	2.2
P(oFS - 0.13 pClS)	0.3	0.8	1.0	2.1
P(oFS - 0.84 oClS)	0	0.3	0.7	1.3
P(oFS - 0.70 oClS)	0.2	0.4	0.8	1.5
P(oFS - 0.61 oClS)	0	0.3	0.6	1.2
P(oFS - 0.49 oClS)	0.3	0.4	0.7	1.1
P(oFS - 0.39 oClS)	0.3	0.4	0.8	1.9
P(oFS - 0.35 oClS)	0	0.3	0.6	1.5
P(oFS - 0.28 oClS)	0.4	0.5	0.9	1.2
P(oFS - 0.20 oClS)	0.3	0.5	0.8	1.9
P(oFS - 0.14 oClS)	0.3	0.5	0.9	1.7

[x] Numbers indicate mole fraction of para chlorostyrene and ortho chlorostyrene in the respective copolymer.

in order to detect the transition from compatibility to incompatibility as a function of copolymer composition. The one Tg criterion was used as a measure of compatibility. The results for the P(oFS-pClS)-PPO blends are present in the Figure 1, and for P(oFS-oClS)--PPO blends are shown in the Figure 2. It is evident that copolymer composition and position of the substituted halogen in benzene ring has very important influence on the compatibility. Only copolymers containing from 14 to about 40% of oClS in the blends with PPO show one Tg at moulding temperatures of 210 C. In the case of P(oFS--pClS) the compatibility with PPO is in much higher range of copolymer compostion, i.e. from 13 to about 70% of pClS in the copolymers. It is evident that copolymers containing pClS are"nearly" mixed with PPO.

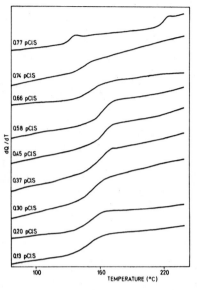

Figure 1. DSC thermograms of 50/50 wt % of P(oFS-co-pClS)-PPO blends. Numbers indicate mole fraction of oClS in the corresponding copolymer.

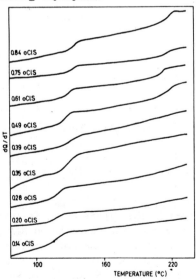

Figure 2. DSC thermograms of 50/50 wt % of P(oFS-co-oClS)-PPO blends. Numbers indicate mole fraction of oClS in the corresponding copolymer.

Table 2. PHASE SEPARATION OF POLY(CHLOROSTYRENE-CO-FLUOROSTYRENE)-
-PPO BLENDS[x]

Mole fraction of para ClS in the copolymers	Glass transitions temperatures of the compatible blends:			
	moulded at 210°C	annealed 15 min at		
		255°C	280°C	325°C
0.74	145	132,218	129,217	-
0.66	157	140,216	-	
0.58	157	157	155	-
0.45	153	153	153	150
0.37	155	153	150	150
0.30	153	153	153	153
0.18	148	145	145	145
0.13	147	147	147	147
Mole fraction of ortho ClS in the copolymers				
0.39	129	121,217		
0.35	125	119,217		
0.28	123	115,216		
0.20	121	112,217		
0.14	121	110,216		

[x] Blend compositions are always one to one by weight.

Phase Separation in the Compatible Copolymer-PPO Blends. Mixture exibiting a single Tg were annealed at temperature of 255, 280 and 325°C respectively. An annealing experiment designed to locate the phase separation temperature, consisted of heating a sample know to be homogeneous, at the fastest available rate (320°C/min) to a selected annealing temperature, and holding at this temperature for an arbitrary period of time. Fifteen minutes was selected as a balance in order to avoid degradation and to allow enough time for equilibrium to be established. Samples were than quenched to ambient temperature as rapidly as possible in the instrument. After that, samples were reheated at a rate of 20°C/min to determine whether one or two phases were present.

The results of the annealing experiments are presented in the

Table 2. It is evident that P(oFS-co-oClS)-PPO and P(oFS-co-pClS)-
-PPO blends exibit quite different behaviour. All PPO-copolymer
blends containing oClS were separated at temperatures of 255°C, a
although blends containing pClS in the copolymer are much more stable. In these systems phase separation was exibited also at 255°C
in the blends containing 74% and 66% of pClS respectively. For all
other compatible blends it was shown that there is no phase separation of the annealing temperatures on the compatibility. Each of
these systems retained its own Tg after annealing.

The conclusion is that in both systems, P(oFS-co-pClS) and
P(oFS-co-oClS), copolymer composition has a pronounced influence
on the compatibility with PPO. It was also shown that the some of
the compatible blends show phase separation by annealing at higher
temperatures, indicating that they are compositionaly dependent.
The range of compatibility and the degree of compatibility is much
wider in the blends of PPO and copolymers containing pClS. Some of
these copolymers were stable even at temperatures as high as 325°C,
and did not show any phase separation after annealing. A possible
explanation of this behaviour can be atributed to the copolymer
structure which may be influenced by the position of the substituted chlorine in the benzene ring. The additional proof of this
statement is considered necessary.

REFERENCES

1. P.Alexandrovich, F.E.Karasz and W.J.MacKnight, Polymer, 1977, 18, 1022.
2. J.R.Fried, F.E.Karasz and W.J.MacKnight, Macromolecules, 1978, 11, 150.
3. P.Alexandrovich, Ph.D.Dissertation, University of Massachusetts, 1978.
4. R.Vuković, V.Kurešević, F.E.Karasz and W.J.MacKnight, The 27th International Symposium on Macromolecules, Strasbourh, July 1981 Vol. 2, p. 1233.
5. R.Vuković, F.E.Karasz and W.J.MacKnight, Hem.ind., 1981, 35, 361.
6. R.Vuković, V.Kurešević, F.E.Karasz and W.J.MacKnight, Proceedings of the Second European Symposium on Thermal Analysis, Ed.D.Dollimore, Hyden and Sons Ltd., London, Philadelphia, Rheine, (1981), p. 243.
7. R.Vuković, V.Kurešević, F.E.Karasz and W.J.MacKnight, Thermochim. Acta, 1982, in press.

THE GLASS TRANSITION OF MESOPHASE
MACROMOLECULES

Bernhard Wunderlich and Janusz Grebowicz
Department of Chemistry
Rensselaer Polytechnic Institute
Troy, New York 12181   USA

INTRODUCTION

For the operational definition of normal liquids and solids it is not possible to ask for a mechanical criterion, since some liquids have, for example, higher shear viscosity than some crystals. We consider it best to require a solid to be either a crystal (stable below its melting temperature) or a glass, which changes at its glass transition temperature to a liquid. The glass is an amorphous solid, usually supercooled relative to the crystalline state. Both, the melting temperature T_m and the glass transition temperature T_g, are easily identified by thermal analysis. The transition temperatures T_m and T_g provide thus for a precise, operational definition for solids and liquids.

Mesophases fit, as the name indicates, between the three limiting states: crystal, glass and liquid. Mesophase materials of linear, flexible macromolecules have gained attention recently when it was found that parallel molecular orientation is possible in some of these mesophases. This orientation can lead to high modulus and tensile strength. [1] Presently there exists a good amount of confusion in the literature about the description, properties, and nomenclature of these macromolecular mesophases and their place in the arrangement of matter. In this paper we will try to resolve this problem and show the importance of the glass transition in the identification of the thermotropic mesophases of flexible, linear macromolecules. [2]

DESCRIPTION OF MESOPHASES

There are three major types of mesophases, namely liquid crystals, plastic crystals, and condis crystals. The first two types represent the positionally and orientationally disordered materials and are well described for small molecules. [3] The term "condis crystal", which is a contraction of the term "conformationally disordered crystal" is coined by us to designate the mesophase most important for flexible, linear macromolecules. We are not aware of prior naming of this class of mesophases. Conformational mobility has been claimed for the high temperature crystals of polytetrafluoroethylene, polyisoprenes, polyethylene and others.

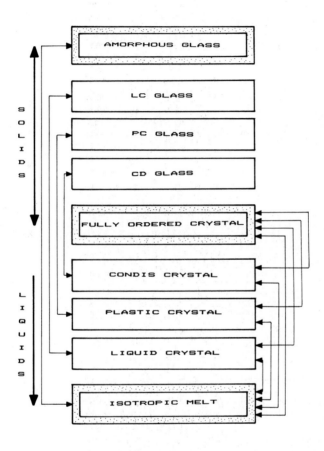

Fig. 1. Schematic diagram of the relationship between the three limiting phases (double outline) and the six mesophases. The top five phases are solid. The bottom four phases show increasing mobility i.e. liquid-like behavior.

While all three mesophases have some degree of long range order in common with the crystal, they also have all some degree of non-vibrational motion in common with the liquid. As a result, all mesophases will show, just as the liquid, a glass transition if full-ordered crystallization can be avoided on cooling. This fact has often been overlooked as a key identifier for the mesomorphic state. Below the respective glass transitions we have three mesophase glasses besides the "normal" amorphous glasses: positionally disordered glasses (LC-glasses), orientationally disordered glasses (PC-glasses) and conformationally disordered glasses (CD-glasses).

THE GLASS TRANSITIONS

Figure 1 illustrates the relationship between the various possible phases. Note that polymorphism of crystals and mesophases has not been considered in this diagram The transitions indicated on the right-hand side are first order transitions which involve a heat of transition and are of little interest to the present discussion. [2] The transitions on the left-hand side are the glass transitions to be discussed. Outermost is the "normal" glass transition from the isotropic melt to the amorphous glass. To distinguish the various glass transitions we add the first letter of the state reached on cooling to T_g, i.e. the "normal" glass transition becomes T_{ga}. The glass formed when a liquid crystal is cooled under conditions which avoid crystallization is an LC-glass and its transition temperature on heating is thus $T_{g\ell}$. Similarly T_{gp} and T_{gc} are the glass transitions to the PC-glass and the CD-glass, respectively.

The vitrification process of a melt at T_{ga} shows no change in entropy, as is illustrated in Fig. 2. Only on decrease in heat capacity occurs at T_g on cooling (curves marked <). The exact temperature range of the change in heat capacity is cooling rate dependent. This reveals the dynamic nature of the glass transition and forces for exact characterization to list T_g and cooling rate.

Empirically, one could observe that molecules with similar size and number of motifs (beads) which start moving on devitrification have similar changes in heat capacity. For small beads we find a change of about 11.3 J/(K mol). [4] For larger motifs such as phenylene or larger ring structures the contribution may be two to three times this amount. [5] In the same way as it is possible to judge from the chemical structure and the entropy increase on fusion the approximate amount and type of disordering, [6] it is possible to judge from the chemical structure and the heat capacity increase on devitrification the approximate amount and type of gained mobility.

For well characterized systems, the glass transitions can also be used for the determination of the weight fraction crystallinity w^c

$$\Delta C_p = (1-w^c)\Delta C_p^\circ \tag{1}$$

where ΔC_p is the measured increase in heat capacity and ΔC_p° is the

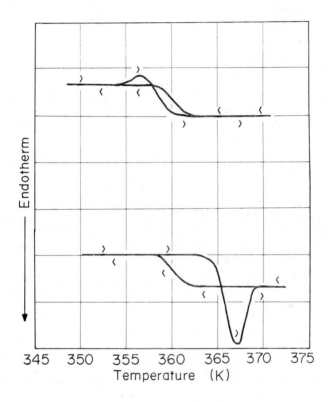

Fig. 2 Schematic of the heat capacity on cooling and heating at largely different rates in the glass transition region. Curves A: Cooling at 5.0 K/min, heating at 0.25 K/min. Curves B: Cooling at 5.0 K/min, heating at 150 K/min. (Hypothetical polymer, time dependence as in polystyrene).

heat capacity increase for the glass transition of the pure amorphous glass or the mesophase glasses. The little information available seems to indicate that there is not much difference between the ΔC_p^o for the amorphous glass and the various mesophase glasses.

Low molecular weight LC-glasses of N-(2-hydroxy-4-methoxybenzylidene)-4'-butylaniline have been described by Sorai and Seki. [9] A cooling rate of 12 K/min was sufficient to go from the liquid crystal to the glass. ΔC_p^o was 107 J/(K mol) at 204 K. Eight or nine beads account for the ΔC_p^o as well as for the chemical structure. In our laboratory we studied the LC-glasses of p-butyl-p'-methoxy-azoxybenzene and its eutectic mixture with p-ethyl-p'-methoxy-azoxybenzene [10]. The $T_{g\ell}$ of the two glasses were 207.6 and 208.1 K, with ΔC_p^o of 129 and 102 J/(K mol), respectively. Again the ΔC_p^o can be rationalized in terms of the mobile parts of the molecules (beads). Tsugi et al. [11] have reported an LC-glass of cholesteryl hydrogen phthalate with a $T_{g\ell}$ of 295 K and a ΔC_p^o of 180 J/(K mol). Again, the magnitude of ΔC_p^o seems reasonable. Petrie [12] found a $T_{g\ell}$ of 283 K and a ΔC_p^o of 170 J/(K mol) for cholesteryl-2,4-dichlorobenzoate (approximately one bead less than cholesteryl hydrogen phthalate). Several other low molecular glasses have been reported without detailed characterization. [12, 13]

For flexible linear macromolecules, liquid crystals are only possible if there is enough conformational freedom to permit pseudopositional mobility of the mesogenic group. Some examples of glass transitions are listed in the review of Takeda. [14] Quantitative data were obtained in our laboratory for poly(acryloyl oxybenzoic acid) [8]. For this polymer it was possible to compare the amorphous glass [T_{ga} = 348 K, ΔC_p^o = 39 J/(K mol)] and the LC-glass [T_{ga} = 408 K, ΔC_p^o = 43 J/(K mol)] which gives direct support to the above contention of small differences in ΔC_p^o between amorphous glass and LC-glass. We also analyzed the LC-glasses of the poly(ethylene terephthalate-co-oxybenzoate) system. [7] Above about 30 mol-% oxybenzoate this copolymer is liquid-crystal-like. The full phase diagram was analyzed and no discontinuity in the change of $T_{g\ell}$ or ΔC_p^o with change from isotropic to anisotropic glass was observed. A final LC-glass we studied was the "flexible spacer" molecule with main-chain mesogenic groups poly(oxy-2,2'-dimethylazoxybenzene-4,4'-diyloxydodecanedioyl) [$T_{g\ell}$ = 288 K, ΔC_p^o = 220 J/(K mol)]. [15]. The similarity of this polymer to the low molecular weight azoxybenzenes is obvious. The 10 spacer CH_2-groups contribute significantly to the glass transition [about 100 J/(K mol)]. It was concluded that the mesophase was a homogeneous material. No partially "liquid crystalline" state seems to exist in the polymeric materials. The mesophase to isotropic state transition shows also the behavior of a close-to-equilibrium transition (418 K, heat of transition 4.1 kJ/mol). A small heat of transition from the completely ordered crystal to the mesophase of 13.6 kJ/mol at 391 K indicates the relatively poor packing in the crystal. A well packed crystal would have been expected to have a heat of fusion of 100 kJ/mol.

Turning to PC-glasses, major work was done by Seki and coworkers. [17,20]. The old heat capacity measurements of Kelly [16] of the glass transition of cyclohexanol were repeated by Adachi et al. [17] and Otsubo and Sugawara [18] and interpreted properly as that of a plastic crystal [T_{gp} = 150 K ΔC_p° = 30 J/(K mol)]. Again the cyclohexanol may easily account for 2 to 3 beads. These data seem to point to a similar ΔC_p° of PC-glasses as in amorphous and LC-glasses. Other examples of thermally fully characterized PC-glasses which fit into the same analysis are cis-1,2-dimethyl-cyclohexane [T_{gp} = 95 K, ΔC_p° = 61 J/(K mol)] [19] and 2,3-dimethylbutane [T_{gp} = 76 K, ΔC_p° = 56.7 J/(K mol)] [20]. PC-glasses of flexible linear macromolecules have not been reported as yet. One expects greater difficulties to give sufficient freedom by conformational motion to permit orientational disordering than is needed in LC-glasses. CD-glasses take the place of the low molecular weight PC-glasses.

To the present there has been no well documented case of a CD-glass transition. The main reason is the close relationship a CD-crystal bears to the fully ordered crystal and the commonly only partial crystallization of linear macromolecules. Rather fast quenching is necessary to avoid the small reorientation necessary to establish the crystal. On heating one finds then first T_{ga} of the remaining amorphous glass. Above T_{ga} where one expects T_{gc}, thermal analysis is complicated by cold crystallization, reorganization and melting. On the other hand, many crystals of linear macromolecules are rather disordered, so that one might assume that these crystals are in reality not fully ordered crystals, but rather represent CD-glasses which have a glass transition temperature somewhat above the melting temperature of the crystal.

In case of polypropylene, for example, a mesophase has been proposed to exist for quenched samples. [21] On heating, these mesophase materials show a glass transition T_{ga} of the normal, uncrystallized portion at about 280 K. On further heating, the mesophase portion changes at about 350 K to the fully ordered crystal with a rather minor exothermic deviation observable in a DSC-trace. Presently we try to find whether this metastable mesophase may have a glass transition at about 350 K. Detailed heat capacity measurements may give the answer. [22]

Looking quantitatively at the glass transition kinetics of amorphous glasses, one finds that the glass transition temperature decreases approximately logarithmically with cooling rate. For poly(methyl methacrylate) [23] and polystyrene [24] the following equations have been found (heating rate q in K/min, T_{ga} in K).

$$T_{ga} \text{ (PMMA)} = 383 + 4.23 \log q \qquad (2)$$

$$T_{ga} \text{ (PS)} = 372.5 + 403 \log q \qquad (3)$$

Since the glass transition is a freezing process on a 3 to 10 mobile unit scale, there are neither phase boundaries nor nucleation processes for the transition.

Measurements on cooling are often difficult to carry out, so that heating experiments are usually substituted. These may lead, however, to hysteresis phenomena if heating and cooling are not carried out at similar rates, or if annealing occurred close to the glass transition temperature before analysis. Figure 2 shows typical apparent heat capacities on heating through the T_g-region at heating rates largely different from the cooling rates (traces indicated by >). No systematic study of the hysteresis effects on mesophase glasses has been reported. It was observed in our laboratory, however, that for liquid crystal type glasses of two small molecules of the azoxybenzene type, [10] of the macromolecular oxybenzoic side chain mesogenic group type, [8] and the main-chain mesogenic type with flexible spacer, [15] the hysteresis was strongly reduced. This can only be interpreted by assuming that the glass transition has a smaller rate dependence than is observed for amorphous glasses. It must be mentioned, in contrast, that the poly(ethylene terephthalate-co-oxybenzoate) liquid crystals showed a reduction in hysteresis only for partial crystallinity of the fully ordered type. [7] Under the condition of partial crystallinity many flexible, linear macromolecules will show also a reduction in hysteresis [25]. Much more detailed experimental information is necessary before an interpretation of these observations will be possible.

CONCLUSIONS

An operational definition for solids is given which allows also to characterize solid mesophases. We recognize five different solid states: crystals, amorphous glasses, and glasses formed on vitrification of liquid crystals, plastic crystals and condis crystals (LC-glasses, PC-glasses and CD-glasses).

The glass transitions of the four types of glasses are compared (designated as T_{ga}, $T_{g\ell}$, T_{gp}, T_{gc}, respectively). One expects in general a higher glass transition for anisotropic glasses than for isotropic glasses, but one example is reported in the literature where both states could be reached. On several examples it seems to be documented that all glasses have similar changes in the heat capacity at the transition. Surprisingly small time dependence seems to be possible in some mesophase glasses.

An important new field seems to have been opened up by the study of mesophase glasses.

Acknowledgements

This work has been supported by the National Science Foundation, Polymers Program, Grant Number DMR78-15279.

REFERENCES

1. see, for example, J. L. White and J. F. Fellers, J. Appl. Polymer Sci., Appl. Polymer Symposium 33, 137 (1978)
2. a more extensive review of this topic is planned by us, to be published 1983 in the Adv. Polymer Sci.
3. see, for example: G. W. Smith, "Plastic Crystals, Liquid Crystals, and the Melting Phenomenon. The Importance of Order." Adv. in Liquid Crystals, Vol. 1, G. H. Brown, ed. Academic Press, New York, 1975.
4. B. Wunderlich, J. Phys. Chem. 64, 1052 (1960).
5. U. Gaur and B. Wunderlich, Polymer Div. Am. Chem. Soc. Preprints, 20, 429 (1979).
6. B. Wunderlich, "Macromolecular Physics, Vol 3, Crystal Melting" Academic Press, New York, 1980.
7. W. Meesiri, J. Menczel, U. Gaur, and B. Wunderlich, J. Polymer Sci., Polymer Phys. Ed., to be published (1982).
8. J. Menczel and B. Wunderlich, Polymer, 22, 778 (1981).
9. M. Sorai and S. Seki, Mol. Cryst. Liq. Cryst. 23, 299 (1973).
10. J. Grebowicz and B. Wunderlich, Mol. Cryst. Liq. Cryst. 76, 287 (1981).
11. K. Tsugi, M. Sorai, and S. Seki, Bull. Chem. Soc. Japan, 44, 1452 (1971).
12. J. Cognard and C. Ganguillet, Mol. Cryst. Liq. Cryst. 49 (Lett), 33 (1978).
13. K. V. Deinz, A. S. Paranjpe, E. B. Mirza, P. S. Parvathanathan and K. S. Patel, J. Physique C3, 40,
14. M. Takeda, these proceedings elsewhere.
15. J. Grebowicz and B. Wunderlich, J. Polymer Sci., Polymer Phys. Ed., to be published 1982.
16. K. K. Kelley, J. Am. Chem. Soc. 51, 1400 (1929).
17. K. Adachi, H. Suga and S. Seki, Bull. Chem. Soc. Japan, 41, 1073 (1968).
18. A. Otsubo and T. Sugawara, Sci. Rep. Res. Inst. Tohoku Univ. A7, 583 (1955).
19. H. M. Huffman, S. S. Todd and G. D. Oliver, J. Am. Chem. Soc. 71, 584 (1949).
20. K. Adachi, H. Suga and S. Seki, Bull. Chem. Soc. Japan, 44, 78 (1971); 43, 1916 (1970).
21. A. Fichera and R. Zanetti, Makromolekulare Chem. 176, 1885 (1975).
22. J. Grebowicz and B. Wunderlich, unpublished.
23. S. M. Wolpert, A. Weitz, and B. Wunderlich, J. Polymer Sci., Part A-2, 9, 1887 (1971).
24. B. Wunderlich, D. M. Bodily, and M. H. Kaplan, J. Appl. Phys. 35, 95 (1964).
25. J. Menczel and B. Wunderlich, J. Polymer Sci., Polym. Letters Ed. 19, 261 (1981).

INFLUENCE OF INTER-/INTRAMOLECULAR ENERGY RATIO
AND CHAIN SEGMENT DENSITY ON THE GIBBS-DIMARZIO
THEORY OF THE GLASS TRANSITION

A.R. Greenberg* and R.P. Kusy[†]
*Department of Mechanical Engineering
 University of Colorado, Boulder, CO 80309
[†]Dental Research Center, University of North
 Carolina, Chapel Hill, NC 27514

INTRODUCTION

Over the last 25 years a number of workers have utilized the Gibbs and DiMarzio (G-DM) statistical mechanical theory of the glass transition [1] to predict the empirically determined dependence of the glass transition temperature (T_g) on polymer molecular weight (MW) [2-10]. This theoretical relationship between T_g and MW is embodied in two parametric equations which contain both intra- and intermolecular contributions to the configurational entropy. If the lattice coordination number, z, is assumed to equal 4, then these equations can be stated for polydisperse systems as follows:

$$\frac{2\beta\exp\beta}{1+2\exp\beta} - \ln[1+2\exp\beta]$$
$$= \frac{\bar{x}}{\bar{x}-3}\left(\frac{1}{1-V_0}\left(\ln V_0 + (1+V_0)\ln\left(\frac{(\bar{x}+1)(1-V_0)}{2\bar{x}V_0} + 1\right)\right) + \frac{\ln[3(\bar{x}+1)]}{\bar{x}}\right) \quad (1)$$

$$\frac{2\alpha S_x^2}{kT_g} = \ln\left(\frac{V_0}{S_0^2}\right) \quad (2)$$

Here β is a dimensionless parameter equal to $-\varepsilon/kT_g$, ε is the energy difference between gauche and trans conformations for the hydrocarbon chain (i.e., the intramolecular or flex energy), $\bar{x}$ is the number average of chain atom segments, V_0 is the fractional free volume at T_g, k is Boltzmann's constant, α is the energy of interaction between a pair of chemically unbonded but nearest neighboring segments (i.e., the intermolecular or hole energy), and S_x is the fraction of these unbonded but nearest neighboring segments (where $S_x = 1-S_0$).

Recently a method for greatly simplifying equation 1 was introduced [11]. This procedure was based upon a plot of the reduced T_g versus the reciprocal number average degree of polymerization, $\bar{P}^{-1}$. When $\bar{x}$ was equated to $2\bar{P}$, a single curve was generated which was independent of ε and only slightly dependent upon V_0 over the range $\bar{P} = 10$ to ∞ for $0.015 \leq V_0 \leq 0.045$.

Comparison of selected empirical results for five different polymers of varying degrees of stiffness with the theoretical predictions of equation 1 indicated that the G-DM relationship failed to adequately describe the T_g vs MW behavior for this representative collection of polymers. These preliminary results suggested that improvements in the accuracy of the theoretical predictions might require a reexamination of the Flory-Huggins lattice model [12] and the associated definition of $\bar{x}$ upon which the G-DM theory is based.

The present effort expands the previous reduced variables plot, which assumed a constant value of V_O over the molecular weight range, to include the parallel development of a constant value for the ratio of the inter-/intramolecular energy (i.e., r). Moreover by setting $\bar{x} = n\bar{P}$ (n = any positive number), a simple modification is introduced which may relate not only the chemical structure but also the intrinsic mobility within a given polymer.

THEORETICAL CONSIDERATIONS

By setting 2α equal to the hole energy [13], E_O, letting $r = E_O/\varepsilon$ [14], and substituting for S_O and S_x, equation 2 can be rewritten as:

$$\beta = -\frac{1}{r}\left[\frac{\ln V_O - 2\ln\left(\frac{2\bar{x}V_O}{2\bar{x}V_O+(\bar{x}+1)(1-V_O)}\right)}{\left(\frac{(\bar{x}+1)(1-V_O)}{2\bar{x}V_O+(\bar{x}+1)(1-V_O)}\right)^2}\right] \quad (3)$$

Equations 1 and 3 can now be solved iteratively by maintaining $\bar{x}$ and r (or $\bar{x}$ and V_O) constant while simultaneously varying V_O (or r) until a unique β, derived from equation 3, satisfies equation 1. The situation will now be considered for several combinations of $\bar{x}$ and r.

Using the reduced variables method, a plot of $T_g/T_{g\infty}$ versus $\bar{P}^{-1}$ is presented in Figure 1. Here theoretical lines have been obtained as a function of r and n, where r has been set equal to 0.8, 1.0, and 1.2, respectively, and n spans the range 0.5-25. The abscissa has been plotted on a logarithmic scale to better resolve the data between the values of $10^3/\bar{P} = 1-10$. The family of lines shown in the figure is quite dependent upon the value of n but relatively insensitive to r. However, as a consequence of the conditions imposed by the solution of equations 1 and 3, this slight dependence of the theory on r requires that the value of V_O vary significantly. This is demonstrated in Figure 2 where V_O is plotted as a function of $\bar{P}^{-1}$ for the parameters, r and n. While the absolute value of V_O varies considerably with changes in $\bar{P}^{-1}$, r, and n, the relative change in V_O depends primarily upon the value of n. For example if $\bar{x} = 2\bar{P}$, V_O varies by ~ 39-46% for r = 0.8-1.2 over the range of $\bar{P}^{-1}$. The magnitude of this change decreases for increasing n.

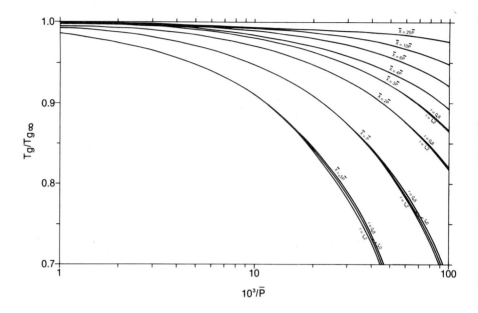

Figure 1. Reduced variables plot indicating the dependence of the glass transition (T_g) upon the logarithmic reciprocal degree of polymerization $(\bar{P})^{-1}$ as a function of constant values of the ratio of hole energy to flex energy (r = 0.8, 1.0, and 1.2) and number average of chain atom segments per $\bar{P}$ (n = 0.5, 1,2,3,4,6,10, and 25). The relationships assume any constant value of ε and require the variation in V_o indicated in Figure 2.

RESULTS

From the available literature the T_g versus molecular weight data of four different polymers [2-7],[15-27] were compared to the theoretical curves derived from equations 1 and 3. This collection included three vinyl polymers, poly(methyl methacrylate) (PMMA) [2-5],[15-17], polystyrene (PS) [18-26], and poly(vinyl chloride) (PVC) [6], and one divinyl polymer, poly-α-methyl styrene (PαMS) [7,27]. These data sets represented an expansion of previous preliminary results [11] to include all available published information for both pure polymers and blends regardless of tacticity, test methodology, physical form, or thermal history. These particular polymers were included in this study because their data sets contained at least 25 data points reasonably distributed over the range $\bar{P}^{-1}$ = 1 to 100.

The 110 point data set for PS is shown in the representative reduced plot of Figure 3. Superposed are the theoretical curves for n = 2,1 and r = 1.0. To a first approximation the former

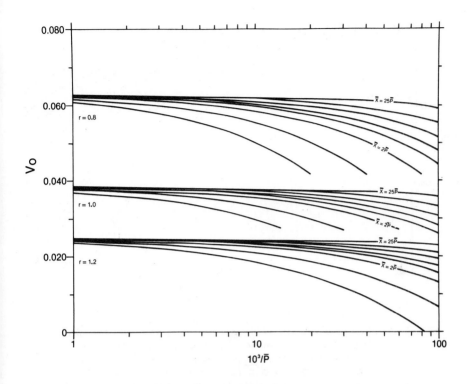

Figure 2. Relationship between the parameter V_o and $(\bar{P})^{-1}$ as determined from equations 1 and 3. Values for r and $\bar{x}$ correspond to those indicated in Figure 1.

delineates an upper bound while the latter designates a lower bound. To adequately describe the entire data set, parametric values intermediate to those shown are apparently required. (For comparison the $n = 2$, $V_o = 0.030$ curve from reference 11 has been included.) Similar plots were constructed for PVC, PMMA, and PαMS which contained 37, 122, and 25 data points, respectively. While upper and lower bounds could be distinguished for the first two polymers, the PαMS results were rather distinctly divided among particular curves depending upon the region of $\bar{P}^{-1}$ under consideration. Within the present context, no one curve could describe that data set.

DISCUSSION

When the reduced variables method was first introduced, the G-DM theory (equation 1) was shown to be reasonably independent of V_o over the range $0 \leq 10^3/\bar{P} \leq 100$ [11]. Moreover when the mean number of chain segments were restricted to a value of $2\bar{P}$, the

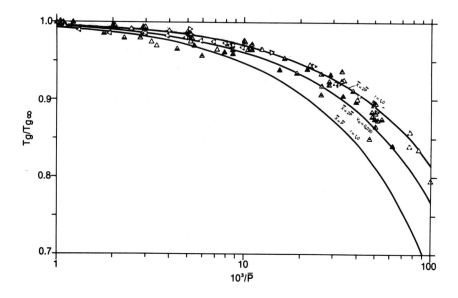

Figure 3. Reduced variables plot for polystyrene: △, Enns et al. [18]; ▲, Fox and Flory [19]; ◁ and ▷, Glandt et al. on blends [20]; △, Krause and Iskandar [21]; △, Richardson and Savill [22]; ▲, Rudin and Burgin [23]; △ and ▲, Stadnicki et al. [24]; △, Ueberreiter and Kanig [25]; and △, Ueberreiter and Kanig on blends [26]. Curves represent solution of the G-DM equations for combinations of n = 1,2 and either r = 1.0 or V_o = 0.030 as indicated.

theory was not applicable to many polymers. This major shortcoming was attributed to the inability of the theory to take into account differences in chemical structure among polymers. The current analysis places these results in better perspective by the inclusion of equation 3 and by the addition of the index n. The first considers the energy ratio (r) as the intermolecular energy per unit of intramolecular energy, while the second defines n as the number of chain segments per repeat unit, $\bar{x}/\bar{P}$. Since each segment is assumed to occupy one lattice site, n represents a chain segment density, i.e., the number of lattice sites occupied per mer. With these modifications the constant parameter r could be plotted against n to generate the complete solution set shown in Figures 1 and 2. This procedure requires that V_o decrease monotonically with decreasing molecular weight (cf Figure 2). The net decrease in free volume as a function of molecular weight occurs because the increased free volume associated with the greater number of chain ends is more than offset by the reduced free volume associated with a lower T_g.

Recognizing the constraints imposed by the parameters r and V_o on the G-DM theory, the experimental results can be reconsidered as a function of n. If for PS, PMMA, and PVC, r is assumed to equal 1.0 (cf Figure 2 for corresponding variations in V_o), then the mean number of lattice sites occupied per mer equals ~ 1.9, 1.2, and 1.5, respectively. This implication, that PS is more flexible than either PMMA or PVC, is contrary to the results reported in the literature [28,29]. In contrast for PαMS, no single combination of n and constant r will adequately represent the empirical results, although the low value of n observed for high $\overline{P}^{-1}$ is consistent with the recognized stiffness of the molecule [28,29]. Additional analysis of several polymers having various stiffnesses [8-10,30-38] suggest that large values of n (n ≳ 10) would be required to describe the most flexible polymers such as polydimethyl siloxane. To attribute any physical significance to such large values of n is difficult, unless n is regarded as being representative of the "effective" mean number of lattice sites occupied per mer. If a dynamic connotation is more correct, then the introduction of n into the theory not only evaluates the segment size on the basis of a specified lattice space size (e.g., a methyl group) [39], but also indicates the inherent flexibility of the molecule within its lattice site, i.e., its jump frequency.

ACKNOWLEDGMENT

We wish to thank E.A. DiMarzio for helpful comments and suggestions.

This investigation was supported by NIH Research Grant No. DE02668 and RCDA No. DE00052 (RPK).

°This text is included here with the permission of the publisher, IPC Science and Technology Press Ltd. The unabridged version will appear in POLYMER.

REFERENCES

1. J.H. Gibbs and E.A. DiMarzio, J. Chem. Phys., 1958, 28, 373.
2. R.B. Beevers and E.F.T. White, Trans. Farad. Soc., 1960, 56, 744.
3. R.P. Kusy and A.R. Greenberg, J. Thermal Anal., 1980, 18, 117.
4. R.P. Kusy, M.J. Katz, and D.T. Turner, Thermochimica Acta, 1978, 26, 415.
5. R.P. Kusy, W.F. Simmons, and A.R. Greenberg, Polymer, 1981, 22, 268.
6. G. Pezzin, F. Zillo-Grandi, and P. Sanmartin, Eur. Polym. J., 1970, 6, 1053.
7. J.M.G. Cowie and P.M. Toporowski, Eur. Polym. J., 1968, 4, 621.
8. J.M.G. Cowie, Eur. Polym. J., 1973, 9, 1041.

9. J.M.G. Cowie and I.J. McEwen, Polymer, 1973, 14, 423.
10. A. Eisenberg, "Advances in Polymer Science," Springer-Verlag, Berlin, (1967), Vol. 5, p. 59.
11. R.P. Kusy and A.R. Greenberg, Polymer, in press.
12. P.D. Gujrati and M. Goldstein, J. Chem. Phys., 1981, 74, 2596.
13. E.A. DiMarzio, J.H. Gibbs, P.D. Fleming, and I.C. Sanchez, Macromolecules, 1976, 9, 763.
14. E.A. DiMarzio, Proc. N.Y. Acad. Sci., in press.
15. S.L. Kim, M. Skibo, J.A. Manson, and R.W. Hertzberg, Polym. Eng. Sci., 1977, 17, 194.
16. G.J. Pratt, J. Mater. Sci., 1975, 10, 809.
17. E.V. Thompson, J. Polym. Sci., 1966, 4, 199.
18. J.B. Enns, R.F. Boyer, and J.K. Gillham, Polym. Preprints, 1977, 18(2), 475.
19. T.G. Fox and P.J. Flory, J. Polym. Sci., 1954, 14, 315.
20. C.A. Glandt, H.K. Toh, J.K. Gillham, and R.F. Boyer, Polym. Preprints, 1975, 16(2), 126.
21. S. Krause and M. Iskandar, Proc. 10th N. Amer. Therm. Anal. Conf., Boston, 1980, 51.
22. M.J. Richardson and N.G. Savill, Polymer, 1977, 18, 3.
23. A. Rudin and D. Burgin, Polymer, 1975, 16, 291.
24. S.J. Stadnicki, J.K. Gillham, and R.F. Boyer, Polym. Preprints, 1975, 16(1), 559.
25. K. Ueberreiter and G. Kanig, Z. Naturforsch., 1951, 6A, 551.
26. K. Ueberreiter and G. Kanig, J. Colloid Sci., 1952, 7, 569.
27. S.L. Malhotra, L. Minh, and L.P. Blanchard, J. Macromol. Sci.-Chem., 1978, A12(1), 167.
28. J.M. O'Reilly, J. Appl. Phys., 1977, 48, 4043.
29. E.A. DiMarzio and F. Dowell, J. Appl. Phys., 1979, 50, 6061.
30. R.B. Beevers, J. Polym. Sci., 1964, A2, 5257.
31. R.B. Beevers and E.F.T. White, Trans. Farad. Soc., 1960, 56, 1529.
32. J.J. Keavney and E.C. Eberlin, J. Appl. Polym. Sci., 1960, 3, 47.
33. J.A. Faucher, J. Polym. Sci., 1965, 3, 143.
34. C.H. Griffiths and A. VanLaeken, Polym. Preprints, 1976, 17(2), 949.
35. B. Ke, J. Polym. Sci., 1963, 1, 167.
36. G. Allen, "Techniques of Polymer Science," Society of Chemical Industry, (1963), No. 17, p. 167.
37. J.B. Enns and R.F. Boyer, Polym. Preprints, 1977, 18(1), 629.
38. R.H. Wiley and G.M. Brauer, J. Polym. Sci., 1953, 11, 221.
39. B. Wunderlich, J. Phys. Chem., 1960, 64, 1052.

THE INFLUENCE OF CROSSLINK DENSITY ON THERMAL PROPERTIES OF
SOME EPOXY RESINS BASED ON 4,4'-DIAMINOPHENYLMETHANE
D. Argyropoylos, Department of Chemistry, McGill University,
Montreal, Canada, HBA 2K6. K.A. Hodd and W.W. Wright,
Department of Non-Metallic Materials, Brunel University, Uxbridge,
Middlesex, England, UB8 3PH.

ABSTRACT

The effects of functionality on the cure behaviour of mono, di, tri and tetra glycidyl amines based on 4,4'-diaminodiphenylmethane and some properties of the cured systems has been studied by thermal methods. To do this the following compounds having mono- di- and tri-epoxy functionality were synthesised.

4-dimethyiamino-4'-methylglycidylamino-diphenyl methane.(I) Monofunctional
4-dimethylamino-4'diglycidylamino-diphenyl methane.(IIA) Difunctional
4-4'-N-methylglycidylamino-diphenyl methane.(IIB) Difunctional
4-N-methylglycidylamino-4'-diglycidylamino-diphenyl methane. (III) Trifunctional

and the tetrafunctional intermediate, the resin MY720 was obtained from Ciba Geigy Ltd.

Tetra-N-glycidyldiamino-diphenylmethane.(MY720)Tetrafunctional

These intermediates, separately and in combination were cured with stoichio metric amounts of 4,4'-diaminodiphenylmethane to produce networks, which possessed identical chemical structural units, and differed only in their crosslink densities.

Linear relationships were observed to obtain between system functionality, and curing exotherm and between system functionality and the glass transition temperature of the fully cured resin.

The relationship between system functionality and thermal stability of the cured resin was less simple and optimal combinations of mono- and tetra-functional or di- and tetra-functional components were identified.

INTRODUCTION

In spite to a variety of studies on epoxy resins, there are few systematic studies on dependence of their properties on the structural parameters of the three dimensional network.

The above compounds, (I),(IIA),(IIB) and (III), were synthesised (1) in order to study the effects of different network parameters on the properties of the cured system.

The tetraglycidyl derivative, MY720, was commercially available, (Ciba Geigy Ltd, Duxford, Cambridge, England).

The synthesised derivatives, were mixed in different proportions with the tetra-glycidyl analogue, to produce systems of systematically varied crosslink density and the properties of the cured resins derived from these systems may then be interpreted in terms of varying network structure.

The following compositions have been studied:

Resins I, IIA, IIB, III, MY720 and mixtures of MY720 with all the resins individually in the ratios 25/75, 50/50, 75/25.

The parameters evaluated for these systems are as follows:
1) Exotherm developed during curing, followed by differential scanning calorimetry, (DSC).
2) Glass transition temperature, determined by thermomechanical analysis (TMA).
3) Thermal stability studies, followed by thermogravimetric analysis, (TGA).

EXPERIMENTAL

Curing and Curing Schedule The resins were cured with the stoichiometric amount of 4-4'-diaminodiphenyl methane (DDM) on the basis that each primary aminic proton reacts with an epoxy group.

The resin/DDM mixture was heated at 95°C, to melt the DDM to effect homogenisation.

The hot low viscosity mixture was poured into circular steel moulds which were sealed with a silicone rubber gasket.

The moulds were clamped and were heated at 165°C for 20 hours. A two hour post-cure at 200°C followed. The cured specimens were kept in small polyethylene envelopes, in a $CaCl_2$ desiccator.

Cured Resin Characterisation The various techniques used to evaluate the curing process and the properties of the resulting specimens are described below:

Differential Scanning Calorimetry DSC was used to determine the amount of heat developed (ΔH) during the curing process. A Perkin Elmer DSC-2 instrument was used at the 5-10m cal/sec sensitivity ranges. An oxygen free nitrogen stream of 20 cm^3/min was maintained through the cell during the measurements. The preweighed liquid sample pans were charged with 2-5 mg of material and dynamic scans were performed at a rate of $10°$ K/min. The instrument was calibrated by using 4-5 mg of indium, ΔH melt = 6.79×10^3 mcal g^{-1}, and the baseline was optimised according to the manufacturers instructions. The exotherms developed were determined by cutting and weighing the areas under the peaks.

Thermomechanical Analysis A loaded-column thermal indentation method was used to determine the glass transition temperature of the specimens using a Perkin-Elmer TMS-1 thermomechanical analyser.

Changes in the specimen were measured using a linearly variable differential transformer through which the probe moved. The probe was a pointed-end quartz rod with a tip diameter of approximately 0.5mm and the probe was loaded with 20g.

Thermogravimetric Analysis The samples were subjected to an increasing temperature at a steady rate of $2°$/min and an air flow of 50 cm^3/min. The change in weight of the sample was monitored by a Du Pont 951 thermogravimetric analyser, module attached to a Du Pont 990 console. Sample size was varied between 3 and 4 mg.

RESULTS AND DISCUSSION

Thermoanalytical Methods
a) Differential scanning calorimetry:

DSC was used to follow and measure the exotherms developed during the curing of the various resin compositions and table 1 summarises the results for the observed peak temperatures and the exotherms which developed.

TABLE I. SUMMARY OF DSC RESULTS FOR ALL RESIN COMPOSITIONS

Resin Composition	Peak Temperature °C	Exotherm Developed Jg^{-1}
100% I	238	128
100% IIA	199	268
100% IIB	237 (187 inflexion)	286
100% III	178	342
100% MY720	166	516
25% I, 75% MY720	197	434
25% IIA, 75% MY720	172	543
25% IIB, 75% MY720	168	555
25%, III, 75% MY720	167	515
50% I, 75% MY720	188	446
50% IIA, 75% MY720	175	417
50% IIB, 75% MY720	167	443
50% III, 75% MY720	170	420
75% I, 25% MY720	207	233
75%, IIA, 25% MY720	185	294
75% IIB, 25% MY720	167	322
75% III, 25% MY720	180	357

(All compositions were cured stoichiometrically with DDM.)

The dynamic scans of the resin systems showed broad reaction exotherms at various peak temperatures.

In two cases (resins I and IIA), due to the pronounced differences in the heat capacities of the cured resins compared to the starting materials, the baseline had to be extrapolated and the heats of reaction estimated approximately.

Assuming that the heat of reaction is directly proportional to the extent of reaction (2)(3), then the position of the exotherm peak may give an indication on the system's reactivity.

The monofunctional derivative of DDM, resin I showed the highest temperature for the onset of the curing reaction and for the maximum peak temperature.

It was found that the temperatures for curing were reduced with increasing resin functionality. ie. I > IIA > IIB > III > MY720. This order probably relates to a function of glycidyl group concentration.

In general the presence of monofunctional species together with MY720 in all proportions (i.e. 100% 75% 50% and 25%) increased the cure temperature requirements.

Curing of resin IIA commenced at a somewhat higher temperature than resin IIB, probably due to steric effects which are more pronounced in resin IIA. Resin IIB shows anomalous behaviour in the DSC dynamic scans.

A peak obtained at $237^{o}C$ is believed not to be the same maximum of the epoxy curing reaction as for all other cases. It may be a maximum of a secondary curing process which follows different kinetics.

The shape of the exothermic profiles obtained for IIB showed an inflection at $187^{o}C$ and a peak at $237^{o}C$. The inflexion implies the onset of a secondary process. The overall heat of reaction for this resin is found to be 286 Jg^{-1}, which is higher by 18 Jg^{-1} than the heat evolved from the isomeric resin IIA. This difference, being within the scatter range of results obtained from successive runs, is insignificant and cannot therefore be attributed to a process other than epoxy ring opening. Additionally, the crosslink density of all systems involving resin IIB, was found to be higher than the crosslink density of all systems involving resin IIA (1).

This, together with the fact that the exotherms developed by both resins were of a similar order of magnitude, indicated that the heat developed in both cases may be attributed to the building of the crosslinked network only.

A common feature of many epoxy resin systems is that the curing process is autocatalytic in nature and is diffusion controlled in its later stages (4).

There is evidence (5) that the rate coefficients for diffusion controlled reactions are inversely proportional to viscosity of the reaction medium.

Due to the symmetric linear geometry of the structure of resin IIB the resulting polymeric species during curing will impose less flow restrictions to the system as compared with resins IIA, III and MY720. The linear polymeric species formed by IIB during the initial stages of the reaction will have the highest mobility of the resins examined.

The isomeric resin IIA will show higher viscosity due to the bulky side groups involved in the initially formed chains, thus restricting flow and consequently lowering the rate coefficients of the diffusion controlled process.

This behaviour will be enhanced with resins II and MY720, due to crosslinking taking place at much earlier stages, because of their increased functionality.

Considering the structures of the multifunctional glycidyl amines studied in this work, resin IIB contains the less sterically hindered nitrogen atoms (6). In fact this is the only resin which will cure if left at $120^{o}C$ for 5 hours, in the absence of curing agent. It is therefore logical to conclude that, the curing process of this resin involves the highest proportion of tertiary nitrogen catalysed reactions.

Resin IIB also contained a higher proportion of chlorine (1) than the other resins and it is known that its presence affects the overall reactivity. If some chlorohydrin structure was still present then it would be expected to dehydrochlorinate at elevated temperatures and the epoxy group formed would then react very rapidly and give a peak in the exotherm.

However the elemental analysis of the cured resin I with the stoichiometric quantity of DDM showed: C 77.6% H 7.8% N 10.0%. The calculated values for the resulting molecule were: C 77.3% H 7.8% N 10.1% (assuming that reaction takes place only between oxirane groups and the primary protons of DDM). The close agreement in the two sets of values showed that in this case no reaction takes place through the tertiary aminic centers.

A plot of resin functionality, (normalizing for epoxy content) versus exotherm developed during curing, shows linear behaviour in all cases.

b) Thermomechanical analysis:

Cuthrell (7) investigated the factors affecting the cure of epoxides using thermal expansion measurements as a means of determining the extent of cure.

Two phase transitions were evident using this technique. Cuthrell demonstrated that these transition temperatures increased as the material's degree of cure increased.

For the densely crosslinked networks formed from polyfunctional glycidyl amines an indentation probe was employed in the TMA apparatus. Indentation measurements were more sensitive than expansion measurements.

Linear relationships between resin functionality and glass transition temperature were obtained for all the resins at various crosslink densities.

It is logical to expect that the presence of mono-di- and tri-functional species in the MY720 matrix will cause chain termination in the following order I > I > III, thus bringing the crosslink density to a minimum in the case of the monofunctional derivative I.

Once again significant differences between resins IIA and IIB were observed. The glass transition temperature of the latter was always found to be higher than the glass transition temperature of the systems resulting from resin IIA, (table 2).

c) Thermogravimetric analysis:

The dynamic, thermal degradation results showed that the tetrafunctional resin MY 720, was not the most thermally stable system, even though it possessed the highest crosslink density amongst the systems under study.

The results are compared in the 200 to 400°C temperature range where an overall 5-10% weight loss was observed. The majority of the TGA results are summarised in table 3.

It may be observed that the introduction of 25-50% of resin I into the MY720 matrix enhances the thermal stability from 314°C (for the 100% MY720) to 367°C for the 25% I 75% MY720 composition. On the other hand, the introduction of 75% resin I into the MY720 system results in a reduction in thermal stability.

A similar enhancement of ther thermal stability of MY720 systems was observed in the presence of 25-75% of resin IIA,

TABLE II. SUMMARY OF TMA RESULTS FOR ALL RESIN COMPOSITIONS

Resin Composition	Glass Transition Temperature, T_g °C
100% of resin I	Low temperature transition.
100% of resin IIA	88
100% of resin IIB	137
100% of resin III	142
100% of resin MY720	229
25% I with 75% MY720	177
25% IIA with 75% MY720	177
25% IIB with 75% MY720	187
25% III with 75% MY720	199
50% I with 50% MY720	102
50% IIA with 50% MY720	156
50% IIB with 50% MY720	202
50% III with 50% MY 720	207
75% I with 25% MY720	73
75% IIA with 25% MY720	116
75% IIB with 25% MY720	142
75% III with 25% MY720	207

(All compositions were cured stoichiometrically with DDM.)

TABLE III. SUMMARY OF TGA RESULTS FOR ALL RESIN COMPOSITIONS

Resin Composition	Temperatures for Areas of Maximum Weight Losses °C	Temperatures at Various Weight Loss Levels °C		
		5%	10%	50%
100% of resin I	260 332 350 490	210	239	418
100% of resin IIA	- 381 409 486	257	361	418
100% of resin IIB	258 300 376 500	259	308	477
100% of resin III	273 379 402 523	293	337	404
100% of MY720	286 351 380 499	274	314	474
25% I 75% MY720	269 367 406 493	314	367	410
25% IIA 75% MY720	282 382 400 500	285	343	480
25% IIB 75% MY720	253 340 381 525	266	311	478
25% III 75% MY720	278 379 - 483	280	329	381
50% I 50% MY720	236 369 - 490	328	366	457
50% IIA 50% MY720	281 341 379 497	290	346	477
50% IIB 50% MY720	275 380 404 509	269	332	474
50% III 50% MY720	274 381 432 488	274	321	432
75% I 25% MY720	248 367 - 500	244	298	444
75% IIA 25% MY720	- 379 434 490	311	370	451
75% IIB 25% MY720	269 384 - 487	283	372	474
75% III 25% MY720	281 323 479 532	285	337	460

(All compositions were cured stoichiometrically with DDM.)

50-75% of resin IIB and 75% of resin III.

The thermal stability of these glycidyl amine systems increases with increasing crosslink density (up to a maximum Mc value, thereafter the trend is reversed).

CONCLUSIONS

1) The glass transition temperatures, for epoxy networks based upon 4,4'-diaminodiphenyl methane glycidyl amines, vary linearly with crosslink density.
2) System reactivity is proportional to overall system functionality.
3) Steric effects play a significant role in the architecture of the cured network. Glycidyl groups attached to the same nitrogen atom generally increase the free volume of the system.
4) The thermal stability of these networks is dependent upon crosslink density. The introduction of monofunctional or difunctional components in the tetrafunctional matrix enhances the thermal stability of the latter.

REFERENCES

1. D. Argyropoylos, M.Phil. Thesis, Brunel University, (1981).
2. H.J. Borchard and J.A. Daniels, Chem Soc., 79, 41, (1957).
3. H.E. Kissinger, Anal. Chem., 29, 1702 (1957).
4. J.M. Barton, Polymer 21, 603, (1980).
5. P.E.M. Allen and C.P. Patrick, "Kinetics and mechanisms of polymerization reactions", Ellis Horwood, Chichester 1974,Chp 2.
6. T. Kakurai and T. Noguchi, J.Soc. Org. Chem. Japan, 18, 485, (1960), ibid., 64 498, (1969), and 65, 827, (1962).
7. R.E. Cuthrell, J.Appl. Pol. Sci., 37, 51, (1959).

GLASS-TRANSITIONS AND MELT-FLOW OF HOMO- AND COPOLYMERS OF ELECTRO-DONOR AND ELECTRO-ACCEPTOR MONOMER PAIRS.

H.-A.SCHEIDER, Institut für Makromolekulare Chemie der Universität, D-7800 Freiburg, Stefan-Meier-strasse 31

INTRODUCTION

According to modern concepts, the elementary act of flow supposes that a molecular-kinetic unit has to overcome an energy barrier between neighbouring equilibrium positions. At the same time there must exist some free space - a "hole" - near the initial equilibrium position, large enough to allow the molecular-kinetic unit to jump into.

Consequently the flow of fluids depend as well on the existence of "holes" near the flow unit as on the accumulation of energy to overcome the potential energy barrier during the transition in the new hole. Therefore the probability of flow may be expressed by the product of the probabilities of formation of new holes and of the overcoming the potential energy barrier.

At higher temperatures where the ammount of fractional free volume is large, the flow of a polymer melt will depend mainly on the accumulation of energy and the temperature dependence of the viscosity will be given by the equation of Eyring, worked out in extension of the theory of absolute rates on transport phenomena.

At lower temperatures, espesially near the glass transition temperature, Tg, the creation of new holes will be the rate determing step of the viscous flow and the temperature dependence obeys the well known expression of Williams, Landel and Ferry (1)

$$\log (\eta/\eta_g) = \log a_T = \frac{-C_{1,g}(T - Tg)}{C_{2,g} + (T - Tg)} \quad (1)$$

derived starting from Doolittle's relation which explains the viscosity in terms of the specific free volume.

The constants $C_{1,g} = B/2.303f_g$ and $C_{2,g} = f_g/\Delta\alpha_f$ are connected with the fractional free volume, f_g and the change of the thermal expansion coefficient, $\Delta\alpha_f$ at Tg. B is a constant and a_T the "shift factor" which allows the supperposition of the flow isotherms to "mastercurves".

Usually the viscosity of polymer melts, defined as the proportionality factor, relating the shear stress, τ, to the shear rate, $\dot{\gamma}$, $\eta = \tau/\dot{\gamma}$ depends not only on temperature, but also on stress and shear rate and is often called "dynamic viscosity".

Its limiting value for zero shear rate is the largest Newtonian viscosity and shows the usual temperature dependence of Newtonian low-molecular liquids in accordance with the equation of Eyring. Corresponding the flow will be characterized by an activation energy of flow.

The temperature dependence of the dynamic viscosity obeys the WLF-equation. Accordingly the flow of polymers is directly related to the glass transition temperature, even if other temperatures insteed of Tg are used as "reference" temperatures for the shifted mastercurves. However, to compare viscosity data the chosen reference temperature has to be correlated to Tg.

RESULTS AND DISSCUSION

In the present paper an attempt is made to corelate viscosity data of atactic polymers, measured with an Instron Rotary Rheometer by the excentric rotating disc (ERD) method, with the glass transition temperatures, measured by differential scanninc calorimetry with the DSC-2 of Perkin-Elmer.

The main attention of this study is concerned with the flow behaviour of amorphous homo- and copolymers containing donor and/or acceptor groups. N-(2 hydroxyethyl)carbazolylmethacrylate (HECM) as electro-donor monomer and methacryloyl-β-hydroxyethyl 3,5-dinitrobenzoate (DNBM) as electro-acceptor monomer are the basic structural elements of the studied polymers, synthesized by radical polymerization(2-4). The obtained data are compared with data of atactic head-to-head (a-hh-PP) and head-to-tail (a-ht-PP) polypropylenes.

The characteristics of the studied polymers are listed in Table 1

TABLE 1. Characteristics of studied polymers

Polymer	$\bar{M}_w$ +)	Tg,K	$\log \eta_o$,KPa.s/rad (at Tg+90)	C_1 (at Tg+90)	C_2
a-ht-PP	23000	260.8	0.42189	5.317	114.06
a-hh-PP	27000	242.3	0.97011	4.911	134.07
PDNBM	12000 ++)	359.65	1.6129	4.73	76.0
Co7DNBM/1HECM	13000	378.25	-	-	-
Co3DNBM/1HECM	13000	389.55	1.3260	13.64	214.6
Co5DNBM/3HECM	15500	398.7	-	-	-
Co1DNBM/1HECM	16000	403.9	3.7883	13.84	164.2
Co3DNBM/5HECM	22000	412.9	-	-	-
Co1DNBM/3HECM	32000	414.3	2.3889	9.17	137.6
Co1DNBM/7HECM	41500	415.5	-	-	-
PHECM	57000	415.85	2.7963	8.01	151.2

Explanations to table 1:

+) The molecular weights were determined by gel permeation chromatography.

++) The figures in the copolymer formulas correspond to the respective content of comonomers in the copolymer.

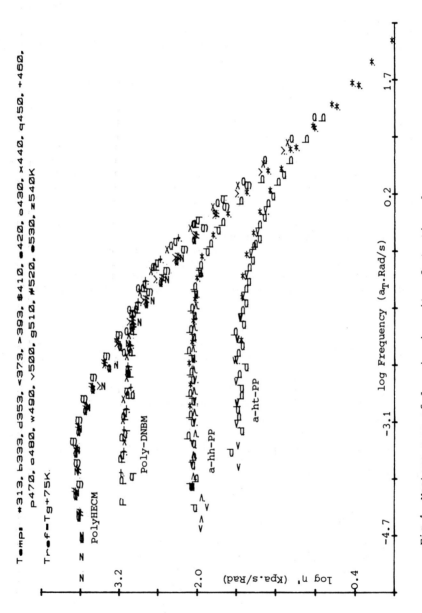

Fig.1 Mastercurves of dynamic viscosity of atactic polymers.

Comparing the Tg-values of the D/A polymers and copolymers with the glass-transition temperature of PMMA (Tg=386.8 for a radicalic PMMA M_w = 45500, polymerized in the same conditions) it seems that the electro-donor HECM-sequence has a higher stiffenes. The Tg behaviour of the DA-copolymers may be expressed in terms of the Gordon-Taylor-Wood relation (5).

The zero shear viscosities of poly-DNBM and poly-HECM are in the normal molecular weight order, but substantially higher than the viscosities of the atactic polypropylenes (Fig.1).

On the contrary the zero shear viscosities of the DA-copolymers deviate from a normal molecular weight order (Fig.2), the 1:1 copolymer showing the highest zero shear viscosity, even much higher than the poly-HECM, although the molecular weight of the latter is twice of that of the copolymer.

This behaviour may be explained by inter- and/or intramolecular interactions due to the presence of donor and acceptor groups. In polymers with predominant donor or acceptor groups, intermolecular interaction will predominat and due to repulsive interactions may loose the encoiling of the polymer coil. These effects will be higher in the homopolymers as in the copolymers where donor and acceptor groups are blocked up partially. In fact the viscosities of the homopolymers are higher than of the corresponding electro-unequilibrated copolymers.

In the electro-equilibrated 1:1 DA-copolymer intramolecular interactions may act against the viscous flow. Both effects will rise the zero shear viscosity of the polymer. Experimental data, however, suggest a more pronounced effect of the intramolecular interaction.

The specific flow behaviour disappears at higher shear rates.

An attempt to use the constants of the WLF-equation, together with the Simha-Boyer rule (6)

$$\Delta\alpha_f \cdot Tg = 0.113 \qquad (2)$$

in order to evaluate fractional free volume data at Tg failed, the obtained values being much to low when compared with the WLF-value of f_g=0.025.

Therefore measurements of thermal expansion coefficients have to complet this study for a better understanding of the correlation between glass-transition, structure and viscous flow of atactic DA-polymers.

Cenerous financial support by AIF is gratefuly acknowledged.

REFERENCES

1. M.L.Williams, R.F.Landel and J.D.Ferry, Journ.Amer.Chem.Soc., 77, 3701 ;1955)
2. V.Percec, A.Natanson and C.I.Simionescu, Polym.Bull., 5, 217, 225 (1981)
3. C.I.Simionescu, V.Percec and A.Natanson, ibid., 3, 535 (1980)
4. C.I.Simionescu, A.Natanson and V.Percec, Ibid., 3, 543 (1980)
5. L.A.Wood, J.Polymer Sci., 28, 319 (1958)
6. R.Simha and R.F.Boyer, J.Chem.Phys., 37, 1003 (1962)

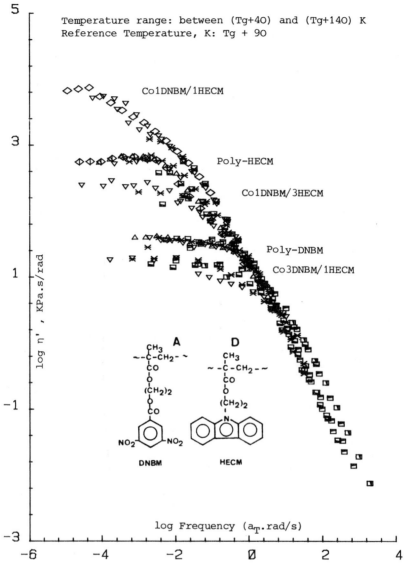

Fig.2 Mastercurves of dynamic viscosity of DA-polymers
squares, temperatures below T_{ref};
rhombs and triangles temperatures above T_{ref};
× temperatures near T_{ref}

KINETIC STUDIES ON THE OXIDATION CHARACTERISTICS
OF CARBONS PREPARED BY THE CARBONISATION OF
VARIOUS POLYMERS.

D. Dollimore and P. Campion,
Department of Chemistry and Applied Chemistry,
University of Salford,
Salford. M5 4WT
England

ABSTRACT

The preparation of carbons by the carbonisation of various polymers at 700°C in nitrogen, followed by further heat treatment at 2000°C and 2700°C in argon is described.

The carbons were oxidised in air and oxygen over the range 400-600°C. The data shows either zero order or a slightly accelerating reaction. Oxidation rates calculated from the Polanyi-Wigner equation are compared with experimental rates. A factor which unifies the theoretical and experimental rates is suggested. It appears that the most probable mechanism is immobile adsorption of the oxygen onto the surface followed by a rate determining dissociation process.

INTRODUCTION

In this study the oxidation of various carbons is reported. An early review on this topic is by Walker et al [1]. In the gasification of carbon by oxygen chemisorption of oxygen is involved [2]. Only a small proportion of the surface is involved in this process [3]. At higher temperatures the chemisorbed oxygen is degassed, when usually more carbon monoxide is released than carbon dioxide; most of the carbon dioxide is lost below 600°, and the carbon monoxide above this temperature [4,5]. In total gasification studies with emphasis on amount of gas produced attention has been drawn to whether carbon dioxide or carbon monoxide is the primary product [6-9]. In this and other studies attention is drawn to the amount of carbon gasified [10,11,12]. An important factor is then the reaction interface [13-18]. There have been attempts made to identify this surface as it could lead to the establishment of the rate of carbonisation per unit area of surface [10, 19-21].

EXPERIMENTAL

<u>Materials</u>. The materials from which the carbons were prepared are lactose, polyfurfuryl alcohol, polyvinyl alcohol, polyphenol-formaldehyde, and polyvinyl chloride. The lactose was supplied by General Chemical Co. Ltd., (specification, sulphated ash - less than 0.1%, copper 3 ppm, arsenic 1 ppm, iron 2 ppm, lead 2 ppm). Polyfurfuryl alcohol was prepared from the monomer by adding 0.4% 2N hydrochloric acid (by vol.) to the monomer when polymerisation took place over several months at room temperature. Polyvinyl alcohol was

supplied to following specifications; specific gravity 1.19∿1.27, ash content less then 1.0%, pH of 4% aq. soln. 6∿8. Polyphenol-formaldehyde was supplied as I.C.I. Mouldrite Resin No. 425. It was heated at 150°C for 24 h. to produce a brittle blood-red solid. Polyvinyl chloride was supplied by British Geon Ltd. as Geon 113.

Bulk Carbonisation. This was carried out in a muffle furnace with the specimens in two 3 in. diam. silica crucibles, in presence of dry oxygen free flowing nitrogen (1 min^{-1}). A heating rate of 5°C min.$^{-1}$ was used up to 700°C, and this maintained for 2h., cooled to room temperature and then removed from nitrogen atmosphere.

Graphitisation. A graphitisation furnace with an argon atmosphere was used. Heating schedules were:

$$20°C \xrightarrow{5h.} 1500°C \xrightarrow{2h.} 2000°C$$

and

$$20°C \xrightarrow{6h.} 2500°C \xrightarrow{2h.} 2700°C$$

Specimens were held at required temperature for $1\frac{1}{4}$h., cooled under flowing argon for 18h. and stored over magnesium perchlorate dessiccant.

Oxidation Unit. Gasification experiments with oxygen at one atmosphere (flow rate 60 cc min.$^{-1}$) utilised an enclosed unit of silica springs. Approximately 0.25 g (accurately weighed) of carbon was used and outgassed at 200°C under vacuum, and temperature rapidly raised to the oxidation temperature. A Stanton thermobalance was used for experiments in flowing air.

RESULTS AND DISCUSSION

Samples were carbonised at 700°C and graphitised at 2000°C and 2700°C. The specific surface areas (calculated from adsorption data) of the carbons were all less than 2 m^2 g^{-1}. Oxidation data is presented firstly as % wt. loss (and hence fraction decomposed, α) with respect to time (t), and on the basis that:-

$$\frac{\text{Specific Reaction Rate}}{\text{per unit time}} = \text{g of carbon gasified (units g/gh}^{-1}\text{)}$$

$$\frac{\text{per unit time}}{\text{g of carbon remaining}} = \frac{d\alpha/dt}{(1-\alpha)}$$

as $\frac{d\alpha/dt}{(1-\alpha)}$ against time

Typical plots are shown in Figures 1 and 2. Table 1 summarises the specific reaction rates in air and oxygen. Samples prepared at 2000°C showed approximately constant reaction rates in air except for the lactose carbon. The estimate of polyfurfuryl alcohol carbon 138 kJ mol^{-1}.

The general behaviour pattern of the carbons is that low temperature carbons showed a decay mechanism (first order kinetics) but for the specimens prepared at higher temperatures the oxidation kinetics were only very approximately zero order and could be described as acceleratory. On the assumption that the rate of reaction per unit area is constant a decay mechanism can only mean that the surface area available to the attaching oxygen is decreased whilst an accelerating reaction rate would indicate an increase in accessible surface area. This is a simplification but the model will be true in the absence of other factors such as the presence of catalytic impurities. Thus a plot of $d\alpha/dt$ versus time (t) effectively describes the changes in surface area that occurs during oxidation and if the initial surface area is known, or if given an arbitrary value of unity, the surface area can be calculated at any other time during the oxidation, and this is illustrated in Table 2. In those carbons where nitrogen BET surface areas have been determined at various burn-offs [10,11,12] then the plots do parallel the variation in $d\alpha/dt$ with time (t).

Gregg [23] and others [24,25] have used the Polanyi-Wigner expression in the form:

$$\frac{-dz}{dt} = N_o \nu e^{-E/RT}$$

where

N_o = Number of molecules per cm^2
 = $\frac{r\rho N}{M}$

ν = frequency of activation

 frequency of vibration of atoms in the solid lattice $10^{13} s^{-1}$

r = distance between atoms of carbon

M = atomic weight

ρ = density of carbon

N = Avogadros Number

E = Energy of Activation

T = Temperature in degrees Kelvin

R = Gas Constant

$\frac{-dz}{dt}$ = Number of molecules decomposing per sec per cm^2

Putting appropriate data to these terms gives:-

$$\log_{10} \frac{-dz}{dt} = \frac{-E}{2.303.RT} + \log_{10} 2.3 \cdot 10^{28}$$

The value for the activation energies found in the present study are of the order 170 kJ mol^{-1}. The temperature dependence of this predicted reaction rate is shown in Figure 3. The experimental reaction rates are listed in Table 3. The predicted rate from the Polanyi-Wigner equation is also shown in these Tables. The rates in these Tables were calculated from a zero order model. The theoretical oxidation rates compared with the experimental rates are a factor of 10^5 too high for the 2700°C carbons and a factor of 10^4 too high for the 2000°C carbons. This could be due to the fact that the Polanyi-Wigner equation does not take into account variations in reactivity caused by previous heat treatment. Apart from the effect of such heat treatment upon the pore geometry and impurity content, Watt and Franklin [26] have shown that the most reactive part of a carbon is that which exists as single graphite layers. A disordered amorphous mass is less reactive than such a graphite layer but is still more reactive than the parallel stacking found in graphite. Therefore, further heat treatment decreases the reactivity by increasing the proportion of parallel-stacked groupings If the Polanyi-Wigner equation is written:-

$$\log \frac{-dz}{dt} = K - \frac{E}{2.302RT} + \log_{10} 2.3 \cdot 10^{28}$$

then
$$K = 10^{-5} \text{ for 2700°C carbons}$$
and
$$K = 10^{-4} \text{ for 2000°C carbons}$$

This can be regarded as a measure of these effects on the basic reactivity of the carbon under consideration.

Eyring et al [27] have put forward a number of models based on the absolute reaction rate theory Gulbransen and Andrew [21] have calculated theoretical rates predicted by these models and compared then with experimentally observed rates. By applying these models to the experimental values found in this study it is concluded that the most probable mechanism is immobile adsorption of oxygen onto the surface followed by a rate-determining dissociation process

REFERENCES

1. P.L. Walker, F. Rusinko and L.G. Austin, Advances in Catalysis (Ed. D.D. Eley, P.L. Selwood and P.B. Weisz), Academic Press, Vol. XI, 1959, p.133.
2. P.J. Hart, P.J. Vastola and P.L. Walker, Carbon, 1967, 5, 363.
3. W.V. Loebenstein and V.R. Deitz, J. Phys. Chem., 1955, 59, 481.
4. R.B. Anderson and P.H. Emmett, J. Phys. Chem., 1952, 56, 753.
5. J.F. Norton and A.L. Marshall, Trans. Am. Inst. Mining Met. Engrs. 1944, 156, 351.
6. R.F. Strickland-Constable, Trans. Farad. Soc., 1944, 40, 333.
7. X. Duval, J. Chim. phys., 1950, 47, 339.
8. T.F.E. Rhead and R.V. Wheeler, J. Chem. Soc., 1913, 103, 461.
9. J.R. Arthur, Trans. Farad. Soc., 1951, 47, 164.
10. D. Dollimore and A. Turner, Trans. Farad. Soc., 1970, 66, 2655.
11. D. Dollimore and A. Turner, 3rd Conf. Indust. Carbon and Graphite (Soc. Chem. Ind. Lond.), 1970, p.65.
12. D. Dollimore and A. Turner, Gas Chemistry in Nuclear Reactors and Large Industrial Plants (Ed. A. Dyer) Heyden, 1980, p.164.
13. B.G. Tucker and M.F.R. Mulcahy, Trans. Farad. Soc., 1969, 65, 274.
14. A.R. Blake, C.A. Hempstead and P.P. Jennings, J. Appl. Chem., 1964, 14, 115.

15. A.R. Blake, J. Appl. Chem., 1964, 14, 382.
16. S.J. Gregg and R.F.S. Tyson, Carbon 1965, 3, 39.
17. D.D. Chadra, R.G. Barradas and M.J. Dignam, J. Colloid and Interface Sci., 1973, 44, 195.
18. E.A. Gulbransen, K.F. Andrews, F.A. Brassart, Vac Microbalance Tech., 1965, 4, 127.
19. G. Blyholder and H. Eyring, J. Phys. Chem., 1957, 61, 682.
20. G. Blyholder and H. Eyring, J. Phys. Chem., 1959, 63, 1004.
21. E.A. Gulbransen and K.F. Andrew, Ind. Eng. Chem., 1952, 44, 1034.
22. P. Campion, D. Dollimore, G. Mason and A. Turner, to be published.
23. S.J. Gregg and R.I. Razouk, J. Chem. Soc., 1949, 536.
24. P. Tobley and J. Hume, Proc. Roy. Soc., 1928, 120A, 209.
26. J.D. Watt and R.E. Franklin, Nature, 1957, 180, 1190.
27. H. Eyring, S. Glasstone and K.J. Laidler, The Theory of Rate Processes, McGraw Hill, 1941.

Table 1 Comparative Oxidation Rates in Air and Oxygen.
Values of Specific Reaction Rate $\dfrac{d\alpha}{dt(1-\alpha)}$

	AIR				OXYGEN	
Heat Treatment Temperature	2000°C		2700°C		2000°C	2700°C
Oxidation Temperature	500°C	550°C	550°C	600°C	500°C	550°C
	Values at $\alpha=0.1$					
Order of Decreasing Reactivity (i)	Lactose 23.0x10^{-3}	Lactose 83.2x10^{-3}	PVA 22.0x10^{-3}		Lactose 177x10^{-3}	PPF 12.2x10^{-3}
	PVA 20.0x10^{-3}	PVA 70.0x10^{-3}	PPF 12.0x10^{-3}	PPF 60.0x10^{-3}	PVA 22x10^{-3}	Lactose 12.1x10^{-3}
	PPF 5.3x10^{-3}	PPF 20.0x10^{-3}	Lactose 12.0x10^{-3}	Lactose 55.0x10^{-3}	PPF 5.5x10^{-3}	PFA 10.0x10^{-3}
	PFA 2.0x10^{-3}	PFA 10.0x10^{-3}	PFA 10.0x10^{-3}	PFA 27x10^{-3}	PVC 5.0x10^{-3}	PVC 5.6x10^{-3}
	PVC 1.7x10^{-3}	PVC 9.0x10^{-3}	PVC 6x10^{-3}	PVC 23x10^{-3}	PFA 4.0x10^{-3}	

Notes: Abbreviations are: PVC - Polyvinyl chloride carbon; PVA - Polyvinyl alcohol carbon; PFA - Polyfurfuryl alcohol carbon; PPF - Polyphenol-formaldehyde carbon; Lactose - Lactose carbon

Table 2 Available Surface Area in Arbitrary Units w.r.t. % burn off for Carbons parpared at 2700°C and Oxidised in Air.

At. 550°C

Time h	PVC carbon		PVA carbon		PFA carbon		PPF carbon		Lactose carbon	
	% Burn off	Surface area	% Burn off	Surface area	% Burn off	Surface area	% Burn off	Surface area	% Burn off	Surface area
5	1.3	1.0	9.2	1.5	2.9	1.2	4.0	1.0	2.9	2.0
10	2.6	1.5	20.1	1.6	5.8	1.8	9.4	1.4	8.4	2.8
15	4.6	2.0	31.2	1.6	10.7	2.0	16.0	1.6	13.1	3.3
20	6.9	2.5	42.3	1.6	15.7	1.8	23.6	1.9	23.2	3.8

At. 600°C

1	1.0	1.0	—	—	1.7	1.0	3.6	1.0	2.5	1.0
2	2.3	1.3	—	—	3.4	1.0	7.8	1.2	5.8	1.3
3	3.8	1.5	—	—	5.2	1.1	13.4	1.5	10.7	2.0
4	5.3	1.5	—	—	7.3	1.2	19.3	1.6	16.0	2.1
5	7.1	1.8	—	—	9.7	1.4	26.6	2.0	—	—
6	9.2	2.1	—	—	12.5	1.6	34.7	2.2	—	—
7	11.6	2.4	—	—	15.1	1.6	43.5	2.4	—	—

Table 3 Experimental Oxidation rates (dz/dt) of carbons

Specimen	Temperature of Oxidation			
	Number of Carbon Atoms Gasified per cm^2 per s		Number of Carbon Atoms Gasified per cm per s	
	500°C	550°C	550°C	600°C
PVC carbon	1.9×10^{12}	7.3×10^{12}	2.1×10^{12}	1.1×10^{13}
PFA carbon	2.5×10^{12}	1.0×10^{13}	6.2×10^{12}	2.1×10^{13}
Lactose carbon	2.5×10^{13}	1.1×10^{14}	4.9×10^{12}	3.1×10^{13}
PPF carbon	1.0×10^{13}	2.5×10^{13}	1.0×10^{13}	4.4×10^{13}
PVA carbon	1.8×10^{13}	8.0×10^{13}	2.2×10^{14}	—
Predicted Rate from Polanyi-Wigner Equation	1.2×10^{17}	5.6×10^{17}	1.2×10^{17}	5.6×10^{17}
	Sample prepared at 2000°C		Sample prepared at 27 0°C	

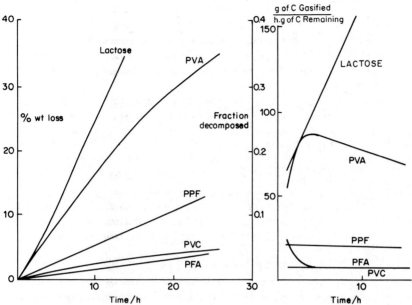

Figure 1. Oxidation of 2000°C heat treated carbons in air at 500°C

Figure 2. Specific reaction rate vs time for 2000°C heat treated carbons at 500°C in air

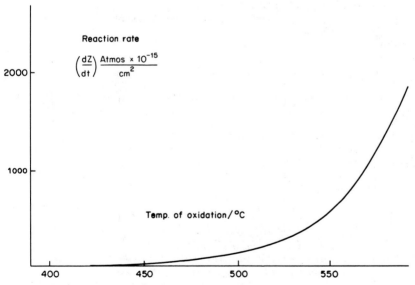

Figure 3. Reaction rate (from Polanyi-Wigner)

THE THERMAL STUDY OF CROSS-LINKED POLYELECTROLYTES
(ION-EXCHANGE MEMBRANES).
Roman Dabek and Rodolfo Morales.
ESIQIE-Instituto Politécnico Nacional.
México 14, D. F.
México.

INTRODUCTION

Thermal analysis under nonisothermal and isothermal conditions in the form of differential scanning calorimetry (DSC) differential thermal analysis (DTA) and thermogravimetric analysis (DTG) have been used successfully in thermal studies of polymers for a number of years. Only a few of these papers, however, have been devoted to cross-linked polyelectrolystes and more exclusively to ion-exchange resin [1-8], i.e. "homogeneous" polyelectrolytes. Up to date no publication dealing with "pseudohomogeneous" ion-exchange membranes has appeared. Ion exchange membranes are used for: electrodialysis systems for desalination of brackish water or seawater, sugar solutions and other aqueous solutions; membranes fuels cells and other important applications [9].

The pseudohomogeneous membrane forms the system of two polymers of which one plays the role of a carrier (for example, polyethylene film is most often used as matrix) and is the continuous phase, while the other is a cross-linked polyelectrolyte (e.g. polystyrenesulphonic acid) dispersed in the continuous phase. In their swollen state they represent a system of two polar phases: the phase of hydro- or fluorocarbon network with functional groups built in and the phase of the internal solution. In their dried state, they represent only one polymeric phase. The structure of this phase was studied by various methods, also by thermal methods.

The present investigation was undertaken to find the correlation between the thermal curves of a cation exchange membrane and the thermal properties of all components of the polymeric phase of the membrane.

The purpose of this paper is to illustrate further information of the DSC, DTA, TG, DTG, for cation exchange membranes available from simultaneous nonisothermal analysis.

EXPERIMENTAL

Two types of cation exchange membranes with $-SO_3X$ groups were examined: a) pseudohomogeneous membrane obtained from polyethylene (low density) chemically modified with the copolymer of styrene and DVB, designated by the symbol PE/PSS in the further text, which was prepared in the Institute of Organic Technology and Plastics, Wroclaw Polytechnical University (Poland)

(Fig. 1a). The preparation of these membranes were described earlier [10-11]; b) membrane obtained on a film of fluorinated polymeric material i.e.: membrane AMF C-311 (American Machine and Foundry Co.); perfluorinated membrane NAFION (Du Pont) fabricated from copolymers of tetrafluoroethylene and monomers such as perfluoro-3,6-dioxa-4-methyl-7-octene sulfonyl fluoride (Fig. 1b).

The examination of other polymers were performed to explain endo- and exothermic effects occuring in the DTA and DSC curves of cation exchange membranes. These polymers were: the initial polyethylene film (PE); the polyethylene film modified with the copolymer of styrene and DVB with different degrees of cross-linking (before sulphonation); tetrafluoroethylene etc.

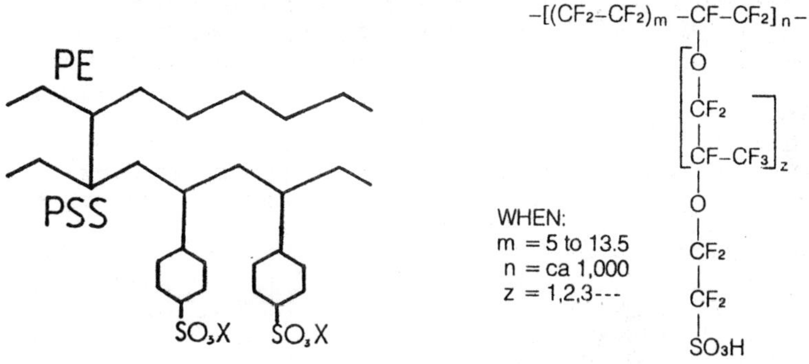

Fig. 1. The chemical structure of: a) cation exchange membrane obtained of the polyethylene film; b) NAFION perflourosulphonic acid-membrane.

The thermal studies were carried out using: a) Differential Scanning Calorimetry (DSC) Model DSC-1B, DSC-2 System and Model TGS-2 Thermogravimetric System made by Perkin Elmer; b) Derivatograph (DTA, DTG and TG measurements) [12-13]. Samples of about 10mg of the polymers foil and membranes in air dried state were taken. The examination was performed with heating rates of 2K to 5K per minute from 173K up to 800K in a 20 litre per hour (argon, helium or air) stream.

RESULTS AND DISCUSSION

Thermal characterization is completed by recording thermograms of the ion-exchange membrane in an inert gas. Some of the characteristic parameters obtainable are illustrated in Fig. 3, providing three sources of information: (i) Transitional: glass transition, melting, (ii) Quantitative: mass change of reactant mass (M), heat of reaction and maximum rate of reaction (R). Also kinetic parameters such as activation energy preexponential factor and reaction order may be obtained by several methods,

(iii) Qualitative (comparative): initial, final and max. rate of reaction temperatures for TGA and DTA with suitable interpolated baselines. Reaction type (i.e., exothermic, endothermic). Collectively, these parameters describe the thermal behaviour.

As the determination of the glass transition temperature (T_g) of studied systems (i.e., PE and PE modified) by the DSC method was proven impossible. T_g's were determined with thermomechanical curves. These measurements were carried out using an apparatus made by the Institute of Polymers, Polytechnical University of Łódź (Poland). T_g of PE and PE modified are 255-268K [14].

On the basis of the DSC or DTA curves obtained it was possible to determine the phase transition, i.e., melting point and ranges, heat of fusion, crystalization etc., of the ion-exchange membranes. (See Fig. 2 and 3).

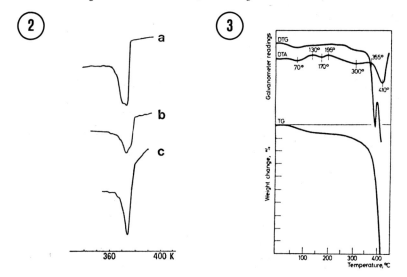

Fig. 2. DSC Thermograms of: a) initial polyethylene film (PE); b) the polyethylene film modified with copolymer of St/DVB; c) cation exchange membrane PE/PSS.

Fig. 3. Thermal curves of the ion exchange membrane AMF C-311, in argon atmosphere.

For the initial polyethylene film (curve 1) it is 384K; for the film modified with the St/DVB copolymer it is 386K (curve 2). The melting point of the matrix (PE) in the ion-exchange membrane is about 386K (curve 3). Compared with pure PE there is a small change of about 3K

The energy of fusion of PE modified is 31.06 cal/gram and the degree of crystallinity of ion-exchange membrane is 47%. The latter value is of the same order as that of the pseudohomogeneous membranes [15] and of the polyethylene carrier. X-rays investigation confirmed this value. The experimental method which we used to study the degree of crystallinity of ion-exchange membrane was based in the work described by B.Kee [16].

As known from the literature, the DTA and DSC curves in the temperature range 384-773K for low density polyethylene and 410-773K for high-density polyethylene show no thermal effects when the heating is performed in a vacuum [17] or in an inert atmosphere (e.g., nitrogen, argon, helium, etc.). The appearance in the temperature range 410-900K of exo- and endothermic effects in the thermograms of ion-exchange membrane can be assigned to the dispersed phase of the polyelectrolyte which is the cross-linked polystyrenesulphonic acid in the case cation exchange membranes. (See Fig. 4 and 5).

This provides a reason, therefore, for using the thermal method to study the thermal behaviour of the cross-linked polyelectrolyte phase in ion-exchange membranes. Analysis of curves (DTA, DSC, DTG, TG, etc.), gives valuable information about the structure of the phase and thermal decomposition.

The three stages of weight loss (see Fig. 5 and 6) can be visibly explained by the occurrence of the following processes [1,2] a) an endothermic dehydration process over the temperature range 300-450K; b) an endothermic desulphonation process over the temperature range 450-650K; c) exothermic processes at thermo-oxidizing polymer decomposition, carbonization and complete com-

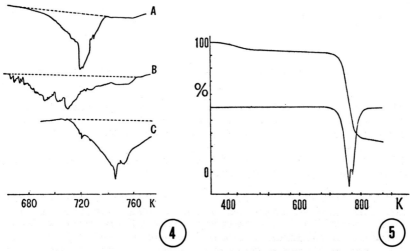

Fig. 4. Thermal decomposition of: a) the initial polyethylene film (PE); b) the polyethylene film modified with copolymer of St/DVB; c) cation exchange membrane PE/PSS (in K^+- form) in nitrogen;

Fig. 5. TG and DTG curves of cation exchange membrane PE/PSS (in K^+-form) in nitrogen.

bustion of crack carbon at temperatures above 650K.

From ambient temperature up to 250K the liberation of differently bound water and the partial decomposition of the ion-exchange membrane take place simultaneously. In the literature a few papers can be found dealing with determination (quantitative results) of the water contents of air-dried or swollen ion-exchangers by simultaneous TG, DTG and DTA measurements (derivatograph) [5-7]. In this investigation cation exchange membrane in air-dried state was examined, e.g. the samples of these membranes contained a low quantity of water on account of the curve TG (see Fig. 5). In the range of temperature 313K-473K could be the weight loss region observed and it is attributed to dehydration. The dehydration process in cation exchange membrane associated with a small endothermic effect and in the same region of temperature occurs also in the region of melting of the polyethylene carrier (in the case PE/PSS membranes).

Thermal decomposition process of the cross-linked polystyrene sulphonic acid in the matrix phase (PE). The threshold temperature of the desulphonation process and the loss of weight in the temperature range of reactions desulphonation are dependent upon the ionic form of ion exchange membrane (e.g. kind of counter-ion in functional group $-SO_3X$). The thermogram behaviour observed for cation exchange membrane in K^+ form is illustrated in Fig 5. The desulphonation process occurs above 690K.
Since thermal decomposition of the cross-linked polyelectrolyte takes place, many gaseous products are formed and they must all diffuse through the matrix phase (PE). In the case of the membranes considered, thermal decomposition of the dispersed phase of the cross-linked polyelectrolyte proceeds in visco-flowing state of PE (matrix phase).

The investigations of ion-exchange membranes were performed also in the oxidizing atmosphere. Threshold decomposition temperature in air for PE is 483K. They formed the basis of conclusions on the thermooxidative destruction of cross-linked polystyrenesulphonic acid and its salts, when the destruction takes place in a polymer matrix in different physical states.

The influence of the physical state of the matrix on the thermal decomposition of cross-linked polyelectrolyte. In order to determine the influence of the physical state of the matrix on the thermal decomposition of cross-linked polystyrenesulphonic acid, the graft copolymer membrane AMF C-311 synthesized on a fluorinated polymeric backboneing was also investigated (Fig.3).

A very high melting point of the matrix (682K), higher than the decomposition temperature of the sodium salt of polystyrenesulphonic acid (603K) [18], is the reason why the thermal decomposition of the polyelectrolyte in the AMF C-311 membrane occurs under quite different conditions when the matrix is in the visco-elastic state. In comparison with the thermal decomposition processes of the cross-linked polyelectrolyte in the PE matrix and in the fluorinated polymer matrix, some differences are revealed.
These differences can be explained for example, by the different technologies of preparation of the membranes.

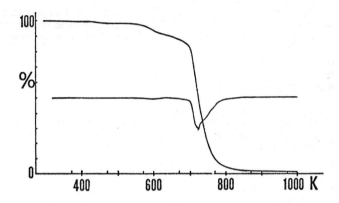

Fig. 6 TG and DTG curves of NAFION membrane. Heating rate of 5K min^{-1} in N_2.

Thermal decomposition process of the perfluorinated membrane NAFION. The investigations of NAFION membranes (See Fig. 6) formed also the basis of conclusions on the thermodestruction perfluorosulphonic acid and its salts. Although NAFION perfluorosulphonic acid membranes are designed for diverse uses, their thermal properties are different and dependent on the thermal and combustion decomposition products of the sulphonic acid analogs of the sulphonyl fluoride copolymers. NAFION membranes are not suitable for melting process. These are compared with the known thermal degradation of the analogous structure for TEFLON [19-20] and fluorocarbon resin.

The endothermic desulphonation process of NAFION N-117 membrane is about 546K in anhydrous systems. The maximum temperature of this process is 597K. Thermal decomposition begins before the products become fluid enough for shaping. In comparison with the thermal decomposition process of the AMF C-311 membrane and in the NAFION N-117 membrane, some differences are revealed.

However, like other fluroplastics, NAFION membranes resist decomposition at higher temperatures than do most other thermoplastics.

ACKNOWLEDGEMENTS

The authors wish to thank Dr. Alejandro Peraza García and Dr. Jorge Ayala. (I.I.E. Cuernavaca). Thanks is also due to Mrs. Ing. Marina E. Rincón (I.I.E. Cuernavaca) for her assistance with part of the experimental work.

REFERENCES

1. V.S. Soldatov, A.I. Pokrovskaya and L.I. Tsukurova, Zh. Fiz. Khim., 5 (1968) 1258.
2. G.F.L. Ehlers, K.R. Fisch and W.R. Powell, J. Polymer Sci. A-1,7 (1969) 2955.
3. I.F. Gleim, V.A. Moichenko and V.S. Doldatov, Ionnyi Obmen Ionity 1970, pp. 58-64. Edited by Samsonov G.V. "Nauka" Leningrad OTD USSR.
4. V.S. Soldatov, Z.R. Pavlovskaya and I.F. Gleim ibid p. 64-70.
5. J. Inczédy, J. Thermal Anal. 13 (1978) 257.
6. J. Kristof, J. Inczédy, J. Paulik and Paulik, J. Thermal Anal. 15 (1979) 151.
7. J. Kristof and J. Inczédy, J. Thermal Anal. 19 (1980) 51.
8. W. Balcerowiak and Terelak K,J.Thermal Anal. 18 (1980) 271.
9. S.T. Hwang and K. Kammermeyer, Membranes in Separations Techniques of Chemistry Volume VII J. Wiley & Sons. 1975.
10. J. Lindeman, H. Czarczynska, Polimery 14 (1969) 601.
11. H. Czarczynska, J. Lindeman, Polimery 15 (1970) 450.
12. F. Paulik, J. Paulik and L. Erdey, Z. Anal. Chem., 160 (1958) 241.
13. F. Paulik, J. Paulik and L. Erdey, Z. Anal. Chem., 160 (1958) 321.
14. R. Dąbek, Polimery 25 (1980) 408.
15. A. Narębska, Proc. First Conference on Ion Exchange Materials "Synthesis, Properties and Application Ion Exchange Resins and Membranes", Toruń 1972 p. 45.
16. B. Ke, The Meaning of Crystallinity in Polymers, Polym. Symp, 18 (1967)
17. F. Danusso and G. Palizzotti, Chim. Ind. (Milan), 44 (1962) 241.
18. V. A. Kargin, T.A. Koretskava, A. M. Kharlamova, G.S. Markova and Yu. K. Ovtchinnikov, Vysokomolekularnyje Soedinenya, 8 (1971) 1811.
19. S.L. Madorsky, V.E. Hart, S. Strauss, V.A. Sedlak, J. Res. Nat. Bur. Stand. 51 (1953) 327-333.
20. S.T. Kosiewicz, Thermal Acta 40 (1980) 319-326.

DSC STUDY OF UNSATURATED POLYESTER RESIN STABILITY
by
Stephen P. Molnar
Armco, Inc.
Research & Technology
Middletown, Ohio 45043

INTRODUCTION

Sheet Molding Compound (SMC) is a complex mixture of a number of components, some reactive and some present in the formulation as an inert diluent. The SMC formulation employed in this study is given in Table 1. Also illustrated are the structural formulae of the initiators studied, tert-butylperbenzoate (TBP), A, and 1,1-bis-(t-butylperoxy)-cyclohexane (USP 400), B. The principle reactive species are resin, a styrenated oligomer of acid and diol, and styrene added as additional crosslinking agent and for viscosity control. The formulation contains both fiberglass and carbon fiber as reinforcement. Problems were encountered in molding the formulation which had been held under appropriate storage conditions for a period of three months. A study of this system was undertaken in an attempt to determine the cause of the poor storage stability of the formulation.

Few studies of the curing reactions of SMC have appeared in the literature[1-4].

EXPERIMENTAL

Differential Scanning Calorimetry data were obtained in air with a Perkin-Elmer DSC II employing standard laboratory practices. SMC pastes were run when new, after aging for three months without carbon fiber, and after being aged for three months in contact with carbon fiber. Aging occurred at ambient laboratory conditions in closed containers.

DISCUSSION

Figure 1 gives a direct comparison of the reactivity of two SMC initiators, in the form in which they are received from the manufacturer. The USP 400 was a 75% solution in an inert phthalate solvent, the TBP was studied as a neat liquid.

Figure 2 gives a direct comparison of TBP and USP 400 in the same SMC resin system. Table 2 summarizes the data obtained from this series of figures.

Although the exothermic enthalpy of reaction of both initiators is approximately three times that of the resin/initiator system, the weighted contribution of the initiator to the evolved energy of the formulations is less than 0.8% and is not considered further. The presence of initiator results in a displacement of the curing exotherm to a lower temperature.

It is reasonable to assume that the rate of evolution of enthalpy is directly proportional to the rate of reaction of double bonds in the polyester and the styrene. The area under the curve is the total exothermic enthalpy of reaction.

Plots of α vs. T, Figure 3 for TBP, are descriptive of the temperature dependence of the amount of cure, in these examples at a fixed heating rate of 10 K/min, where α is the extent of reaction.

Let us turn to a consideration of the extraction of kinetic parameters from DSC studies of these two systems. Kinetic models for unsaturated polyester curing reactions can provide insight into the reaction mechanisms and provide predictability. As a minimum, such models should allow interpolation of kinetic data within the experimental temperature region. Classically, the isothermal rate law for the species A may be written:

$$-\frac{dA}{dt} = k A^n \qquad (1)$$

in which k = rate constant,
A = concentration of reactive species (double bond)
n = "reaction order", and
t = time

Separating the variables, we obtain

$$-\int_0^A \frac{dA}{A^n} = k \int_0^t dt \qquad (2)$$

Integrating equation (2) gives rise to two cases

$$\ln\left(\frac{A_o}{A}\right) = kt \quad n = 1 \qquad (3)$$

and

$$\frac{1}{A^{n-1}} - \frac{1}{A_o^{n-1}} = (n-1)kt \quad n \neq 1 \qquad (4)$$

Let

$$A = 1 - \alpha \qquad (5)$$
$$A_o = 1$$

These equations (3) and (4) become

$$\ln\frac{1}{(1-\alpha)} = kt \qquad (6)$$

and

$$(1-\alpha)^{1-n} - 1 = (n-1) kt \qquad (7)$$

Where is obtained from DSC experiments conducted at constant temperatures, and $(1 - \alpha)$ is the concentration of reactant at time t.

Equations (6) and (7) can be transformed into equations of a straight line. If we then plot the left hand side of the equation vs. time, the magnitude of the slope of the line is equal to the rate constant. A principal problem in the extraction of rate data from thermal analysis experiments is the selection of the appropriate model. Many workers in this field employ the correlation coefficient resulting from a linear least squares regression analyis of the experimental data. The danger inherent in the use of this statistic is illustrated by the work of Chen and Fong[5]. Employing artificial data with an activation energy of 26.0 Kcal/mole, these authors employed 17 different models for the calculation of the activation energy, E, and the Arrhenius pre-exponential factor, A. Twelve of these models gave a correlation coefficientof 0.99+. Unfortunately the calculated values of E ranged from a low of 10.6 Kcal/mole to a high of 100.7 Kcal/mole. A better statistic for the selection of an appropriate kinetic model is the standard error of the estimate, a measure of the "goodness" of the best straight line fit to the experimental data. The standard error of the estimate is a measure of the scatter of the experimental points about the linear least squares fitted regression line. It is the square root of the sum of the squares of the residuals divided by the square root of the number of experimental points. The smaller the numerical value of the standard error of the estimate, the better the fit of the model to the data. We iteratively search for a value of n which minimizes the statistic.

Figure 4 shows the effect on the standard error of the estimate of changing the value of n assumed in calculating the straight line fit of isothermal data to equations (6) and (7). The discontinuity in the plot at a value of n = 1 results from the fact that two different models are being compared. The salient point is that the value of the standard error of the estimate employing the model of equation (7) is tending to a minimum as n approaches a value of one.

Figure 5 which is a plot of Y = left hand side of equation (4) vs. (n-1) illustrates the point that improper selection of a value for n results in a non-straight fitted line for the experimental data. The data used to generate this figure had a minimum value for the standard error of the estimate for the model n = 1 (equation 6).

The kinetics of both initiators evaluated, TBP and USP400, were found to approach a minimum in the standard error of the estimate when $\ln (1-\alpha)^{-1}$ vs. t was plotted, i.e. a classical first order model. The results of these calculations for 10-90% reaction are summarized in Table 3. Figure 6 illustrates the first order behavior of the SMC. The percent reaction range for the calculations was selected to eliminate the possibility of anomolous end effects from the results. Figure 7 shows the temperature dependence of the rate constants according to equation (8), the Arrhenius equation:

$$\ln k = \ln A - \frac{E}{R}\frac{1}{T} \qquad (8)$$

where E = activation energy
A = Arrhenius pre-exponential factor
R = gas constant
T = absolute temperature
k = rate constant (slope of the regression line).

Table 4 summarizes the Arrhenius rate parameters for the curing reactions of the two systems. In this application the use of the correlation coefficient, r, to measure the "goodness" of fit of the model to the experimental data is acceptable because we are not attempting to select between mathematical models but rather to judge how well a model fits the experimental situation. As can be seen from Table 4 all of the samples, with the exception of New/TBP, showed excellent agreement with the Arrhenius model. TBP decreased in reactivity when allowed to age for 88 days in comparison to USP400. The USP400 showed only a slight increase in activation energy during the same time period. Indeed, the difference in activation energies between new and aged USP400 may be within experimental error. However, insufficient experimental information is available to allow us to address this question.

The reactivity of the TBP and USP400 in contact with Carbon Fiber suffered inhibition, reflected in marked increase in activation energy. The activation energy of the TBP system increased by 67% on aging while that of the USP400 system increased by a smaller amount, 42%. It has been shown that the storage stability of SMC system has been affected by the presence of this particular carbon fiber, which had been assumed to be a chemically inert constituent the formulation.

REFERENCES

1. M. R. Kamal and S. Sourour, Polym. Engr. and Sci., $\underline{13}$, 59 (1973).
2. S. Sourour and M. R. Kamal, SPE Technical Papers, $\underline{18}$, 93 (1972).
3. M. R. Kamal, S. Sourour, and M. Ryan, ibid., $\underline{19}$, 187 (1973).
4. L. J. Lee, Polym. Engr. and Sci., $\underline{21}$, 483 (1981).
5. P. H. Fong and D.T.Y. Chen, Thermochimica Acta., $\underline{18}$, 273 (1977).

TABLE 1
SMC FORMULATION

Component
Resin
Styrene
Initiator
Zinc Stearate
Cab-O-Sil
MgO
Carbon Fiber
Fiberglass

Initiators:

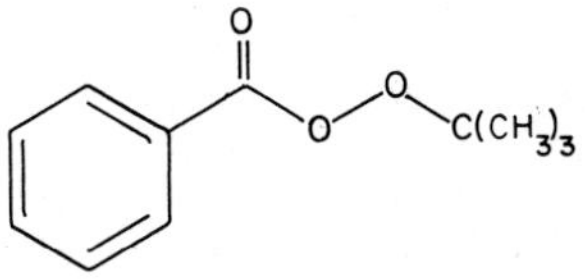

A

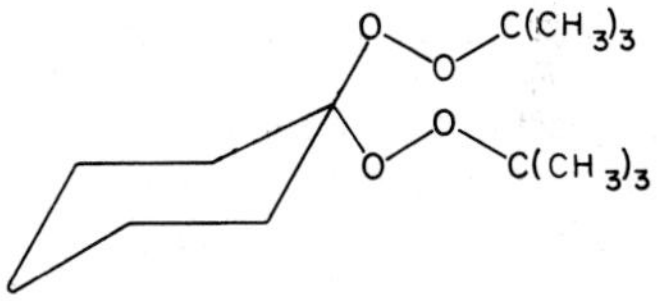

B

TABLE 2
TBP VS. USP400

	T(K) Initial	Max.	Final	ΔH (mcal/mg)
TBP	382.5	427.5	464.0	282.935
TBP/Resin	390.0	407.3	475.2	87.721
USP400	367.8	422.8	454.2	277.473
USP400/Resin	382.5	405.5	470.0	89.923
Resin	461.1	477.5	507.0	51.003

(a) As received, no initiator.

TABLE 3
SMC CURE RATE CONSTANTS

Sample	T (K)	k (sec^{-1})	e^a
New/TBP	405	3.116 x 10^{-2}	7.94 x 10^{-2}
	410	5.252 x 10^{-2}	5.25 x 10^{-2}
	422	1.409 x 10^{-1}	9.75 x 10^{-2}
Aged/TBP	405	3.602 x 10^{-2}	3.60 x 10^{-2}
	415	8.441 x 10^{-2}	1.94 x 10^{-1}
	422	1.972 x 10^{-1}	1.08 x 10^{-2}
Inhibited/TBP	405	1.350 x 10^{-2}	5.35 x 10^{-2}
	410	2.880 x 10^{-2}	6.02 x 10^{-2}
	415	3.850 x 10^{-2}	2.36 x 10^{-1}
New/USP400	393	1.586 x 10^{-2}	1.59 x 10^{-2}
	405	3.346 x 10^{-2}	2.22 x 10^{-2}
	422	1.557 x 10^{-1}	1.09 x 10^{-1}
Aged/USP400	393	1.131 x 10^{-2}	6.94 x 10^{-2}
	413	8.837 x 10^{-2}	1.73 x 10^{-2}
	422	1.123 x 10^{-1}	2.27 x 10^{-1}
Inhibited/USP400	393	1.022 x 10^{-2}	2.34 x 10^{-2}
	405	4.246 x 10^{-2}	1.24 x 10^{-1}
	413	1.040 x 10^{-2}	1.45 x 10^{-1}

(a) Standard error of estimate.

TABLE 4
SMC Cure Arrhenius Parameters

Sample	E(Kcal/mole)	E(%)	r^a
New/TBP	27.6_9	---	0.9255
Aged/TBP	33.7_1	21	0.9941
Inhibited/TBP	47.1_4	67	0.9800
New/USP400	26.1_7	---	0.9926
Aged/USP400	27.3_4	4	0.9816
Inhibited/USP400	37.1_8	42	1.0000

(a) Correlation coefficient.

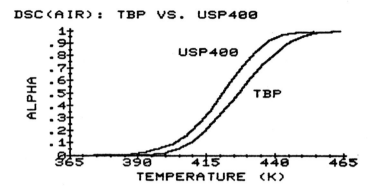

FIGURE 1 - Initiator Extent of Reaction as a Function of Temperature.

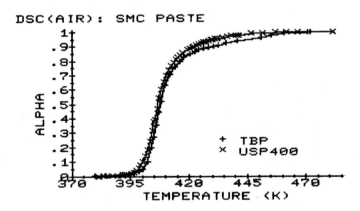

FIGURE 2 - SMC Paste Extent of Reaction as a Function of Temperature.

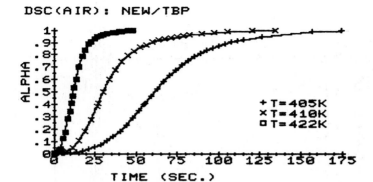

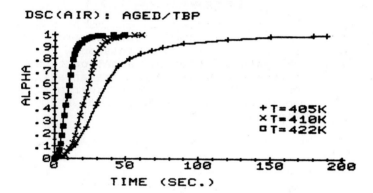

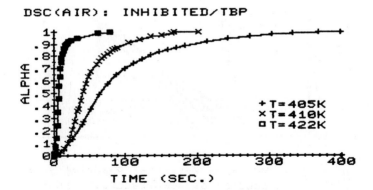

FIGURE 3 - TBP Extent of Reaction as a Function of Temperature.

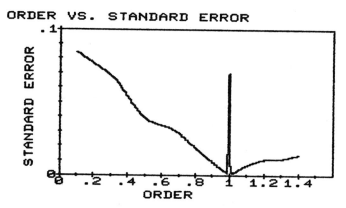

FIGURE 4 - Standard error of the estimate as a function of assumed reaction order in a classical kinetics expression. (Note: the parallel vertical lines centered around the value of n=1 are an artifact of the computer plotting program employed in the production of the figure).

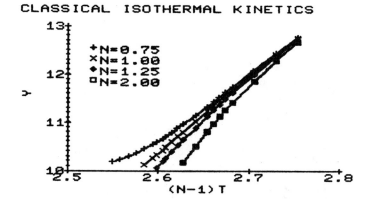

FIGURE 5 - Effect of assumed reaction order on a classical kinetic plot (Note: Abscissa is in unit of time for n=1).

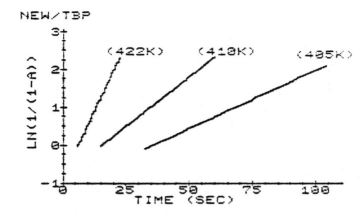
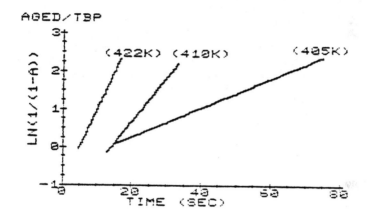
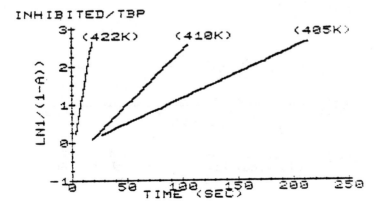

FIGURE 6 - First Order Model Plots - TBP

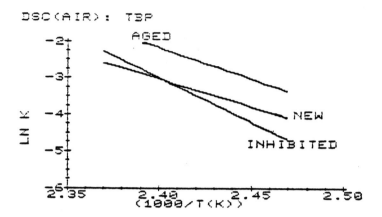

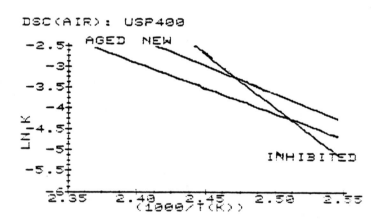

FIGURE 7 - Arrhenius Plots

THERMAL DEGRADATION OF PHENOL-FORMALDEHYDE POLYCONDENSATE

by

G. Camino, M.P. Luda di Cortemiglia, L. Costa and L. Trossarelli
Istituto di Chimica Macromolecolare dell'Università di Torino
V. G. Bidone, 36 - 10125 Torino, Italy

INTRODUCTION

It has been suggested that the thermal degradation of crosslinked phenol-formaldehyde (PF) resins primarily occurs by an oxidative mechanism regardless of whether the degradation is carried out in air or in inert atmosphere /1/. This has been attributed to the fact that at high temperature the resin itself could act as an oxygen source for the oxidative process /1-3/. However, it has also been shown that partial oxidation of the polymer, possibly by absorbed oxygen, generally occurs during the crosslinking step, even when it is carried out in an inert atmosphere /1/. Therefore the oxygen promoting the oxidative degradation of the polymer in previous studies might well have been introduced during the crosslinking step.

In order to avoid interference by external oxygen, we have studied the thermal degradation of a PF novolac type polycondensate, in absence of crosslinking agents and in dynamic high vacuum.

EXPERIMENTAL

Materials. A sample of PF novolac type polycondensate (SIR SpA, Italy) was used ($\bar{M}n$: 830). NMR characterisation showed that in the polymer predominate o,p substituted diphenylmethane structures:

Gel permeation chromatography and liquid-liquid chromatography showed a continuous distribution of molecular weights starting from the dimer. Unreacted phenol and formaldehyde, as well as hydroxymethyl substituents, were below detectable amounts.

Thermogravimetry (TG). Thermogravimetry and isothermal mass change determinations were carried out under vacuum (p<0.1 Pa) on a DuPont 951 thermobalance on 15 mg samples. For comparison purposes, a TG has also been carried out under nitrogen flow (4.5 l/h).

Differential thermal analysis (DTA). The DuPont DTA module was used under nitrogen flow (4.5 l/h). The cell was repeatedly evacuated and filled with nitrogen prior to each experiment, to accurately exclude atmospheric oxygen.

Thermal volatilisation analysis (TVA). Thin layers of powder samples of 100-150 mg were degraded under dynamic high vacuum (p<0.001 Pa) in a TVA apparatus provided with differential condensation of the degradation products (DC-TVA) /4-5/. The rate of evolution of products gaseous at room temperature which are volatile at -100°C or at -196°C (not condensable), were measured as a function of the temperature of the sample in dynamic experiments (10°C/min) or of time in isothermal ones. Condensable gaseous products were collected as obtained from degradation or after separation by subambient TVA (SA-TVA) /5-7/ and identified by IR spectroscopy.

The fraction of the degradation products which are volatile at the degradation temperature but condensable at room temperature were condensed in the water cooled upper part of the degradation vessel (cold ring fraction, CRF). This fraction was quantitatively collected at the end of experiments by dissolution in acetone and evaporation of the solvent, weighed and examined by IR as a film casted on KBr from ethanol solution. Evaporation of solvents were carried out under nitrogen stream to avoid oxidation.

By using removable glass sample holders, the residue of degradation could be easily recovered, weighed and examined by IR (KBr pellets).

Molecular weight measurements. Number average molecular weights of the original novolac, of CRF and of residues were measured in acetone solutions at 37°C using a Mechrolab 302 vapor pressure osmometer.

RESULTS AND DISCUSSION

Overall thermal behaviour. The relevant features of the thermal behaviour of the PF polycondensate are summarised in figure 1, while the weight of the various fractions of the degradation products are listed in table 1. The TG curves (figure 1A) show that the degradation process occurs with a continuous weight loss at a rate which is strongly dependent upon the pressure above the sample. In particular a larger weight loss is observed under vacuum than under nitrogen. This behaviour is typical of degradations in which a large amount of CRF is formed in TVA as in the present case (table 1). The evolution of products which are gaseous at room temperature occurs through several steps as shown by the TVA curve of figure 1B. This complexity

TABLE 1. COMPOSITION AS WEIGHT PERCENT OF THE DEGRADATION PRODUCTS OF PF POLYCONDENSATE HEATED TO 500°C AT 10°C/MIN.

Condensable gaseous products	1.5 (H_2O, CO_2, benzene)
Cold ring fraction (CRF)	60
Residue	38

of the degradation process does not appear in the TG curve since the gases represent only 1.5 percent of the original weight of the polymer (table 1). On the contrary, TVA is very sensitive to the evolution of small amounts of gases. Two main peaks at about 100°C and 420°C respectively, concerning gases completely condensable at -100°C, characterise the TVA curve. At 500°C, 38% of the original polymer remains as an insoluble charred residue (table 1).

The DTA curve of figure 1C shows an endothermic process at 50°C which is the softening temperature of the PF polycondensate, followed by a broad endothermic process extended over the whole range of temperatures examined, with an ill-defined maximum rate at about 400°C.

Degradation products: condensable gases. By using the TVA curve of figure 1B as a guideline, successive degradation experiments were carried out up to temperatures at which successive steps in the formation of gaseous products are completed. The separation by SA-TVA and IR spectra of the products collected after each step, showed that the first group of peaks of the TVA curve (up to about 280°C) mainly concerns the elimination of water and of trace amounts of other impurities deriving from the synthesis of the PF novolac.

The presence of benzene and carbon dioxide besides water in the degradation products collected above 300°C, shows that the second peak of the TVA curve (max. rate 420°C) concerns the fragmentation of the polymer structure. Formation of carbon dioxide indicates that fragmentation partially involves an oxidation process.

"Cold ring fraction". CRF represents the largest fraction of products formed on heating the PF novolac. Its formation begins at about 100°C and continues up to 500°C. By comparing weight increasing of this fraction and weight loss in TG as a function of the temperature, one can see that the shape of the TG curve (figure 1A) is mostly determined by the formation of the CRF. The chemical structure of CRF which is a readily acetone soluble white crystalline solid is very similar to that of original PF novolac, as shown by the close similarity of the significant absorptions in their IR spectra (figures 2A, 2B). The most evident difference between the two spectra is a decreasing of the absorptions at 1445 and at 1485 cm^{-1}

in the case of the CRF. On the other hand, these two absorptions show a corresponding increasing in the spectrum of the residues (figure 2C). It is important to note that IR absorptions related to oxidised species (1650 cm^{-1}) /1_/, are absent in the spectra of both CRF and residues. The attribution of the absorptions at 1445 and 1485 cm^{-1} to specific structures is rather difficult owing to the complexity of the IR spectrum of PF polycondensates in this region /8_/. However, the data of table 2 show that at low temperature (150°C) only evaporation of oligomers present in the original PF novolac takes place whereas at increasing temperature (270°C) also a condensation reaction (branching) between polymer molecules occurs. Further heating leads to a crosslinked structure which by fragmentation (>350°C) produces volatile chain fragments further increasing the weight of the CRF. Thus, although the chemical structure of the CRF does not change sensibly throughout the degradation, as shown by IR, the oligomers by which it is formed, originate from two different mechanisms namely evaporation and fragmentation. Therefore the absorptions at 1445 and 1485 cm^{-1} could be tentatively attributed to tetrasubstituted units /8_/ which accumulate in the residue as branching and crosslinking proceed. This assignment implies the existence of some tetrasubstituted units also in the original novolac. Once reacted phenolic nuclei pendant from the main chain of novolacs to which they would be linked by means of a single methylene bridge, have indeed been previously suggested /9-11/.

TABLE 2. $\overline{Mn}$ OF CRF AND RESIDUE OBTAINED IN ISOTHERMAL HEATING TO CONSTANT WEIGHT.

Temperature, °C	CRF		$\overline{Mn}$ of residue	
	$\overline{Mn}$	Yield%	experimental	calculated (a)
150	150	4	1000	1000
270	450	33	2900	1400
500 (b)	400	60	insoluble	-

(a) calculated assuming only evaporation of oligomers to the CRF
(b) dynamic TVA.

Residue. At 500°C a charred residue is obtained which shows roughly the same characteristic IR absorptions of the PF polycondensate although less resolved and with different relative intensities (figure 2D). Even at this temperature, the residue does not show spectroscopical evidence of oxidation.

Mechanism of degradation. The thermal behaviour of the PF novolac is the result of the competition between several processes partially overlapping, namely evaporation of oligomers, branching and crosslinking and chain fragmentation. The mechanism proposed for the thermal crosslinking reactions in novolac type PF polycondensates in the absence of curing additives, involves the cleavage of the single bonds linking the pendant groups to the main chain /9 /. Although our results do not allow any conclusive inference on the mechanism of thermal crosslinking, they show very clearly that the process occurs without detectable modification of the main chemical structure of the polymer and in particular it does not involve oxidation reactions. This is shown by the absence of additional absorption at 1650 cm^{-1} (=CO) in CRF and residues, whereas this absorption was generally found in crosslinked PF resins obtained from resol type polycondensates on which the most extensive degradation studies have been carried out /1 /. Moreover, CO which would be eliminated by degradation of oxidised structures /1 /, has not been found among the degradation products in our case. However, the fragmentation process of the crosslinked structure, occurring above 350°C, produces carbon dioxide implying that in this step an oxidation process takes place. Thus it is confirmed that the resin itself acts as a source of oxygen for the oxidation process /1-3/. Since benzene is formed during the degradation, it has been suggested that loss of OH radicals must occur which in turn would represent a source of oxygen for oxidation reactions /1 /. The following scheme of reactions, which is one of those proposed by Conley /1 / in the case of PF resol type polycondensates, could also explain the products of degradation of our novolac:

Successive chain scissions can originate short chain fragments condensable in the CRF. Hydrogen abstraction in the above scheme occurs from methylene bridges and when abstraction is performed by OH radicals, water is formed /1 / which is also a degradation product.

ACKNOWLEDGEMENT The authors wish to thank Dr. S. Pezzoli and Dr. L. Cavalli of SIR for supplying characterised samples.

REFERENCES

1. R.T. Conley, "Thermal Stability of Polymers", R.T. Conley, Ed., Dekker, New York, (1970).

2. W.M. Jackson and R.T. Conley, J. Appl. Polym. Sci., 1964, 8, 2163
3. H.W. Lochte, E.L. Strauss and R.T. Conley, J. Appl. Polym. Sci., 1965, 9, 2799.
4. I.C. McNeill, Europ. Polym. J., 1970, 6, 373.
5. I.C. McNeill, "Developments in Polymer Degradation", N. Grassie, Ed., Appl. Sci. Publ., London, (1977), Vol. 1, p. 43.
6. I.C. McNeill, L. Ackerman, S.N. Gupta, M. Zulfiqar and S. Zulfiqar, J. Polym. Sci. Polym. Chem. Ed., 1977, 15, 2381.
7. W.J. McGill, L. Payne and J. Fourie, J. Appl. Polym. Sci., 1978, 22, 2669.
8. D.O. Hummel and F. Scholl, "Infrared Analysis of Polymers, Resins and Additives an Atlas", Wiley, New York, (1971) Vol. 1.
9. E.L. Winkler and J.A. Parker, J. Macromol. Sci. Revs. Macromol. Chem., 1971, C5(2), 245.
10. J.J. Gardikes and F.M. Konrad, Rept. Org. Coatings and Plastics Chem., 1966, 26, 131.
11. R.W. Lenz, "Organic Chemistry of Synthetic High Polymers", Interscience, New York, (1967), p. 139.

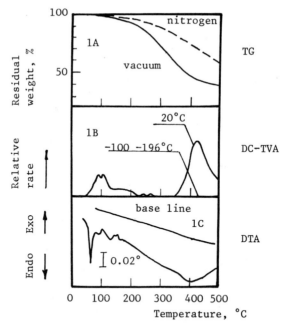

FIGURE 1. Thermal analysis of novolac. Thermogravimetry (1A), under nitrogen (---), under vacuum (——); thermal volatilisation analysis (1B) rate of evolution of products volatile at room temperature, at -100 and -196°C; differential thermal analysis (1C).

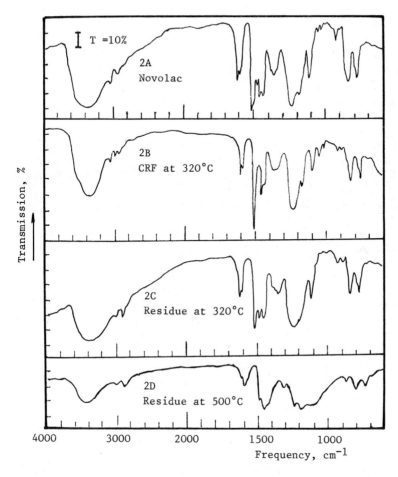

FIGURE 2. IR spectra of: original novolac (2A); CRF from degradation at 320°C (2B); residue at 320°C; residue at 500°C (2D).

THERMAL DEGRADATION OF UREA-FORMALDEHYDE
POLYCONDENSATE
by
G. Camino, L. Operti, L. Costa and L. Trossarelli
Istituto di Chimica Macromolecolare dell'Università
V. G. Bidone, 36 - 10125 Torino, Italy

INTRODUCTION

Very little has been reported on the thermal degradation of urea-formaldehyde (UF) polycondensates. Most of the literature on this topic /1, 2/ concerns thermal oxidative degradation processes rather than the purely thermal one. Besides the interest in extending the knowledge on the thermal behaviour of chemical structures in polymers, the thermal degradation of urea-based polymers acquires a particular importance in the light of the growing interest for polymeric fire retardant materials. UF polymers are indeed characterised by a considerable flame resistance and have been used in conjunction with other substances to impart fire retardancy for example to various cellulose materials /2 /. In the complex process by which a polymer burns, its thermal degradation supplying combustible volatile products to the flame and possibly producing a charred residue which prevents the underlying polymer from further burning, is of paramount importance. Therefore the knowledge of the thermal degradation mechanism of UF polycondensates is essential to the understanding of their fire retardant action.

EXPERIMENTAL

Materials. A curing catalyst-free UF polycondensate moulding powder (F/U = 1.6, Resem-Montedison, Italy) was used. The chemical structure of the polymer was characterised by combination of proton NMR and chemical analysis /3 / (table 1). A qualitative test showed the presence of free formaldehyde in the sample.

Thermal analysis. Thermogravimetry (TG) was carried out either under air or nitrogen flow (60 cm^3/min) or under vacuum (p<0.1 Pa) at 10°C/min on samples of 10 mg using a DuPont 951 thermobalance.
Differential thermal analysis (DTA) was carried out under nitrogen flow (60 cm^3/min) at 10°C/min using a DuPont DTA module. The cell was repeatedly evacuated and filled with nitrogen before each experiment to exclude atmospheric oxygen in order to obtain reprodu-

TABLE 1. CHEMICAL STRUCTURES IN UF POLYCONDENSATE, MOLES PER 100 G SAMPLE

-CO-NH-	-CO-NH$_2$	-CO-N=	-CH$_2$-OH	-CH$_2$-O-CH$_2$-	=N-CH$_2$-N=
1.2	0.2	0.6	0.6	0.3	0.8

cible results.

Thermal volatilisation analysis (TVA). Thermal degradations were carried out under dynamic vacuum (p < 0.001 Pa) in a TVA apparatus provided with differential condensation of the volatile degradation products /4, 5/. Rates of volatilisation of products volatile at room temperature and of those volatile at -100°C or at -196°C, were recorded as a function of temperature which was raised at 10°C/min to 500°C. The volatile degradation products were collected at the end of the run, separated by the subambient thermal volatilisation technique (SA-TVA) /6 / and identified by infrared spectroscopy (IR). Mass spectrometry of gases evolved in an experiment carried out in static vacuum, allowed the identification of non condensable gases. The fraction of the degradation products volatile at degradation temperature but condensable in the water cooled upper part of the degradation vessel (cold ring fraction, CRF) was quantitatively collected by dissolution in ethanol, weighed after evaporation of the solvent under nitrogen stream and examined by IR. By using a removable glass sample holder, the residue at 500°C could be easily collected and weighed.

RESULTS AND DISCUSSION

Effect of atmosphere on degradation. The TG curves of figure 1 show the effect of pressure or of the oxygen content in the atmosphere surrounding the sample upon its thermal degradation behaviour. In particular, below 300°C the three curves are almost coincident within the experimental error. Between 300-500°C, the most noticeable difference is a somewhat faster rate of weight loss in the TG carried out under vacuum as compared to those carried out at atmospheric pressure. As discussed below, this is due to the formation of high boiling fragments from the degradation of the polymer structure beginning at about 300°C, which, under vacuum, are more efficiently removed from the degrading sample. Oxygen seems to take part directly to the degradation process only above 500°C leading to further volatilisation of the residue of thermal degradation which is otherwise stable in inert atmosphere to 700°C. Thus, below 500°C, the UF polycondensate seems to undergo purely thermal degradation

TABLE 2. PRODUCTS OF THE THERMAL DEGRADATION TO 500°C UNDER VACUUM OF FU POLYCONDENSATE, WEIGHT PERCENT.

Gases	74
CRF (high boiling)	12
Residue	14

even when heated in air.

The thermal degradation of the UF polycondensate is shown to occur in two main steps by the differential TG curve of figure 1 (DTG) with maximum rate at 180 and 300°C respectively.

Degradation process. A detailed picture of the thermal degradation process of UF polycondensate is obtained by comparing TG, DC-TVA and DTA curves (figure 2). The DTA curve shows the occurrence of several endothermal phenomena between 70-260°C corresponding to the first stage of the TG curve which involves a relatively moderate overall weight loss (20-25%). The main step of weight loss (from 25 to 85%) is due to a process which tends to be exothermal at first (260-280°C) and then endothermal again (280-400°C).

Throughout the thermal degradation of the UF polycondensate the major part of the products evolved is volatile at room temperature in the TVA apparatus (table 2) and hence the TVA curve of figure 2B gives a deepest insight into the degradation process than any other single technique in this case. Indeed it shows that overlapping steps which cannot easily be resolved for example by monitoring weight loss, are well evidentiate by the more sensitive rate of gas evolution measurements. Figure 2B clearly shows that the thermal degradation of the UF polycondensate occurs through five steps with rate maxima at: 75, 125, 155, 260 and 290°C respectively. It is interesting to note that the differential condensation of the degradation products is very much helpful in recognising the existence of truly separated processes in correspondence of peaks 2-3 and 4-5 respectively. This would have been somewhat questionable on the basis of the total rate of gas evolution alone, whereas the evolution of products condensable at -100°C, at -196°C and not condensable in correspondence of peak 2 and the predominant evolution of only products condensable at -100°C for peak 3, clearly indicate the occurrence of two separate processes. Similarly, steps 4 and 5 are well distinguishable if one considers the rate of production of volatiles condensable between -100°C and -196°C.

The composition of the volatile products of degradation (table 3) indicates that while below 100°C impurities (free CH_2O, H_2O) are eliminated from the sample, the range of temperature 100-200°C should essentially concern branching and crosslinking ractions, for

TABLE 3. VOLATILE PRODUCTS FORMED IN THE THERMAL DEGRADATION UNDER VACUUM OF UF POLYCONDENSATE.

Temperature, °C	Step	Products	Condensation temperature, °C
100	1	H_2O, CH_2O	--100
100-200	2-3	H_2O, CH_2O	--100
		CO_2	-196
200-500	4-5	NH_2CH_3, $N(CH_3)_3$	--100
		NH_3, CO_2	-196
		CO, CH_4	not condensable
		chain fragments	20

example /7/:

$$\begin{array}{c} \diagdown N \diagup \\ | \\ CH_2OH \\ H \\ | \\ \diagup N \diagdown \end{array} \xrightarrow{-H_2O} \begin{array}{c} \diagdown N \diagup \\ | \\ CH_2 \\ | \\ \diagup N \diagdown \end{array} \quad (1)$$

$$\begin{array}{c} \diagdown N \diagup \\ | \\ CH_2OH \\ CH_2OH \\ | \\ \diagup N \diagdown \end{array} \xrightarrow[-H_2O]{-CH_2O} \begin{array}{c} \diagdown N \diagup \\ | \\ CH_2 \\ | \\ \diagup N \diagdown \end{array} \quad (2)$$

The overall process is endothermal probably owing to the heat required to evaporate H_2O and CH_2O.

Above 200°C the fragmentation of the polymer structure should begin as shown by elimination of ammonia, amines, CO_2 etc. However, as discussed above, both crosslinking and fragmentation processes occur in two steps probably involving structures characterised by different thermal stability. For example, as far as crosslinking is concerned, the reaction occurring at lower temperature (step 2), could involve mainly CH_2O elimination from $-CH_2-O-CH_2-$ groups which are known to be more thermally labile /8/. As to the polymer fragmentation, the exothermal step 4 could involve a cyclisation reaction developing structures more thermally stable than the original ones. Extensive endothermal fragmentation of the polymer occurs indeed only in the last step of degradation as shown by the fact that high boiling chain fragments are only produced in this step.

The study of the chemical reactions taking place in each of the four degradation steps (2,3,4,5 figure 2B) will be carried out by analysing the degradation products and the modifications of the structure of the polymer in stepwise isothermal degradations at temperatures selected on the basis of the DC-TVA curve.

ACKNOWLEDGEMENT The authors wish to thank Dr. N. Lupi of Resem-Montedison for supplying characterised samples.

REFERENCES

1. R.T. Conley, "Thermal Stability of Polymers", R.T. Conley, Ed., Dekker, New York, (1970).
2. N.B. Sunshine, "Flame Retardancy of Polymeric Materials", W.C. Kuryla and A.J. Papa, Eds., Dekker, New York, Vol. 2 (1973).
3. M. Chiavarini, N. Del Fanti and R. Bigatto, Angew. Makromol. Chem., 1975, 46, 151.
4. I.C. McNeill, Europ. Polym. J., 1970, 6, 373.
5. I.C. McNeill, "Developments in Polymer Degradation", N. Grassie, Ed., Appl. Sci. Publ., London, 1977, Vol. 1.
6. I.C. McNeill, L. Ackerman, S. Gupta, M. Zulfiqar and S. Zulfiqar, J. Polym. Sci. Polym. Chem. Ed., 1977, 15, 2381.
7. G. Widmer, "Encyclopedia of Polymer Science and Technology", H.F. Mark, N.G. Gaylord and N.M. Bikales, Eds., Interscience, New York, (1965).
8. K.J. Saunders, "Organic Polymer Chemistry", Chapman and Hall, London, (1973).

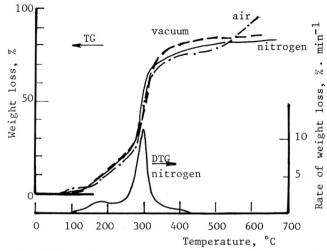

FIGURE 1. TG of UF, under air (-··-), nitrogen (——), vacuum (---).

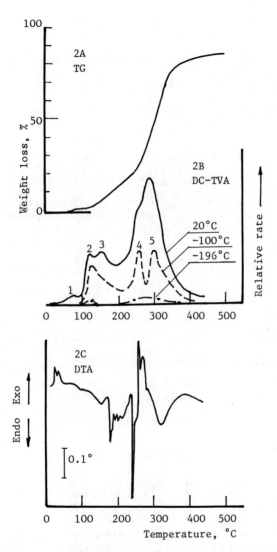

FIGURE 2. Thermal analysis of UF polycondensate. Thermogravimetry (2A); thermal volatilisation analysis (2B), rate of evolution of products volatile at room temperature (——), at -100°C (- -), non condensable (-··-); differential thermal analysis (2C).

CURING KINETICS OF POLY(ETHYLENE ADIPATE) AND
TOLUENE DIISOCYANATES

Rosalie G. Ferrillo, Albrecht Granzow and
Volker D. Arendt
American Cyanamid Company
Chemical Research Division
Bound Brook, NJ 08805 USA

ABSTRACT
The curing kinetics for the reaction of poly(ethylene adipate) and either 2,4-toluene diisocyanate or 2,4-2,6-toluene diisocyanate have been studied by dynamic and isothermal differential scanning calorimetry. Energy of activation and enthalpy of curing values for both systems have been calculated.

QUANTITATIVE DTA OF EPOXY ADHESIVES AND PREPREGS
by
Alberto Schiraldi and Paola Rossi
Istituto di Chimica Fisica e di Elettrochimica
Università di Pavia. 27100 PAVIA (ITALY)

INTRODUCTION

Epoxy resins have been recognized as a technologically important class of polymers and are widely employed as adhesives or, when mixed with suitable fillers, as prepregs.

The composition of such polymers, tentatively and empyrically planned by the producer, is extremely complex and remains largely unknown. Accordingly, the detailed mechanism of the cure reaction cannot be easily understood.

Nonetheless, an important check of the material, before its use, can be easily carried out by means of DTA or DSC investigations.

DTA or DSC thermograms typically show endothermic shifts of the base line at the glass transition temperature, T_g, and an exothermic peak corresponding to the heat, Q, delivered by the cure reaction occurring in the meanwhile of the DTA scanning.

Generally, the heat Q is supposed to be proportional to the amount of uncured material undergoing DTA and is employed for the computation of the cure degree, α,

$$\alpha = 1 - Q/Q(0) \qquad (1)$$

where $Q(0)$ is the heat delivered in the DTA scanning by the so called prepolymer, i.e. by the uncured material.

In the present work we discuss the validity of the above way of working out the experimental data and we suggest a more correct procedure to attain α and the phenomenological law obeyed by the cure reaction.

EXPERIMENTAL

Details concerning the preparation of the samples, the conditions of their isothermal cure before the DTA investigation, as well as the heating rate and the sensitivity employed in the DTA scanning, are given in a previous paper [1].

Here we summarize the main aspects:
(i) weight of samples about 50 mg;
(ii) T_{cure} = 100 °C;
(iii) DTA apparatus, Du Pont Thermoanalyzer mod. 900;
(iv) T_g values are taken at the interception between base line and tangent at the flexus point of the endothermic shift;
(v) investigated materials are: (1) AF 163 WT06; (2) AF 163-2 WT06; (3) FM 53; (4) FM 123-2; (5) NARMCO 1113; (6) NARMCO 5216; (7) CIBA 920. Materials (1)-(5) and (6)-(7) are employed as adhesives and as prepregs, respectively.

GENERAL CONSIDERATIONS

It is well known that the integration of a peak in a DTA or DSC thermogram allows to evaluate the amount of heat delivered in the T range where the peak occurs, i.e. in non isothermal conditions.

When such a heat is the the thermal effect accompanying a definite chemical reaction, the usual way of attaining its phenomenological kinetic parameters, through the evaluation of the derivative dQ/dT at various points of the peak profile, may be reasonably justified [2].

This is not the case for the polymerization of complex materials such as epoxy resins, since the process may not be considered a definite chemical reaction with a unambiguous final product.

It is easy to realize and to experimentally verify (by checking the T_g of various samples fully cured at different T_{cure}: the higher T_{cure}, the higher T_g) that the conditions of the cure, viz. T_{cure} or heating rate and pressure, strongly affect the nature and the properties of the final polymer. Presumably they act on the length of chain and on the extention of the crosslinks.

Furthermore eq.(1) may not be safely employed, as it implies $\alpha = 0$ for the prepolymer, which is not at all the case of epoxy adhesives and prepregs.

Therefore if one still wants to employ a non isothermal analysis of these materials, it is necessary to circumvent the above scrapes.

We suggest to put

$$\alpha = 1 - Q(t)/Q_0 \qquad (2)$$

with

$$Q(0) \neq Q_0 ,$$

where $Q(t), Q(0)$ and Q_0 are the heat delivered in the DTA scanning by a sample isothermally cured at a given T_{cure} for a cure time t, by the prepolymer and by the corresponding mixture of monomers, respectively.

A special care is required to realize the actual meaning of eq.(2) where:
(i) α is defined in terms of non isothermal quantities;
(ii) $Q(t)$ is not the amount of heat delivered at the time t of the DTA scanning, but the overall heat delivered in the DTA scanning by a sample partially cured in isothermal conditions for a time t.

Therefore eq.(2) must be considered only a formal expression connecting to one another the experimental $Q(t)$'s, so that α, given there, may not be referred as to an actual function of state.

However, although this means that point derivatives, such as dQ/dt, checked along the profile of the peak are misleading, no objection may be raised against the use of the overall values of $Q(t)$, delivered by samples isothermally cured at the same T_{cure} for different t_{cure}'s.

As a matter of fact, if N is the initial number of moles of the mixture of monomers, N_i will be that of the moles undergone isothermal polymerization after a given cure time, while $N_{ni} = (N-N_i)$ will be still unlinked.

The overall thermal effect, $Q(t)$, observed in the DTA scanning corresponds to the extent of the non isothermal cure of N_{ni} moles of the mixture of monomers, no matter how complex is its mechanism, as it will be the same for any value of N_{ni}.

Accordingly,

$$Q(t)/Q_o = N_{ni}/N = 1 - \alpha =$$
$$= 1 - N_i/N = 1 - \alpha_i$$

where α_i is the isothermal cure degree.

Therefore α may be treated as an actual polymerization degree for an isothermal cure reaction, viz.

$$d\alpha/dt = K_n(1-\alpha)^n \qquad (3)$$

where n and K_n play the role of a reaction order and of a kinetic constant, respectively, provided that eq.(3) is employed <u>only</u> in the integrated form.

For $n > 0$ and $n \neq 1$, integration of eq.(3) gives

$$[1 - \alpha]^{(n-1)} = [1 + (n-1)K_n t]^{-1} = [Q(t)/Q_o]^{(n-1)} \qquad (4)$$

Eq.(4) is numerically solvable through an iterative procedure reported elsewhere [1], which gives the best n, K_n and Q_o for the fit of the experimental data.

This procedure allows also to recognize the case $n_{best} = 1$, when integration of eq.(3) leads to the peculiar form requiring the linear regression $ln\ Q(t)$ vs t.

Typically, n=1 and n=2 are found for epoxy resins, i.e.

$$\alpha = 1 - exp(-K_1 t) \qquad \text{for n = 1}$$
$$\alpha = K_2 t/(1+K_2 t) \qquad \text{for n = 2}.$$

It must be born in mind that the parameters Q_o, n and K_n, allowing eq.(4) to fit the experimental data, refer to a definite cure reaction, i.e., for a given material, to a definite T_{cure}.

This means that, for example, it is a nonsense to evaluate an activation energy from K_n values obtained for different T_{cure}'s, as they actually correspond to different reactions leading to different final polymers.

Accordingly, also the value of the polymerization degree of the prepolymer, $\alpha(0) = Q(0)/Q_o$, does not correspond to a definite state of the system: it may be used only as a qualitative index to compare differents lots of the same prepolymer, e.g. to roughly evaluate the relative extent of their aging: the higher $\alpha(0)$, the more aged the prepolymer.

RESULTS

TAB.1 summarizes the results of the investigation. For each material, indicated with the same number as in section EXPERIMENTAL, are reported Q_o, $\alpha(0)$, n, K_n and the lowest Tg observed for the prepolymer.

TAB.1 Q_o, T_g AND K_n ARE IN cal gram^{-1}, °C AND minutes^{-1}, RESPECTIVELY. THE Q_o'S OF THE PREPREGS (6) AND (7) REFER TO THE EPOXY RESIN CONTENT, 33% AND 20%, RESPECTIVELY.

RESIN	Q_o	$\alpha(0)$	n	K_n	T_g
(1)	40.75	0.14 ±0.04	2	$3.8\ 10^{-2}$	-11
(2)	26.97	0.13 ±0.03	2	$1.8\ 10^{-2}$	-14
(3)	30.91	0.11 ±0.04	1	$1.3\ 10^{-2}$	-13
(4)	32.23	0.13 ±0.04	1	$2.9\ 10^{-2}$	- 4
(5)	37.56	0.14 ±0.04	1	$2.6\ 10^{-2}$	-11
(6)	47.02	0.0 ±0.04	1	$3.6\ 10^{-2}$	+13
(7)	55.15	0.13 ±0.04	1	$4.9\ 10^{-2}$	+11

Due to their hybrid nature, these materials show at least two T_g signals, at about or below 0°C and at higher T, respectively.

Generally, the former corresponds to the component acting as elastomer, the latter to that providing resistance to compressive strengths.

The material (4), viz. FM 123-2, has been more extensively investigated. The results obtained from a series of samples cured at 130°C allow to recognize again that n = 1 and give K_n=1.16 10^{-1}min^{-1}.

The final polymers, obtained after a suitable cure period at 100°C and at 130°C, show different T_g values, viz. close to 100°C and to 130°C, respectively.

This means that they are quite different from each other, so that is a nonsense to attempt the evaluation of an activation energy from the corresponding K_n values.

Two lots of the material (4) have been investigated: the former giving some surprising results [1], the latter, more recently purchased, showing a "normal" behaviour.

With respect to the isothermal cure at 100°C, the former gives $\alpha(0)$ = 0.20 ± 0.05, the latter $\alpha(0)$ = 0.13 ± 0.04, which is consistent with the different age of the two lots and might explain the poor adhesivity sometimes observed for this material [3].

REFERENCES

1. A.Schiraldi, V.Wagner, G.Samanni and P.Rossi, J.Therm.Analysis, 1981, 21, 299
2. J.Sestak and J.Kratochvil, J.Therm.Analysis, 1973, 5, 193
3. P.Rossi, G.Samanni, V.Wagner and A.Schiraldi, Proc. II Nation. Conf. of A.I.C.A.T. (Associazione Italiana Calorimetria Analisi Termica) 1981, p.75.

CURING OF A POLYIMIDE RESIN

G. A. Pasteur
H. E. Bair
F. Vratny

Bell Laboratories
Murray Hill, NJ 07974

INTRODUCTION

Because of their excellent high temperature stability, polyimides are currently considered and in some cases used in the fabrication of integrated circuits. They are supplied as solvated polyamic acids. To be suitable for IC technology, the cured polymer should not bleed solvent or polymerization products. The purpose of this study is to evaluate the curing process of the polymer through thermogravimetry, infrared spectroscopy and calorimetry.

EXPERIMENTAL

The polyimide tested was polyimide iso - Indro quinazolinedione (PIQ).

Thermogravimetric analysis (TG) was carried out on liquid samples of PIQ weighing several milligrams. A Perkin-Elmer thermogravimetric system (TGS-2) was used.

Differential scanning calorimetric (DSC) measurements were carried out on samples weighing about 15 mg each. The details of these measurements have been reported elsewhere [3].

A duPont Moisture Analyzer (26-321A) was used to determine the amount of water eliminated during the imidization process. This device uses a coulometric technique to measure water quantitatively.

The glass transition temperature, Tg, of a cured, crosslinked film of PIQ was measured by thermal mechanical analysis on a Perkin-Elmer TMA-1 instrument. The penetration of a small (0.03 in. dia.) hemispherical quartz tip under 5 grams load was monitored as a function of temperature. The sample was heated in a helium atmosphere at a rate of 10°/min.

Infrared spectra were recorded on a Perkin-Elmer 282 double beam spectrophotometer from 4000 to 400 cm^{-1}. The prepolymers were cast on salt plates and baked in air up to 400°C.

RESULTS AND DISCUSSION

Polymides are synthesized in a two stage process [2]. The first stage involves an addition reaction between a dianhydride and a diamine carried out in a polar solvent to produce a high molecular weight polyamic acid. This is the prepolymer supplied by the manufacturers. The solvents are typically n-methylpyrrolidone (NMP); N, N-dimethyl acetonamide; or cellosolve acetate. The second stage involves cyclization to the polyimide by heating above 150°C with evaporation of the solvent and dehydration. The second stage of polymerization can be followed using several techniques: (1) TG measurement of the weight loss due to solvent and/or water evaporation, (2) water analysis and (3) infrared spectroscopic determination of the decrease of peaks due to solvent solvent and the increase of peaks due to imide rings.

In the first TG analysis, PIQ was heated from room temperature to 700°C in air at a heating rate of 10°/min. The solvated, uncured thermal resist lost weight in several stages (Figure 1):

1. The sample began to volatilize rapidly above 65°C and continued at this rate until about 72 wt.% was volatilized and 140°C was attained.
2. Between 140° and 225° the sample lost additional weight at a comparatively slow rate until a loss of 85 wt.% was reached.
3. Continued heating to 310° liberated one additional weight percent of volatile components (85 to 86 wt.%).
4. From 310° to 515° the residue was stable and lost no weight.
5. Beyond 515° the residue began to degrade and volatilize. Complete volatilization occurred near 670°.

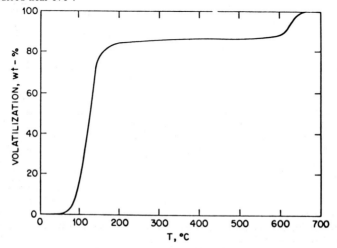

1. Volatilization of PIQ when heated in air from 23° to 700°C at 10°C/min.

In separate thermogravimetric runs, four PIQ samples were heated at 1°/min to 90°C and then held at this temperature for one hour. All samples lost approximately 75 wt.% in 30 minutes. After one hour the total volatilized had only increased from 75 to 77 wt.%. Following this treatment the first sample was heated abruptly to 215° (Figure 2), the second to 250°, the third to 300° and finally the fourth to 350°. Within several minutes the weight loss of each sample had stabilized at 86.0 ± 0.1 wt.%. Heating these samples to 400°C only yielded a few tenths of a weight percent more of volatile material.

For the direct determination of water produced by the imidization reaction, a sample of PIQ was heated to 90°C in an air oven for 1 hour to eliminate most of the solvent and any absorbed water [3]. IR analysis indicated no significant imidization occurred at 90°C. The sample was then placed in a coulometric water measuring device. When it was heated from 23 to 225°C, it began to give off water rapidly above 150°C. The amount of water vaporized was equivalent to about 20% of the material volatilized between 90° and 215° in the previous thermogravimetric study. The remaining volatile substances are shown in the subsequent IR study to be due to the solvent, NMP.

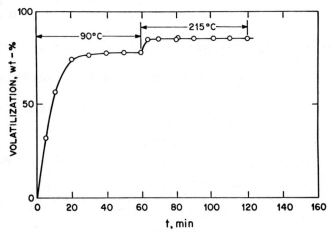

2. Isothermal weight loss of PIQ at 90°C for 60 minutes followed by 60 minutes at 215°C.

Figure 3 shows the IR spectrum of PIQ at different states of the curing process: as cast (3a), air dried over night (3b), after 150°C bake for 40 min. (5c), after 200°C bake for 1 hour (3d), after 400°C bake for 1 hour (3e). In Figure 3a the presence of NMP is confirmed by the shape of the C-H absorption between 3000 and 2800 cm^{-1} and by the shoulder at 1680 cm^{-1} due to C=O. However the solvent evaporation can only be followed by the one band which can be singled out in the spectrum of the freshly cast film at 1410 cm^{-1}. The two other possible solvents (n,N-dimethylacetonamide and cellosolve acetate) are probably not present because their strongest bands do not appear in the spectra. By comparing spectra 3a through 3c of PIQ, it can be seen that the overnight solvent evaporation is incomplete (3b) and that after the first bake at 150°C the solvent removal is almost complete (3c); a weak shoulder due to NMP is still present at 1410 cm^{-1}. After a 200°C bake (3d) the band at 1410 cm^{-1} is not visible indicating the solvent evaporation is complete.

The prepolymer form of the imide (polyamic acid) is characterized by amide and acid groups which react during the curing stage to form imide groups. Therefore the imidization reaction can be followed by the decrease of bands due to the amide and acid groups. These are: sharp small bands due to N-H stretching between 3000 and 3300 cm^{-1}, the amide I band around 1650 cm^{-1} (C=O), the amide II band at 1540 cm^{-1}, weak bands on a broad background (2500 to 3500 cm^{-1}) due to bonded OH stretching of the acid group, the carbonyl band at 1730 cm^{-1} superimposed on an imide band and C-O stretching between 1200 and 1300 cm^{-1}.

The imidization reaction can also be followed by the appearance and increase of the bands due to the imide ring. The planar configuration of the imide structure gives rise to two carbonyl bands as in an anhydride. These bands can be seen in Figure 3c through 3e at 1730 and 1780 cm^{-1}. The use of the band at 1730 cm^{-1} would be controversial because it could also be due to the acid carbonyl of the polyamic acid. Two other imide bands can also be used: C-N stretching at 1380 cm^{-1} and deformation vibration of the imide ring at 730 cm^{-1} [2]. In spectrum 3a, a weak band at 1780 cm^{-1} indicated that some imidization has occurred in the prepolymer phase. The band at 1500 cm^{-1}, attributable to the C=C stretching vibration in the aromatic ring, would not be affected by the heat treatment. This band can therefore be used as an internal standard. The imidization can be followed by plotting the ratios of the absorbances at 1780 cm^{-1}, 1380

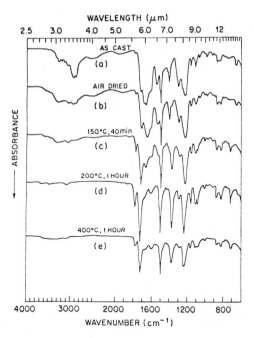

3. Infrared spectra of PIQ films: 5a, as cast; 5b air dried over night; 5c, baked at 150°C for 40 min.; 5d, baked at 200°C for 1 hour; 5e baked at 400°C for 1 hour.

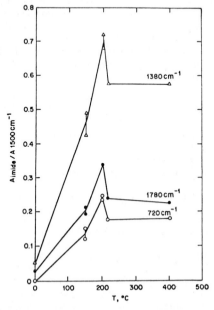

4. Intensities of imide ring absorptions at 1780, 1380 and 720 cm^{-1} normalized to the aromatic absorption at 1500 cm^{-1} versus temperature.

cm^{-1}, and 730 cm^{-1} to that at 1500 cm^{-1} as a function of temperature. Figure 4 shows that these ratios increase with temperature up to about 200°C and then decrease. This implies that the amount of imide rings decreases above 200°C. TG analysis showed the weight of the polymer was stable within a few tenths of percent between 215 and 400°C. Therefore the reduction in the concentration of the imide rings is probably not due to the decomposition of the polyimide. A better assumption is that the imide ring will rupture and recombine to form an interchain imide bond leading to a crosslinked network [2-3]. A. P. Rudakov, et. al., [4] did not detect any absorption at 1600 cm^{-1} characteristic of the newly created non-cyclic -OC-N-CO- group. However, in our study this band is present as a shoulder after 1 hour at 200°C (Figure 3d) and has increased after 1 hour at 230°C (spectrum not shown) and 1 hour at 400°C. Thus, at 200°C the crosslinking mechanism has started as indicated by the decrease in the IR absorption at 730, 1380 and 1780 cm^{-1} relative to the imide ring at 1500 cm^{-1}. At 230°C the crosslinking process is obviously terminated as the intensities of imide ring absorptions are stable (Figure 4).

The level of crosslinking can generally be followed by calorimetric measurements. Moy and Karasz [5] have observed a marked reduction in ΔC_p at Tg for an epoxy system as the degree of crosslinking was increased. Similarly, ΔC_p was found to decrease non-linearly as the ratio, of styrene acrylonitrite copolymer grafted to polybutadiene was increased [1]. Therefore, ΔC_p at Tg was measured for PIQ in the uncrosslinked and crosslinked state. Sufficient liquid PIQ was added to an aluminum pan to yield a film which weighed approximately 10 milligrams after one hour at 90°C. This film was heated at 40°/min from room temperature to 220°C. Above 130° Cp rose rapidly and discontinuously with the center of the glass transition near 168°C. Unfortunately, in this initial run, a broad endotherm occurs and masks Tg so that ΔC_p at Tg could not be measured accurately. However, after cooling the sample to 23° it was reheated to 240° at 40°/min and Tg occurred at 197° with ΔC_p equal to 0.12 cal °C^{-1}g^{-1}. Once again the film was cooled rapidly to 23°C and reheated to 280° at 40°C/min. Under these conditions Tg was shifted upwards to 260° and ΔC_p was reduced to 0.05 cal °C^{-1}g^{-1}. Subsequent cooling to 30°C and reheating to 300° yielded Tg at 287° and ΔC_p was equal to about 0.01 cal °C^{-1}g^{-1}. One last cycle to 400°C showed Tg at 316° and ΔC_p was approximately 0.01 cal °C^{-1}g^{-1}. These results are summarized in Table 1. Thermomechanical analysis of this same film in a penetration mode revealed that significant softening did not occur until 310°C. Thus, these findings support our contention that IR data shows crosslinking occurred as films were aged at 200° and higher.

TABLE 1

PIQ Heated to x°C at 40°/m	Tg, °C	ΔC_p cal °C^{-1}g^{-1}
220	168	(0.50)
240	197	0.12
280	260	0.05
300	287	0.01
400	316	0.01

CONCLUSION

PIQ, a solvated polyamic acid used in the fabrication of ICs, was found to contain about 84 wt.% of n-methylpyrrolidone (NMP). In this system, NMP volatilizes completely at 150°C or higher. In addition, near 150°C imidization was detected spectroscopically by the development of several characteristic imide ring absorption bands at 1780, 1380 and 730 cm^{-1} and coulometrically by the rapid evolution of water. At 200°C a crosslinked resin started to form. The development of crosslinks was followed by IR and DSC measurements. The IR imide ring bands intensity reaches a maximum at 200°C and then decreases between 200 and 230°C to a steady value up to at least 400°C; this indicates the imide ring has opened. A band at 1660 cm^{-1} detected after 200°C shows that crosslinking has taken place. ΔCp at Tg was found to decrease from 0.08 to 0.01 cal °C^{-1}g^{-1} as the crosslink density increased. The curing process which accompanied the high temperature aging of the material caused Tg to shift from 168 to 316°C. These results lead to the conclusion that when PIQ is used in ICs it should be cured by heating to at least 230°C to insure completion of the imidization and the initiation of crosslinking processes.

REFERENCES

1. H. E. Bair, L. Shepherd and D. J. Boyle, in Thermal Analysis in Polymer Characterization, E. A. Turi, ed., Heyden and Son, Inc. Philadelphia-London-Rheine, 1981 pp. 114-124.

2. Polyimides, A New Class of Thermally Stable Polymers. N. A. Adroua, M. I. Bessenov, L. A. Laius, A. P. Rudakov. Technomic Publication, Stanford, Conn. 1970.

3. W. Wrasido, P. M. Hergenrother and H. H. Levine, Am. Chem. Soc., Polymer Preprints 5(1), 141 (1964); CA, 64, 3499g (1966).

4. A. P. Rudakov, M. I. Bessonov, M. M. Koton, E. I. Pokrovskii and E. F. Fedorova, Dokl. Akad, Nauk SSSR 161(3), 617-619 (1965); C.A. 63, 740 d (1965).

5. P. Moy and F. E. Karasz in "Water in Polymers", S. P. Rowland, ed., ACS Symposium Series, Washington, D. C., 1980, 127, 505.

Section V Applied Science and Industrial Applications

Plenary Lecture   COAL, OIL SHALE AND THERMAL ANALYSIS

S. St. J. Warne
Department of Geology, University of Newcastle
New South Wales, 2308, Australia

INTRODUCTION

The predicted exhaustion of the world's plentiful and cheap resources of petroleum, the restricted nature of the present sources of supply and the marked lack of oil self sufficiency in many major industrialised countries, coupled with a wholesale price increase, $\approx$ 1600%, between 1970-1981 has led to a search for alternative types and sources of energy.
The results of this have been:-
a) a large upsurge in the actual use and planned use of coal for electricity generation.
b) a duplication (and proposed triplication) of the only commercial full scale oil from coal plant in the world (SASOL), parts I and II of which now produce 35% of South Africa's current petrol needs from 27 million tonnes of poor quality coal per annum.
c) a very serious look at oil from coal processes at the multi-million dollar pilot and demonstration plant level, in several parts of the world.
d) the ongoing critical, appraisal, basic and proving up research investigations of a number of large oil shale deposits and related recovery methods, in a number of different countries.
Thus interest and research in the fields of coal and oil shale are very much to the fore at the present time.
Most of the coal currently produced is either burnt directly as a fuel (mainly for power generation) or carbonised to form coke (for steel smelting). In the case of oil shales, the material, usually as mined, is heated to drive off from it recoverable liquid and gaseous hydrocarbon products. Thus some form of heat treatment is commonly involved so that the application of thermal analysis methods is of immediate and potential use in characterisation and utilisation studies of these two fuels.
Additionally, thermal analysis methods have the advantage of direct application to whole coal/shale samples which (except for crushing) have not been pretreated in any way. Thus such problems as the loss, alteration and imperfect concentration/removal of particular constituents are obviated and the resultant curves represent all of the original constituents in the sample. It is not implied however, that thermal analysis of specifically prepared products would not yield useful results. In fact, such applications, e.g. to coal washery or spent shale retorting products represent additional fields of sound research.

THERMAL ANALYSIS TECHNIQUES.

The thermal analysis techniques applicable to solid hydrocarbon fuels are differential thermal analysis (DTA), Thermogravimetry (TG), derivative thermogravimetry (DTG), evolved gas analysis (EGA), differential scanning calorimetry (DSC), thermomagnetometry (TM), and variable temperature and atmosphere X-ray diffraction. Of these DTA followed by TG has been most commonly applied.

PREVIOUS WORK - GENERAL

The most recent reviews of applications of thermal analysis to coal, coal minerals and coal ashes have appeared in specialist books [1,2,3,4 and 5] which in turn contain references to detailed and older works.

In view of the previous work and the recent aspects described herein, it would appear that thermal analysis has an established and potential rôle to play in at least the following areas of research, development and economic assessment of solid fossil fuels.

1. Rank evaluation.
2. Identification of specific minerals.
3. Evaluation of mineral contents.
4. Improved detection of carbonates in flowing CO_2.
5. Iron isomorphous substitution in carbonates.
6. Mineral thermal stability and decomposition rates.
7. Production of mineral decomposition products including gases.
8. Mineral decomposition product reactions.
9. The establishment and magnitude of the endothermic/exothermic nature of mineral and mineral component decompositions and or reactions and the temperatures at which they occur.
10. Determination of ignition temperatures.
11. Carbonisation under various atmospheric conditions.
12. Proximate analysis.
13. Product characterisation.
14. Retorting conditions and practice e.g. effects of heating rates, gas atmosphere conditions, and heat balance.
15. Residual carbon determinations of process end products.
16. Carbonaceous contents of other rocks.
17. Fluidised bed reactor studies.
18. The effects of catalysts on coal/oil shale temperature dependent processes.
19. Improved economic assessment of exploration and production samples.
20. The correlation of stratigraphic units.

The applications of thermal analysis are clearly of too wide a range to be considered adequately in a single address. Thus rather than attempt to cover all things for all people, it would appear realistic to restrict the coverage generally to mineralogical aspects and in particular to the lines of research undertaken in my own laboratory at the University of Newcastle (Australia).

MINERALOGICAL ASPECTS

The DTA curves of coal and oil shale determined in air show a large, rather ill defined exothermic feature, which varies somewhat in temperature (dependant on the type of material [6]), within the temperature range 250 to 900°C [4,7 and 8]. This feature, typically shows two broad, exothermic maxima (Fig.1 curves 1 and 2). These peaks represent firstly the oxidation/combustion of the volatile hydrocarbons released during heating, while the second is due to the later rapid oxidation of the remaining fixed carbon.

Within the temperature range of this large exothermic feature lie the much smaller peaks of the minerals commonly found in coal. These occur in relatively low amounts with the organic component (cf. curves 1 and 2 with 4 to 12, Fig.1). With the exception of siderite, the superposition of these opposed thermal effects is negating. Thus the mineral caused peaks are suppressed and rarely appear in recognisable form on the resultant DTA curves.

This masking effect may however, be alleviated by determinations in flowing inert gas atmospheres (see below under "Differential Thermal Analyses").

For siderite ($FeCO_3$) the endothermic decomposition reaction $\approx 600°C$) releases FeO which immediately oxidises to give a marginally larger exothermic effect. Due to the high oxidation potential of FeO, these two reactions become superimposed in conditions of readily available oxygen (i.e. in flowing O_2, or for sample contents of less than 30% if determined in air [9]. Such DTA curves show only a small resultant exothermic peak. This is completely overwhelmed by the huge coal feature, and is not detected (cf curves 1 and 2 with 8, Fig.1).

The decomposition of pyrite (FeS_2) occurs in the same temperature region as siderite, and is endothermic. It is also accompanied by rapid oxidation, to form iron sulphates. The effects of heating in air are analagous to siderite but the oxidation effects are larger and in air tend to give unsatisfactory reproducibility as O_2 availability fluctuates due to localised O_2 depletion [10].

For these reasons a "typical" DTA curve of pyrite has not been included in Fig.1. Likewise its effects are also indistinguishable from those of the much greater coal reactions.

Conversely, the combustion of organic matter causes such a strong exothermic reaction, in the above temperature range, that its presence may be used to confirm the existence of small amounts of organic matter in sediments, soils [11] and in fly ash [12]. Also in relation to coal rank [12] and to the degree of graphidisation of organic matter fragments (phytoclasts) in rocks as a function of their metamorphic grade [13].

Detection limits may be improved by determinations in flowing O_2 instead of air, which promotes a more vigorous reaction.

Currently a number of low carbon content materials are available in large quantities as natural deposits or waste products e.g. some carbonaceous shales and coal cleaning/washing and oil shale retorting residues, are being carefully assessed as commercial sources of low grade fuels. This is because they are now becoming economically exploitable due to the very high cost of competing fuels and the fact that their production costs, have often already been met for other purposes.

The fuel value of these materials is directly relatable to their moisture content, hydrocarbon yield and residual carbon content, all of which may be assessed by thermal analysis methods.

Such determinations provide information in the area where standard laboratory methods for coal analysis have been formally grouped under the heading of proximate analysis.

Thermogravimetry and Proximate Analysis. The proximate analysis of coal and oil shale involves the determination (weight percent) of

the following components and is covered by standard methods i.e. ASTM D 3172-75
1. Moisture
2. Volatiles
3. Fixed carbon
4. Ash.

All these determinations may be obtained by TG, the results of which agree well with the standard methods [14,15].

Thermogravimetry has the following advantages:
a) The four determinations may be carried out on the same sample as a continuous process.
b) They are therefore determined under identical conditions.
c) Fixed carbon content is obtained by direct measurement instead of by difference.
d) The full proximate analysis can be completed in under 30 minutes.
e) The thermobalance can be fully automated, including shut down on completion. Thus operator involvement is restricted to loading and starting the TG run.
f) By interfacing with suitable computer and plotting/ print-out facilities, the final data can be presented in the most suitable form.
g) Experimental control difficulties are minimised.

Descriptions of the method and suitable advanced equipment packages have been published recently [14,15 and 16].

One method [14] involves heating in purging N_2 (for N_2 read oxygen free nitrogen throughout) followed by O_2 (Fig.2), i.e.
1. heating the sample rapidly (50 to 60°C) per minute to 110°C where it is held isothermally for 5 minutes by which time constant weight has been achieved. The weight loss to this point represents the moisture content;
2. heating rapidly (100°C per minute) to 950°C, where it is again held isothermally for 10-15 minutes to constant weight. This second weight loss represents the loss of volatiles;
3. at 950°C the flowing N_2 gas atmosphere is replaced by flowing O_2. Under these conditions the residual carbon burns. The sample is held under these conditions to constant weight. This last weight loss represents the fixed carbon content.
4. The remaining sample weight represents the ash content.

Either by the standard or TG methods the volatile matter weight loss is considered to be essentially due to the release of hydrocarbon gases. It is accepted that minerals which decompose in this temperature range would contribute volatile components, such as CO_2 (carbonates), H_2O (dehydroxilation of clays) and SO_x (sulphides). However, their proportion of the total gas yield is comparatively small, due to the relatively low mineral matter content of coals, and with the exception of SO_x, their presence is accepted as inevitable and is not considered particularly significant.

This is not the case however, with oil shales. These, in many cases, could be classified as impure carbonate rocks. Irrespective of the actual rock composition, the organic kerogen fraction is usually dominated by the mineral content which is commonly in excess of 50% [17].

Furthermore, it has been stated that, "the inorganic constituents of oil shales differ substantially from one deposit to another" [18].

With the exception of the Green River oil shales (U.S.A.), little

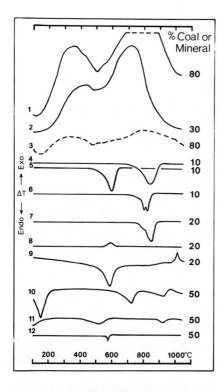

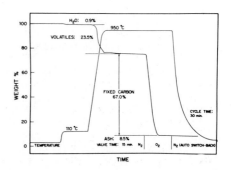

Fig.2. Illustration of Proximate Analysis of Coal by TG. After Earnest and Fayans [14].

Fig.1. Comparison of DTA curves of a low ash sub-bituminous coal and minerals commonly in coal, diluted with Al_2O_3. Curves (1 to 3) 80, 30 and 80% coal, (4) 10% calcite, (5) 10% magnesite, (6) 10% dolomite, (7) 20% ankerite, (8) 20% siderite, (9) 20% kaolinite, (10) 50% montmorillonite, (11) 50% illite, (12) 50% quartz. Heating rate 15°C/minute (static air), except curve 3 (flowing N_2). Additional equipment and experimental details [10].

detailed mineralogical information is available [19]. In this deposit, a large number of inorganic constituents have been identified. These include some 70 authigenic minerals, which range from common to rare and include several new minerals and a surprising range of carbonates [20].

With the recent upsurge of interest in oil shales, papers on the mineralogy of other deposits are beginning to appear. Although as yet too few, these tend to indicate that the "typical" oil shale mineralogy may be much simpler. This certainly is the case for many of the large E.Australian deposits such as at Rundle in Queensland [21]. Here calcite, siderite and dolomite (includes Fe substituted members, ferroan dolomite-ankerite), together with the clay minerals, montmorillonite, illite and kaolinite occur with quartz and some pyrite.

Initially, it is the effects this limited number of minerals have on the TG and DTA curves of oil shales, which is under current study. In fact these effects can be so marked, that one scientist worked

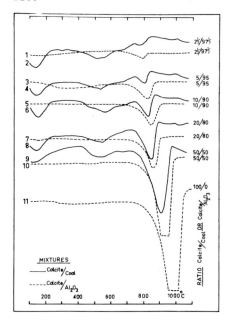
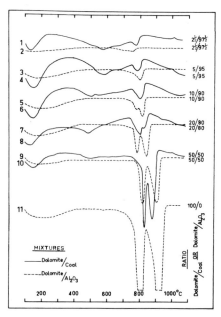

Fig.3. DTA curves obtained in flowing N_2 from mixtures of calcite with Al_2O_3 (– – –) or the coal (———). After Warne [2 and 10].

Fig.4. DTA curves obtained in flowing N_2 from mixtures of dolomite with Al_2O_3 (– – –) or the reference coal (———). After Warne [2 & 10].

only on demineralised oil shale samples [22].

In order to accurately locate where these mineral weight loss reactions occur, DTA was used as it gives a clearer result than TG.

Differential Thermal Analysis.   Previous work [2,10 and 23] established that if DTA runs were made in inert furnace atmosphere conditions of flowing N_2, the coal curve could be suppressed (cf curves 1 and 3, Fig.1). Under these conditions the specific mineral caused peaks could be clearly recorded and identified.

In order to establish in detail the diagnostic curve configurations of each mineral, at various concentrations, when diluted with inert Al_2O_3 or coal, a series of artificial mixtures were made and subjected to DTA in flowing N_2. The results demonstrated the identity of the diagnostic mineral peaks and their detection limits (e.g. Figs.3 and 4). Also that the presence of much smaller amounts of anhydrous carbonates and sulphides could be detected compared to the clays [2] (e.g. Fig.1 curves 4 to 12), where larger contents of some minerals have been needed to produce peaks of comparable size.

Under these inert conditions, the superimposed negating effects of the immediate oxidation of the FeO, released on the decomposition of siderite, does not occur. Thus the single endothermic

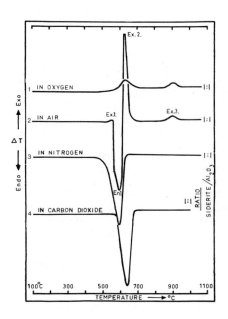
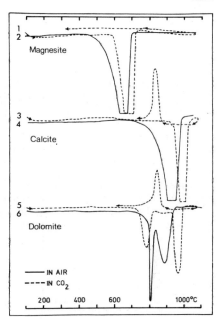

Fig.5. DTA curves of siderite showing the effects of different furnace atmospheres. All samples diluted 1:1 with Al_2O_3.

Fig.6. DTA curves of magnesite, calcite and dolomite, determined in flowing CO_2 and N_2. Diagnostic peak movements and recarbonation peaks (from the $CaCO_3$ components) shown by curves determined in CO_2. Featureless cooling curves in air omitted for clarity. All samples diluted 1:1 with Al_2O_3.

decomposition peak is recorded in its entirety (cf curves 1 to 3, Fig.5). This results in much improved detection limits i.e. comparable to calcite and magnesite (Fig.1 curves 4 and 5).

A previous investigation [24] produced much improved DTA peak resolution and identification when cerussite ($PbCO_3$) was determined in flowing CO_2. This led to the consideration that the substitution of flowing CO_2 for N_2 would still act as an inert gas as far as coal combustion was concerned, would not affect other minerals, but would provide improved carbonate peak definition and detection.

This technique proved most successful [2] and led to several valuable applications.

Firstly, for single peaked carbonate decompositions (i.e. magnesite and calcite) there is a marked increase in peak height as the increased partial pressure of CO_2 delays the dissociation reaction. Thus when it does occur, it takes place, more rapidly, over a smaller temperature range and with an increased peak height, as individual peak areas remain constant. The detection limit is therefore considerably increased i.e. down to contents of $\approx 0.25\%$ [25].

Secondly, this delayed dissociation means that it now occurs at a higher temperature. There is therefore a marked diagnostic up scale movement of such single endothermic peaks when the DTA is determined in CO_2 as compared to N_2 or air (cf curves 1 to 4, Fig.6).

Thirdly the two and three peaked curves of dolomite ($CaMg(CO_3)_2$) and ankerite ($Ca(MgFe)(CO_3)_2$) behave somewhat differently. In both cases their initial endothermic peak is displaced down scale to a lower temperature, while the higher temperature dolomite peak and ankerite peaks move to higher temperature positions (cf curves 5 and 6, Fig.6). The increased parting of these diagnostic peaks, remains right down to the limits of their detection, $\approx 0.25\%$ [25].

This feature is of considerable importance because in N_2 or air, as the dolomite or ankerite sample content falls, the typical multipeaked DTA curves progressively fuse (Fig.4). Thus for contents of <10% both minerals are represented by similar single much less diagnostic endothermic peaks. These are difficult to identify specifically and distinguish from the comparable peak of calcite.

Fourthly, DTA in flowing CO_2 provides a method for assessing the presence of "calcium carbonate" contents from the re-carbonation peak of CaO which occurs on the cooling curve (Fig.6, curves 3 and 5). Furthermore, any Fe, Mg, Mn, Zn and Pb oxides produced during heating will not re-carbonate in this way.

Here it should be noted that strontianite ($SrCO_3$) and witherite ($BaCO_3$) show one ($\approx 875°C$) and two (≈ 750 and $970°C$) crystallographic re-inversion peaks, respectively, on their DTA cooling curves, irrespective of furnace atmosphere conditions [26]. Their presence on cooling curves produced under flowing N_2 conditions distinguishes them from the calcite re-carbonation peak which will be absent.

The presence of Fe as a component of carbonates from which it can be released on heating is of importance because:-
a) It lowers the ash fusion temperatures and has important implications for boiler tube deposits and corrosion.
b) In the dolomite-ankerite series the Mg content is progressively diminished by increasing substitution by Fe. A lowering of the decomposition temperature results [27]. In turn this makes available earlier the MgO component (but in lesser amounts, which is important for potential evolved SO_x complexing and retention [29]. This aspect has been investigated by thermal analysis methods [30], and the process may be reversed later to reclaim the sulphur [31].
c) It acts as a catalyst in oil liquifaction processes [32].
d) The iron may be present also in the form of siderite. This has economic implications because the forms in which siderite and ferroan dolomite-ankerite are present usually differ [33], which effects the methods and success with which these minerals may be removed by coal preparation processes.

It has been established that siderite, ankerite and dolomite can be clearly identified, individually by DTA, in flowing CO_2. This led to three other important questions concerning the further application of DTA, i.e.
1. Could ankerite be identified when present with dolomite?
2. Could siderite be identified when present with ankerite?
3. Could the degree of Fe substitution in the dolomite-ankerite series be determined?

In the first case, artificial mixtures of dolomite with ankerite in various proportions were subjected to DTA in flowing CO_2 furnace atmosphere conditions. The resultant curves (Fig.7) demonstrate that

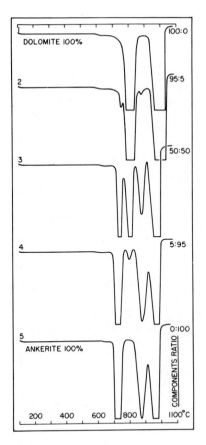
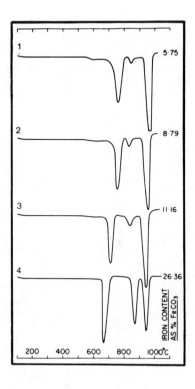

Fig.7. DTA curves of artificial mixtures of dolomite and ankerite determined in flowing CO_2. Detection limits of ≃2% of each with the other are indicated.

Fig.8. DTA curves of members of the dolomite/ankerite series whose Fe contents range from 5.75 to 26.36 weight % expressed as $FeCO_3$. Illustrates the relationship of first peak temperature and second peak size to $FeCO_3$ content. Detection limit ≃1% $FeCO_3$. All samples diluted 1:1 with Al_2O_3, determined in flowing CO_2. Details, Warne, Morgan and Milodowski, [25].

detection of the contents of each mineral to at least 2% is possible.

The second and third points have been the subject of detailed studies of which part I, "Iron Content Recognition and Determination by Variable Atmosphere DTA" has recently been published [27] and part II, "Thermogravimetry in CO_2 Atmospheres" is in manuscript [28] (with Dr. D.J. Morgan and Mr. A.E. Milodowski of the Institute of Geological Science, London).

In essence, from the DTA (in flowing CO_2) of members of the dolomite-ankerite series it has been shown that, the size of the middle endothermic peak depends on the amount of Fe substitution which has occurred (Fig.8). Furthermore, the minimum detectable amount of Fe present (expressed as $FeCO_3$) is in the order of 1%.

In addition, the decomposition of siderite occurs at a temperature sufficiently below the first ankerite peak, that the peaks of both minerals stay resolved down to the detection limits of approximately 1% (Fig.9).

As the decomposition of dolomite starts at a somewhat higher temperature than ankerite (cf curves 1 and 5, Fig.7), the separation of the siderite and the first dolomite peak is greater and its identification also clear.

Complementary DTA and TG Determinations. With the decomposition temperatures and ranges of these carbonates clearly located by DTA, the equivelant portions of the proximate analysis related TG curves could be examined for evidence of the complementary weight losses caused by the CO_2 released during such decompositions (Determinations in flowing N_2).

This phase of the work was carried out using a Stanton Redcroft simultaneous DTA-TG thermobalance. It gave excellent DTA-TG results, from the same sample under strictly controlled atmosphere and heating rate conditions (heating rate $15°C$/minute with 30mg samples at -150# B.S. sieve).

The results of this aspect of our oil shale research project are typified by dolomite [Fig.10] as detemined in flowing N_2 and CO_2. Full details of this work are in manuscript [34] (with Mr. D.H. French, University of Newcastle, Australia).

The upper part of Fig.10 shows four curves obtained from a low mineral content oil shale from the Rundle deposit in Queensland (Australia). Curves 1 and 2 show the DTA and TG curves produced by heating to $1000°C$, a mixture of 80% of this oil shale with 20% inert Al_2O_3. The TG curve shows three distinct weight losses which are mirrored by endothermic peaks on the DTA curve. In order of increasing temperature ranges these are caused by, initial water losses (to $150°C$), gaseous hydrocarbon release ($350-550°C$) and a small carbonate decomposition CO_2 loss ($650-700°C$).

Curves 3 and 4 were obtained from artificial mixtures of the same amount of the oil shale with 20% of dolomite. They clearly show the modifications to both DTA and TG curves caused by the presence of this mineral (cf curves 1 and 2 with 3 and 4, Fig.10).

Similar modifications at marginally different temperatures are produced on curves from mixtures containing the same oil shale content and 20% ankerite or calcite.

Such weight loss modifications to TG curves of oil shales caused by ankerite, dolomite and calcite are collectively easily recognised. They occur in the approximate temperature range 625 to $775°C$, which is clearly separated from the hydrocarbon release weight loss at 350 to $550°C$. However, the identification of which of these three carbonates is actually present is not clear using the flowing N_2 method.

Here it should be noted that the much improved parting of the DTA peaks of dolomite in flowing CO_2 is also reflected on its TG curve. The result is a marked increase in the definition and separation of the two decompositional weight losses as compared to determinations in N_2 (Fig.11) or air. (Discussed in detail in the forthcoming publication [28]. This also applies to oil shale containing dolomite.

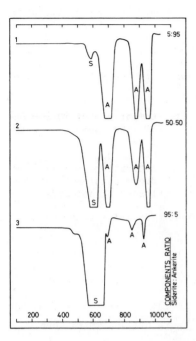

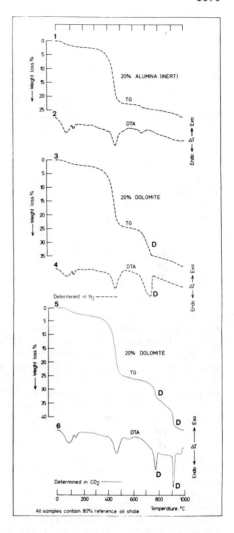

Fig.9. DTA curves of mixtures of siderite with ankerite. Illustrate detection and identification limits of ≈1%. After Warne, Morgan and Milodowski, 1981 [25].

Fig.10. DTA and TG curves of a low mineral content oil shale (Rundle Australia), determined in flowing N_2 and CO_2. Samples contain 80% oil shale mixed with 20% Al_2O_3 (curves 1 & 2) or 20% dolomite (curves 3,4,5 & 6). Portions of curves due to dolomite decomposition are marked D.

A comparison of the curves 5 and 6 with 3 and 4 in Fig.10, shows that the single DTA peak and TG weight loss caused by the decomposition of dolomite in flowing N_2, is replaced by two such features on both the DTA and TG curves when determined in flowing CO_2.
Furthermore, when determined under the same conditions,
a) ankerite is recorded as three separate DTA peaks TG weight losses,

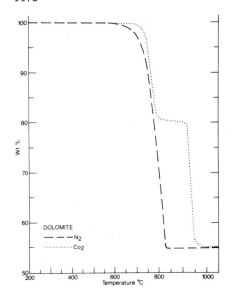

Fig.11. Comparison of TG curves of the same dolomite determined in flowing CO_2 or N_2, shows marked separation of the two decompositional weight losses of dolomite.

b) calcite decomposition remains as a single DTA and TG feature, but both move up scale some $75^\circ C$.

Thus by using simultaneous DTA and TG determinations in flowing N_2 and CO_2 the individual presence of these carbonates, the magnitude of their thermal effects and their CO_2 contribution to the overall volatile matter yield can be ascertained.

In contrast, the decomposition temperatures of siderite and magnesite are considerably lower. In flowing N_2 their decomposition effects are partially superimposed on the hydrocarbon yield features on DTA and TG curves of oil shale. Of these siderite is superimposed on the higher temperature portion of the hydrocarbon yield feature, which is almost complete before the decomposition of magnesite starts.

Again, determinations in flowing CO_2 cause the single decomposition feature of siderite and magnesite to move up scale (occur at higher temperatures). The movement of magnesite is somewhat greater than for siderite. As a result siderite is only marginally superimposed on and magnesite is well clear of the hydrocarbon yield feature.

In this way determinations in flowing N_2 compared to flowing CO_2 show up scale movement of the siderite and magnesite features on the DTA and TG curve which,
a) confirms their individual presence,
b) allows for their full recording without interference,
c) indicates the true magnitude of their thermal effects,
d) isolates and delineates the effects of their CO_2 evolutions, which allows;
e) the assessment of their CO_2 contribution to the volatile matter yield from TG curves.

In contrast the detection limits of the clay minerals, kaolinite, montmorillite and illite, together with quartz are not as good (Fig.1

curves 9 to 12). For these minerals considerably larger amounts of each are required to produce comparable sized DTA peaks. Furthermore, the modifications to the DTA and TG curves of oil shale, caused by their presence, are much less evident and cannot be improved or diagnostically moved up scale by determinations in flowing CO_2.

At this stage therefore we see less applications, as described above, of DTA and TG to this group of minerals. However, a profitable line of investigation, not available to us, would be to make determinations in atmospheres of much increased water vapour pressure. In this way the definition and position of water producing reactions could be diagnostically manipulated.

CONCLUSIONS

In the fields of coal and oil shale research, characterisation, development and utilisation, the work to date indicates that important contributions can be made with variable atmosphere DTA and TG.

Direct applications are to mineral identification and content evaluation, with very good detection limits for anhydrous carbonates. The effects of carbonate minerals in relation to their CO_2 contribution to the volatile matter yield may be determined. Decomposition temperatures and rates together with the relative size of these dominantly endothermic reactions may be assessed. These parameters are of particular relevance to the production and maintenance of the required heat balance in retorting technology.

Such diagnostic information appears also to be directly applicable to a number of economic aspects e.g. deposit characterisation, establishment of variation trends, quality control, blend and beneficiation product assessment and has significance in relation to conditions of deposition, preservation, diagenesis, degree of coalification and probably metamorphism.

ACKNOWLEDGEMENTS

I am fortunate in having collaborated with several stimulating scientists, with whom working has been a great pleasure, viz., R.L. Mackenzie and B.D. Mitchell, P.K. Gallagher, P. Bayliss, D.J. Morgan and A.E. Milodowski, A.C. Cook and D.H. French.

The oil shale research, except where acknowledged, forms part of the Universities of Newcastle and Wollongong (Australia) project "Low Rank Oil Shales", currently supported under the "National Energy Research, Development and Demonstration Program", administered by the Department of National Development and Energy (Australia).

Permission to republish several Figures is much appreciated and has been kindly granted as follows: Fig.2, Perkin-Elmer Corp., Figs.3 and 4, the Institute of Energy and Fig.9 Thermochimica Acta.

REFERENCES

1. G.J. Lawson, Solid Fuels, Chapt.25, in Differential Thermal Analyses, VII, Ed. R.C. MacKenzie Acad. Press, (1970), p.705-726.
2. S.St.J. Warne, Chapt.52, in Analytical Methods for Coal and Coal Products, V.III, Ed. C.Karr Jr., Acad. Press, (1979), p.447-477.

3. D. Schultze, "Differentialthermoanalyse", VEB Deutscher Verlag der Wissenschaften, E. Berlin, (1971), p.338.
4. N.I. Voina and D.N. Todor, Chapt.37, in Analytical Methods for Coal and Coal Products, V.II, Ed. C. Karr Jr., Acad. Press, London, (1978), P.619-648.
5. M.I. Pope and M.D. Judd, "Differential Thermal Analysis - a Guide to.....", Chapt.15, Heyden and Sons, London, (1980), p.114-123.
6. S.St.J. Warne, Jour. Therm. Anal., 1981, $\underline{20}$, 225.
7. J.B. Stott and O.J. Baker, Fuel London, 1953, $\underline{32}$, 415.
8. S.St.J. Warne, unpub. Ph.D. thesis, Univ. New South Wales (Australia), (1963), p.367.
9. S.St.J. Warne, Chem. de Erde, 1976, $\underline{35}$, 251.
10. S.St.J. Warne, Jour. Inst. Fuel, 1965, $\underline{38}$, 207.
11. W. Smykatz-Kloss, "Differential Thermal Analysis, Applications and Results in Mineralogy", Springer-Verlag, Berlin, (1974), p.185.
12. D.J. Swaine, Proc. 2nd Internat. Conf. Therm. Anal., 1969, $\underline{2}$, 1377.
13. C.F.K. Diessel and R. Offler, N. Jb. Miner. Mh., 1975, $\underline{I}$, 11.
14. C.M. Earnest and R.L. Fyans, Perkin-Elmer, Thermal Analysis Application Study No.32, (1981), p.8.
15. J. Elder, Proc. 10th NATAS. Conf. (Boston) U.S.A. (1980), 247.
16. M.R. Ottaway, Fuel, in press.
17. S.St.J. Warne, "Role of Determinative Mineralogy in Relation to Oil Shale Petrology", in, Oil Shale Petrology Workshop, Eds. A. A.C. Cook & A.J. Kantsler, Keiraville Kopiers, Wollongong, (1980).
18. D.C. Duncan, Chapt.2, in Oil Shale, Eds. T.F. Yen and G.V. Chilingarian, Elsevier New York (1976), p.13-26.
19. W.C. Shanks, W.E. Seyfried, W. Craig Meyer and T.J. O'Neil, Chapt.5, in Oil Shale, Eds. T.F. Yen and G.V. Chilingarian, Elsevier New York (1976), p.81-102.
20. C. Milton, Wyoming Univ. Contrib. Geol., 1971, $\underline{10}$(1), 57.
21. A. Cook, A. Hutton, A. Kantsler & S. Warne, Sci. Aust.,1980, $\underline{4}$,6.
22. J.D. Saxby, Thermochim. Acta., 1981, $\underline{47}$, 121.
23. S.St.J. Warne, Jour. Inst. Fuel, 1970, $\underline{43}$, 240.
24. S.St.J. Warne and P. Bayliss, Amer. Mineral., 1962, $\underline{47}$, 1011.
25. S.St.J. Warne, Nature, 1977, $\underline{269}$, 678.
26. K.H. Wolf, A.J. Easton, and S.St.J. Warne in Carbonate Rocks Pt.B. Eds. G.V. Chilinger, H.J. Bissell and R.W. Fairbridge, Elsevier New York (1967), p.253-341.
27. S.St.J. Warne, D.J. Morgan and A.E. Milodowski, (Part I, in) Thermochim. Acta, 1981, $\underline{51}$, 105.
28. D.J. Morgan, A.E. Milodowski and S.St.J. Warne, (Part II, Thermogravimetry in CO_2 Atmospheres), in manuscript for submission to Thermochim. Acta.
29. L. Fuchs, E. Nielsen & B. Hubble, Thermochim. Acta, 1978, $\underline{26}$, 229.
30. C. Sun, E. O'Neill & D. Keairns, Thermochim. Acta, 1978, $\underline{26}$, 283.
31. R. Snyder, W. Wilson & I. Johnson, Thermochim. Acta,1978,$\underline{26}$, 257.
32. S.B. Alpert and R.H.H., Chapt.28, in Chemistry of Coal Utilization (Ed. M.A. Elliott), 2nd Suppl. Vol., John Wiley, New York.
33. E. Stach, M-Th. Mackowsky, M. Teichmuller, G.H. Taylor, D. Chandra and R. Teichmuller, "Stach's Textbook of Coal Petrology", Gebruder Bointraeger, Berlin, (1975), p.428. (1981), p.1919-1990.
34. S.St.J. Warne and D.H. French, (The Application of DTA and TG to Some Aspects of Oil Shale Mineralogy), in manuscript for submission to Thermochim. Acta.

DETERMINATION OF REACTION RATE AND KINETICS
OF THE NICKEL CATALYSED METHANATION OF CO,
MEASURED BY DSC

G. Hakvoort and L.L. van Reijen
Delft University of Technology
Department of Chemistry
The Netherlands

INTRODUCTION

At our laboratory the preparation of high-temperature resistent Ni/Al_2O_3 catalysts is investigated [1,2,3]. These catalysts are to be used for the methanation of CO, according to the reaction:

$$CO + 3H_2 \rightarrow CH_4 + H_2O$$

Since this reaction is highly exothermic, with a heat effect of about 215 kJ/mol CO at 200°C, it may be investigated by DSC [4].

This paper describes two kinds of experiments with coprecipitated nickel-alumina catalysts, namely activity measurements and determination of the kinetics.

EXPERIMENTAL

The preparation and the composition of the catalysts are described elsewhere [1,2,3]. For the DSC measurements a Dupont 910 apparatus is used. A certain amount of reduced catalyst, passivated with water vapour, is placed in an open DSC cup, made of aluminum. First the catalyst is rereduced by heating in 1 atm H_2 to 400°C with a heating rate of 10°C/min. Once at 400°C, this temperature is kept constant during 15 minutes. After this the sample is cooled down to about 120°C.

In all kinetic experiments the heating rate was 10°C/min and the total gas flow rate is kept constant at 39 ml/min at a pressure of 1 atm. First a baseline is measured by heating in pure hydrogen up to 400°C (see Figure 1). After cooling down to 120°C, a small percentage of CO is admitted, together with an excess of H_2, and the heating is repeated.

The formation of methane causes a strongly exothermic DSC-signal. This signal reaches a maximum value (v_{max}) upon heating. For low initial CO-concentrations there is a linear relation between the calculated heat flow, corresponding to 100% conversion of CO, and the value of v_{max}. The ratio of these two values corresponds to the DSC-calibrating factor C. Within our experimental conditions C varies between 3.5 and 4, depending on the amount of catalyst and the degree of contact with the DSC cup. For exact measurements this calibration factor should be determined for every fresh sample, with help of a standard experiment. From the value of the DSC-signal v the reaction rate r, the fraction converted and the remaining CO-pressure p_{CO} are calculated.

Isothermal experiments with varying initial CO-pressures and hydrogen pressures were made in order to study the influence of hydrogen and CO-pressure on the reaction kinetics.

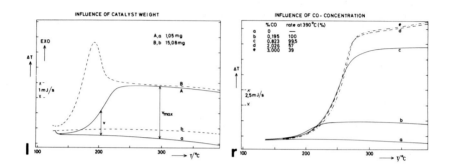

Figure 1. Influence of reaction parameters; 1: influence of the amount of catalyst, a and b are baselines (100% H_2) and A and B are reaction lines (0,21% CO - 99,79% H_2), v_{max} corresponds to the maximum reaction rate; r: influence of (initial) CO-concentration for 1,05 mg of catalyst. The rate at 390°C is given as a percentage of the rate, corresponding to complete conversion of the CO into methane.

RESULTS AND DISCUSSION

Activity measurements. In the first place the activities of different types of catalyst were determined with DSC and compared with results obtained with a flow reactor.

In both cases standard conditions were applied. For the flow reactor, these conditions were: gasflow 71/hr, consisting of 15% CO and 85% H_2, pressure 1 atm and temperature 300°C. For each experiment a mixture of 25-100 mg of catalyst with 4 g of α-Al_2O_3 powder was used [1].

At these conditions DSC experiments showed fast catalyst deactivation. Therefore the DSC testing conditions were chosen as follows: catalyst weight 2 mg, gasflow 39 ml/min, consisting of 2% CO and 98% H_2. The reaction rate r at 250°C was calculated from a linear heating rate experiment (5 or 10°C/min).

As seen in Figure 2, there is a good linear relationship between both kinds of measurements. So it can be concluded that the activity of these kinds of catalysts can be determined with DSC in a fast and easy way.

Determination of kinetics. The kinetics of the catalytic reaction were determined by measurements at different temperatures and CO/H_2 ratios. Several preliminary investigations were made, in order to find the optimal conditions for this kind of measurements. The

results are as follows:
- The maximum heating rate is $10°C/min$. Higher rates give less reproducible results.
- Deactivation of the catalyst is a serious problem at the DSC experiments. Cleaning of the DSC cell, low CO-concentrations and low temperatures are needed for a longer catalyst life. For this reason linear heating rate experiments, which can be made in shorter time, are preferred in comparison with isothermal experiments. Regular repetition of a standard experiment is needed to control whether the activity of the catalyst is still unchanged.
- The amount of catalyst is limited to a few milligrams (see Figure 1). With higher amounts a peak is shown in the DSC curve, followed by a decreasing DSC signal at higher temperatures. Probably this is caused by the formation of a thermal lag between the upper part of the catalyst layer, in which, at high temperatures, total conversion occurs, and the DSC-sensor.
- At high CO-concentrations the reaction becomes diffusion-limited from a certain degree of conversion (see Figure 1).
- Variation of the rate of gasflow influences the DSC calibration factor C. Substitution of a part of the hydrogen by helium shows no distinct variation of C.

For these reasons the measuring conditions of the kinetic experiments were chosen as follows: catalyst weight (having a Ni/Al ratio of 2): about 1 mg, CO-concentration less than 1%, total gasflow 39 ml/min, total pressure 1 atm.

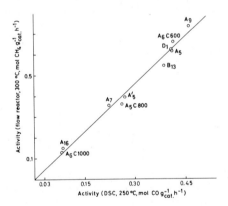

Figure 2. Comparison of the activities of catalysts, measured with DSC or with a flow-reactor. The activities are expressed in: mol CO/g cat. hr.

Results of kinetic measurements. Figure 3a shows a set of isothermal curves, constructed by making sections through linear heating rate experiments with different initial CO-concentrations. As in really isothermal experiments, a maximum in the reaction rate is clearly visible. The exact position of this maximum depends on the reaction temperature and the remaining CO-pressure p_{CO}.

In a search to the reaction mechanism, a great number of kinetic equations is tested [5]. The best fit, found at a nearly constant H_2 pressure of about 1 atm, corresponds to the equation:

$$r = kp_{CO}^{\frac{1}{2}}/(1 + k_c p_{CO}^{\frac{1}{2}})^2 \qquad (1)$$

This can be converted into:

$$(p_{CO}^{\frac{1}{2}}/r)^{\frac{1}{2}} = 1/k^{\frac{1}{2}} + (k_c/k^{\frac{1}{2}}) \, p_{CO}^{\frac{1}{2}} \qquad (2)$$

As shown in Figure 3b, there is a good fit, according to equation (2).

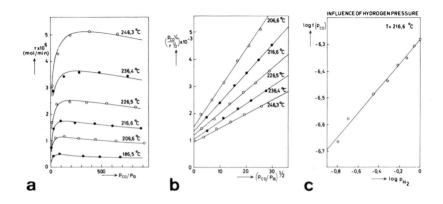

Figure 3. Isothermal curves; a and b: constructed from linear heating rate experiments at 1 atm H_2; c: determination of the influence of the H_2-pressure, according to eq. 7, p_{H_2} is expressed in atm.

The temperature dependence of k and k_C can be described by:

$$k = k_o \exp(-E_A/RT) \qquad (3)$$

and

$$k_C = K_o \exp(+\lambda/RT) \qquad (4)$$

This is shown in Figure 4. In Table 1 the results of two sets of experiments are summarized.

To investigate the hydrogen dependence of the methanation rate, two sets of isothermal experiments have been made, both at the same temperature:
- At first the value of k_c was determined at about 1 atm H_2 and varying CO-concentrations.
- Then the hydrogen pressure was varied by substitution of hydrogen

by helium at a constant initial CO-pressure.

Assuming that k_c is independent of the hydrogen pressure, the hydrogen dependence is given by the equation:

$$k = k' \, (p_{H_2})^a \tag{5}$$

Combination of equations (1) and (5) then leads to:

$$k = f(p_{CO}) = r \, (1 + k_c \, p_{CO}^{\frac{1}{2}})^2 / p_{CO}^{\frac{1}{2}} = k' \, p_{H_2}^a \tag{6}$$

and

$$\log f(p_{CO}) = \log k' + a \log p_{H_2} \tag{7}$$

In Figure 3c the results are given for one experiment. It appears that the exponent a of eq. (5) for all experiments equals about 0,5: a = 0,47 ± 0,02.

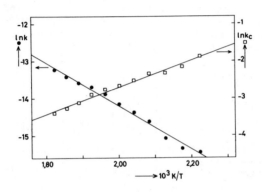

Figure 4. Arrhenius plot of the kinetic parameters k and k_c.

TABLE 1. RESULTS OF KINETIC MEASUREMENTS

Sample weight (mg)	0,90	0,92
Temperature region (°C)	176-276	166-295
k_o (mol/min Pa$^{\frac{1}{2}}$)	$7,91.10^{-2}$	$3,76.10^{-2}$
K_o (1/Pa$^{\frac{1}{2}}$)	$5,15.10^{-5}$	$7,43.10^{-5}$
E_A (kJ/mol)	48,5±3,8	45,4±4,3
λ (kJ/mol)	29,8±2,9	28,5±5,6

Mechanism. A possible mechanism for explaining the rate equation found, could be the dissociative adsorption of CO and H_2, followed by surface reactions:

$$CO(g) + 2s \rightleftarrows C_a + O_a \quad ; K_1 = \Theta_C \Theta_O / \Theta_s^2 p_{CO}$$

$$H_2(g) + 2s \rightleftarrows 2H_a \quad ; K_2 = \Theta_H^2 / \Theta_s^2 p_{H_2}$$

$$C_a + H_a \rightarrow CH_a + s \quad ; r_3 = k_3 \Theta_C \Theta_H$$

$$O_a + H_a \rightarrow OH_a + s \quad ; r_4 = k_4 \Theta_O \Theta_H$$

$$1 = \Theta_s + \Theta_C + \Theta_O + \Theta_H$$

$$r_3 = r_4 \qquad \text{stoichiometric condition.}$$

In these equations C_a means a carbon atom, adsorbed on the catalytic surface; s is an unoccupied catalytic surface site; Θ_i is the fraction of catalytic surface sites, covered with i. Further it is assumed that the first and second reaction are fast equilibria. The third or fourth reaction is rate determining. This means that CH_a and OH_a will react very fast with formation of methane and water vapour. So the surface concentration of CH, CH_2, CH_3 and OH is very low, compared with Θ_s, Θ_O, Θ_C and Θ_H, resulting in the fifth equation. A further calculation with these equations gives:

$$r = k' \, p_{CO}^{\frac{1}{2}} p_{H_2}^{\frac{1}{2}} / (1 + k_c p_{CO}^{\frac{1}{2}} + K_2 p_{H_2}^{\frac{1}{2}}),$$

in which

$$k' = (K_1 K_2 k_3 k_4)^{\frac{1}{2}} \text{ and } k_c = K_1^{\frac{1}{2}} \{(k_4/k_3)^{\frac{1}{2}} + (k_3/k_4)^{\frac{1}{2}}\}.$$

Since we have found that r varies with $(p_{H_2})^{\frac{1}{2}}$, it may be assumed that the term $K_2 \, p_{H_2}^{\frac{1}{2}}$ is much smaller than $(1 + k_c p_{CO}^{\frac{1}{2}})$. This is valid when the adsorption of CO is much stronger than the adsorption of H_2, within our experimental conditions.

CONCLUSIONS

The methanation reaction can be very precisely measured with DSC, provided that the experimental conditions are carefully chosen.

The activity of methanation catalysts can be determined with DSC in a fast and easy way.

The rate equation found corresponds to the results obtained by other investigators, who used more conventional apparatus [5].

ACKNOWLEDGEMENT

The authors are indebted to mr. Y. Timmerman for the performance of a great part of the experiments.

REFERENCES

1. E.C. Kruissink, "Coprecipitated nickel-alumina methanation

catalysts", Thesis, Delft (1981).
2. E.C. Kruissink, L.L. van Reijen and J.R.H. Ross, J. Chem. Soc., Faraday Trans. 1, 1981, 77, 649.
3. L.E. Alzamora, J.R.H Ross, E.C. Kruissink and L.L. van Reijen, J. Chem. Soc., Faraday Trans. 1, 1981, 77, 665.
4. T. Beecroft, A.W. Miller and J.R.H. Ross, J. Cat., 1981, 40, 281.
5. R.Z.C. van Meerten, J.G. Vollenbroek, M.H.J.M. de Croon and J.W.E. Coenen, Applied Catalysis, to be published.

THERMOANALYTICAL INTERPRETATION OF THE REACTION
MECHANISM OF PROPENE OXIDATION ON THALLIUM(III)
OXIDES (1)

C.Mazzocchia (2), P.Cardillo (3), F.Di Renzo,
R.Del Rosso, P.Centola.

Dipartimento di Chimica Industriale e di Ingegneria Chimica del Politecnico di Milano
Piazza Leonardo da Vinci, 32 - I20133 Milano

INTRODUCTION

Several selective oxidation reactions of olefins on metal oxides based catalysts follow a redox mechanism.
The determining role played in selective oxidation by lattice oxigen - and not adsorbed oxygen - was pointed out by Bielanski and Haber [1]. Mars and van Krevelen had already determined the mathematic shape of the redox mechanism [2].
In recent years thallium(III) oxides were increasingly used in heterogeneous catalysis. Above all they were used in dimerization [3] [4] and selective oxidation reactions [5].
In a previous work [6] the authors pointed out that on thallium(III) oxide it is possible to perform propene epoxidation.
The authors observed that, in tests carried out in a pulse reactor, lattice oxygen oxidizes propene to propene oxide and to carbon dioxide.
From an industrial point of view the heterogeneous epoxidation of propene could present so many advantages to justify the study of the reaction mechanism.
Therefore this paper aims to check the redox mechanism proposed for the epoxidation of propene on thallium(III) oxide catalysts.

EXPERIMENTAL

In a previous work [6] the authors detailed the preparation and the characterization of the catalysts used in thermogravimetric tests. In this paper we report only some properties (Table 1), which are useful to a best understanding of the thermogravimetric data. The O/Tl atomic ratio was checked by atomic absorption using a Perkin

(1) Research supported by Consiglio Nazionale delle Ricerche (CNR) with a CNR-CNRS joint program.

(2) To whom all correspondence should be sent.

(3) Stazione Sperimentale per i Combustibili - S.Donato Milanese

Elmer 303 spectrophotometer. The surface areas were measured with a Ströhleim Areameter. Density data were determined by picnometry in toluene.

Thermogravimetric analyses were carried out on a Cahn-Ventron C100 thermobalance and on a Mettler TA2000C thermoanalyser. The former was used to define rate of reduction and re-oxidation of catalysts; the latter was used for simultaneous evaluations of gravimetric data and heats of reaction.

Reduction tests were carried out with both pure propene or hydrogen and mixtures of propene and nitrogen at different propene pressure values. In re-oxidation tests both pure oxygen and oxygen diluted with nitrogen were used.

All tests of reduction and re-oxidation on the Cahn-Ventron C100 were carried out at isothermal conditions, using 1 g of catalyst with the particle size showed in Table 1. The gas program we used was a cycle of cleaning by nitrogen, reduction by propene or hydrogen, cleaning by nitrogen, and re-oxidation by oxygen. Gases were fed at a flow of 20 Nlph. In the quantitative evaluation of all tests buoyancy effects due to changes of gas were deduced by comparison with blank tests.

Methods used in tests of catalytic activity were detailed in a previous work [6]. Conversion was limited to remain in condition of differential reactor. Then it was possible to obtain initial reaction rates.

RESULTS AND DISCUSSION

Tables and figures show the thermal evolution of catalysts in different reaction conditions and the data of reduction and re-oxidation rates based on these results.

Study of the thermal evolution of catalysts. A first survey on the shape of the reduction and re-oxidation curves showed that the higher weight loss rates occur at the beginning of the reaction. This fact allowed an easy evaluation of the initial reaction rates.

In the tables initial rates of reduction or re-oxidation are expressed as

$$K_i = \frac{\text{moles}}{\text{g sec atm}},$$

Were moles are referred to the O_2 emitted by the catalyst or reacted with them and grams are referred to the original amount of unreacted catalyst.

Reduction tests with pure propene and hydrogen. From the data reported in Tables 2 and 3 the changes of the reduction and re-oxidation rates with temperature can be remarked.

It is important to underline that the three studied compounds presented differences in reactivity threshold. The catalyst A reacted with propene only from 250°C, while the catalyst B already reacted at 160°C (Table 2). Catalyst C, in spite of several activation conditions used (Table 1), didn't react at 250°C. Catalyst C(I), activated at 300°C, began to react at 300°C, while the catalyst C(II), activa

TABLE 1. PHYSICO-CHEMICAL PROPERTIES OF THE CATALYSTS

Catalyst	Conditions of thermal activation	Mesh	Surface area (m^2/g)	Density (g/cm^2)	Atomic ratio O/Tl
A	4h, 350°C	120	3.2	9.2	1.64
B	2h, 300°C	180	2.7	9.0	1.62
C(I)	2h, 300°C	80	0.11	10.3	1.52
C(II)	2h, 400°C	80	0.04	10.3	1.52

TABLE 2. REDUCTION RATES (Moles/sec atm g) IN PROPENE ATMOSPHERE VERSUS TEMPERATURE (°C)

Catalyst	160	200	250	300	400
A			$3.6\ 10^{-8}$		
B	$1.4\ 10^{-8}$	$2.8\ 10^{-8}$	$1.1\ 10^{-7}$		
C(I)				$9.4\ 10^{-9}$	
C(II)					$1.6\ 10^{-7}$

TABLE 3. RE-OXIDATION RATES (moles/sec atm g) IN OXYGEN ATMOSPHERE VERSUS TEMPERATURE (°C)

Catalyst	160	200	250	400
A			$2.0\ 10^{-5}$	
B	$1.2\ 10^{-7}$	$2.8\ 10^{-6}$	$4.6\ 10^{-6}$	
C(II)				$1.15\ 10^{-5}$

ted at 400°C, only reacted at higher temperature (Table 2).

Catalyst C(II) also underwent reduction tests with hydrogen. This time the compound already reacted at 270°C. Generally reduction tests with hydrogen showed an higher reduction rates than reduction tests with propene, e.g. catalyst C(II) presents an initial reduction rate constant $1.0\ 10^{-5}$ at 300°C in hydrogen and $1.5\ 10^{-6}$ at 400°C in propene.

<u>Stoichiometry, thermal activation and reduction rates</u>. A previous work [6] related preparation methods of thallium(III) oxides

to some physico-chemical properties: stoichiometry, colour, density, surface area, and acidity.

Another work [7] pointed out different behaviours in the thermal activation in air of precursors prepared through different methods of precipitation. Above all authors underlined that compound C presented no weight loss over 280°C, while compound A and B lost weight until 420°C.

It is possible to relate differences in stoichiometry and thermal behaviour to differences in reduction rate mentioned in the former paragraph. Table 2 shows that, at constant O/Tl ratio (Table 1, see compounds C(I) and C(II)), the different thermal activation can explain the lower reactivity of C(II), compared with C(I).

The higher O/Tl ratio of the compound B compared with the compound A (Table 2) is probably related to its different thermal activation and it can explain the higher reduction rate of compound B. Moreover compound C, whose O/Tl ratio is near to the stoichiometric ratio, showed a lower reduction rate than compound A and B, which present a significant excess of oxygen.

Confirmation of the redox mechanism. The feature of the initial re-oxidation rate K_{io} greater than the initial reduction rate K_{ir} (Tables 2 and 3) is presented from all studied catalyst. Generally K_{io} was much greater than K_{ir}. The gap between these rates decreased only when less than 1 % of oxygen has been depleted in the reduction phase. However K_{io} remained greater than K_{ir}, e.g. on catalyst B at 160°C K_{ir} is $1.4 \cdot 10^{-7}$, K_{io} after 1% oxygen depletion is $1.1 \cdot 10^{-6}$.

Tests showed the repeatibility of the reduction-oxidation cycles.

It is useful to recall that catalytic activity tests carried out in pulse reactor suggested the intervention of the lattice oxygen in the formation of propene oxide, carbon monoxide and carbon dioxide on thallium(III) oxides [6].

On the base of these data a redox reaction mechanism was proposed for the oxidation of propene in flow reactor, in the same temperature range used for thermogravimetric tests.

The redox mechanism requires that the reduction of catalyst performed by propene is the slow stage of the reaction and it requires also that the catalyst is in its more oxidized state.

Re-oxidation tests of the catalyst B, performed at different oxygen pressure values, verify the latter condition when the ratio between oxygen pressure and propene pressure in not lesser than 0.15 (Figure 2 and 3).

On the ground of these results it is possible to propose a reaction mechanism composed of two stages:

$$\text{Propene} + Tl_2O_3 \xrightarrow{K'_{ir}} \text{oxidation products} + Tl_2O_3 \text{ red}$$

$$Tl_2O_3 \text{ red} + O_2 \xrightarrow{K'_{io}} Tl_2O_3$$

Rate constants marked with a superscript refer to catalytic data. Assuming some hypotheses it is possible to write the expression of initial reduction and oxidation rates in the shape:

$$r_{ir} = K'_{ir} \, P_{C_3H_6} \, [Tl_2O_3],$$

$$r_{io} = K'_{io} \, P_{O_2} \, [Tl_2O_3 \text{ red}],$$

where $[Tl_2O_3]$ and $[Tl_2O_3 \text{ red}]$ are the surface ratios between oxidized or reduced catalyst and the total surface catalyst. In all tests in which reaction conditions allow a steady state

$$r_i = r_{ir} = r_{io},$$

where r_i is the initial reaction rate of the catalytic test. Then it is possible to write

$$\frac{1}{r_i} = \frac{1}{K'_{io} \, P_{O_2}} + \frac{1}{K'_{ir} \, P_{C_3H_6}}$$

Thermogravimetric data show that K_{io} is much greater than K_{ir}, and then the oxygen reaction order is zero and we can write

$$r_i = K'_{ir} \, P_{C_3H_6}$$

Catalytic activity data on catalyst B were related in this framework to thermogravimetric data. The influence of the propene pressure on the initial reduction rate of the catalyst B is reported in Figure 3.

On the ground of these results, we assumed that, in a range of propene pressure from 0.8 to 0.4 atm, the reaction order of propene in thermogravimetric tests was in good agreement with the reaction order in catalytic activity tests. From catalytic activity data, using a feed of propene (60%) and oxygen (40%), we obtained a value of $7 \cdot 10^{-7}$ moles/sec atm for K'_{ir} at 250°C. From thermogravimetric data a calculated value of $1.2 \cdot 10^{-7}$ moles/g sec atm for K_{ir} at 250°C at a propene pressure of 0.6 atm can be given.

Value of K_{ir} and K'_{ir} are calculated supposing reaction order one. However the ratio between the rate constants is not affected by a different reaction order, if the assumption of equal reaction order is maintained.

The likeness of K_{ir} and K'_{ir} further supports the proposed redox mechanism for the reaction of propene epoxidation.

Clearly in this paper experimental data were used in the direct way, according to Barret [8], to give the most useful preliminary results. From these results it will be possible to devise a further experimental program to obtain the entire kinetic mechanism. Studies of the stages of absorption and desorption and of the all-products kinetic seems above all interesting.

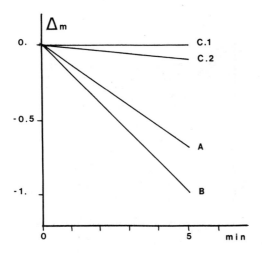

FIGURE 1. Reduction under propene: weight loss percent on catalyst oxygen. Test temperature: Catalyst A, B, and C(I) test 1 250°C; catalyst C(I) test 2 300°C.

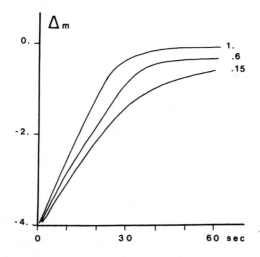

FIGURE 2. Re-oxidation under propene in nitrogen at different propene/total flow ratios; weight recovery percent on catalyst oxygen. Test temperature: 250°C. Initial rate of weight recovery (moles/g sec atm) $2.0 \cdot 10^{-5}$ at propene 15%; $5.6 \cdot 10^{-6}$ at propene 60%; $4.6 \cdot 10^{-6}$ at propene 100%.

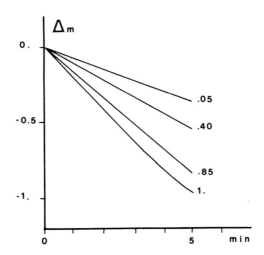

FIGURE 3. Reduction under propene in nitrogen at different propene/
/total flow ratios: weight loss per cent on catalyst
oxygen. Test temperature: 250°C. Initial rate of weight
loss (moles/g sec atm) $7.9 \cdot 10^{-7}$ at propene 5%; $1.6 \cdot 10^{-7}$
at propene 40%; $1.1 \cdot 10^{-7}$ at propene 85%; $1.1 \cdot 10^{-7}$ at pro̲
pene 100%.

REFERENCES

1. A.Bielanski and J.Haber, Catal.Rev.Sci.Eng., 1979, 19, 1.
2. P.Mars, D.W.van Krevelen, Chem.Eng.Sci.Suppl., 1954, 3, 41.
3. D.L.Trimm and L.A.Doerr, J.Catal., 1971, 23, 49.
4. P.Centola, R.Del Rosso and C.Mazzocchia, Chimica e Industria (Milano), 1978, 60, 91.
5. L.A.Petrov, J.Catal., 1976, 43, 367.
6. C.Mazzocchia, F.Di Renzo, P.Centola and R.Del Rosso, in press, proceedings "VIII Simposio Iberoamericano de Catalisis", Huelva, 1982.
7. C.Mazzocchia, R.Del Rosso and P.Centola, proceedings "Journées de Calorimétrie et d'Analyse Thermique" 3, 13-1, Ed. AFCAT Barcelona (1980).
8. P.Barret, "Cinétique hétérogène", Ed. Gauthier-Villars, Paris, (1973).

CALORIMETRIC STUDY OF THE INTERACTION OF HYDRO-
GEN WITH CoMo HYDRODISULPHURISATION CATALYTS

by M.B. POLESKI[*], A. AUROUX and P.C. GRAVELLE

Institut de Recherches sur la Catalyse, CNRS,
69626 Villeurbanne, France

[*]Present address : Institute of Petroleum,
Mining and Gas Technology, 01224, Warsaw,
Poland

ABSTRACT

It is well established that cobalt promotes the hydrodisulphurisation and hydrogenation activities of molybdenum-based sulfide catalysts. The origin of this beneficial effect has been the subject of much attention but is not yet conclusively established.

The object of the present study was to determine whether cobalt only stabilizes (or enhances) surface properties already present in molybdenum sulfide or whether the cobalt-promoted catalysts exhibit new surface properties. Hydrogen was selected as a test molecule since it participates in both hydrogenolysis and hydrogenation reactions. Differential heats of its interaction with a series of CoMo catalysts were recorded at 483 and 593K by means of a heat-flow microcalorimeter.

The unsupported catalysts were prepared by codigesting mixtures of molybdenum and cobalt oxides, in different proportions, with an aqueous solution of ammonium sulfide at 350-360K during 6-8hrs. The dried batches were further sulfided in gaseous mixtures of either H_2-H_2S or H_2-dibenzothiophene at 673K and kept in an argon atmosphere before the calorimetric measurements. Some nickel-containing catalysts were also prepared.

The calorimetric results indicated that, in the absence of cobalt, the affinity of the molybdenum catalyst towards hydrogen is low. However, a small proportion of cobalt (0.02) in the catalyst causes a very important enhancement of the heat of adsorption of hydrogen (from 15 to 100 kJ.mol^{-1}). The catalyst affinity towards hydrogen remains high when the proportion of cobalt increases to ~ 0.5. For further additions of cobalt, the heats of adsorption of hydrogen decrease and, finally, a catalyst only containing cobalt ions does not react with hydrogen. Nickel-containing catalysts qualitatively exhibited a similar behaviour.

These calorimetric data were compared to the results obtained from alumina-supported Co-Mo catalysts and an interpretation of the promoting action of cobalt (or nickel) ions in the molybdenum catalysts was proposed.

EFFECT OF ACIDIC & BASIC CATALYSTS ON PYROLYTIC
DECOMPOSITION OF CELLULOSE STUDIED BY COMPUTER
CURVEFITTING OF THERMOGRAVIMETRIC DATA.

Virginia Garcia Randall, M. Sid Masri & Attila
E. Pavlath, Western Regional Research Center,
Agricultural Research Service, U.S. Department
of Agriculture, Berkeley, CA 94710

INTRODUCTION

Dwindling natural oil and gas reserves and concern with reliability of their supply and delivery have stimulated research for alternate energy sources. Recovery of the energy content of biomass is such an alterate, albeit of limited potential overall contribution to projected energy needs [1]. Biomass has the advantage of being a renewable resource, being dependent ultimately on capture of solar energy. We discuss here results of experiments on pyrolytic decomposition of cellulose and effect of inorgnic base and acid catalysts on the decomposition, using thermogravimetric methods. Thermal behavior and degradation of cellulosic materials and effect of additives have been extensively studied in the literature [2-15], yielding considerable discussion and mechanisms of pyrolytic endothermic and exothermic reactions, effect of additives, and identification of pyrolytic products. In general, pyrolysis yields fixed gases, tars (mainly levoglucosan) and a carbonaceous char residue. Variations of pyrolysis conditions and of substrate pretreatment modify the course of the pyrolysis and influence the proportion of conversion to these products and the specific composition of the volatile components. It is likely that both concurrent and consecutive reactions take place, complicating interpretation of the results. Recently Pavlath and Gregorski [16] proposed a new approach for interpreting thermogravimetric pyrolytic results through curvefitting of the first derivative thermograms to provide means of comparing decomposition patterns. The assumptions and limitations of the method were discussed [16]. In the present study we used this approach to compare the effect of different acidic, neutral and basic inorgnic salts on the pyrolytic decomposition of cellulose.

MATERIALS AND METHODS

Sample Preparation. Catalysts, 1-20% (W/W), were added to 10 g aliquots of microcrystalline reagent grade cellulose. The samples were prepared by adding the catalyst dissolved in 20 ml water to the cellulose. The samples were mixed well, dried and ground in a Wiley Mill to pass a 20 mesh screen. The inorganic additives tested were $ZnCl_2$, $MgCl_2$, LiCl, NaCl, NaOH, KOH, LiOH, $Ca(OH)_2$ and Na_2CO_3.

Pyrolysis. Thermogravimetric analyses were carried out under nitrogen in a Perkin-Elmer TGS-2 unit equipped with a microprocessor-controlled programmer. The pyrolysis conditions were kept constant, namely 1-2 mg sample size, 40°C/min heating rate, and sweeping of the pyrolsis chamber with N_2 gas at constant flow. Constant pyrolysis conditions are necessary for valid comparison of the results. For example, changing the rate of heating of cellulose from 40°C/min to 4°C/min lowered the pyrolysis temperature about 60°C. Constancy of sample size and N_2 sweeping rate are also important from mass transfer consideration and for controlling the extent of secondary reactions by removing primary products from the pyrolysis zone. Thermogravimetric weight loss and the first derivative curves as well as the char residue were recorded. First derivative thermograms from TGA were deconvoluted as described [16]. The assumption was based on best fitting of the experimental thermogram with a maximum of 10 individual peaks, each having a Gaussian distribution and a half width of 22°C. The curvefitting of the first derivative was done by the least square and weighted chi-square method. The resulting data, expressed as the number of components, percentage of contribution of each component to the total decomposition curve, and the corresponding temperatures at which the maximum rate of weight loss for each component occurs, were computed. Some of the data were also processed to yield graphic presentation of the decomposition patterns to facilitate visualization of the decomposition envelopes.

RESULTS AND DISCUSSION

The results show that the deconvoluted patterns obtained with the additives are different, not only from those with neat cellulose but also from each other, as shown in Tables 1 and 2. In a sense, each condition (particular additive and concentration) has its own unique decomposition envelope or "fingerprint", providing a quantitative profile useful for comparing the course of pyrolysis for the different conditions. Thus, regardless whether the individual component curves represent actual chemical reactions in a physical sense, the profiles nonetheless, provide a quantitative mathematical meaning relating to temperature regions where changes in the rate of weight loss occur during pyrolysis. Comparision of the decomposition envelopes brings out the following points.

Lewis Acids. Addition of $ZnCl_2$ at 1,3 and 10% lowered the pyrolysis temperatures appreciably with the extent of shift being related to the concentration of the Lewis acid. Also the weight loss was spread out over a wider temperature range. Char production was also increased. Maximal effect appears to have been reached at 10% $ZnCl_2$, with the 20% level showing no further shift. The weaker Lewis acids, magnesium and lithium chlorides, at the 1-10% levels, also lowered the pyrolysis temperatures, spread the weight loss and increased char. The effect was less pronounced than with $ZnCl_2$. These results are shown in Table 1 and Fig. 1.

Surprisingly, even the neutral non-Lewis acid sodium chloride also lowered the pyrolysis temperature and increased char, although

Table 1. PEAK TEMP. AND WT. LOSS OF COMPONENTS DECONVOLUTED FROM FIRST DERIVATIVE OF TGA OF CELLULOSE WITH ADDITIVES AT VARYING CONCENTRATIONS[a].

Additive	Wt %	Deconvolution Components Peak temp., °C and (% Wt loss)								
$ZnCl_2$	1	274 (4)	292 (6)	308 (9)	324 (12)	340 (19)	356 (24)	365 (7)	381 (2)	400 (5)
	3	241 (4)	261 (8)	280 (16)	298 (20)	316 (15)	334 (9)	356 (5)	376 (4)	397 (3)	417 (10)
	10	214 (2)	235 (1)	251 (3)	268 (7)	283 (12)	298 (9)	316 (7)	332 (4)	348 (3)	365 (14)
	20	217 (5)	236 (12)	252 (12)	269 (8)	285 (4)	304 (2)	325 (2)	344 (2)	364 (2)	382 (14)
$MgCl_2$	1	257 (1)	277 (1)	297 (1)	316 (1)	335 (3)	353 (12)	369 (35)	383 (29)	405 (2)	422 (3)
	10	239 (11)	256 (6)	274 (3)	295 (3)	314 (7)	334 (12)	352 (16)	362 (6)	384 (3)	404 (12)
LiCl	1	268 (2)	289 (4)	310 (7)	330 (12)	348 (20)	363 (20)	375 (7)	391 (1)	413 (1)	428 (2)
	10	241 (4)	264 (11)	282 (16)	302 (19)	311 (8)	331 (2)	352 (2)	373 (2)	397 (2)	420 (4)
NaCl	10	297 (4)	313 (3)	329 (8)	345 (16)	361 (22)	370 (28)				

[a] Thermogravimetric analysis (TGA): Heating rate, 40°C/min; atmosphere, N_2; temp. range of deconvoluted curve, 120-480°C; char ranged from 8 to 33%.

the effect was less pronounced than with the Lewis acids. While the effect with the Lewis acids might be explained in part by catalyzed dehydration of cellulose (and consequent effects of generating water, hydrolysis of glycosidic bonds, and promotion of intermolecular interaction [5-7]), the action of NaCl is obscure. Since water which is generated during pyrolysis may play a significant role as a solvent, promoting ionic charge transfer, hydrolysis and intermolecular interactions [8], the effect of NaCl may be mediated through the solvent water rather than through a direct action of the solid salt.

Inorganic Bases. Sodium, potassium and lithium hydroxides at the 1 to 20% levels shifted the pyrolysis to lower temperatures and enhanced char formation. The magnitude of the effects was related to the base concentration. The decomposition envelopes for sodium and potassium hydroxides were strikingly similar, and maximal effect was reached by the 10% level. With the weaker base, LiOH, a further shift (about 30-40°C) to lower temperature was obtained at the 20% level as compared to the 10% level. Results with sodium carbonate between 1-20% were strikingly similar to those obtained with sodium and potassium hydroxides. Although lower pyrolysis temperatures were obtained with $Ca(OH)_2$, the effect was less pronounced than with the other bases. Char formation was only

Table 2. PEAK TEMP. AND WT. LOSS OF COMPONENTS DECONVOLUTED FROM FIRST DERIVATIVE OF TGA OF CELLULOSE WITH ADDITIVES AT VARYING CONCENTRATIONS[a].

Additive	Wt %	Deconvolution Components Peak temp., °C and (% Wt loss)								
NaOH	1	227	245	264	283	305	326	344	360	376	393
		(3)	(3)	(3)	(4)	(5)	(6)	(9)	(15)	(15)	(12)
	3	182	204	226	245	267	289	310	331	347	364
		(1)	(3)	(5)	(6)	(6)	(8)	(13)	(16)	(9)	(7)
	10	201	219	240	259	277	295	314	329	344	363
		(3)	(3)	(4)	(8)	(11)	(12)	(12)	(7)	(1)	(8)
	20	212	232	250	271	289	305	319	330	347	365
		(7)	(2)	(4)	(7)	(13)	(14)	(7)	(2)	(1)	(10)
KOH	10	190	212	233	253	272	291	309	327	347	370
		(3)	(4)	(7)	(8)	(9)	(11)	(13)	(7)	(3)	(13)
	20	204	223	243	263	281	299	317	335	353	370
		(5)	(5)	(7)	(11)	(13)	(13)	(8)	(2)	(1)	(7)
LiOH	1	228	252	275	297	317	336	356	372	385	410
		(4)	(3)	(4)	(4)	(6)	(10)	(20)	(24)	(4)	(4)
	3	230	250	270	290	309	328	346	362	375	394
		(3)	(3)	(3)	(4)	(6)	(10)	(18)	(18)	(6)	(4)
	10	221	242	262	282	303	322	338	353	370	390
		(4)	(3)	(4)	(7)	(8)	(11)	(13)	(10)	(2)	(7)
	20	180	200	226	250	273	291	308	324	339	360
		(5)	(2)	(2)	(2)	(6)	(11)	(13)	(13)	(7)	(10)
$Ca(OH)_2$	1	295	320	337	356	372	386	398			
		(3)	(1)	(2)	(2)	(10)	(36)	(36)			
	3	251	270	290	310	330	347	364	378	390	402
		(2)	(1)	(1)	(2)	(2)	(3)	(5)	(10)	(37)	(22)
	10	250	280	305	327	350	368	388	395	420	440
		(2)	(2)	(3)	(6)	(9)	(21)	(31)	(7)	(2)	(6)
	20	218	237	254	272	291	312	330	345	355	380
		(4)	(1)	(2)	(2)	(4)	(9)	(22)	(30)	(9)	(4)
Na_2CO_3	1	247	266	286	306	325	342	359	376	392	410
		(30)	(3)	(4)	(4)	(4)	(5)	(11)	(20)	(19)	(6)
	3	220	240	258	277	297	318	336	354	371	388
		(5)	(3)	(4)	(5)	(6)	(9)	(14)	(14)	(8)	(8)
	10	183	204	224	245	265	285	305	325	342	362
		(2)	(2)	(3)	(4)	(8)	(10)	(12)	(12)	(8)	(11)
	20	203	223	246	266	284	304	320	336	355	375
		(5)	(5)	(6)	(8)	(12)	(12)	(8)	(6)	(2)	(10)

[a] Thermogravimetric analysis (TGA): Heating rate, 40°C/min; atmosphere, N_2; temp. range of deconvoluted curve, 120-480°C; char ranged from 8 to 28%. For neat cellulose: peak temperatures are 340, 358, 373, 387 and 400 °C with corresponding weight losses of 10, 31, 37, 15 and 2%; char: 5%

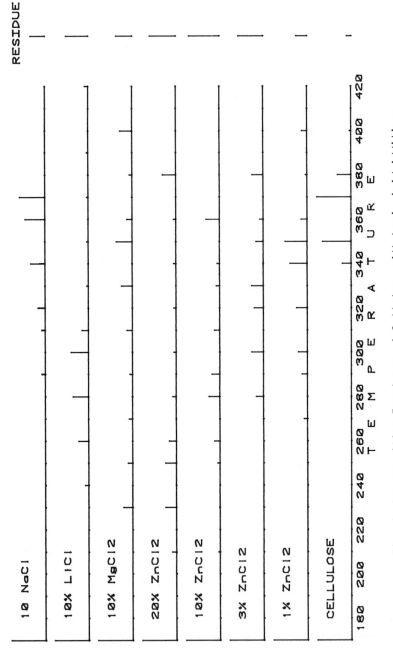

Fig. 1. Decomposition Envelopes of Cellulose with Lewis Acid Additives. (Maximum dimension of y-axis is 37%)

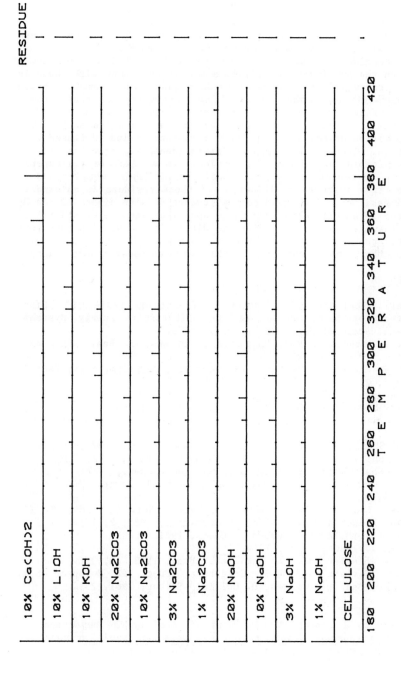

Fig. 2. Effect of Added Inorganic Bases on Decomposition Envelopes of Cellulose. (Maximum dimension of y-axis is 37%)

slightly increased. The results with $Ca(OH)_2$ may be in part related to sample preparation. Because of its limited solubility, $Ca(OH)_2$ was added as a slurry, and thus intimate contact and homogenous distribution of the base may not have been achieved. These results with the bases are shown in Table 2 and Fig. 2.

In conclusion, our results on the effect of bases and Lewis acids on the pyrolysis of cellulose are in agreement with those in the literature. They further provide a quantitative profile based on a curvefitting method of the first derivative TGA thermograms. These profiles are useful in comparative studies on the effect of additives in modifying pyrolysis. The bases and acids shifted the decomposition of cellulose to lower temperatures and increased char formation. These results may assist in identifying practical conditions suitable for biomass conversion. As an example, in experiments to be reported elsewhere, addition of KOH to wheat and safflower straw improved gasification of these residues in an experimental gasifier. The improved gas production with higher CO and H_2 resulted in a self-sustaining flame upon ignition. Gasification of these residues without additives is difficult and usually gave only marginal results (frequent bridging and caking of biomass, stoppage of gas flow, poor gas composition and non-sustainable flame).

REFERENCES

1. Natl. Res. Council, "Energy in Transition 1985-2010", Final Report of Comm. on Nuclear and Alternative Energy Systems, Natl. Acad. of Sci., Wash., D.C., 1979.
2. S. L. Madorsky, V.E. Hart and S. Straus, J. Res. Natl. Bur. Std., 1956, 56, 343.
3. R. F. Schwenker, Jr. and L.R. Beck Jr., J. Polym. Sci.; Part C, 1963, 2, 331.
4. C.H. Mack and D.J. Donaldson, Text. Res. J., 1967, 37, 1063.
5. F.A. Wodley, J. Appl. Polym. Sci., 1971, 15, 835.
6. F. Shafizadeh, C.W. Philpot and N. Ostojic, Carbohyd. Res., 1971, 16, 279.
7. F. Shafizadeh and Y.Z. Lai, J. Org. Chem., 1972, 37, 278.
8. F. Shafizadeh and Y.L. Fu, Carbohyd. Res., 1973, 29, 113.
9. F. Shafizadeh and Y.Z. Lai, Carbohyd. Res., 1973, 31, 57.
10. H.A. Schuyten, J.W. Weaver and J.D. Reid, Advan. Chem. Ser., 1954, 9, 7.
11. W.K. Tang and W.K. Neill, J. Polym. Sci. C, 1964, 6, 65.
12. F.J. Kilzer and A. Broido, Pyrodynamics, 1965, 2, 151.
13. F.H. Newth, Advan. Carbohyd. Chem., 1951, 6, 83.
14. E.F.L.J. Anet, ibid., 1964, 19, 181.
15. R.L. Whistler and J.N. BeMiller, ibid., 1958, 13, 326.
16. A.E. Pavlath and K.S. Gregorski, Proc. II European Sym. Therm. Anal., Aberdeen, D. Dollimir Editor, 1981, 251.

Reference to a company and/or product named by the Department is only for purposes of information and does not imply approval or recommendation of the product to the exclusion of others.

A THERMOGRAVIMETRIC STUDY OF THE CATALYTIC
DECOMPOSITION OF NON-EDIBLE VEGETABLE OILS

K.N.NINAN, K.KRISHNAN AND K.V.C.RAO
Chemicals Group, Vikram Sarabhai Space
Centre, Trivandrum (India).

INTRODUCTION

In pursuit of energy from replenishable sources, attempts to convert non-edible vegetable oils to hydrocarbon fuels have been made during the last few decades. The Indian Space Research Organisation (ISRO) has developed some catalysts for the production of hydrocarbons by the thermolysis of vegetable oils [1]. The utilisation of thermal analysis techniques has recently been shown to be a specially suitable method of investigating catalytic systems [2]. Thermogravimetry has been specifically used to evaluate the catalytic degradation of organic compounds [3,4]. In the present investigation, it is attempted to study the effect of some of the ISRO developed catalysts on the thermal decomposition of six vegetable oils abundantly available in India.

EXPERIMENTAL

Materials. Commercial samples of six vegetable oils (Neem, Punna, Groundnut, Castor, Karanj and Rubber seed oils) obtained by the conventional extraction of the seeds were used in the study. Five ISRO developed catalysts (TISIAL - 1, MOSIAL, ZIRSIAL, PLATSIAL and VASIAL) were used in the form of extruded strands.

Instruments. The TG experiments and isothermal mass-loss measurements were carried out using DuPont 990 Thermal Analyser in conjunction with 951 Thermogravimetric Analyser.

Procedure. For TG measurements, the heating rate was 10°C min^{-1}. Isothermal mass-loss measurements were conducted at 420°C (Good yields of hydrocarbons were obtained at this temperature in our pilot plant set up). Sample mass was 10 mg. and atmosphere was nitrogen (flow rate 50 cm^3 min^{-1}) in both the cases. TISIAL-1 was chosen as a typical catalyst for studying the catalytic decomposition of all the oils. The catalyst was socked in the oil and the size of the catalyst was adjusted to absorb about 10 mg. of the oil. For comparing different catalysts Rubber seed oil was chosen and isothermal and non-isothermal experiments were carried out with all the five catalysts.

RESULTS AND DISCUSSION

The values of the temperature of inception of reaction (T_i), the temperature of completion of reaction (T_f) and % mass loss obtained from the TG experiments of the oils without the catalysts are given in Table 1. (Typical TG curves are shown in Fig.1). The results (duration of reaction and % mass loss) from the corresponding isothermal experiments are also given in Table 1. (Typical isothermal mass loss curves are shown in Fig.2).

Table 1. RESULTS OF THERMAL DECOMPOSITION OF N.E.O. WITHOUT CATALYST

Oil	TG Results			Isothermal Results	
	T_i °C	T_f °C	% mass loss	Duration (min)	% mass loss
Neem	175	465	100	5.2	90
Punna	170	480	100	5.1	89
Groundnut	290	465	100	5.7	88
Castor	175	475	100	3.3	88
Karanj	175	480	100	5.5	88
Rubber	175	485	100	5.7	80

In absence of the catalyst, all the oils give a single step TG curve with inception temperature around 175°C, except in the case of groundnut oil for which T_i is 290°C. In spite of this difference, T_f values of all oils fall in the range of 460-480°C. From isothermal experiments, it can be seen that the mass loss takes place rapidly upto around 90%, followed by a very slow mass loss. The duration of the former step is more or less same in all the cases except in the case of castor oil whose decomposition seems to be faster.

In presence of the catalyst (TISIAL - 1) all the oils

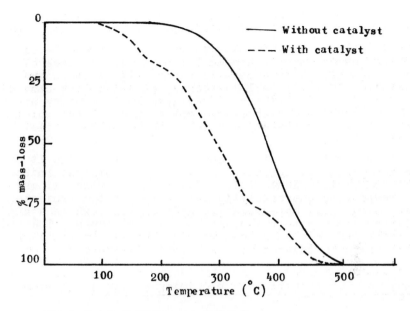

Fig.1. Typical TG Curves of N.E.O.

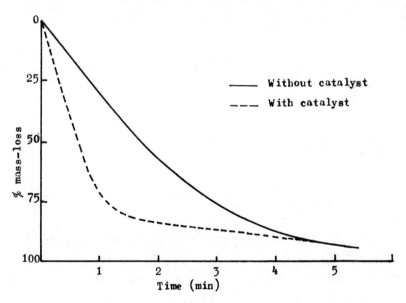

Fig.2. Typical Isothermal Mass-loss Curves of N.E.O.

show a three stage degradation pattern in their TG curves. The values of T_i, T_f and % mass loss for each stage are given in Table 2. The catalyst brings down the decomposition temperature to around 70°C even in the case of groundnut oil. The % mass loss for stage I is in the region of 12-16% in all the cases, except in the case of Castor oil (23%). The oils employed in this study are triglycerides of molecular weight around 900, for which decarboxylation corresponds to 14.7% mass loss while the loss of $3CO_2$ and $3H_2O$ leads to a mass loss of 20.7% (Castor oil is a typical case of the latter). So it may be inferred that the first stage is the catalyst induced decarboxylation/dehydration of the oils. Stage II may correspond to the catalytic conversion of the products from stage I to more volatile hydrocarbons, while stage III corresponds to the higher boiling residues. The above observations are in agreement with the results from the pilot plant studies conducted in our laboratory.

Table 2. TG RESULTS OF N.E.O. WITH TISIAL-1 CATALYST

Oil	Stage I			Stage II			Stage III		
	T_i°C	T_f°C	% mass loss	T_i°C	T_f°C	% mass loss	T_i°C	T_f°C	% mass loss
Neem	65	200	15	200	375	59	375	500	26
Punna	75	215	16	215	390	59	390	510	25
Groundnut	65	225	13	225	390	59	390	510	25
Castor	65	240	23	240	385	47	385	500	30
Karanj	70	225	15	225	380	60	380	510	23
Rubber	65	210	12	210	390	59	390	510	29

The results of isothermal studies of all oils with TISIAL-1 catalyst are given in Table 3.

Table 3. ISOTHERMAL RESULTS OF N.E.O. WITH TISIAL-1 CATALYST

Oil	Neem	Punna	Groundnut	Castor	Karanj	Rubber
Duration (min)	2.1	2.3	1.9	1.7	2.3	2.0
% mass loss	80	85	75	70	80	80

The catalyst brings down the duration of reaction from 5.3 minutes to about 2 minutes. However, the decarboxylation/dehydration step is not distinctly seen here.

This step occurring at low temperatures (T_i = 70°C) may not be getting resolved at high temperature (420°C). Table 4 summarises the TG results of Rubber seed oil with five different catalysts and Table 5 gives the corresponding results from isothermal measurements. From these tables it can be observed that the decomposition pattern is more or less same with all the catalysts.

Table 4. TG RESULTS OF RUBBER SEED OIL WITH CATALYSTS

Catalyst	Stage I			Stage II			Stage III		
	T_i°C	T_f°C	%mass loss	T_i°C	T_f°C	%mass loss	T_i°C	T_f°C	% mass loss
TISIAL-1	60	210	13	210	390	59	390	510	28
MOSIAL	65	200	13	200	365	55	365	500	32
ZIRSIAL	65	200	12	200	385	56	385	510	32
PLATSIAL	60	195	12	195	405	60	405	525	28
VASIAL	65	200	12	200	385	60	385	520	28

Table 5. ISOTHERMAL RESULTS OF RUBBER SEED OIL WITH CATALYSTS

Catalyst	TISIAL-1	MOSIAL	ZIRSIAL	PLATSIAL	VASIAL
Duration (min)	1.8	1.9	2.0	2.3	1.8
% mass loss	80	80	80	75	80

Kinetic parameters. The efficiency of a catalyst system is best evaluated from its effect on the kinetic parameters governing the reaction rate. So the kinetic parameters (energy of activation, E and pre-exponential factor, A) were calculated from all the TG curves using the Coats-Redfern equation [5]. The order parameter 'n' was found to be near unity in all the cases from the best-fit value of 'n'. Tables 6 and 7 give the values of E and A along with the correlation coefficient, r, of the corresponding curves for the catalytic and non-catalytic decomposition of the oils. Table 6 shows the effect of TISIAL-1 catalyst on all oils, while Table 7 depicts the effect of different catalysts on Rubber seed oil. From these tables, it can be observed that the activation energy for the decomposition reaction is considerably reduced by all the catalysts.

Table 6. KINETIC PARAMETERS FOR DIFFERENT OILS

Oil	Without Catalyst			With TISIAL-1 Catalyst		
	E (kJ mole^{-1})	A (sec^{-1})	r	E (kJ mole^{-1})	A (sec^{-1})	r
Neem	136.6	1.943×10^8	0.9990	76.40	2.312×10^4	0.9975
Punna	113.6	2.681×10^6	0.9980	94.34	7.441×10^5	0.9986
Groundnut	202.7	1.173×10^{13}	0.9998	88.76	3.390×10^5	0.9955
Castor	128.8	5.345×10^7	0.9985	50.76	1.212×10^3	0.9997
Karanj	151.8	2.331×10^9	0.9993	84.25	8.610×10^4	0.9913
Rubber	121.1	8.190×10^6	0.9986	86.31	1.315×10^5	0.9990

Table 7. KINETIC PARAMETERS FOR RUBBER SEED OIL WITH CATALYSTS

Catalyst	E (kJ mole^{-1})	A (sec^{-1})	r
NIL	121.1	8.190×10^6	0.9986
TISIAL	86.31	1.315×10^5	0.9996
MOSIAL	76.20	3.340×10^4	0.9993
ZIRSIAL	82.25	6.125×10^4	0.9992
PLATSIAL	75.00	9.301×10^4	0.9965
VASIAL	73.12	9.163×10^3	0.9994

ACKNOWLEDGEMENT

We thank Mr.N.S.Madhavan for help in computer programming and Dr.Vasant Gowariker, Director, Vikram Sarabhai Space Centre for his keen interest and encouragement.

REFERENCES

1. The Indian Space Research Organisation, British Patent 1,524,781 (1978).
2. D.Dollimore, Thermochim. Acta, 1981, 50, 123.
3. J.Scheve and K.Heise, "Thermal Analysis," Vol.3, Birkhauser Verlag, Basel (1971), p.71.
4. D.Dollimore, B.W.Krupay and R.A.Rose, "Proc. First European Symposium on Thermal Analysis," Heyden,(1976)p.455.
5. A.W.Coats and J.P.Redfern,Nature,1964, 201, 68.

A TG-DTA STUDY OF THE ADSORPTION OF SMALL HYDRO-
CARBON MOLECULES BY VARIOUS MODIFIED ZSM-5
ZEOLITES.

Zelimir GABELICA, Jean-Pierre GILSON, Guy DEBRAS
and Eric G. DEROUANE,

Facultés Universitaires de Namur
Département de Chimie, Laboratoire de Catalyse,
61, rue de Bruxelles, B-5000 - Namur - Belgium

INTRODUCTION

Zeolite ZSM-5 shows unique molecular shape selective catalytic properties which strongly depend on the free pore volume and openings and on the tortuosity of its framework channel-system [1]. The internal pore characteristics of (pentasil) zeolites can be characterized by the so-called "constraint-index" test-method [2] or by their sorption selectivities and capacities for different hydrocarbons[3-8]. Recent studies have indicated that the effective pore dimensions of zeolite ZSM-5 can be slightly altered by chemical treatments such as ion-exchange with larger cations, by controlled deposition of carbonaceous residues, or by impregnation with P or B - containing chemicals, thus leading to drastic modifications of its shape selective properties [9-13].
Preliminary work [14] has shown that small hydrocarbon molecules, such as n-hexane, are reliable probes to monitor slight changes in the free pore volume of chemically modified materials. The present paper reports further investigation using two hydrocarbon adsorbates, namely n-hexane and its methyl-branched isomer, 3-methylpentane to characterize various ZSM-5 zeolites.

EXPERIMENTAL

Table 1 lists the zeolites used in the present investigation and their characteristics. Si/Al ratios were evaluated by proton induced γ-ray emission [15-16]. Syntheses of ZSM-5 zeolites have been described and discussed earlier [16]. The chemical and structural identity of all the materials was ascertained by X-ray diffraction, i.r. spectroscopy, thermogravimetric and chemical analyses. Zeolites were conventionally acidified by treatment with HCl or NH_4NO_3 followed by calcination at 550°C [17]. Modifications include the introduction of cesium ions in the zeolite during its synthesis [21], the deposition of carbonaceous residues by reaction with CH_3OH at 370°C for 16h, the impregnation with H_3BO_3 or trimethylphosphite solutions with, as a result, an incorporation of boron [19] or phosphorus [10] respectively.

A STA-780 thermal analyser (TG-DTA-DTG) from Stanton Redcroft was used to study the n-hexane (n-C_6) and 3-methylpentane (3M-C_5) adsorptions and their subsequent removal, by following the sample weight gain (loss) as a function of time or temperature. About 0.05 g of zeolite was preactivated in situ in a dry N_2 flow (30 ml.min^{-1}) at 500°C for 2h and then cooled at 20°C prior to hydrocarbon adsorption. The zeolite, held in a Pt crucible within the thermobalance, was exposed to n-C_6 and 3M-C_5 vapors, by diverting the N_2 flow (30 ml.min^{-1}) through a double stage saturator containing the liquid hydrocarbon held at 20°C (n-C_6, p/p_o=0.18) or at 0°C (3M-C_5, p/p_o=0.15). Weight variations are referred to the weight of zeolite after heating at 500°C.

RESULTS

A typical hydrocarbon (HC) adsorption experiment is schematized in fig. 1.

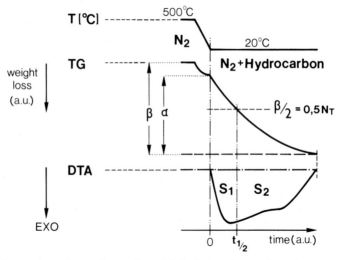

Figure 1. Adsorption of small hydrocarbon molecules with N_2 as gas carrier : schematic TG-DTA pattern and definition of the various parameters.

It consists initially of a slow adsorption of N_2 while the zeolite is cooled from 500 to 20°C (β-α). Exposure to the HC vapor in the thermobalance at 20°C, results in a subsequent important weight gain related to its rapid adsorption (α). As demonstrated previously [14], weakly-physisorbed N_2 is completely displaced by the HC molecules so that the total HC uptake, expressed as the number of hydrocarbon molecules per unit cell of zeolite, N_T, is equal to the entire weight gain β when equilibrium is reached.

While N_T is sensitive to pore volume modifications, pore mouth restrictions and/or channel tortuosity changes by various modifiers are expected to affect essentially the HC adsorption rates or the shape of the DTA response during adsorption.

In our experimental conditions, the linear $n-C_6$ molecule diffuses through the whole channel system of zeolite ZSM-5, thus rapidly filling its entire void pore volume [5-8]. This adsorption is accompanied by an exothermic DTA peak which presents a typical shape [14].

By contrast, $3M-C_5$, one of the branched $n-C_6$ isomers, adsorbs more slowly into the zeolite [8]. Because of steric hindrance effects imposed by the ZSM-5 channel structure and other steric constraints due to the side methyl-groups, the $3M-C_5$ packing into the zeolite is less favored than that of n-paraffins [8]. As a consequence, the DTA peak presents a different shape from the one observed in the case of $n-C_6$ [14].

For zeolite (H)-ZSM-5, at 20°, equilibrium state is almost reached after ca. 0.7h. The loading then corresponds to the adsorption of N_T=5.5-6.6 molecules per unit cell of zeolite, value which is in good agreement with those previously reported [5,8].

Values of N_T and $t_{1/2}$ may be expected to vary slightly from one H-ZSM-5 or (Na,H)-ZSM-5 zeolite to another (samples 1, 3 and 6 - table 1). Indeed, it was reported [18] that specific equilibrium uptakes and/or the form of the corresponding isotherms in the case of HC adsorptions on pentasil zeolites (ZSM-5) could markedly depend on the method of preparation of the zeolite (or on its exchange treatment) but to a minor extent on the size of the crystallites. A reduction in adsorption capacity may occur because of the presence of exchange cations or that of extraneous species present in the pores for Si/Al ratios below 23 [12].

The time required for the adsorption of an amount of HC corresponding to $0.5 \ N_T$ ($t_{1/2}$), is of course related to the rate of the adsorption and it can be measured directly from the TG curve. Two effects must be distinguished in the HC adsorption process :
  i. An initial sorption step which affects N_T and thus $t_{1/2}$. The resulting sorbate-zeolite interactions affect the shape and the intensity of the corresponding DTA exothermic peak.
  ii. A subsequent intracrystalline rearrangement of the sorbate molecules which has little or no influence on N_T but however gives an exothermic DTA response because of additional sorbate-sorbate interactions.

As for our sorption measurements the total DTA peak stems from both effects, it is interesting to compare the proportionality of its intensity (surface) with the HC loading during the sorption process, for example by measuring the surface S_1 after an uptake of $0.5 \ N_T$ (fig. 1) and by comparing the various $S_1/S_1+S_2 = S_1/S_T$ values at $t_{1/2}$. Values of S_1/S_T close to 0.5 indicate the occurence of a simultaneous pore filling and internal rearrangement of the HC molecules during the whole sorption process. Values higher than 0.5 are found for high zeolite-sorbate interactions when the heat of adsorption decreases with loading, that is if molecule-zeolite interactions predominate over sorbate-sorbate interactions only at low loading. By contrast, values smaller than 0.5 indicate a filling of the pores followed by a molecular rearrangement to achieve an optimum packing of the sorbate molecules until maximum loading is reached. This situation will depend on steric modifications of the pore volume of the zeolite.

Table 1

ADSORPTION OF $n-C_6$ AND $3M-C_5$ ON ZSM-5 ZEOLITES : TG/DTA PARAMETERS

	Zeolite[a]	Chemical Treatment	Ads of $n-C_6$			Ads of $3M-C_5$		
			N_T[b]	$t_{1/2}$[c]	S_1/S_T[d]	N_T[b]	$t_{1/2}$[c]	S_1/S_T[d]
1	(Na,H)-ZSM-5(44.5)	—	7.8[e]	360	0.40	5.9[f]	310	0.36
2	(Na,Cs,H)-ZSM-5(45.2)	Cs^+ incorp.	7.0	310	0.26	5.8	242	0.19
3	(H)-ZSM-5(25.9)	Acidif. (HCl)	7.7[e]	335	0.42	6.6[f]	324	0.40
4	(H)-ZSM-5(25.9) + C	Coke deposit.	7.4	505	0.61	6.45	410	0.53
5	(H)-ZSM-5(27.6) + B	Impregn. H_3BO_3	—	—	—	6.55	504	0.55
6	(H)-ZSM-5(14.5)	Acidif. (NH_4NO_3)	8.05[e]	324	0.41	5.5[f]	256	0.34
7	(H)-ZSM-(12.6) + P	Impregn.P$(OCH_3)_3$	—	—	—	5.1	660	0.53

(a) (Cation)-ZSM-5 (Si/Al) ratio. For all the samples but sample 2, the particle sizes, as measured by SEM, are in the range 1-5 μ (platelets or spherical agregates).

(b) molec./u.cell zeolite, measured from β (fig.1) after flushing 20 min. with $N_2 + n-C_6$ ($p(n-C_6) = 110$ Torr) or 45 min. with $N_2 + 3M-C_5$ ($p(3M-C_5) = 140$ Torr).

(c) in seconds ($\pm$ 10).

(d) S_1 = surface fraction (a.u) of the DTA peak measured after $t_{1/2}$ (fig.1) ; S_T = total surface of the DTA peak.

(e) literature values : 8.14 mol/u.cell [5] and ∿ 8.3 mol/u.cell (from graphical extrapolation [8]).

(f) literature values : 6.74 mol/u.cell [5] and ∿ 5.5 mol/u.cell (at 30°C, from graphical extrapolation [8]).

Values of N_T, $t_{1/2}$ and S_1/S_T are reported and compared for various reference and modified ZSM-5 zeolites (table 1).

DISCUSSION

Substitution of Na^+ ions by protons and vice versa, does not influence the total pore volume of zeolite ZSM-5, as evidenced by comparable values of N_T in the case of samples 1, 3 and 6 (n-C_6 adsorption). By contrast, smaller N_T values (for n-C_6) are observed in the case of (Na,Cs,H)-ZSM-5, indicating that the large Cs^+ ions affect the internal free volume of the zeolite. The small S_1/S_T values for both n-C_6 and 3M-C_5 indicate rapid diffusion of the sorbates into the channel system, followed by their progressive molecular rearrangement, their optimal packing being retarded because of the presence of the Cs^+ ions.

A pronounced restriction of the pore mouths is obtained by carbonaceous residues deposition on the zeolite surface [13]. Consequently, both n-C_6 and 3M-C_5 show reduced adsorption rates (higher $t_{1/2}$ values for sample 4 than for sample 3) but reach finally similar equilibrium sorption values (similar N_T values), confirming that only a minor amount of coke is deposited in the intracrystalline volume of ZSM-5 materials [13]. The higher S_1/S_T values account for a progressive internal rearrangement of the sorbate molecules during the (slower) filling process.

By impregnation of ZSM-5 with H_3BO_3, a zeolite containing 0.3 wt % of B (sample 5) was obtained. Such a small amount of B does not affect, as expected, the total sorption capacity of the material (comparable N_T (3M-C_5) values are observed for sample 5 and for sample 3, its precursor). By contrast, the rate of adsorption of 3M-C_5 is reduced and the S_1/S_T value increased, thus suggesting that boron affects the external surface of the zeolite near the pore mouths, as also observed and just discussed for the coke deposition. This proposal has recently been substantiated by catalytic tests conducted on B-containing zeolites [19].

The addition of phosphorus to a ZSM-5 zeolite (sample 7 which contains 2.03 wt % P) leads to additional constraints in the external shell of the zeolite crystallites and limits to some extent the free space available at the channel intersections. It thus decreases the capacity and increases the tortuosity of the channel network, as evidenced by the lower N_T values, the higher S_1/S_T ratio and the higher $t_{1/2}$ values observed for sample 7, relative to its precursor, sample 6. Our conclusions agree with the postulated role and the location of phosphorus in modified ZSM-5 zeolites [10,11,20].

CONCLUSION

The present investigation demonstrates that a careful analysis of the complementary TG and DTA results, obtained upon adsorption of two related sorbate molecules differing by their molecular shape, provides an easy means for the description of the adsorption and packing of hydrocarbons in zeolites of intermediate pore size.

REFERENCES

1. For example : E.G. Derouane in "Catalysis by Zeolites" (B. Imelik et al., eds.), Elsevier, Amsterdam (1980) ; p. 5, and references therein.
2. V.J. Frilette, W.O. Haag and R.M. Lago, J. Catal., 1981, 67, 218.
3. N.Y. Chen and W.E. Garwood, J. Catal., 1978, 52, 453.
4. R.M. Dessau in "Adsorption and Ion Exchange with Synthetic Zeolites", ACS Symp. Series 1980, 135, 123 ; ibid. Eur. Patent, 0,031,676 (1981).
5. E.G. Derouane and Z. Gabelica, J. Catal., 1980, 65, 486.
6. E.G. Derouane, Z. Gabelica and P.A. Jacobs, J. Catal., 1981, 70, 238.
7. P.A. Jacobs, H.K. Beyer and J. Valyon, Zeolites, 1981, 1, 161.
8. D.H. Olson, G.T. Kokotailo, S.L. Lawton and W.M.Meier, J. Phys. Chem., 1981, 85, 2238.
9. N.Y. Chen, W.W. Keading and F.G. Dwyer, J. Amer. Chem. Soc., 1979 101, 6783.
10. W.W. Keading and S.A. Butter, J. Catal., 1980, 61, 155.
11. W.W. Keading, C. Chu, L.B. Young, B. Weinstein and S.A. Butter, J. Catal., 1981, 67, 159.
12. E.G. Derouane, P. Dejaifve, Z. Gabelica and J.C. Vedrine, "Selectivity in Heterogeneous Catalysis", Faraday Disc. Chem. Soc., n°72 (1982), in press.
13. P. Dejaifve, A. Auroux, P.C. Gravelle, J.C. Vedrine, Z. Gabelica and E.G. Derouane, J. Catal., 1981, 70, 123.
14. Z. Gabelica, J.P. Gilson and E.G. Derouane, Proc. Second Eur. Symp. Thermal Analysis, Aberdeen 1981, Heyden, London, (1981), p. 434.
15. G. Debras, E.G. Derouane, G. Demortier, J.P. Gilson and Z. Gabelica, submitted for publication.
16. E.G. Derouane, S. Detremmerie, Z. Gabelica and N. Blom, Appl. Catal., 1981, 1, 201.
17. D.H. Olson, W.O. Haag and R.M. Lago, J. Catal., 1980, 61, 390.
18. H.J. Doelle, J. Heering, L. Riekert and L. Marosi, J. Catal. 1981, 71, 27.
19. J.P. Gilson, Dr. Sc. Thesis, Facultés Universitaires de Namur, 1982.
20. J.C. Vedrine, A. Auroux, P. Dejaifve, V. Ducarme, H. Hoser and S. Zhou, J. Catal., 1982, 73, 147.
21. E.G. Derouane, Z. Gabelica and N. Blom, submitted for publication.

THE EFFECT OF DIFFERENT TYPES OF DISTRIBUTIONS ON THERMODESORPTION KINETICS

V. Dondur, D. Fidler, D. Vučelić

Institute of General and Physical Chemistry, Belgrade
Institute of Physical Chemistry, Faculty of Sciences and Mathematics
University of Belgrade, P.O.Box 551, Belgrade, Yugoslavia

INTRODUCTION

The basic characteristics of desorption are that the rate constant in general depends upon the degree of coverage and that it is a first or second order reaction |1|. Only under ideal conditions, when the surface is energetically homogeneous, is it possible to apply classical kinetic methods. However, energetic surface heterogeneites will change kinetic equations due, to the dependence of activation energy $E(\theta)$ and the preexponential factor $A(\theta)$ on the degree of coverage |2|. Some authors say that the value of the preexponential factor is constant |3,4|; others believe the preexponential factor to change according to the same function as activation energy |5,6|. Functions in wich a change in activation energy is expressed may be elementary |7| (linear, exponential, etc.), although more complex forms do also exist |8,9|.

Generally, adsorbed molecules are not in the same energetic state. Statistically there is a molecular energy distribution. The influence of the distribution on the rate of desorption and what form there is between a particular type of distribution and the $E(\theta)$ are not clearly understood.

RESULTS AND DISCUSSION

The energy required for molecules to leave a heterogeneous surface is not constant. This is illustrated in Fig. 1 by a simple potential diagram. E_g represents the potential energy of molecules in a gas. Activation energy for desorption changes from $E_g - E_{min}$ to $E_g - E_{max}$, with the average value $E_{\bar{s}}$. For unassociated types of desorption the first order of the reaction will be valid. From the total number of molecules, δN has an energy between W and $E + \delta E$. The rate of desorption may be expressed by an equation in which the kinetic parameters are constant, since δE is taken to be small.

$$- \frac{d(\delta N)}{dt} = k \delta N \qquad (1)$$

The number of molecules changes with time according to the following equation:

$$N(t) = N_0 e^{-kt} \qquad (2)$$

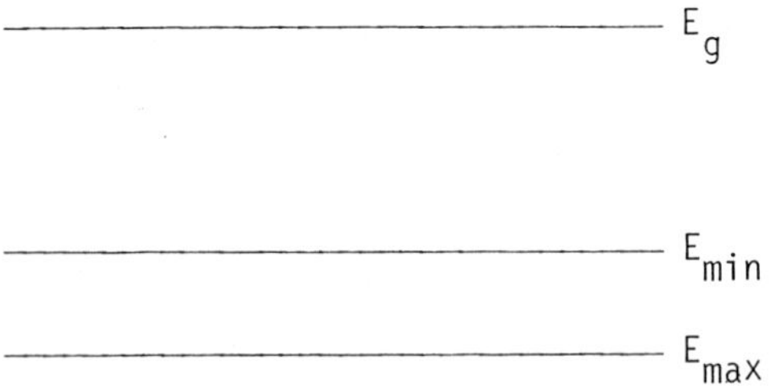

Fig. 1.
Simplified potential diagram for desorption reaction from a heterogeneous surface

The number of molecules δN_0 which at time $t = 0$ has an energy between E and $E + \delta E$ is equal to the product of the corresponding probability and total number of molecules N_0 |10|.

$$\delta N_0 = N_0 P(E,0) \delta E \qquad (3)$$

$$\delta N(t) = N_0 P(E,0) e^{-kt} \delta E \qquad (4)$$

By substituting (3) in equation (2), an equation is obtained according to which the number of molecules δN changes with time. The total number of molecules which after time t remain on the surface is equal to the sum of all N molecules which are in an energy range from E_{min} to E_{max}. This may be described by the following equation:

$$N(t) = N_0 \int_{E_{min}}^{E_{max}} P(E,0) e^{-A e^{-(E_g-E/RT)}} dE \qquad (5)$$

For a given temperature the kinetic curve $N(\theta)$ will depend upon the density function $P(E,0)$.

The TG curve ($N(T)$) or its differential-from DTG curve (dN/dT) plays the role of kinetic curve under dynamic conditions. With a given heating rate, $N(T)$ will also depend upon the density function:

$$N(T) = N_0 \int_{E_{min}}^{E_{max}} Q(E,T_0) \, e^{-A/\beta \int e^{-(E_g-E/RT)}dT} \, dE \qquad (6)$$

The density function $Q(E,T_0)$ is equal to the function $P(E,0)$. The shape of the kinetic curve under isothermal and dynamic conditions depends on the type of distribution. If the $P(E,0)$ function is of the Dirac type, then equation (5) represents the kinetic curve for a first-order reaction. Kinetic parameters of this process may be determined by applying non-isothermal methods |11| The existence of adsorption complexes is proof that the distribution on the surface is of the Dirac type.

The effect of the Gauss distribution on the kinetic curve is presented in Fig. 2. In Fig. 2a distribution functions for different parameters are presented, while in Fig. 2b appropriate thermodesorption peaks are shown. With an increase in average energy the peaks, shift and an increase in dispersion leads to a proadening of peaks. Similar results were obtained for lorentz's distribution. The symmetrical distribution of the Gauss type is characteristic in the case of water desorption from silica gel. As can be seen from Fig. 3, the agreement between the experimentally obtained and calculated curve is satisfactory. Distribution parameters are $\bar{E} = 44$ kJ/mol, $\sigma = 4$ kJ.

Desorption peaks in the case of a surface which corresponds to the asymmetrical type of distribution will also be asymmetrical functions. Fig. 3 shows results obtained for Weibul distribution functions 12 . A Weibul distribution with different parameter values is presented in Fig. 4a and appropriate thermodesorption peaks in Fig.4b. The shape and position of the peaks changes depending on the value of the function parameter. The other asymmetrical functions have an effect on the change in the shape of peaks in a similar way.

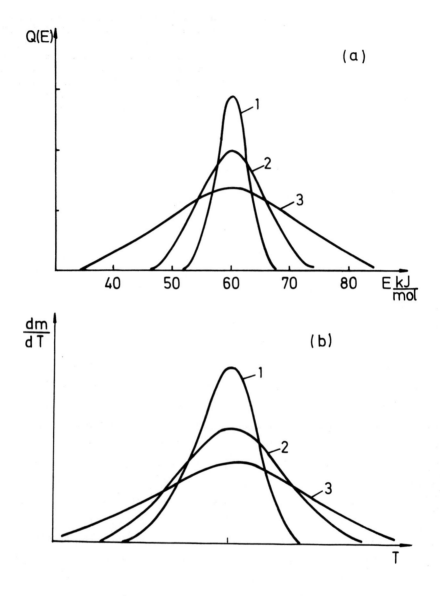

Fig.2. Example of the effect of a Gauss distribution on the appearance of thermodesorption peaks

Fig.2a. Different Gauss distributions; Fig.2b. Appropriate thermodesorption peaks

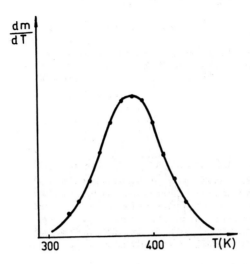

Fig. 3. Water desorption from silica gel. The experimental curve is indicated by a solid line, while the broken line indicates the calculated curve for the Gauss distribution

The fact that one peak in thermodesorption always appears is common to all peaks displayed. Broad thermodesorption peaks are the result of the great dispersion of energy.

Complex thermograms may appear as the result of the existence of several adsorption complexes (two or three Dirac distribution functions) or several active centers on which the energy distribution comes.

A typical example of a complex thermogram is with zeolites. In water desorption from zeolites several distorted peaks always appear. In complex processes it is difficult to assume in advance what a suitable type of distrubution would be and predict the possibile number of sorption centers.

Results of numerical analysis in determining the distribution function $Q(ET_o)$ in the case of the zeolite water system are presented in Fig. 5. In Fig. 5a thermograms of type A zeolite with monovalent cations LiA, NaA, and KA are presented. The ageement betwen experimen-

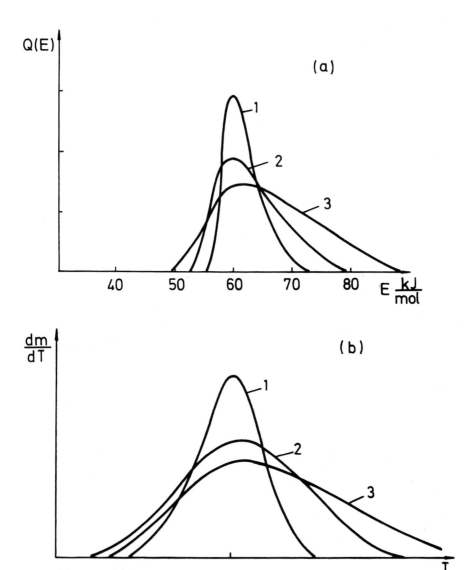

Fig. 4. Example of Weibull distribution with the appearance of a thermodesorption peak
Fig. 4a Different Weibull distributions
Fig. 4b Appropriate thermodesorption peaks

tally obtained and the calculated curve is satisfactory. Appropriate energy distributions are given in Fig. 5b. It was observed that the energy dispercion is the lowest for KA, and for LiA and NaA there are energy levels (of 80 - 100 kJ/mol). It is evident that water on zeolite represents a system with a complex distribution. The dependence of activation energy on the coverage degree calculated on the basis of the distribution function is very similar to the experimentally obtained curve of adsorption heat, as can be seen in Fig. 6.

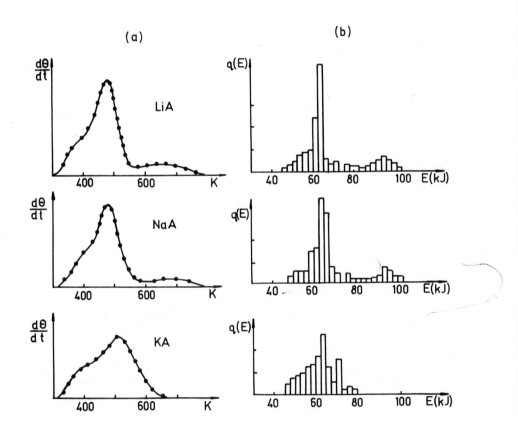

Fig. 5 Water desorption from type A zeolite with a monovalent cation
Fig. 5a The solid line indicates the experimental curve; the points indicates calculated results
Fig. 5b Corresponding numerically obtained distribution functions

The complex dependence of activation energy on the degree of coverage appears as the result of the complex distribution function on the surface. The method applied took the preexponential factor as a constant value, which may be considered a shortcoming.

Also in the case of other complex systems (e.g. aluminum trioxide), results were obtained which were in agreement with the experiment.

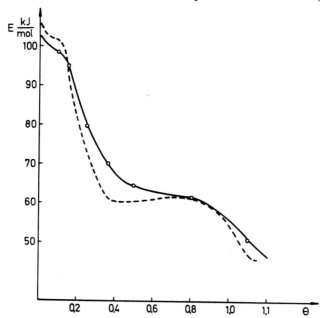

Fig. 6 Change in energy activation calculated on the basis of the distribution function. The broken line indicates differential heats of adsorption |13|

REFERENCES

1. J.R. Cvetanović, Y. Amenomiya, Advan. Catal., 1976, 17, 103
2. D. King, Surf. Sci., 1975, 47, 384
3. G. Carter, G. Armour, Vacuum, 1969, 19, 460
4. P.A. Redhead, Trans. Faraday, Soc., 1961, 57, 641
5. V.I. Yakerson, V.V. Rozanov, Fiz. hem. kinet., Vol. 3, Moskva,1974
6. V.I. Yakerson et al., Surf. Sci, 1968, 12, 221
7. V.I. Yakerson et al., Izv.Acad.Nauk USSR, ser.him.,1968,7,1468
8. E. Bauer, et al., J. Appl. Phys., 1974, 45, 5164
9. J. Tokoro, T.Uchijima, Y.Yonenda, J. Catal., 1979, 56, 110
10. J.N. Giba Probality and Statistical Inferences for Scientists and Engineers, Prentice-Hall, Engelwood Cliffs. NJ, 1973
11. M.Smutek, S. Cerny, F. Buzek, Advan. Catal., 1975, 24, 343
12. W.Weibull, J. Appl. Mech., 1951, 18, 293
13. M.M.Dubinin, et al. Izv.Akad.Nauk. USSR, Ser.Him., 1969, 11,2355

THERMAL PROGRAMMED REDUCTION (TPR): A NOVEL THERMAL ANALYSIS TECHNIQUE FOR THE INVESTIGATION OF CATALYSTS

A. Bossi, A. Cattalani and N. Pernicone

Istituto Guido Donegani S.p.A.
Centro Ricerche Novara
Via Fauser 4, 28100 NOVARA (Italy)

INTRODUCTION

Reduction is an important step in the preparation of many industrial catalysts. However it is often not easy to reproduce on a laboratory scale the actual reduction process, owing to frequent overlapping of other chemical and physical processes. The usually employed thermogravimetric techniques give meaningful data only in favourable cases, when the reduction process is well separated from other chemical processes occurring with weight change (e.g. dehydration).

The Temperature Programmed Reduction Technique, based on the continuous determination of hydrogen consumption during reduction, allows to neglect the effect of any other reaction simultaneous with the reduction [1,2].

Several advantages of TPR can be envisaged in comparison with DTG, as for instance:

- higher sensitivity in the detection of reducible species
- reactions overlapping the reduction process are not recorded
- the reducing gas flows through the sample bed, not over as occurs in thermogravimetry
- the apparatus is less expensive.

Disadvantages are lower precision in quantitative measurements and some delay in the detection of the phenomenon, which depends on the volume of the tubing between the sample and the TG detector and on the gas flow rate.

In this paper the evaluation of the TPR technique as a tool for studying catalyst reduction processes is carried out by comparing TPR and DTG data in the reduction of CuO and of high and low temperature CO conversion catalysts.

EXPERIMENTAL

The following catalysts have been investigated:
- pure CuO prepared by decomposition of $Cu_2(OH)_2CO_3$;
- mixed CuO-ZnO, CuO-Al_2O_3, CuO-ZnO-Al_2O_3 prepared with the procedures used for the Montedison-Ausind low temperature CO conversion catalyst;
- Fe_2O_3-Cr_2O_3 MHTC Montedison-Ausind high temperature CO conversion catalyst.

The DTG curves were recorded by means of the Dupont 990-951 thermogravimetric equipment.

The TPR apparatus is shown in Fig. 1.

The technique is based on the following principle: a reducing mixture hydrogen-inert gas (except helium) flows through: thermal conductivity detector, sample, trap to condense the reduction (and decomposition) products, measurement side of the T.C. detector.

During the reduction experiment the hydrogen concentration is recorded while the sample is heated at constant rate by a suitable temperature programmer. The T.C. detector displays a signal proportional to the difference of hydrogen concentration between the inlet and the outlet of the sample tube.

At constant gas flow rate this concentration difference is proportional to the reaction rate.
In such a way at the temperature considered, while the peak area is proportional to the hydrogen consumption, a TPR spectrum can be obtained with one or several peaks corresponding to reducible species.

A typical TPR peak for the reduction of CuO is shown in Fig. 2 in comparison with the corresponding DTG profile.

For quantitative measurements of the amount of hydrogen consumed previous calibration is needed. An electronic mass flow controller is necessary to have a reproducibility of $\pm$ 5% in such measurements.

RESULTS AND DISCUSSION

Several parameters may influence both the shape of TPR peak and the maximum peak temperature, T_M, namely gas flow rate, sample mass, hydrogen partial pressure.

In Table 1 the influence of these parameters on T_M in CuO reduction is shown. Our results are in good agreement with those recently published by Gentry et al. [3].

TABLE 1 - Influence of operative parameters on maximum peak temperature T_M

Parameter variation	Effect on T_M
Hydrogen concentration increases	T_M decreases
Hydrogen flow rate increases	T_M decreases
Sample mass increases	T_M increases

This behaviour is very similar to that usually found in DTG experiments. The experimental conditions (mainly internal diameter of the sample holder, gas flow rate, sample mass) must be carefully controlled to avoid the influence of external diffusion on reaction kinetics. The effect of the sample mass, m, can be eliminated by employing different amount of catalyst and extrapolating to m = 0 [3].

In practice the peak profile represents a description of the reduction kinetics. The treatment of the experimental data can be carried out by integrating the well known differential equation for non isothermal reactions [4]

$$\frac{d\alpha}{dT} = \frac{z}{\phi} \exp(-\Delta E/RT)(1-\alpha)^n$$

where:

α = reduction degree
ϕ = heating rate
ΔE = activation energy of the reduction
z = preexponential factor
n = reaction order

Several approximations can be applied for the solution of this equation [4]. The value of the activation energy of the reduction, apart from any assumption on the reaction mechanism, can be obtained according to the approximation of Ozawa (or Kissinger), by plotting log ϕ vs $1/T_M$ [5,6].

In Fig. 3 our experimental data for the CuO reduction are plotted, showing good agreement between TPR and DTG techniques.

To obtain some information on the reaction mechanism from these dynamic thermal analysis techniques our usual procedure is to calculate the activation energy by the Ozawa (or Kissinger) method and to check the reaction mechanism by the Satava approximation [7].

In our case, for CuO reduction, straight lines (Fig. 4) were obtained using an equation on the type

$$g(\alpha) = \left[-\ln(1-\alpha)\right]^n$$

The most probable reaction mechanism, giving an activation energy close to that calculated by the Ozawa method, corresponds to $n = 1/3$ (random nucleation as rate-determining step).

The potentialities of TPR for catalyst reduction studies are well evidenced in Fig. 5 and 6. In Fig. 5 DTG and TPR spectra of the Montedison-Ausind high temperature CO conversion catalyst are compared. The reduction pathway is:

$$\text{hydrous } \alpha\text{-FeOOH} \xrightarrow{-H_2O} \text{anhydrous } \alpha\text{-FeOOH} \xrightarrow{-H_2O} Fe_2O_3 \xrightarrow[-H_2O]{+H_2} Fe_3O_4 \xrightarrow[-H_2O]{+H_2} Fe$$

(the last reduction step is to be avoided).

The first and the second DTG peaks are due to dehydration steps. This is clearly demonstrated by the TPR profile where only the reduction steps are recorded. Important practical informations on the beginning of the reduction were obtained by TPR.

In Fig. 6 the TPR spectra of the $CuO-ZnO-Al_2O_3$ system are shown. A partial interaction between CuO and ZnO seems to occur (probably solid solution of CuO or Cu_2O in ZnO).

Much stronger is the interaction between CuO and Al_2O_3, with probable formation of highly disordered $CuAl_2O_4$ not evidenced by X-ray diffraction. Both these interactions make more difficult the nucleation of metallic copper, which was shown to be the most probable rate-determining step.

REFERENCES

1. I.W. Jenkins, B.D. Nicol, S.P. Robertson, CHEMTECH, 316 (1977).
2. N. Pernicone, F. Traina, Pure Appl. Chem., 50, 1169 (1978).
3. S.J. Gentry, N.W. Hurst, A. Jones, J. Chem. Soc. Faraday Trans. I, 75, 1688 (1979).

4. E. Koch "Non-Isothermal Reaction Analysis", Academic Press., 1977.
5. T. Ozawa, J. Thermal Anal. 2, 301 (1970).
6. H.E. Kissinger, Anal. Chem. 29, 1702 (1957).
7. V. Satava, F. Skvara, J. Am. Ceram. Soc. 52, 591 (1969).

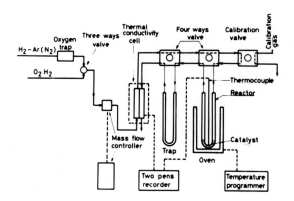

Fig. 1 - TPR apparatus.

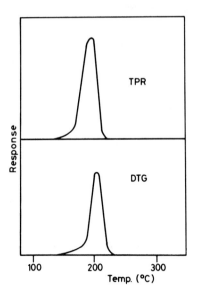

Fig. 2 - Comparison between TPR and DTG peaks for the CuO reduction.

Fig. 3 - Ozawa plot for the TPR and DTG reduction of CuO.

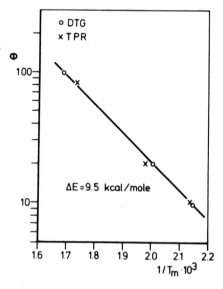

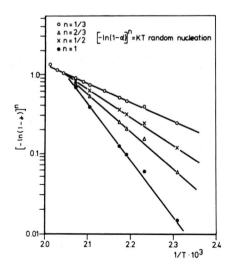

Fig. 4 - Satawa plot for the TPR of CuO.

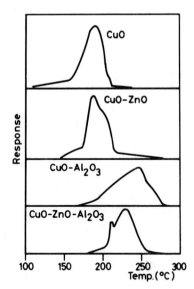

Fig. 5 - Comparison between TPR and DTG spectra in the reduction of a Montedison high temperature CO conversion catalyst.

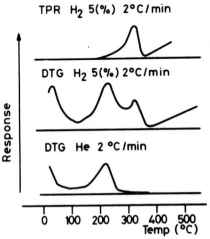

Fig. 6 - TPR spectra of the $CuO-ZnO-Al_2O_3$ system.

CHARACTERIZATION OF SUPPORTED METAL CATALYSTS BY THERMAL PROGRAMMED REDUCTION

R. Bertè, A. Bossi, F. Garbassi and G. Petrini

Istituto Guido Donegani S.p.A.
Centro Ricerche Novara
Via Fauser 4, 28100 NOVARA (Italy)

INTRODUCTION

The complexity of the interaction of ruthenium compounds with oxide supports has been shown by chemico-physical techniques [1-5] and their different catalytic behaviour [6]. In particular, it is widely agreed that the $Ru-SiO_2$ interaction is weak and easily destroyed in reducing conditions [1,3,4], while that of ruthenium with alumina is quite stronger, giving rise to species reducible with difficulty [1,4]. Recent studies give a theoretical support to this picture [7]. Finally, the interaction of Ru with magnesia does not appear very strong and seems to be of a different kind. Its occurrence is dependent on the preparation procedure, which can give rise to samples showing a well different catalytic behaviour [6]. Recent literature shows that Ru supported on MgO is partially proof against reduction [8], although ultimately reducible [9]. The existence of a stable Ru-Mg-O surface complex has been already suggested on the basis of Ru loss measurements by high temperature volatilization [10]. However, its stability against reduction was not measured.

It is aim of this work to characterize the different ruthenium species that can be stabilized on silica, alumina and magnesia supports with the Temperature Programmed Reduction (TPR), a promising technique for supported metal catalysts [11,12].

EXPERIMENTAL

Samples were prepared by impregnation using as precursor Ru chloride in the case of SiO_2 and Al_2O_3, and Ru nitroso-nitrate for MgO. The chemico-physical characteristics of the final samples are reported elsewhere [1].

The measurements were carried out with the TPR system described in another communication at this Conference [13].

The reducing agent was a mixture of 5% H_2 in argon, flowing at a rate of 2400 cm^3 hr^{-1}. The gas flow was regulated by a Matheson mod. 8240 Mass Flow Controller. The use of nitrogen as carrier gas was avoided for the possible catalytic formation of ammonia in the presence of Ru-containing samples.

The reactor had a diameter of 4 mm and was heated at a linear rate of 4 K min^{-1} from room temperature to 1073 K. The employed catalyst amount was taken in the range 0.03-0.1 g. A trap for water and other undesired compounds was placed between the reactor and the thermal conductivity detector. The response of this last was calibrated before each run by means of an injection of pure argon in the gas mixture.

RESULTS

TPR spectra of Ru/SiO_2, Ru/Al_2O_3 and Ru/MgO samples are reported in Figs. 1-3 respectively.

The reduction behaviour and hydrogen consumption of the catalyst precursors after impregnation and drying appear markedly different on the various supports. (A spectra in Figs. 1-3). On the silica- and alumina- supported samples, major differences are the position and intensity of the peaks at T > 523 K, while the peaks centered near 453 K are quite similar.

The hydrogen consumptions, expressed as H/Ru atomic ratios, are about 4 on silica, a value near to that expected for the reduction of RuO_2, and somewhat larger (mean value: 5.5) on alumina.

In the case of MgO, a very different situation was observed, with the occurrence of a main peak at 553 K, and a minor one at 673 K. On this sample, very high H/Ru = 18 values were found and confirmed in repeated experiments.

A series of TPR experiments was carried out on MgO samples after pretreatments in different conditions (see caption of Fig. 3), with the aim to identify the reducible ruthenium species arising from the interaction with the support. For the same reason, some pure compounds were also submitted to TPR: RuO_2, giving a single narrow peak at 453 K, Ru nitrate, with a broad peak near 573 K and Mg nitrate, showing a maximum at 703 K.

DISCUSSION

It is quite clear that the presence of precursor salt anions (Cl^- and NO_3^-) and the nature of the support strongly influence the appearance of the TPR diagrams. The complete elimination of chloride ions gives rise to a single peak diagram on SiO_2 (Fig. 1-curve B), with disappearance of the high temperature feature and shift of the maximum under 373 K. On alumina, the elimination of the high temperature peak is very difficult, and obtainable after reduction and reoxidation at 1073 K, while a single heating at 973 K is not sufficient (Fig. 2-curves B and C).

Two peaks are generally obtained, always at temperature higher than 373 K.
Differences observed on the two supports suggest that already at the impregnation stage an interaction of higher strenghtening is established between the precursor compound and the support surface on alumina rather than on silica.

Final peaks (1-B and 2-C) indicate also that a stronger interaction survive between Ru and alumina. The occurrence of two peaks on curve 2C suggests that two different Ru species are present, one similar to that found on silica, the other one more difficulty reducible.

The presence of Ru species reducible only at high temperature is consistent with XPS results on the same samples, where the persistency of Ru species with a binding energy higher than that of the metal was found also after reduction in H_2 at 873 K [1].

A different situation appears on MgO (Fig. 3-A): the two detected peaks are easily assigned, by comparison with

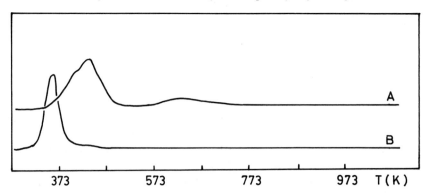

Fig. 1 - TPR diagrams of Ru/SiO_2 catalysts
(A: dried at 383 K; B: after Cl^- elimination)

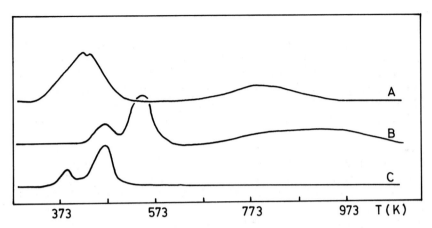

Fig. 2 - TPR diagrams of Ru/Al$_2$O$_3$ catalysts
(A: dried at 383 K; B: calcined at 973 K;
C: reduced at 673 K and re-oxidized at 1073 K)

TPR diagrams of pure compounds, to ruthenium and magnesium nitrate respectively.

With this assignement also the high hydrogen consumption (H/Ru = 18) is justified, taking into account the reduction of the nitrate ions, according to the following reactions

$$Ru(NO_3)_3 + 9H_2 \rightarrow Ru + 1.5N_2 + 9H_2O \qquad (H/Ru = 18)$$

$$Ru(NO_3)_3 + 13.5H_2 \rightarrow Ru + 3NH_3 + 9H_2O \qquad (H/Ru = 27)$$

Traces of ammonia were actually detected inserting a Nessler bottle in the TPR system, just after the TPR reactor.

The TPR diagram after washing (Fig. 3-B) shows a remarkable lowering of the Ru nitrate peak, the persistency of Mg nitrate and the occurrence of a new peak, possibly due to Ru hydroxide.

As the catalytic behaviour of the Ru/MgO catalyst in the CO activation is quite different whether reduced "as prepared" or after a heating in vacuo at 673 K [6], a sample was pretreated in He flow at this temperature before reduction, to check the occurrence of new Ru species. The correspondent TPR spectrum shows a single small peak near 520 K (Fig. 3-C), with a very low H/Ru value of 1.

The persistency of a noticeable amount of nitrate in the sample (about 1/3 of the initial amount) after such treatment suggests the formation of a nitrate containing surface complex, stable in reducing conditions.

A subsequent TPR experiment on the same sample after air exposure gives the diagram of Fig. 3-D, showing two peaks and a higher hydrogen consumption (>4).

Finally, after repeated oxidation-reduction cycles, TPR diagrams gradually show a single peak similar to that observed on silica, consistent also for the hydrogen consumption.

Although TPR cannot give a complete information on the chemical nature of the species present in the samples, it can be envisaged from the whole of results, accompanied by nitrate analysis after each step, that a very stable

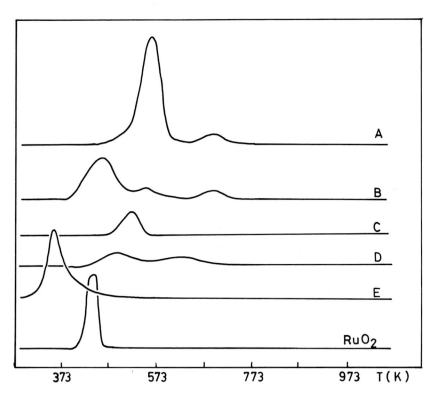

Fig. 3 - TPR diagrams of Ru/MgO catalysts
(A: dried at 343 K; B: washed in H_2O; C: heated at 673 K in He; D: as C after air exposure; E: cyclically reduced at 1073 K and re-oxidized at 423 K).

Ru-MgO-NO$_3$ surface complex can form in particular conditions, namely in absence of humidity, onto the samples. This complex appears very stable against reduction, since no hydrogen consumption up to 1073 K has been measured, apart some residual Ru nitrate.

TPR after reoxidation in air, compared to reoxidation with He + 5% O$_2$ in the TPR system, indicate that this complex is on the contrary very sensitive to humidity.

REFERENCES

1. A. Bossi, F. Garbassi, A. Orlandi, G. Petrini and L. Zanderighi, "Preparation of Catalysts II", Elsevier, Amsterdam (1979), p. 405.
2. E.B. Prestridge, G.H. Via and J.H. Sinfelt, J. Catal., 1977, 50, 115.
3. F.W. Lytle, G.H. Via and J.H. Sinfelt, J. Chem. Phys., 1977, 67, 3831.
4. C.A. Clausen, III.and M.L. Good, "Characterization of Metal and Polymer Surfaces", Academic Press, New York, (1977), vol. I, p. 65.
5. E. Guglielminotti, A. Zecchina, A. Bossi and M. Camia, J. Catal., in press.
6. A. Bossi, F. Garbassi, G. Petrini and L. Zanderighi, J. Chem. Soc., Faraday Trans., I, in press.
7. S. J. Tauster, S.C. Fung and R.L. Garten, J. Catal., 1978, 55, 29.
8. I.W. Bassi, F. Garbassi, G. Vlaic, A. Marzi, G.R. Tauszik, G. Cocco, S. Galvagno and G. Parravano, J. Catal., 1980, 64, 405.
9. J. Schwank, G. Parravano and H. Gruber, J. Catal., 1980, 61, 19.
10. S.J. Tauster, L.L. Murrell and J.P. DeLuca, J. Catal., 1977, 48, 258.
11. S.D. Robertson, B.D. McNicol, J.H. de Baas, S.C. Kloet and J.W. Jenkins, J. Catal., 1975, 37, 424.
12. J. Lemaistre and M. Houalla, C.R. Acad. Sci. Paris, 1981, 292, 1977.
13. A. Bossi, A. Cattalani and N. Pernicone, this Conference.

STRONG METAL-SUPPORT INTERACTION ON NICKEL CATALYSTS

Ling Yuan Chen, Chih Yung Lin
Department of Chemical Engineering
National Cheng Kung University
Tainan, Taiwan, Republic of China

INTRODUCTION

The observation of interaction between metal and support on the surface of support-metal catalysts has been reported by many investigators (1)(2)(3)(4). The suggestion of a strong metal-support interaction on titania support moble metal catalyst made by Tauster and his coworkers (3)(4) is base on the observation that the capability of noble metal for adsorbing hydrogen and carbon monoxide is greatly suppressed by using titania as the support. The same suppression has also been reported for using Nb_2O_3 or Ta_2O_5 as the supports. Many modern techniques are applied to the investigation of the metal-support interaction. A study using x-ray photo-electron spectroscopy made by Vedrine (5) revealed that the interaction between nickel oxide and support retards the reduction of nickel to its metallic state. Houalla and Delmon (6) also use x-ray photo-electron spectroscopy to study the nickel catalysts with silica-alumina as the support. According to their report, the increase of alumina in the support makes the carystallinity of nickel oxide, or metallic nickel on the catalysts changed. The reduction of nickel oxide becomes more diffcult.

Several suggestions have been made for the origin of the metal-support interaction in the catalyst system. Houlla and Delmon (6) think that a new compound could be formed by the combination of nickel oxide and alumina in alumina support nickel catalyst. Many investigators have reported the formation of nickel aluminate on this catalyst (7). For the noble metal catalysts with titania as the support, Tauster et. al. (3)(4) suggested that the origin of strong metal-support interaction could be either the direct conbination of noble metal and titania, or the formation of intermetallic compound between noble metal and titanium that is formed by the partial reduction of titania.

In this investigation, metal-support interaction on the nickel catalysts with different supports was studied. Titania, alumina, silica, and graphite were used as the supports. Graphite is chosen for its inertness. The interaction between nickel and supports was observed in the experiments of thermal analysis and chemisorption of hydrogen on the catalysts. The thermograms of the nickel catalysts were compared, and discussed in order to study the origin of the metal-support interaction.

EXPERIMENTS

The catalysts were prepared by impregnation of supports with the aqueous solution of nickel nitrate. Nickel contents of the catalysts prepared are 18.36% in Ni/Al_2O_3, 18.35% in Ni/TiO_2, 17.95% in Ni/SiO_2, and 18.35 in Ni/graphite by weight. BET surface areas of the supports used are 94 m^2/g for alumina, 15 m^2/g for titania, 45 m^2/g for silica, and <1 m^2/g for graphite.

Hydrogen chemisorption at room temperature in the catalysts were measured volumetrically in a conventional high vacuum system capable of 10^{-6} torr. The catalysts were sealed in a flow through cell, and reduced in hydrogen stream at 500°C prior to the chemisorption experiments. Oxygen up-take on the reduced catalysts was also measured to estimate the degree of reduction of the catalysts.

The x-ray diffraction patterns were taken as the catalysts in fresh, calcined at 500°C, and reduced by hydrogen at 500°C, respectively. The reduced catalysts were contacted to oxygen at the pressure of 10 torr before the XRD was measured to protect the metallic nickel particles on catalyst surface from further oxidation.

The thermal analysis experiments were conducted by a DTG system of the 3M type differential thermal analyzer made by Shimatze Co., Japan. In this system, DTA and TG curves of the samples are recorded simutaneously. Gases flowed through the sample cell at 30 ml/min during the measurements. Nitrogen was used as the flowing gas for studying thermal decomposition of the catalysts, and hydrogen for studying the reduction. In both series of experiments, the cell temperature was raised at 5°C/min.

STRONG METAL-SUPPORT INTERACTION

Hydrogen up-take on the catalysts after the reduction and the degree of reduction of the catalysts reduced in the indicated conditions are shown in Table 1. The metallic nickel on the reduced catalysts adsorbs hydrogen dissociatively. It can be seen from Table 1 that Ni/TiO_2 and Ni/graphite show very low capability of adsorbing hydrogen after they have been of reduced at 500°C for five hours. The low adsorption up-take of Ni/graphite catalyst is expected because of its very low BET surface area. For Ni/TiO_2 catalyst, low hydrogen up-take in chemisorption is coincident with the results of the catalysts of noble metals on TiO_2 reported by Tauster (3)(4). X-ray diffraction analysis shows that the nickel particles on Ni/TiO_2 are as small as those on Ni/Al_2O_3 or Ni/SiO_2 while the nickel particles on graphite support are pretty large. Then, nickel particles on Ni/TiO_2 should have surface area large enough to adsorb hydrogen as those on Ni/Al_2O_3 or Ni/SiO_2 catalysts. Moreover, the degree of reduction of Ni/TiO_2 shown in Table 1 is the same as that of Ni/SiO_2 which has large hydrogen up-take in chemisorption. Thus, the low hydrogen up-take on Ni/TiO_2 catalyst is also not due to the difficulty of reduction. Therefore, the suppression of the hydrogen chemisorption on Ni/TiO_2 catalyst shown in Table 1 strongly indicates the existence of strong metal-support interaction suggested by Tauster (3)(4).

The interaction between nickel and alumina in Ni/Al_2O_3 catalyst is also observed. The exceptional low degree of reduction of Ni/Al_2O_3 catalyst is the consequence of this interaction. X-ray diffraction patterns of Ni/Al_2O_3 catalyst were analyzed purposely in order to confirm the formation of nickel aluminate. Several weak diffraction lines were found in the diffraction patterns of Ni/Al_2O_3 catalyst after the calcination. The d values of these lines are very close to the d values of the diffraction lines of nickel aluminate

TABLE 1 HYDROGEN UP-TAKE AND DEGREE OF REDUCTION OF THE NICKEL CATALYSTS

	Ni/Al_2O_3	Ni/SiO_2	Ni/TiO_2	Ni/graphite
Reduction Time		2 hours		
Hydrogen Up-take mole/g x 10^6	95.57	89.51	19.03	2.61
Degree of Reduction %	73.76	99.81	98.86	62.21
Reduction Time		3 hours		
Hydrogen Up-take mole/g x 10^6	86.50	56.90	8.61	1.41
Degree of Reduction %	80.11	100.00	100.00	65.32
Reduction Time		5 hours		
Hydrogen Up-take mole/g x 10^6	78.58	52.90	0.00	0.00
Degree of Reduction %	95.92	100.00	100.00	77.51

*Reduction Conditions: 500°C; hydrogen flow rate: 97 ml/min. g-catalyst.

given in the index.

THERMAL DECOMPOSITION OF NICKEL NITRATE

When nickel nitrate is heated in nitrogen, it decomposes to $Ni(OH)(NO_3)$ and, then, to NiO (8). The thermograms of the supported nickel catalyst heated in nitrogen are shown in Figure 1 (TG) and Figure 3 (DTA). For nickel nitrate crystal and nickel with supports of TiO_2, SiO_2, and graphite, all the TG curves and DTA curves clearly show the two stages of thermal decomposition. The thermal decomposition temperatures of nickel nitrate crystal are 214°C and 306°C for the two stages respectively. The decomposition temperatures decrease to 208°C and 283°C as the nickel nitrate crystallites are dispersed on the surface of silica and titania. For Ni/Al_2O_3 catalyst, the thermograms shown in Figure 1 and Figure 3 are so different from the others. No step changes are seen in the TG cuves and the endothermic peaks in DTA curves are broadened and even diminished. The different thermal behaviors of Ni/Al_2O_3 catalyst from the other's during the decomposition could come from the formation of a new compound, such as nickel aluminate, on the surface that has different mechanism of thermal decomposition from the mechanism of nickel nitrate. Since Ni/TiO_2 catalyst has the same thermogram as those of Ni/graphite catalyst during the thermal decomposition, and graphite is an inert material, the formation of a new compound between nickel nitrate and

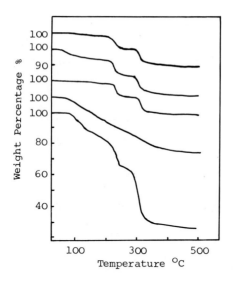

Figure 1. TG Curves of the Samples in Nitrogen Stream

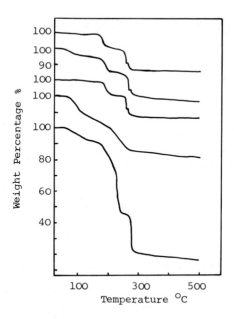

Figure 2. TG Curves of the Samples in Hydrogen Stream

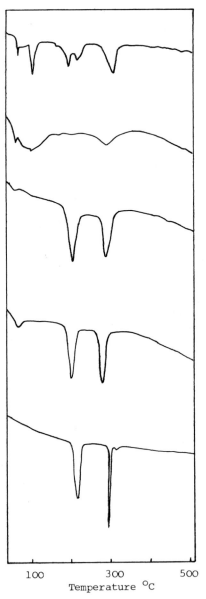

Figure 3. DTA Curves of the Samples in Nitrogen Stream

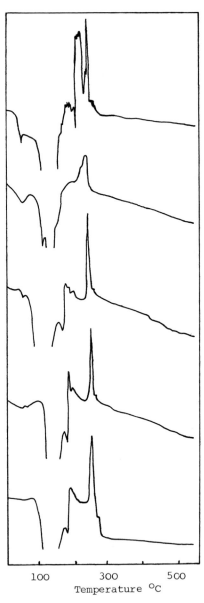

Figure 4. DTA Curves of the Samples in Hydrogen Stream

titania is very unlikely.

THE REDUCTION OF NICKEL CATALYSTS

The reactions taking place during the reduction of the catalysts should include the decomposition of nitrate, the reactions between hydrogen and nitrogen oxides produced in decomposition, and the reduction of nickel oxide to metallic nickel. No attampt is made on the deduction of the reaction scheme for the complicated reduction reactions. The comparison of the thermograms obtained by heating the samples in hydrogen, however, can be used as the basis to discuse the metal-support interaction on the catalysts.

The behaviors of the catalysts during the reduction are shown in the thermograms of Figure 2 (TG) and Figure 4 (DTA). For the supported nickel catalysts except Ni/Al_2O_3, TG curves show two step changes at 180°C, and 245°C respectively, and in DTA curves, a strong and broad endothermic peak at 65 to 150°C with a shoulder at 163 to 170°C, and two exothermic peaks at 180°C, and 245°C are shown. For Ni/Al_2O_3 catalyst, again, no step changes are seen in the TG curve, and the DTA curve is basically the same as the others, but the peaks become broad and weak. The different behavior of Ni/Al_2O_3 from the other catalysts during the reduction can be clearly reconized from the comparison. The formation of a new compound on Ni/Al_2O_3 catalyst could be the reason for this different.

In the DTA curve of Ni/TiO_2 catalyst, the exothermic peak at 180°C is very weak. For Ni/SiO_2 and Ni/graphite catalysts, this peak is strong and accompanies the step change at 180°C in TG curve. The diminish of this exothermic peak in DTA curve of Ni/TiO_2 catalyst could be an indication of the interaction between nickel and titania. If this is the case, the interaction should occur before the nickel is completely reduced to its metallic state at 250°C.

CONCLUSIVE SUMARY

1. Thermal analysis shows that dispersion of the nickel nitrate crystallites on inert supports, such as titania, silica, or graphite can lower its decomposition temperature.
2. Strong metal-support interaction in Ni/TiO_2 catalyst as well as in Ni/Al_2O_3 catalysts is observed in chemisorption experiments and thermal analysis.
3. The thermal behaviors of Ni/Al_2O_3 catalyst strongly indicate the formation of a new compound such as nickel aluminate on catalyst surface. The formation of new compound on the other catalyst is not observed.
4. The origin of strong metal-support interaction in Ni/TiO_2 cannot be concluded, but the suggestion on formation of new compound between nickel and titania which causes the interaction cannot be comfirmed by this investigation.

REFERENCE

1. M.A. Vannice, J. Catalysis, 1975, 37, 449
2. M.A. Vannice, R.L. Garten, J. Catalysis, 1979, 56, 236

3. S.J. Tauster, S.C. Fung, and R.L. Garten, J. Am. Chem. Soc., 1978, 100, 170
4. S.J. Tauster, S.C. Fung, J. Catalysis, 1978, 55, 29
5. J.C. Vedrine, G. Hollinger, and T.M. Duc, J. Phy. Chem., 1978 82, 1515
6. M. Houalla, B. Delmon, J. Phy. Chem., 1980, 84, 2194
7. K. Morikawa, J. Shirasaki, and M. Okada, Adv. Catalysis, 1969, 20, 97
8. J.W. Mellor, "A Comprehensive Treatise Inorganic and Theroretical Chemistry", (1922), Vol. 15, p.487

CATALYST CHANGE INVESTIGATION OF THE SYSTEM ZnO-CuO-Cr_2O_3 DURING ITS REGENERATION IN THE INDUSTRIAL REACTOR

G.Rasulić, Lj.Milanović, S.Jovanović
Control and Research Department
Chemical Industry "Pančevo"- Pančevo
Yugoslavia

INTRODUCTION

It is known that the catalyst for low temperature carbon monoxide conversion on the basis of zinc, copper and chrome oxides is sensitive to raised temperature effects and sulphur and chlorine compounds influence, which are poisons for this catalyst. Under its influence during catalyst life in the industrial reactor precrystallization of active component and growth of the elementary copper crystals appear, which cause the catalyst deactivation/1/. By regeneration process, which is anticipated for this catalyst, its activity is improved and the catalyst life is prolonged. The catalyst, used in our ammonia plant, has been exposed to regeneration three times. According to manufacturers'instructions, the catalyst regeneration is carried out into two stages, the first of which includes the catalyst charge oxidation, by which the sulphur poisoning effects are decreased, and the second - demineralized water washing, by which dissolved chlorides are removed/2/.

The aim of this paper is to investigate active component changes of low temperature shift catalyst during its regeneration process by thermal analysis.

THE EXPERIMENTAL PART

The change investigations of the low temperature shift catalyst during its regeneration process have been carried out on the sample from the top of catalyst bed of the two parallel industrial reactors after different stages of regeneration.

It enables to follow the active component changes of catalyst after oxidation stage and catalytic charges washing in the first regeneration, compared to the new, not used sample. The changes of the catalyst taken from the same point in the industrial reactor after II and III regeneration that have included only the stage of catalytic charge oxidation have been also followed. The following of "catalyst ageing" has been carried out compared to the new, not used sample to the samples from the same point in the industrial reactor after the same stage of I, II and III regenerations.

The review of the investigated samples, their history and internal marks have been given in Table 1.

TABLE 1. THE REVIEW OF THE INVESTIGATED SAMPLES

internal mark	sample history	reactor
1	The new, not used catalyst	–
2	After oxidation in I regeneration	II
3	After oxidation in I regeneration	I
4	After washing in I regeneration	I
5	After washing in I regeneration	II
6	After oxidation in II regeneration	I
7	After oxidation in II regeneration	II
8	Before III regeneration	I
9	Before III regeneration	II
10	After oxidation in III regeneration	I
11	After oxidation in III regeneration	II
13	End of catalytic life, 28 months	I
14	End of catalytic life, 28 months	II
15	End of catalytic life, 28 months from overheating layer, bottom	II

The investigations have been carried out on the Derivatograph in the static atmosphere of air, in the temperature interval from $30°-1000°C$, at heating rate of $10°C/min$. and with alumina as the reference, as well as on the Differential Scanning Calorimeter DSC-1B in the temperature interval from $30°-500°C$ in the dynamic atmosphere of hydrogen.

THE RESULTS AND THE DISCUSION

All investigated samples present a commercial catalyst from the moment of putting it into the reactor to its taking out and replacement by a new one. The samples have been previously subjected to differential thermal analysis with the aim to investigate the thermal behaviour of the samples and the tendency of changes, connected to active component.

The differential thermal analysis has shown, on the basis of the exotermic peak in the temperature interval from $180°$ to $320°C$, the presence of the elementary copper in the samples taken from the catalytic industrial reactor. With the exploitation time increase, the intensity of this peak grows i.e. the content of the elementary copper grows in the samples.

Considering that the content of the elementary copper grows with the catalyst exploitation time in the samples, while at the same time their activity decrease is evident, this copper in the form of enlarged crystals does not present the catalyst active component any more. The catalyst active component is oxidated at taking the catalyst out of the reactor and we have been determining it by measuring the thermal effect of copper oxide reduction process in the investigated samples.

Peak sizes that follow the reduction of the present cupric oxide in the investigated samples, on the basis of which the reduction energy is determined, can be considered as a measure of the active component in the catalyst. Supposing that cupric oxide reduction energy is proportional to its specific area, and knowing both the cupric oxide p.a. specific area and its reduction energy, the cupric oxide specific area has been calculated in the catalyst samples by their reduction energy determination.

Having in view that the catalyst activity decrease is followed by the cupric oxide specific area, the calculated specific area of cupric oxide in the catalyst can be considered as the catalyst active area/3/. The value of cupric oxide reduction energies and cupric oxide specific area, obtained in the catalyst samples, have been shown in the Table 2. for I regeneration.

TABLE 2. THE CHANGE OF CATALYST ACTIVE AREA DURING BOTH STAGES IN I REGENERATION

intern. mark	cupric oxide reduction energy J/g cat.	cupric oxide specific area $m^2 CuO$/g cat.	reactor
1	169,66	2,94	-
3	111,96	1,94	I
4	116,45	2,02	I
2	114,36	1,98	II
5	130,79	2,26	II

During oxidation process, as the first stage of catalyst regeneration, a part of copper stays in the nonactive form. The results obtained show the decrease of cupric oxide reduction energy, in other words, the decrease of cupric oxide specific area, for about 33% according to the new, not used catalyst. By applying the second stage of regeneration the active area increases for 5-14% according to the samples after the first stage of regeneration.

The second regeneration after 16 months of catalyst exploitation included only the oxidation stage. The changes of the active component have been observed separately from the top of both the first and second of industrial reactor after the catalyst oxidation.

The catalyst active area after II regeneration is larger for about 4% according to the active area after the first stage of I regeneration/Table 3/.

In the Table 4. the results of cupric oxide specific area of the samples before and after III regeneration, that has included only the oxidation stage, have been shown.

TABLE 3. THE CATALYST ACTIVE AREA AFTER II REGENERATION

intern. mark	cupric oxide reduction energy J/g.cat.	cupric oxide specific area $m^2CuO/g.cat.$	reactor
1	169,66	2,94	-
2	118,08	2,04	I
3	116,82	2,02	II

TABLE 4. THE CATALYST ACTIVE AREA BEFORE AND AFTER III REGENERATION

intern. mark	cupric oxide reduction energy J/g.cat.	cupric oxide specific area $m^2CuO/g.cat.$	reactor
1	169,66	2,94	-
9	80,93	1,40	I
11	68,97	1,19	I
10	82,72	1,43	II
12	102,85	1,78	II

In the samples before III regeneration considerably lower active area has been registered according to the new, not used catalyst.

The effect of the oxidation in III regeneration has been registered only in the second reactor and it represents the increase of the catalyst active area for 24% according to the sample before regeneration.

In the sample from the top of the first reactor, the decrease of active area, after III regeneration, of about 15% has been obtained by our measuring. Such behaviour can be explained by the difficulties at taking the sample from catalytic reactor and it is possible to consider that it does not always present the representativ sample of the investigated catalytic bed.

In the Table 5. the parallel results of the active areas of the investigated samples, taken from the top of the first and second parallel industrial reactors, after oxidation, as I regeneration stage in all three regenerations, have been shown.

The results point out that the catalyst regeneration after the oxidation in I and II regeneration has been practically carried out with the same effect. The samples have aproximately equal active areas. The values are for 31-34% lower than the same for the new, not used catalyst which points out that a part of copper remains even after the regeneration continuously non-active.

After III regeneration, further decrease of active area according to the new, not used sample, for 40-60% is obtained.

TABLE 5. THE CHANGES OF THE CATALYST ACTIVE AREA AFTER THE OXIDATION IN I,II AND III REGENERATION COMPARED TO THE NEW CATALYST

intern. mark	cupric oxide reduction energy J/g.cat.	cupric oxide specific area $m^2CuO/g.cat.$	reactor
1	169,66	2,94	–
3	111,97	1,94	I
7	118,08	2,04	I
11	68,97	1,19	I
2	114,36	1,98	II
8	116,82	2,02	II
12	102,85	1,78	II

The samples have been taken from the reactor after 19 and 28 months of its exploitation time and the change of its active area with time has been determined. The results obtained have been shown in the Fig.1.parallel to the effect results of the regenerations carried out. We can only suppose the behaviour of the first part of the active area decrease curve with the catalyst exploitation time in industrial reactor, and for that reason this part is presented by a broken line.

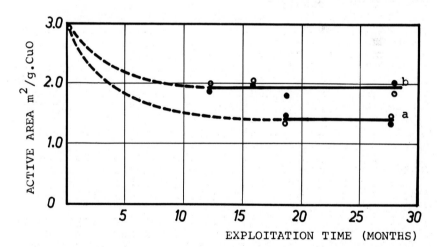

Fig.1. The Catalyst active area change in function of exploitation time in the industrial reactor :
a/. Before regeneration
b/. After regeneration

In order to estimate the efficiency of the regeneration carried out, we have compared the values of the active area after regeneration to the values read out from the diagram before regeneration. Table 6. shows the active area of the investigated samples before regeneration, read out from the diagram in Fig.1., and the active area after regeneration defined by experimental procedure.

TABLE 6. THE CHANGE OF THE CATALYTIC ACTIVE AREA CAUSED BY REGENERATION IN FUNCTION OF EXPLOITATION TIME IN THE FIRST AND SECOND INDUSTRIAL REACTOR

intern. mark	regeneration	exploit. time /months/	active area of CuO $m^2 CuO/g.cat$ before reg.	after reg.	reactor
1	-	0	2,94+	-	-
3	after I	12	1,60	1,94	I
7	after II	16	1,50	2,04	I
9	after III	19	1,40	1,19	I
13	end of life	28	1,45	1,87	I
2	after I	12	1,60	1,98	II
8	after II	16	1,50	2,02	II
10	after III	19	1,43	1,78	II
14	end of life	28	1,35+	2,01	II
15	end of life	28	0,24+	1,97	II

+ The values obtained by the experimental procedure

By regeneration processes the catalyst active area increase for 21 to 36% according to the state before regeneration has been attained. Our results have not pointed out to the decrease of regeneration effect with exploitation time increase as it could be expected. The second regeneration effect is better than the first, and the third is equal to the first. The catalyst shows practically the same changes in both catalytic reactors.

The Table 6., has been supplemented by the samples from the catalyst end life from the top of the first reactor, from the top of the second reactor and from the bottom of the second reactor that has undergone overheating in the process.

These samples have been regenerated in the laboratory reactor with simulation of conditions in the process. The active area defined after the regeneration points out to the increase of 29-49% for the samples from the catalyst end life. The fact that suprised us very much is a very good regeneration of the overheated sample as well, which practically shows the same value of the active area as the non-overheated one.

Another interesting fact is that the samples, regenerated in the laboratory reactor, having previously been used 28 months in the process, have shown for about 12% bigger active area according to the samples after III regeneration in the industrial reactor. It would mean that the catalytic charge active area value can be possibly influenced by regeneration conditions.

CONCLUSION

Thermal investigations of the changes for the low temperature shift catalyst during regeneration process, have been carried out by the methods of differential thermal and microcalorimetric analysis. On the basis of the results obtained it is possible to conclude the following:

1. During the first stage of regeneration - oxidation stage , active area of low temperature shift catalyst increases for about 20-30% according to the same catalyst before regeneration.
2. By the second regeneration stage - washing stage, further active area increase of 5-14% according to the catalyst active area after the first regeneration stage, is attained.
3. On the basis of the results obtained, the catalyst active area, after regeneration, regardless of the catalyst explotation time, amounts about 70% according to the active area of the new, not used catalyst. It would mean that the active area changes have not the predominant influence on the catalyst life.

REFERENCES:

1. P.M.Young,C.B.Clark, Chem.Eng.Progr.1973,<u>69</u>,No.5,69.
2. R.S.Collard, LSK Regeneration Experience in an American Ammonia Plant, AIChE Ammonia Safety Meeting, Boston,sept. 1975
3. G.Rasulić,S.Jovanović,Lj.Milanović, Catalyst change investigation of the system $ZnO-CuO-Cr_2O_3$ during its exploitation in an industrial reactor, by thermal analysis, Aberdeen,Scotland,spet.1981

CALORIMETRIC STUDY OF WATER VAPOUR INTERACTION
WITH BISMUTH MOLYBDATE (2:1) REDUCED AND REOXIDIZED

Luigi Stradella and Giovanni Venturello

Istituto di Chimica Generale ed Inorganica,
Facoltà di Farmacia, Università di Torino,
Via Pietro Giuria, 9 - 10125 TORINO, Italy

INTRODUCTION

Bismuth molybdate are largely used in industry either with or without inert matrix as a catalyst for partial oxidation and ammoxidation |1|, |2| of olephines: their catalytic activity and selectivity have been principally related to bulk characteristics as crystal structure, chemical composition, reducibility and oxidizability |3|.

There are three principal phases of bismuth molibdate:α (Bi_2O_3-3MoO_3);γ (Bi_2O_3-MoO_3);β (Bi_2O_3-2MoO_3), even if there is no general agreement for the β phase composition |4|, |5|. An important role in the oxidation process has been assigned to the lattice oxygen, that can diffuse in the bulk and resaturate the reduced site |6|.

We are carrying out systematic research |7|, |8|, using thermal methods (TGA, DSC, adsorption calorimetry) with the aim of clarifying the nature of the oxidation active site. This paper, in particular, concerns the role of the preadsorbed water vapour; there is still a good deal of controversy among researchers as to the role of this adsorptive on the catalyst.

Matsuura and Schuit |9| think preadsorbed water does not influence propene interaction with bismuth molybdate, while Krivanek et alii |10| have shown that water vapour inhibits active sites in olephine adsorption.

EXPERIMENTAL

The samples of $Bi_2O_3 \cdot MoO_3$, prepared according to Batist et alii |11|, were characterized before and after reduction and reoxidation treatment by X-ray analysis: in both cases the X-ray spectrum was practically coincident with that reported in literature for γ phase |12|.

The reduced sample (for which we use the symbol Red.) was obtained submitting the standard oxidized sample to a H_2 pressure of 665 Pa at 623 K for 30 minutes. The characteristics of such a sample are summarized in table 1: it may be noted that the reduction percentage is very low (<1%), that there is about one anionic vacancy every five unit cells and that the reduced sample can be completely reoxidized.

TABLE 1

Reduction temperature	Reduction time	Reduction %	Anionic vacancies n° in unit cell	Reoxid. %
623 K	30'	0.91	0.21	100

The reoxidation was obtained introducing 665 Pa of oxygen at 673 K on the evacuated sample (for such a sample we use the symbol Ox.).

The water vapour adsorption measurements were performed at 32°C using a Tian Calvet microcalorimeter associated to a vacuum line free from grease (10⁻³ Pa), following an already described static method of gas volumetry |13|. This method allows simultaneous determination of adsorbed water moles and emitted heats.

The equilibrium pressures were determined using a differential pressure transducer.

RESULTS AND DISCUSSION

In fig. 1 we have reported an example of a typical calorimetric experiment of water adsorption, respectively on the reduced sample and the oxidized one. The heat emission peak is quite different in the two cases: the peak of the reduced sample shows slower heat emission and a long tail; this fact is associated with a slow, clearly activated, irreversible adsorption, which is also apparent from the analogous slow decrease in the gas pressure on the adsorbent.

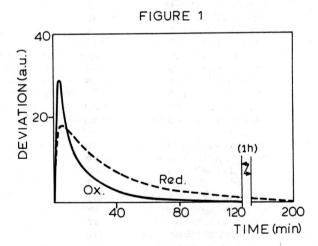

Fig. 1 - Typical calorimetric peak. Heat evolution represented by the calorimeter deviation (expressed in arbitrary units) as a function of time.

Fig. 2 and 3 show the trend of the adsorption and calorimetric isotherms respectively for the two kinds of samples. The remarkable influence of the reduction treatment for water adsorption on the γ phase is very evident. The reduced sample shows greater adsorbed amount of water vapour and greater evolved heat at each equilibrium pressure.

FIGURE 2

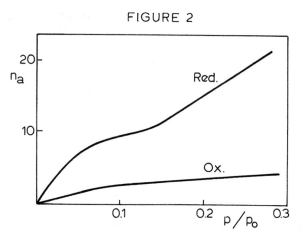

Fig. 2 - Volumetric isotherms. Number (n_a) of adsorbed water moles (μmol/g) as a function of equilibrium relative pressures (p/p_o).

More detailed information can be deduced from the analysis of the histogram of the integral molar heats of adsorption, reported in fig. 4.

FIGURE 3

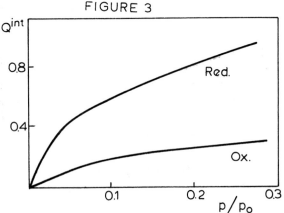

Fig. 3 - Calorimetric isotherms. Adsorption integral heats (Qint)(J/g) as a function of equilibrium relative pressures.

From such a graph we can draw these considerations:
i) there is a small fraction of chemisorbed water, characterized by very high energy of adsorption (>200 kJ/mol). This type of very strongly bounded water has been well recognized on molybdenum oxides by other authors too with independent techniques |14|, |15|. This high energy of interaction might be explained assuming a dissociative chemisorption on a surface cation, according to a mechanism well defined for other metal oxides |16|, |17|, |18|. There is, howevern some spectroscopic evidence |19| that molybdenum oxide cannot dissociate water due to the predominant covalent character of the Mo-O bond.
Nevertheless these high energy sites could be represented by Mo surface atoms: actually there is good agreement in the case of the reduced sample between the number of water moles that are very strongly adsorbed (i.e. 1,3 µmol/g) and the number of Mo atoms present on an ideal surface of the γ phase (1,5 µmol/g) |20|. The reduction treatment slightly increases the number of these active sites.
ii) A weaker interaction follows, which may be assigned to molecular adsorption (energy of the order of 60 kJ/mol), irreversible to vacuum outgassing.
iii) The third adsorption is constituted by physically adsorbed water, which is characterized by an energy of about 15 kJ/mol, is completely reversible to evacuation and is probably associated with Van der Waals interaction. For the oxidized standard sample the second and the third types of adsorbed water appear to be in competition.

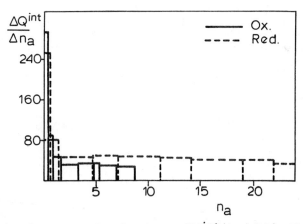

Fig. 4 - Integral molar heats ($\Delta Q^{int}/\Delta n_a$)(kJ/mol) as a function of number (n_a) of adsorbed water moles (µmol/g).

REFERENCES

1. G.C.A. Schuit, J. Less Common Metals, 1974, 36, 329.
2. G.W. Keulks, T.L. Hall, C. Daniel and K. Suzuki, J. Catal., 1974, 34, 79.
3. J. Haber and B. Grzybowska, J. Catal., 1971, 20, 19.
4. P.A. Batist, J.F.H. Bouwens and G.C.A. Schuit, J. Catal., 1972, 25, 1.
5. Tu Chen and G.S. Smith, J. Solid State Chem., 1975, 13, 288.
6. A.W. Sleight, W.T. Linn and K. Aykan, Chem. Tech., 1978, 8, 235.
7. L. Stradella and V. Bolis, Atti XIV Congr. Naz. Chimica Inorganica, Torino, 14/9/81, p. 349.
8. L. Stradella and G.F. Vogliolo, Atti III Convegno AICAT, Genova, 14/12/81, p. 59.
9. I. Matsuura and G.C.A. Schuit, J. Catal., 1971, 20, 40.
10. M. Krivanek, P. Jiru and J. Strnad, J. Catal., 1971, 23, 259.
11. P.H. Batist, J.F.H. Bouwens and G.C.A. Schuit, J. Catal., 1972, 25, 1.
12. A.C.A.M. Bleijenberg, B.C. Lippens and G.C.A. Schuit, J. Catal. 1965, 4, 581.
13. B. Fubini, G. Della Gatta and G. Venturello, J. Coll. Interface Sci., 1978, 64, 470.
14. N. Sotani, S. Masuda, S. Kishimoto and H. Hasegawa, J. Colloid Interface Sci., 1979, 70, 595.
15. J. Matsuura and G.C.A. Schuit, J. Catal., 1971, 20, 19.
16. H. Knozinger, Adv. Catalysis, 1976, 25, 184.
17. M.P. Boehm, Adv. Catalysis, 1966, 16, 179.
18. A. Zecchina, S. Coluccia, L. Cerruti and E. Borello, J. Phys. Chem., 1971, 75, 2783.
19. P. Vergnon, D. Bianchi, R. Benali Chaudri and G. Coudurier, J. Chim. Phys., 1980, 77, 1043.
20. R.W.G. Wickoff, "Crystal Structures", Interscience Publ., New York, (1968), p. 368.

THE BEHAVIOR OF IRON OXIDES IN REDUCING ATMOSPHERES
S. Soled, M. Richard, R. Fiato and B. DeRites
Corporate Research Science Laboratories
Exxon Research and Engineering Co.
P.O. Box 45
Linden, NJ 07036

INTRODUCTION

Bulk iron oxide (α-Fe_2O_3) is a common catalyst precursor for the Fischer-Tropsch synthesis of hydrocarbons from CO and H_2 [1]. In the atmosphere of the hydrocarbon synthesis reaction, mixtures of iron oxides and carbides form, and the actual nature of the catalytic species is still debated [2]. Addition of promoters such as K or Al enhances the activity or improves the physical integrity of the catalyst [3]. The present TG/DTA study investigates the influence of different reducing environments such as H_2, CO, or H_2-CO mixtures on iron-based catalysts. We have studied α-Fe_2O_3, with and without K or Tl promoters. The largest differences among the three catalysts occur in CO-containing atmospheres, where the onset of carbide formation and the rate of carbon growth varies. Relative to the unpromoted α-Fe_2O_3, the K promoter enhances whereas the Tl promoter retards the growth of surface carbon on the catalyst. The growth of carbon on the catalyst affects the amount of methane produced in the reaction and can create problems of reactor plugging as well. Consequently, it is a significant parameter to follow in describing the catalyst systems.

EXPERIMENTAL

A Mettler TA2000C recorded simultaneous TG/DTA patterns in H_2 (Linden, extra dry, 99.95%), CO (Matheson, 99.5%) and a 1:1 H_2/CO mix (Matheson, 99.5%). A molecular sieve drier and deoxo purifier scrubbed the gases prior to use. Seventy-five cc of each gas (at atmospheric pressure) were passed over 100 mg samples of catalyst. A temperature program of $8°/min^{-1}$ to 500°C was adopted as a standard heating condition. Powder x-ray diffraction spectra, taken before and after the runs, identified the phases present. The promoted systems contained 5 gm atoms of K (as K_2CO_3) or 3 gm atoms of Tl (as $TlNO_3$) per 100 gm atoms of Fe (as α-Fe_2O_3). The Fischer Tropsch reactions, run under commercial synthesis conditions (270-300°C, 2.0 MPa) in a tubular reactor with 5 cc of catalyst, provided a comparison with the thermogravimetric data.

RESULTS

H_2 Treatment. The reduction of α-Fe_2O_3 in hydrogen proceeds via two stages as shown in Fig. 1: at ~290°C the Fe_2O_3 begins to reduce to Fe_3O_4 (magnetite); and then at ~350°C, the Fe_3O_4 begins to reduce to α-Fe. A slight inflection in the TG curve indicates the formation of Fe_3O_4. The reduction of Fe_2O_3 to Fe_3O_4 is mildly exothermic whereas the reduction Fe_3O_4 to Fe is strongly endothermic. The addition of either the K_2CO_3 or $TlNO_3$ promoter has no visible effect on the temperature at which reduction begins or on the rate of iron reduction in H_2. With K_2CO_3, the thermogram differs from pure α-Fe_2O_3 only in the appearance of an initial low temperature (~100-150°C) loss of the water of hydration. The thermogram of the $TlNO_3$-promoted Fe_2O_3 is more complex: at ~145°C, a crystallographic ($\beta \rightarrow \alpha$) transition occurs (endothermic); at ~205°C, the $Tl(NO_3)$ melts (endothermic); at ~280°C the $Tl(NO_3)$ reduces to Tl metal (exothermic); and, at ~305°C, the thallium metal melts (endothermic). Consequently, the fully reduced material consists of Fe^0 with a Tl^0 promoter. On cycling the temperature, Tl recrystallizes and remelts: no bulk Tl-Fe alloy has formed.

CO Treatment. Again, the α-Fe_2O_3 (with or without promoters) reduces in a two-step sequence as shown in Figure 2. Qualitatively, the iron oxide reduces faster in CO than in H_2. To demonstrate this point further, we compared the isothermal reduction at 270°C in H_2 versus CO for each of the three samples. In a H_2 atmosphere, no measurable reduction occurred: in CO, all the samples reduced. Before the reduction proceeds to iron, the sample gains weight rapidly and continuously, as first, iron carbide (principally Fe_5C_2) forms (exotherm) and then an overlayer of carbon grows. With the addition of a K promoter, the carbon growth begins at a lower temperature (i.e. after a lesser degree of reduction), even though the rates (at 500°C) are similar. On the other hand, Tl slows the rate of carbon growth.

H_2/CO. Having established that CO is a stronger reducing agent than H_2, we examined the reduction of the three samples in a 1:1 H_2/CO blend. These results, shown in Figure 3, indicate that this mix behaves similarly to a diluted stream of CO. In the unpromoted Fe_2O_3 and the K-promoted α-Fe_2O_3, carbide forms at a higher temperature than in pure CO. In addition carbon grows at a slower rate in the H_2/CO blend compared to the same samples in pure CO. With the Tl promoter, although carbide still forms, the growth of carbon is suppressed dramatically.

H_2 Followed by H_2/CO. Since iron oxides are often reduced in H_2 prior to exposure to synthesis gas, our final treatment involved a prereduction in H_2 to 500°C, cooling to room temperature, and a treatment in a 1:1 H_2/CO mix to 500°C (Figure 4). As described previously, in the initial H_2 reduction iron oxides are reduced to metallic iron. In the K-promoted sample, K_2CO_3 remains as an anhydrous surface phase. In the Tl-promoted sample, Tl metal (as a liquid above 305°C) forms on the surface. The iron phases form carbides in the H_2/CO blend at ~300°C (via an exothermic Tl reaction): 150 to 200°C lower than without the prereduction step. A carbon overlayer also forms. With K-promotion, the onset occurs earlier than without K, whereas in the Tl-promoted system, only a minimal amount of carbon forms.

DISCUSSION OF RESULTS

The reduction of iron oxide proceeds in two steps, with an Fe_3O_4 intermediate. In the promoted systems, K_2CO_3 forms an anhydrous phase whereas the $Tl(NO_3)$ reduces to Tl metal. Carbon monoxide is a stronger reducing agent than H_2. In the presence of H_2, Fe_2O_3 reduces to Fe metal: in CO, the final product is iron carbide. Prereduction of the materials facilitates the formation of iron carbide and a carbon overlayer. The K promotes the formation of carbon on the catalysts whereas Tl suppresses it. High temperature pretreatment of K-promoted Fe_2O_3 catalysts can lead to the growth of excessive surface carbon on the catalyst in the subsequent Fischer Tropsch reaction in CO/H_2 blends. This can lead to reactor plugging in extreme cases. In addition, the presence of carbon on the surface of the catalyst is often associated with the formation of undesirable amounts of CH_4 in the product gas. The novel Tl-promoted systems because of the limited amount of surface carbon growth, do not exhibit this problem.

REFERENCES

1. M. E Dry: The Fischer-Tropsch Synthesis. In: Catalysis-Science and Technology. (J. R. Anderson; M. Boudart, ed.) New York: Springer-Verlag 1981, pp. 159-255.
2. M. A. Vannice, Catal. Rev. Sci. Engr. 14, 153 (1976).
3. H. P. Bonzel and H. J. Krebs, Surf. Sci. 109, 527 (1981).

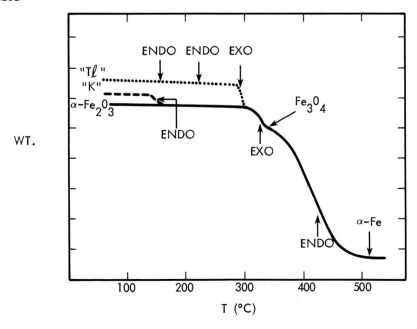

Fig. 1. Reduction in H_2.

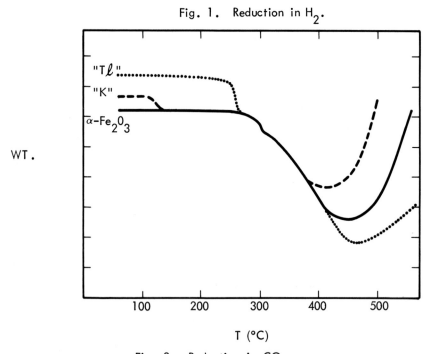

Fig. 2. Reduction in CO.

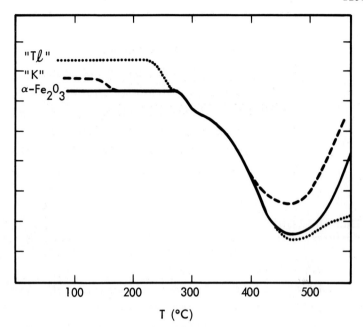

Fig. 3. Reduction in 1:1 CO/H_2.

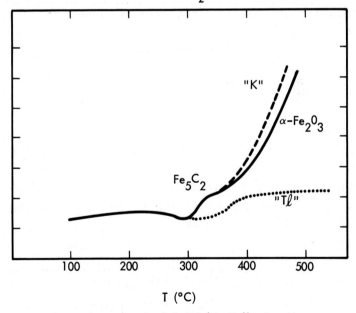

Fig. 4. Treatment in 1:1 CO/H_2 Following H_2 Prereduction.

THE EVALUATION OF FOSSIL FUELS
BY THE DTA AND PY-GC METHODS
V. Dobal, P. Šebesta and V. Káš
Institute of Geology and Geotechnics,
Czechoslovak Academy of Sciences,
V Holešovičkách 41, CS-182 09 Prague 8
Czechoslovakia

INTRODUCTION

Coal is a multicomponent heterogenous system (MCHS). A general feature of these systems is their multivariety with regard to both morphology and behavior. In characterizing these systems we are always confronted with considerable difficulties. For classification purposes an exhaustive analysis would at best be overflowed with information, and for the system behavior description has proved to be insufficient. So far we have not been able to decide whether the properties of such systems are given by properties of the individual components, or the system as a whole forms new properties which cannot be derived from the behavior of its components.

More significant for the characterization of MCHS appear to be methods that work with the sample as with the whole. Furthermore, the most reliable procedures in evaluating the behavior of these systems in technological processes are still those that imitate the process itself. TA methods undoubtedly satisfy the first criterion, and in many cases even the second one.

MCHS have one characteristic peculiarity, which is their history. In their structures they have coded paths along which they were getting to the form we observe. Without knowledge of this history an unabiguous interpretation of the behavior of these systems does not seem to be possible in our deterministic description. On the other hand, decoding the past of the material may provide a considerable amount of new, often unexpected, information. A nice example of this is excellent paper of H.G. Wiedemann [1], who by means of DTA evaluates the dating of production, the material used, and even the technologies of ancient paper.

The aim of our paper is to show the possibilities of using DTA in oxidative atmosphere in combination with Py-GC in the evaluation of some structures of coal related to their past.

EXPERIMENTAL

The results given in this paper have been obtained from two ranks of coals whose basic parameters are in Table 1. They come from a set of metamorphic low volatile bituminous coal and from two samples of brown coal. Each set of samples was obtained from the same mine and from the same seam. The samples in sets differ mainly in the degree of oxidation. Sample 1 is the least metamorphic coal, while samples 2 and 3 belong to a group of highly metamorphic coals, so called crop coal. The second set of samples represents brown coal where sample 4 reflects properties of the original coal and sample 5 represents weathered coal after discovering the seam. Thus each sample underwent a different historical development.

The DTA experiment were carried on using a "Derivatograph". A 30 - 40 mg portion of the sample was spread in a uniform thin layer across the platinum cascade crucible. In all experiments the heating rate was 3.5 deg. min^{-1} from room temperature to $600°C$. Air was used as the oxidation atmosphere in dynamic conditions. In preliminary experiments the determined flow rate of 200 ml. min^{-1} was sufficient for an excess of oxygen during the entire experiment. That is in good agreement with the

TABLE 1   ANALYTICAL DATA OF SAMPLES CHARACTERIZED

Coal No.	Proxim. analysis (wt.%)				Ultim.anal. (wt. % daf)				
	Water	VM	FC	Ash	C	H	N	S	$O_{dif.}$
1	2.9	20.5	74.6	2.0	84.9	4.1	1.2	0.3	9.4
2	15.3	24.9	48.6	11.2	75.8	2.9	1.9	0.5	18.9
3	14.0	26.0	41.0	19.0	71.1	2.8	4.2	0.5	21.5
4	16.7	38.7	37.6	7.0	74.4	5.7	1.3	0.7	18.0
5	17.0	37.7	42.5	2.8	72.5	5.1	1.0	0.3	21.1

value given by Stott and Baker [2].

Pyrolysis gas chromatography was performed in the apparatus described by Romováček and Kubát [3], consisting of a quartz reactor with a molten tin bath heated to 575°C. The gas chromatographic conditions used in this work have been described previously [4].

RESULTS AND DISCUSSION

The oxidative degradation of the samples under discussion occurs in five regions as is evident from the DTA curves in Fig. 1 or from the peak onset temperatures in Table 2. The first two regions are typical of brown coals, while the fifth one is typical of the oxidative decomposition of bituminous coals. The third and the fourth regions represent the degree of metamorphosis. This is in agreement with the generally known experience that with increasing rank the main oxidative maximum (the second) shifts to the right and the first disappears [2]. The DTA curves of the anomalous samples 2 and 3 seem to show that inversion of bituminous coal towards brown coal came about due to oxidation. This is further confirmed relatively high yield of alkali soluble substances (15 % for coal 2 and 25 % for coal 3, both on daf basis). The absence of the first maximum might be explained be the observed fact that as a consequence of previous oxidation this maximum decrea-

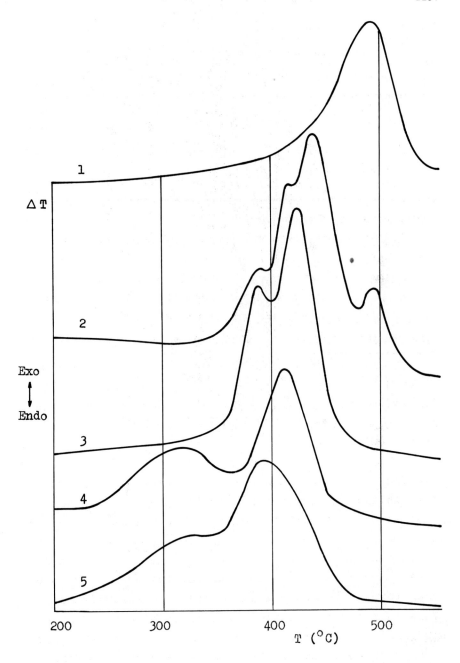

Figure 1 DTA curves of bituminous coal (1), crop coals (2,3), brown coal (4) and weathered coal (5).

TABLE 2 PEAK ONSET TEMPERATURES (°C) AND WEIGHT LOSS

Coal No.	Region in DTA spectrum				
	I	II	III	IV	V
1	-	-	-	-	418 (76)[+]
2	-	350 (17)	400 (20)	435 (40)	475 (14)
3	-	365 (36)	425 (38)	-	-
4	235 (36)	380 (57)	-	-	-
5	220 (24)	355 (71)	-	-	-

[+]weight loss in parentheses, (%)

ses even in brown coal samples, see Table 2.

Nevertheless, it is this shift that the Py-GC results presented in Table 3 exclude. With increasing oxidation the sample pyrograms decrease very markedly, which is essentially in good qualitative agreement with the observed influence of oxidation upon the pyrograms. Total decreasing of all peaks in the spectrum follows, decrease of peaks corresponding to aliphatic strustures being much faster than decrease of aromatics [3]. On the other hand, brown coals, even highly weathered ones, always yield much richer pyrograms than sample 1. The pyrograms of samples 2 and 3 very strongly recall those of graphites with small admixtures of bituminous or humic substances. Devolatilization of sample 2 and 3 has shown that about 50 weight % of volatile matter is formed by carbon oxides with a maximum speed of development between 600 and 700°C. S.Matsumoto et al. [5] ascribe this behavior to oxidized graphite structures.

DTA method in an oxidative atmosphere together with Py-GC allows to study of the material structures with respect to their past. Some very different structures on the other hand may give the similar results and for these reasons our interpretation have to be careful.

TABLE 3 RESULTS OF PY-GC FOR CROP COALS

Retention time (min)	Coal 1	Coal 2	Coal 3	Compounds
	(Integration units per 1 mg of dry sample)			
0.96	40476	3129	1138	CH_4, C_2-hydrocarbons
1.38	23456	1373	484	C_3-hydrocarbons
2.38	7683	357	207	C_4-hydrocarbons
5.51	190	-	-	n-pentane
10.10	3524	1397	633	benzene
12.09	120	-	-	n-heptane
14.78	3840	261	178	toluene
19.46	1813	59	81	m- and p-xylene
20.42	1132	-	-	o-xylene
23.54	1967	-	-	phenol, pseudocumene
32.89	1512	211	99	naphtalene
37.43	654	-	-	methylnaphtalene

REFERENCES

1. H.G. Wiedemann, "The State-of-the-Art of Thermal Analysis", NBS Special Publication 580, Washington (1980), p. 201.
2. J.B. Stott, O.J. Baker, Fuel 1958, 32, 415.
3. J. Romováček, J. Kubát, Anal. Chem. 1968, 40, 1119.
4. P. Šebesta, V. Dobal, J. Kubát, J. Anal. App. Pyr. 1981/1982, 3, 263.
5. S. Matsumoto, H. Kanda, Y. Sato, N. Setaka, Carbon 1977, 15, 299.

A THERMOGRAVIMETRIC METHOD FOR THE RAPID
PROXIMATE AND CALORIFIC ANALYSIS OF COALS AND
COAL PRODUCTS

C. M. Earnest and R. L. Fyans
Perkin-Elmer Corporation
Norwalk, CT 06856
USA

INTRODUCTION

A method for the rapid determination of moisture, total volatiles, fixed carbon, and ash content of coals was presented by Fyans [1] at the 28th Annual Pittsburgh Conference on Analytical Chemistry and Applied Spectroscopy in 1977. The method employed the Perkin-Elmer TGS-2 Thermogravimetric Analyzer and exploited the ability of the low mass furnace employed by this system to achieve both rapid heating rates and short cool down time. The method itself involved the rapid programming of the low mass furnace to $110°C$ and holding isothermally for 5 minutes while the moisture was lost from the coal sample. The furnace was then heated at $80°C/min$ to $950°C$ and held for approximately 7 minutes under flowing nitrogen until all volatile matter was expelled from the coal. The purge gas was then switched to either air or oxygen and the fixed carbon content of the char was oxidized leaving the ash content as the residue. Due to both the rapid analytical procedure and the short cool down time of the furnace, he was able to achieve two complete analyses per hour. Despite this achievement, no microcomputer programmer was yet available and the analysis, though rapid and precise, required total operator attention with the heating rates and upper temperature limit being manually changed along with the manual purge gas switching.

PROXIMATE ANALYSIS BY MICROCOMPUTER CONTROLLED TG

With the introduction of the System 4 Microcomputer Controller (shown in Figure 1 along with the TGS-2 Thermogravimetric Analyzer and Perkin-Elmer 056 Strip Chart Recorder) and Gas Selector Accessory in 1978, such multi-step programs as that employed in the proximate analysis of fossil fuels may now be performed both automatically and unattended. The multi-step analytical method is easily developed through an automatic set-up routine which guides the operator through the entry sequence on the System 4 keyboard. Furthermore, the method may be stored along with other routine analytical procedures and recalled when needed. The microcomputer controller keyboard contains a "Valve Time" key which allows the operator to enter the time (in minutes) at which purge gas switching is to occur when using the Perkin-Elmer Gas Selector Accessory. The Gas Selector Accessory is employed using a two arm furnace tube which allows the active oxidizing gas to enter near the top of the furnace rather than flowing through the balance mechanism. Figure 2

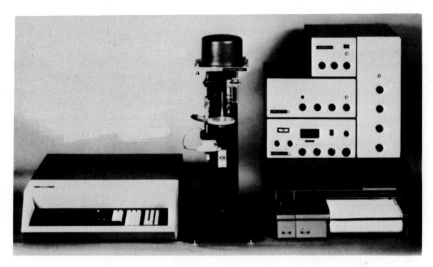

Fig. 1. Perkin-Elmer TGS-2 Thermogravimetric Analysis System with System 4 Microcomputer Controller and Model 056 Strip Chart Recorder.

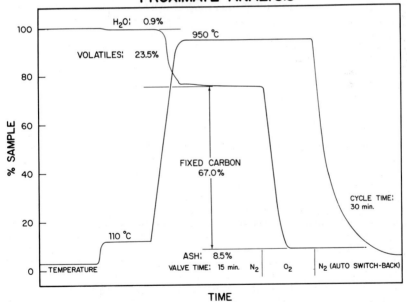

Fig.2. A Typical Proximate Analysis of Coal by Microcomputer Controlled Thermogravimetry.

shows the proximate analysis of coal using the Perkin-Elmer microcomputer controlled thermogravimetric analysis system. The ascending temperature profile has been added to demonstrate the System 4 Microcomputer program. One will notice that once the program has been executed, the purge gas automatically switches back to nitrogen and the system automaticaly cools back down to "load temperature." Furthermore, the total time elapsed for both the execution of the proximate analysis program and cooling back to load temperature in this case is only 30 minutes.

REDUCTION OF ANALYSIS TIME BY OPTIMUM SELECTION OF HEATING PROGRAM

It has been found in our studies that the optimum heating program will vary with the rank of coal under study. For example, in the lower rank specimens such as the American lignites and the Australian brown coals where larger amounts of water are encountered, a longer time of isothermal hold is required at $110°C$ than in coals of higher rank. On the other hand, antracitic coals require a longer isothermal period at $950°C$ for complete evolution of total volatiles than do coals of lower rank. These observed variations in the heat-hold program are all directly related to differences in coal structure and degree of coalification.

In the case where coals of the same rank, and often from the same source, are analyzed on a day to day basis such as in coal burning power plants, the System 4 analytical method may be minimized with respect to time for maximum sample throughput. A good starting point with any proximate analysis is given by the program listed below.

SYSTEM 4 KEY Parameter		Function
"P_1TEMP"	$50°C$	Initial Isotherm
"P_1TIME"	0 Min	Hold Time
"P_1RATE"	$60°C$/Min	Heating Rate
"P_2TEMP"	$110°C$	Isothermal Temp.
"P_2TIME"	5 Min	Hold Time
"P_2RATE"	$100°C$/Min	Heating Rate
"P_3TEMP"	$950°C$	Isothermal Temp.
"P_3TIME"	10-15 Min	Hold Time*
"VALVE TIME"	22 Min	Elapsed Time before gas switching

*Depends on whether purge gas is switched to air or oxygen.

After obtaining the TG thermogram using the above program, one may shorten the time of analysis by determining the point (in time) at which all volatile matter has been expelled from the coal.

For example, Figure 3 describes the proximate analysis of an Australian Bituminous coal from the Greta Seam in Newcastle, New South Wales. On the initial run using the program given above, the recorder pen leveled after 15.0 minutes (total analysis time) and no further loss of volatiles was observed after holding isothermally at 950°C for an additional 7.0 minutes. In this case, the "valve time" may be reduced to 16 minutes representing a time savings of 6 minutes per run. The "valve time" was not reduced the full 7.0 minutes in order to allow for variation in sample size, etc.

Additional time may be cut from the analytical procedure by using oxygen as the active gas instead of air. As can be seen in Figure 3, the time required for ashing was 7.5 minutes using air. The same flow rate of pure oxygen (Figure 4) requires approximately 2.0 minutes to ash the coal sample. Thus, an additional reduction of 5.5 minutes may accrue from the use of oxygen rather than air. It should also be mentioned that the time of analysis may be further reduced by increasing both heating rates (P_1Rate and P_2Rate) given in the above System 4 program. The maximum heating rate of the TGS-2 microfurnace is 200°C/min when using the System 4 Microcomputer Controller.

After the operator establishes both the minimum "valve time" and time of ashing with the active purge gas of choice, the "P_3 Time" parameter is established as the minimum time of isothermal hold at 950°C. This optimized program may now be stored into the memory of the System 4 Microprocessor Controller.

RELIABILITY OF THE METHOD

Although there have been many convincing studies, recently Elder [2] reported the results of the proximate analyses of ten bituminous coals, four biomass species, and the thermogravimetry of several Devonian shales using the TGS-2 Thermogravimetric Analyzer, System 4 Microcomputer Controller and Gas Selector Accessory. In his work, a study involving twenty-five repetitive analyses of a medium volatile bituminous coal standard (Alpha Resources #216) was included. This study shows that when compared to results obtained from standard ASTM methods [3], the microcomputer controlled TGS-2 results were within intra-laboratory ASTM specifications for the determination of moisture, total volatiles, fixed carbon, and ash content. Furthermore, when the results were compared on a moisture free (dry weight) basis, agreement within ASTM inter-laboratory specifications were obtained.

INDIRECT DETERMINATION OF THE CALORIFIC VALUE OF FOSSIL FUELS FROM THERMOGRAVIMETRY

Whereas the rapid determination of moisture, total volatiles, fixed carbon, and ash content in coals and coal products by thermogravimetry has been widely practiced in recent years, the calorific value has been directly determined by either the Oxygen Bomb Calorimeter (ASTM D3173) or Differential Scanning Calorimetry [1] and indirectly from elemental analysis data [4]. Since the heat of combustion of fossil fuels is primarily the result of the oxidation of

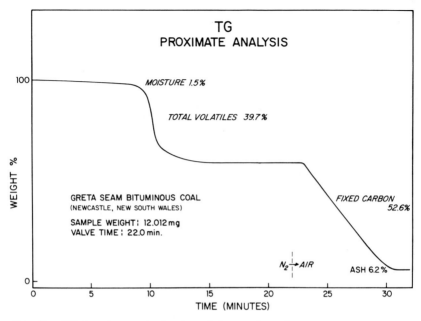

Fig.3. TG Proximate Analysis of Greta Seam Bituminous Coal.

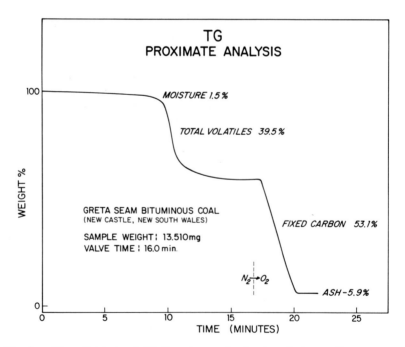

Fig.4. Time Optimized TG Proximate Analysis of Greta Seam Bituminous Coal.

the volative matter and fixed carbon, an equation relating these two variables to the calorific value should be possible.

Realizing both the value and convenience that such a correlation would lend to those involved in coal analyses, the task of developing such a relationship was undertaken in this laboratory. During our period of empirical experimentation, a literature search revealed such a relationship which was published by Goutal [5] some thirteen years prior to the origin of the thermobalance [6]. Although Goutal's Equation was published to apply to all ranks of coal, our experimentation has found that the relationship gives excellent results for bituminous coals but must be modified in order to apply to other ranks

THE GOUTAL EQUATION

As a result of the experimentation with 600 coal specimens of various ranks, Goutal [6] arrived at the empirical relationship

$$P\ (cal/g) = 82C + aV \qquad \text{(Equation 1)}$$

where P = Calorific value in calories per gram
C = % fixed carbon (moisture free basis)
V = % volatile matter (moisture free basis)
a = A coefficient which varies with the percentage volatile matter, V, in the coal.

Thus, the equation states that a constant correlation of 82 calories per percent of fixed carbon and its contribution to the calorific value of the coal exists. It is interesting to note that this value is within 4 calories per percent of that for the standard heat of combustion of graphite. The contribution of the volatile matter to the total heat of combustion will vary with the amount of volatiles and it is here that the greatest empiricism lies.
Table 1 lists some of the values for the Goutal coefficient "a" versus the percentage volatile matter expressed as a dry-ash free basis (V'). These values should be used by the reader to construct a large graph, such as shown in Figure 5, to assist in selecting the proper value of "a".

TABLE I

SOME VALUES OF THE GOUTAL COEFFICIENT "a" FOR SELECTED VALUES OF THE VOLATILE CONTENT (V')

V'(%)	a(Calories/%)
5	145
10	130
15	117
20	109
25	103
30	98
35	94
38	85
40	80

NOTE: In anthracitic coals, a value of a = 82 is employed and is considered constant. For coke samples, the value of a = 0 is recommended.

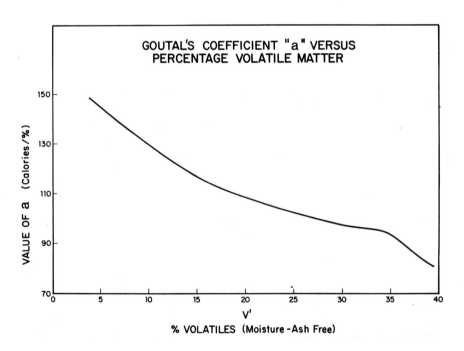

Fig.5. Goutal's Coefficient "a" versus Percentage Volatiles (DAF).

Obviously, the method works best for coals having a DAF volatile content between 5 and 35% as can be deduced from the plot in Figure 5. After experimenting with over 100 bituminous coals, we have found that the calculated calorific value will seldom vary more than 2% from that obtained by the bomb calorimeter. The reader should satisfy himself by applying the equation to any of the results for bituminous coals published by the Penn State Data Base or those appearing on commercial standards. Obviously, the prerequisite for this indirect determination of the calorific value is an accurate proximate analysis.

MODIFIED EQUATION FOR ANTHRACITIC COALS AND COKE

The correlation chart for the assignment of the coefficient "a" should not be used for anthracitic coals, coke, sub-bituminous coals, or lignites. Anthracitic coals and coke have low volatile content and high fixed carbon plays the predominant role in determining the calorific values. Since anthracitic coal seams are few in number when compared to those of other ranks, a representative correlation should be easily made. It was found in our studies that the best overall performance of the Goutal Equation is obtained through the use of a constant value for "a" of 82. Thus, for anthracitic coals

$$P \text{ (cal/g)} = 82C + 82V$$

which reduces to

$$P \text{ (cal/g)} = 82 \text{ (V + C)} \qquad \text{(Equation 2)}$$

Table III lists some of the results of the Goutal Equation. As can be seen from the values in Table II, the calculated values for these six anthracitic specimens agree within 1.0% of that obtained from the ASTM Bomb Calorimeter Method. No general statement of accuracy will be made however until more anthracitic samples are available to us.

Coke is a very common coal product produced by the destructive distillation of organic substances in the absence of air from coals of certain rank. Its major uses are as a reducing agent in the iron and steel industry and as a fuel. The "carbonization" process of coke production represents the second major use of coal. Since high quality coking coals have been found to be those with a DAF volatile range of 20-32% [7], the proximate analysis procedure by thermogravimetry is commonplace in this industry. In the carbonization process both the "reactive" and "inert" maceral components undergo thermal alteration although to a different extent. This thermal alteration accompanied by the evolution of tar and gas would suggest that the coefficient "a" in Goutal's Equation should be less meaningful when applied to this carbonized coal product. With this in mind, and through our empirical experimentation, we recommend the value of a = 0 for the use of Goutal's equation for the indirect determination of the calorific values of coke samples from thermogravimetric data. Thus, for coke samples, the calorific value is calculated by

   $P \text{ (cal/g)} = 82C$ (Equation 3)

or

   $P(\text{BTU lb}) = 147.6C$ (Equation 4)

and is determined from the fixed carbon value only. Table II lists the results of this indirect determination for 5 commercially available coke standards. As can be seen in the Table, the calculated values show a relative agreement ranging from 0.06% for the AR-114 Standard to 4.30% for the AR-772 Standard. The average relative agreement for these five coke standards is 1.82%.

References

1. Fyans, R.L., "Rapid Characterization of Coal by Thermogravimetric and Scanning Calorimetric Analysis," Thermal Analysis Application Study #21, Perkin-Elmer Corp., Norwalk, CT (1977).

2. Elder, J., "Thermogravimetry of Bituminous Coal and Oil Shales," Proc. of the 10th NATAS Conference, October, 1980, Boston, MA.

3. ASTM Standards, Gaseous Fuels, Coal, and Coke, Part 26, Methods D3172-75, 1973.

4. Culmo, R.F., "CHN Analysis of Coals with the Perkin-Elmer Model 240 Elemental Analyzer," Elemental Analysis Applications Study #2, Perkin-Elmer Corp., Norwalk, CT (1977).

5. Goutal, M., "Sur le Pourvoir Calorifique de la Houille," Compt. Rend., 135, 477 (1902).

6. Honda, K., Sci. Rep. Tohoku Univ., 4, 97 (1915).

7. Gibson, J., "Carbonization and Coking," in Coal and Modern Coal Processing: An Introduction. G.J. Pitt and G.R. Millward, Editors, Academic Press, New York (1979).

TABLE II

COMPARISON OF CALORIFIC VALUES OBTAINED FROM MODIFIED
GOUTAL CALCULATION AND ASTM (BOMB CALORIMETER)
METHOD FOR ANTHRACITIC COALS

Coal	% Volatiles (Moisture Free)	% Fixed Carbon (Moisture Free)	Calorific Value (BTU/lb) ASTM	Calorific Value (BTU/lb) Modified Goutal
PSOC-80	5.77	80.29	12,603	12,702
PSOC-81	5.65	86.52	13,741	13,604
PSOC-82	5.76	85.07	13,349	13,406
PSOC-177	4.72	90.94	14,149	14,119
PSOC-178	5.22	85.40	13,448	13,376
PSOC-179	4.78	77.57	12,097	12,155

TABLE III

RESULTS OF MODIFIED GOUTAL CALCULATIONS FOR COKE
SAMPLES BASED ON FIXED CARBON CONTENT

Coke Specimen	% Fixed Carbon	Calorific Value (BTU/lb) ASTM	Calorific Value (BTU/lb) Modified Goutal	% Relative Difference
AR-112	89.58	13,601	13,223	2.78
AR-113	88.38	13,000	13,045	0.35
AR-114	88.29	13,024	13,032	0.06
AR-771	90.36	13,552	13,337	1.59
AR-772	90.01	12,737	13,285	4.30

THERMO-MAGNETO-GRAVIMETRIC ANALYSIS OF PYRITE IN
COAL AND LIGNITE

D. Aylmer and M. W. Rowe
Department of Chemistry
Texas A&M University
College Station, TX 77843

INTRODUCTION

Industrial analysis of coal is ordinarily divided into three parts: (1) Proximate analysis, which is the determination of moisture, volatile matter, fixed carbon and ash contents; (2) Ultimate analysis, which is the determination of carbon, hydrogen, sulfur, nitrogen, sometimes phosphorous, ash and by difference, oxygen contents; and (3) Calorific value, which is the determination of the heat developed by combustion of a unit weight of the fuel. For many coals and lignites, the economically important calorific value can be accurately determined (with $\pm$ 1% uncertainty) by the use of proximate analysis only.

Lignite in Texas, and elsewhere, is an important long-term energy-source for utilization. However, sulfur is a well known and serious contaminant in coal and lignite. To properly utilize the lignite one must remove the sulfur. The effectiveness of sulfur removal and the choice of methods of sulfur removal are dependent upon the chemical form of the sulfur, generally present as organic sulfur and iron compounds, principally pyrite, FeS_2. Thus it is important to have good, fast and accurate methods of analysis for sulfur in lignite and coal.

Hyman and Rowe [1] recently developed a new technique for the measurement of pyrite in coal and lignite. Furthermore the proximate analysis is obtained as a preliminary step in the analysis of the pyrite content. Several laboratories have emphasized the advantage of proximate analysis determination with commercial electrobalances [1-3]. This new thermogravimetric technique has several possible advantages over the ASTM method, both for pyrite analysis and proximate analysis. For example, coals often contain the oxidation product of pyrite, $FeSO_4$, especially if they are not analyzed soon after being exposed to the atmosphere after grinding. The $FeSO_4$ is recorded as FeS_2 by our technique so that no error is introduced by oxidation of the pyrite. Thus thermo-magneto-gravimetric analysis (TMGA) gives the percent FeS_2 prior to oxidation. In some cases in which the pyrite grains are completely surrounded by acid-insoluble organic material, the ASTM method may yield an estimate of the pyrite which is lower than the actual value. Since the organic material is completely oxidized in our method, it will accurately record even those pyrite grains totally surrounded by organic material. Thus in these two possible situations, TMGA should yield superior results compared to the ASTM method.

In addition there are advantages in operation of TMGA compared to the ASTM methods of proximate analysis and pyrite determination as follows: (1) In TMGA both the proximate analysis and the pyrite content can be obtained in only about one and a half hours. (2) Unskilled technicians can be trained easily. With about 10 minutes instruction, several different graduate students at Texas A&M University were able to reproduce the pyrite content in a coal sample within the estimated precision. The ability to use unskilled technicians for routine operation makes the method cost effective. (3) The necessary equipment is fairly widespread in industrial and larger university laboratories. (4) Computer-controlled operation of proximate analysis is presently available on commercial thermobalances and could be rather easily accomplished for pyrite analysis as well. However, the currently used electrobalances suffer from two drawbacks: (1) They are moderately expensive at $\sim$ \$30,000 and (2) They are designed to handle relatively small samples of $\sim$ 50 mg.

Although our present system uses a Cahn RG-2000 electrobalance, but we intend to construct a new system using an electronic analytical balance. Microprocessor controlled operation of the heating schedule and automatic calculation of the proximate analysis and pyrite content are intended. It is anticipated that the entire system can be built for less than \$7500 and will allow determination of both pyrite content by the thermo-magneto gravimetric method [1] and the proximate analysis [1-3]. Thus the two drawbacks stated above for the present TMGA technique will be overcome. A schematic diagram of the apparatus proposed for the rapid, accurate determination of pyrite and proximate analysis of coal and lignite is shown schematically in Figure 1.

The objective of this report is to further test of TMGA for pyrite analysis. To that end a representative suite of coal and lignite samples have been obtained from Dr. P. Dolsen of the Coal Research Division of the Pennsylvania State University. Measurements of the pyrite content and the proximate analysis will be presented and compared to the previously determined ASTM values. The calorific value will be calculated from the proximate analysis for those coals which are appropriate and these will also be compared to values previously determined.

EXPERIMENTAL PROCEDURE

Proximate Analysis by Thermogravimetry - The procedure for determining the proximate analysis has been described earlier. It is perhaps best understood from an examination of a schematic representation of the thermogravimetric curve shown in Figure 2. The coal sample is prepared as specified in the ASTM Standard book (1974 Annual Book of ASTM Standards, Part 26, Gaseous Fuels, Coals and Coke, ASTM, 1916 Race Street, Philadelphia, PA 19103, D 271-70). The experiment is begun by weighing a sample on the balance (region A on Figure 2). For later use in the pyrite determination, a magnet with Faraday poles is momentarily moved into place to see whether the sample initially contains magnetic iron oxides. The system is flushed with dry nitrogen for ten minutes at a flow rate of 80 ml/minute. The heater is then set at 105°C and held for 8-10 minutes which drives off the moisture and the dry weight, W_d, is recorded. The temperature is then raised to 950°C for seven minutes and another weight reduction is noted as the volatiles are removed. The ASTM procedure prescribes a heating rate of 80°/minute to 950°C and held at 950°C for 7 minutes. The temperature is reduced to 700°C and then the system is opened to air which oxidizes the organic material, the so-called fixed carbon, resulting in another weight loss until only ash is left and the weight becomes constant. At this point the furnace is turned off and allowed to cool to 150°C. We are investigating the use of lower temperatures for proximate analysis. Then, 20% H_2 in 80% nitrogen carrier is flowed through the system for 10 minutes at 80 ml/minute in preparation for the determination of pyrite. The proximate analysis is calculated in a straightforward way from the thermogram in Figure 2 by the use of the following equations:

$$\% \ H_2O = (W_i - W_d) \times 100/W_i \quad \ldots \ldots \ldots \quad (1)$$

$$\% \ \text{Volatile matter} = (W_d - W_e) \times 100/(W_i \text{ or } W_d) \quad . \quad (2)$$

$$\% \ \text{Fixed C} = (W_e - W_a) \times 100/(W_i \text{ or } W_d) \quad \ldots \ldots \quad (3)$$

$$\% \ \text{Ash} = W_a \times 100/(W_i \text{ or } W_d) \quad \ldots \ldots \ldots \quad (4)$$

where W_i or W_d is used in equations (2) - (4) depending on whether the percentage of the component is sought in the original sample or for dry weight, respectively; normally, W_d is used. Accurate results have been obtained [1-3]. There seems to be little doubt that thermogravimetric analysis provides a viable, precise alternative to the more tedious ASTM methods for proximate analysis.

Pyrite Content by Thermo-Magneto-Gravimetry. Measurement of the pyrite in coal and lignite by our new method combines the techniques of thermogravimetry and thermogravimetry and thermomagnetometry. Oxidizing and reducing gases, as well as an inert atmosphere is utilized. To understand the method we propose for estimation of the pyrite in coal, previously described [1,4], it is necessary to realize the combined effect of temperature and a strong magnetic field on the apparent weight of a ferromagnetic material suspended from the balance. Figure 3 illustrates that the apparent weight (the saturation magnetization) slowly decreases as the temperature is increased until the apparent weight finally becomes virtually the actual weight of the iron at 770°C, the Curie point of the iron. The Curie point, also called the Curie temperature, is characteristic for the particular ferromagnetic material under question. For example, it is 390°C for metallic Ni, 595°C for Fe_3O_4 (magnetite), 770°C for metallic Fe, etc.

The method for determination of pyrite is perhaps best demonstrated by examination of Figure 4 which begins where Figure 2 ended; that is, we begin at (G) where ash only remains. Incorporated in this ash is the oxidation product of the pyrite, namely, Fe_2O_3, hematite. When we ended the proximate analysis, we stated that 20% H_2 in nitrogen carrier was flushed through the system. After at least 10 minutes at 80 ml/minute flow, the next phase of the experiment is ready to begin. A permanent magnet is inserted which results in a weight increase due to the saturation magnetization, J_s, of the Fe_2O_3 (H). The heater is then raised to a temperature of 400°C with the H_2 continuing to flow and the Fe_2O_3 begins to be reduced to metallic Fe, measured as a large increase in apparent weight (I) due to the iron becomes constant indicating that the reaction is complete (J) and the furnace is turned off. Further increase in apparent weight is noticed as the sample cools due to the increase of the saturation magnetization of the iron with decreasing temperature (Figure 3) as exhibited by region (K) in Figure 4. Once again the apparent weight, W_{fm}, will become constant as the temperature approaches room temperature (L). The magnet is then removed and the final weight, W_f, of the residue is recorded (M). For determination of the pyrite, only three regions of Figures 2 and 4 are of critical importance. They are: (1) either the initial weight, W_i, of region (A), or the dry weight, W_d, of region (D), both from Figure 2, (2) the final apparent weight with the magnet in place, W_{fm}, of region (L) from Figure 4, and (3) the final weight, W_f, of region (M) also from Figure 4. The pyrite content is then calculated using stoichiometry and the saturation magnetization of the iron, J_s = 218 emu/g, as:

$$\% FeS_2 = [(W_{fm} - W_f) \times M.Wt.FeS_2 \times 100]/[(218) \times M.Wt.Fe \times (W_i \text{ or } W_d)] \quad (5)$$

where M.Wt.Fe is the molecular weight of Fe (55.85 g/mole) and M.Wt. FeS_2 is the molecular weight of FeS_2 (119.97 g/mole).

Using the procedure described schematically in Figure 4, Hyman and Rowe [1] reported the pyrite contents of five samples of coal and lignite and compared them with those obtained by the ASTM method. Agreement between the pyrite content as measured by the thermo-magneto-gravimetric analysis (TMGA) proposed here and the ASTM values was encouraging. We will determine whether further measurements support this early optimism.

ACKNOWLEDGEMENTS

This work was supported in part by the Center for Energy and Mineral Resources of Texas A&M University. We are grateful to Dr. P. Dolsen, Coal Research Division of the Pennsylvania State University, for supplying us with samples for study. Discussion with M. Hyman was helpful.

REFERENCES

[1]. M. Hyman and M. W. Rowe (1981) A New Method for Analysis of Pyrite in Coal and Lignite, ACS Symposium Series No. 169, New Approaches in Coal Chemistry, Chp. 22, 389-400.
[2]. R. L. Fyans (1977) Rapid Characterization of Coal by Thermogravimetric and Scanning Calorimetric Analysis, Perkin-Elmer Reprint, Thermal Analysis Application Study 21, presented at the 28th Pittsburgh Conference in Cleveland, Ohio, March, 1977.
[3]. M. B. Harris and J. P. Elder (1981). DSC and TG Studies of Kentucky Coals, Proc. NATAS II, 287-293.
[4]. M. W. Rowe (1981) Determination of the Pyrite Content and the Proximate Analysis of Coal and LIgnite, Proc. NATAS II, 311-317.

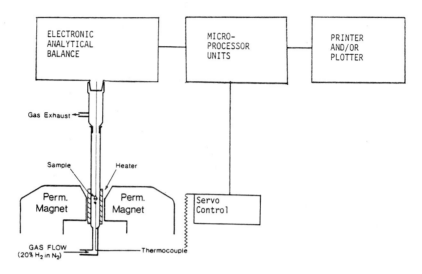

Figure 1. Schematic diagram of proposed instrument for pyrite determination and proximate analysis of coal or lignite.

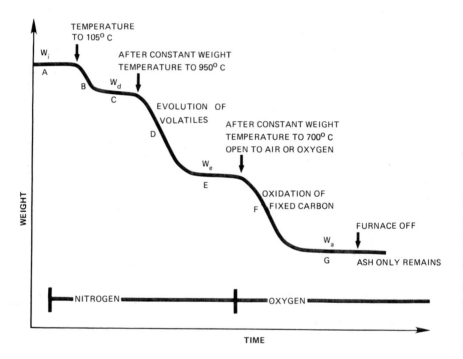

Figure 2. Schematic thermogram showing procedure for determining proximate analysis of coal or lignite.

1275

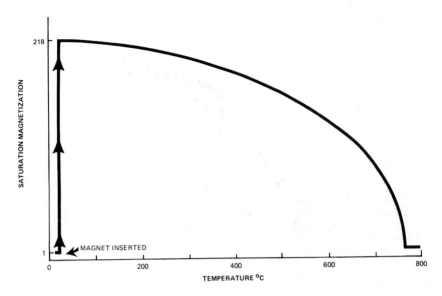

Figure 3. Schematic thermomagnetic analysis of iron. Note large increase in apparent weight with insertion of magnet due to saturation magnetization.

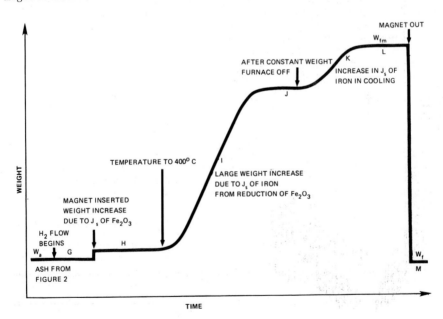

Figure 4. Schematic thermomagnetogram showing procedure for analyzing the pyrite content of coal or lignite.

THERMAL ANALYSIS OF MIXTURES OF POLYVINYLCHLORIDE WITH VARIOUS COKE CHEMICAL AND PETROL PRODUCTS

D.RUSTCHEV, T.GANTCHEVA, O.ATANASSOV

THE HIGHER INSTITUTE OF CHEMICAL TECHNOLOGY DARVENITSA, SOFIA, BULGARIA

The methods of thermal analysis are successfully employed in investigating polyvinylchloride and its mixtures with various additions. Murphy, Hill and Schaecher (1) proved that the characteristic endothermal peak for polyvinylchloride at 300°C is linked mostly with the separation of hydrochloride, while during the next exothermal plateau at 400°C, they do not discover this disintegrating product of the thermal destruction of the polymer. Kinney (2) with the aid of the curve DTA established that the temperature of vitrification of polyvinylchloride is 82°C. Nagy and his co-workers (3) skilfully make use of the thermal analysis to investigate the stabilizing effect of such additions to the polyvinylchloride as stearic acid, lead stearate, etc., in amounts from 1 to 5%.

The thermal analysis, during our investigations, was carried out with suspension polyvinylchloride, mark K 68, manufactured at the State Chemical Combine at the town of Devnya; therefrom are prepared laboratory mixtures of 0.5 kg., which contain 78 to 93% of polyvinylchloride, 5 to 20% of the investigated coke-chemical and petrol products (anthracene fraction, coal pitch, extracts of petrol distillates and bitumens) and the necessary amount of calcium stearate and stabilizers. These mixtures are made homogeneous on a mixer and are rolled at 423°C K in the course of 12 minutes.

Samples of 0.4 gr. were subjected to simultaneous differentially thermal and thermogravimetric analysis. There was used for this purpose a Hungarian apparatus, system Paulic-Paulic-Erdei, model OD-102. We worked at a speed of heating 6 and 10°C/min. inert substance dialuminium trioxide and uncontrolled air atmosphere.

In Fig.1 are given DTA- and TG curves of the initial polyvinylchloride and its mixtures with 5, 10 and 15% anthracene fraction. The pure polymer is characterized with two endo- and two exoeffects (curve a). DTA and TG are changed considerably only when the added anthracene fraction reaches up to 15% (curves d, d'). With these mixtures the primary endothermal effect, linked with the separation of hydrochloride and evaporating the addition, is registered in the temperature interval 315–330°C, which does not differ essentially from the one established by Murphy and co-authors (1). Parallel with the increase of the participation of the anthracene fraction, the maximum of the first exothermal effect is 410°C and moves off at higher temperatures — 420, 430 and 445°C, while the second exothermal effect is registered at lower temperatures of 600 to 610°C, instead of 640°C — curves b, c, d, correspondingly a. Anthracene fraction does not change the beginning of the intensive reduction of the mass, which begins at 235–240°C — curves a', b', c' and d'.

The differentially-thermal and the thermo-gravimetric curves, we obtained, are close to the curves received by Gorshkov (4) for polyvinylchloride rolled in a mixture with plasticizers and fillers. He explains the exothermal effects in the temperature interval of 400–600°C with the formation of double links, reviving the chains of dehydrochlorated polyvinylchloride and oxidation of the products of the thermal destruction. The pyrolysis of polyvinylchloride is accompanied by an endothermal effect, which is deposited with the exothermal effects. One observes, at the thermal destruction of the initial polyvinylchloride, a clearly expressed endothermal effect at 545°C, which in its mixtures with anthracene fraction settles at lower temperatures — 490–500°C (curves a of Fig.1, respectively curves b, c, d).

Coal pitch of various composition and temperature of melting was used in the next experiments as additions. On Fig.2 are given DTA- and TG curves of mixtures of polyvinylchloride with 15% of these coke-chemical products. In these mixtures is preserved the typical endothermal effect at 300–330°C and the exothermal peak at 400–430°C — curves b, c. The second endothermal effect, which, at the initial polyvinylchloride has a clearly outlined maximum at 545°C (curve a) is registered in a broader interval — 490 to 550°C, wherein is observed the going on of the above mentioned processes with a maximum at 520°C — curve b.

Experiments were made at the end with additions of various petrol products, too. The type of DTA and TG-curves does not change in mixtures of polyvinylchloride with an extract of low viscosity dstillate. A series of experiments were made at a lower speed of heating — 6°C/min., in order to study in greater detail the thermal effects in the temperature interval up to 400°C. For the initial polyvinylchloride, the first endothermal effect linked with the separation of hydrochloride, was in this case, registered at 278°C. The addition of 5, 10, 15 and 20% of extract of high viscosity petrol distillate moves the outpointed peak to higher temperatures — 275 to 295°C (Fig.3, curves b, c, d, e). These data of ours quite well coincide with the results of Gorshkov (5), who finds that the maximum of the exsamined endothermal effect in the suspension and the latex polyvinylchloride is registered in the interval from 280 to 290°C.

Interesting results were obtained with mixtures of polyvinylchloride with 5% of petrol bitumen (road and hydro) — Fig.4. DTA- and TG curves are analogous to the mixtures with the rest of the coke-chemical and the petrol products, but the intensive gas separation in this case begins at higher temperatures — 250, respectively 255°C, which shows that petrol bitumens improve the thermal stability of the polyvinylchloride compositions — curves b', c'.

CONCLUSIONS:

1. There have been carried out differentially — thermal and thermogravimetric investigations on mixtures of suspension polyvinylchloride (mark K 68) with various coke-chemical and petrol products (anthracene fraction, coal pitch, extracts of petrol fractions and bitumens).

2. It has been established that the coal pitch broadens the temperature interval (490–550°C) of the second characteristic, for the polyvinylchloride endothermal effect (545°C) and brings about the appearance of an exothermal peak at 520°C.

3. Petrol bitumens increase the beginning of the intensive gas separation from 240 to 250 and up to 255°C and improve the thermal stability of the polyvinylchloride compositions.

REFERENCES:

1. C.B.Murphy, J.A.Hill, G.P.Schaecher. Analyt. Chem. 32 1374, 1960.
2. P–Y.Kinney, J. Appl. Polymer. 11, 193–209, 1967.
3. J.Nagy, T.Gabor, E.Brandt-Petrik. Thermal Analysis — vol.2. Producing Fourth ICTA Budapest, 1974.
4. V.S.Gorshkov, Termographia stroitelnih materialov, M. 1968.
5. V.S.Gorshkov, Zh.P.H. No.2, 448–451, 1966.

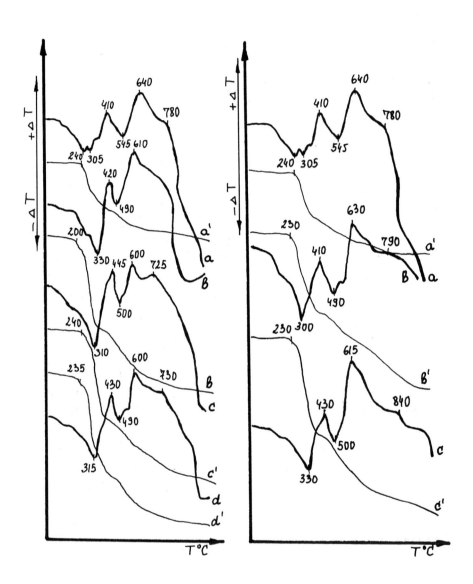

Fig.1.

Fig.2.

Fig.1 — DTA (a, b, c, d) and TG (a', b', c', d') curves of polyvinylchloride (a and a') and its mixtures with 5, 10, 15% anthracene fraction (b, c, d — b', c', d').

Fig.2 — DTA (a, b, c) and TG (a', b', c') curves of polyvinylchloride (a and a') and its mixtures with coal pitch with a melting point of: 78°C (curves b and b') and — 115°C (curves c and c').

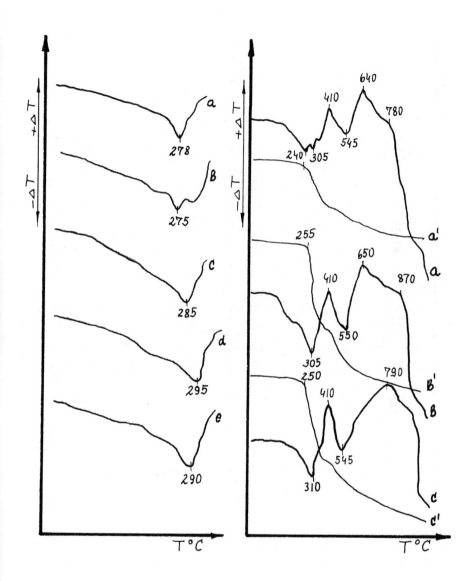

Fig.3.

Fig.4.

Fig.3 — DTA curves of polyvinylchloride (a) and its mixtures with 5, 10, 15 and 20% high viscosity petrol distillate (b, c, d, and e).

Fig.4 — DTA (a, b, c) and TG (a', b' and c') curves of polyvinylchloride and its mixtures with petrol bitumen — road (curves b and b'), and -hydro (curves c and c').

INVESTIGATION OF CALCIUM CARBONATE - SULFUR
TRIOXIDE REACTION BY THERMOGRAVIMETRY

R.K. Chan and K. Ser
Department of Chemistry
University of Western Ontario
London, Ontario, Canada   N6A 5B7

INTRODUCTION

The sulfur-bearing gases discharged by coal-fired electrical power plants and metal-refining activities aggravate the problem of air-pollution and acid rain. Many processes have been developed to control them. Most of them are based on the socio-economic and political expediency. This investigation is an attempt to investigate the reaction of calcined calcium carbonate with sulfur trioxide.

EXPERIMENTAL

Calcination. The pure precipitated calcium carbonate powder is Baker Analyzed reagent grade. Approximately 100 mg of calcium carbonate were carefully placed in the quartz bucket and weighed accurately on the Cahn electrobalance to the nearest 0.01 mg. Note that the sample has to be spread out evenly to avoid incomplete calcination before the weighing. The sample was flushed with dry nitrogen at a rate of 78 mL/min.

A Fisher Model 360 linear temperature programmer was used to control the furnace temperature. The sample was heated in the furnace to 1020 K as rapidly as possible (about 15 min.). During the course of calcination

$$CaCO_3(s) \rightarrow CaO(s) + CO_2(g)$$

the weight loss of the sample (44.0%) and the variation of temperature were measured simultaneously by the Cahn electrobalance and a 26-gauge chromel-alumel thermo-couple, and recorded by a 2-channel strip chart recorder.

Surface Area Measurements. After calcination the product (CaO) was cooled to room temperature and degassed overnight. About 15 hours were required to achieve a vacuum of the order of 10^{-4} mm Hg suitable for surface area measurements. This vacuum was achieved by a rotary mechanical pump in conjunction with a mercury diffusion pump. The surface area measurements were done by the N_2 adsorption isotherm method [1].

After degassing, the system reached a reading around 10^{-4} mm Hg in a McLeod Gauge. A liquid nitrogen cold trap was used to prevent Hg vapor from entering the concentric flow-through tube and thereby condensing on the calcined sample. About 20 min. later, a second liquid nitrogen cold trap was placed around the concentric flow-through tube. Dry nitrogen at a pressure of about 4 cm Hg was let into the system. Since the heat capacity of the nitrogen gas was much smaller than that of the sample and the quartz bucket, it should attain the liquid nitrogen temperature of 78 K in less than 5 min. while the sample and quartz bucket were still considerably above 78 K. When the nitrogen gas gradually cooled towards 78 K, the buoyancy effect increased and the sample showed an apparent weight loss. The weight reached a minimum corresponding to the maximum buoyancy effect, and then increased gradually as the sample was cooled towards 78 K and started to adsorb. It took usually 15 to 20 min. before the amount of nitrogen adsorbed became constant (Figure 1).

The actual amount of nitrogen adsorbed was AC. AB was the buoyancy correction equal to 1.87×10^{-3} mg/mm Hg of N_2 at pressure P. Therefore, the weight of sample corrected to vacuum at liquid nitrogen temperature was B. Additional doses of N_2 were admitted until the pressure was about 250 mm Hg. Each equilibrium required about 5 min.

Figure 2 is the nitrogen adsorption isotherm. The surface area is calculated according to the BET equation for multilayer adsorption [1]. The area of a nitrogen molecule is taken as 1.62×10^{-19} m^2.

Sulfur Trioxide Absorption by Thermogravimetric Analysis.

A flow of dry nitrogen at 78 mL/min. containing 3% of SO_3 is used. The corresponding partial pressure of SO_3 is 23 mm Hg. The rate of heating is 10 K/min.

RESULTS AND DISCUSSION

A more reactive sample of CaO was produced if it was evacuated at the completion of calcination while the sample was still warm. Presumably, the enhanced reactivity is due to the removal of last traces of CO_2 from the interior of sample, thus preventing its recombination to yield an inert coating of $CaCO_3$ on CaO. From the various surface area measurements obtained in this investigation it is obvious that the surface area was greatly affected by the calcination temperature, percentage of calcination, and the size of the sample. For example, prolonged heating of the calcined product at elevated temperature or increasing calcination temperature will generally lead to an appreciable reduction of surface area [2].

The TGA curve (Figure 3) can be explained qualitatively as follows. The calcined sample absorbed SO_3 rapidly at room temperature. When the temperature started to rise, the rate of absorption rose gradually until it levelled off at about 500 K. When the temperature increased further, a fast absorption took place at 600 K. As the temperature reached the isothermal zone of about 1020 K, the absorption again tapered off, this presumably due to the dropping off in rate and saturation.

The initial absorption at room temperature was due to the chemisorption of SO_3 forming a monolayer and a stable complex of $CaSO_4$ with the CaO.

$$CaO + SO_3 \longrightarrow CaSO_4.$$

The intermediate absorption region from 350 to 500 K was due to the increase in the rate of SO_3 diffusion into the micropores between crystallites (Figure 4). The plateau from 500 to 600 K represented the saturation when the micropores were filled and reached the end of this activated diffusion stage.

Further increase of temperature caused appearance of fast absorption region, which was due to the disproportionation of $CaSO_3$ and other high temperature reactions (3).

The proposed mechanism of the process is as follows.

Calcination $\qquad CaCO_3 \rightarrow CaO + CO_2$

Absorption $\qquad CaO + SO_3 \rightarrow CaSO_4$ at room temp.

At higher temp. (> 600 K), SO_3 begins to decompose into SO_2 and O_2 noticeably

$$SO_3 \rightarrow SO_2 + 1/2\ O_2$$

and $\qquad CaO + SO_2 \rightarrow CaSO_3$

$$4CaSO_3 \rightarrow 3CaSO_4 + CaS.$$

The final products are $CaSO_3$, $CaSO_4$ and CaS.

When $CaSO_3$ started to disproportionate the surfaces of $CaSO_3$ were disrupted to a very large extent, effectively breaking down the original CaO structure and exposing fresh layers of CaO, hence more SO_3 or SO_2 was absorbed [4]. Therefore, the TGA curve increased further in weight until it reached the temperature limit which was set at about 1020 K.

CONCLUSION

Calcination of $CaCO_3$ in a flow of 78 mL/min of dry N_2 was a rapid process at a temperature of about 1020 K. Lower temperature would cause incomplete calcination. The surface area measured by the N_2 adsorption isotherm method before the absorption of SO_3 at room temperature was 17 m^2/gm. After absorbing about 20% of its weight of SO_3 the surface area decreased to 5 m^2/gm. For the final product with 120% of its weight of SO_3, the surface area was reduced further to about 2 m^2/gm. The TGA curve was explained qualitatively by the key reactions with the aid of calculated thermodynamic data.

REFERENCES

1. S. Brunauer, P.H. Emmett and E. Teller, J. Amer. Chem. Soc., 1938, 60, 309.
2. R.S. Boynton, "Chemistry and Technology of Lime and Limestone", Interscience, (1966), p. 147.
3. F. Foerster and K. Kubel, Z. Anorg. Allg. Chem., 1924, 139, 261.
4. R.K. Chan, K.S. Murthi and D. Harrison, Can. J. Chem., 1970, 48, 2979.

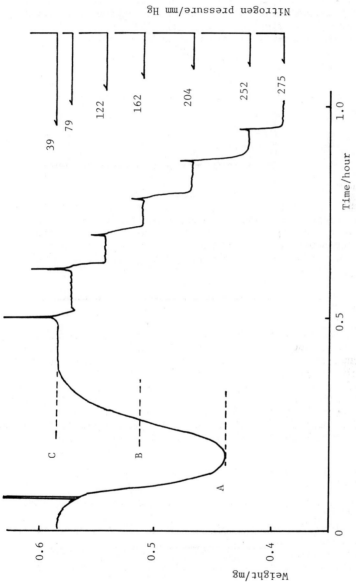

Figure 1. Apparent weight of sample at various pressures of nitrogen

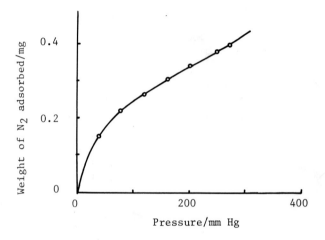

Figure 2. Nitrogen adsorption isotherm at 78 K

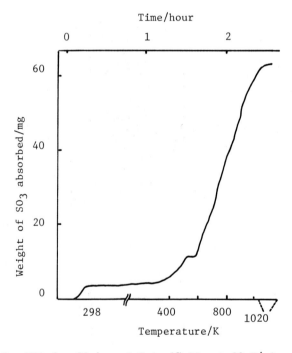

Figure 3. TGA for 55.4 mg CaO in 3% SO₃ at 10 K/min

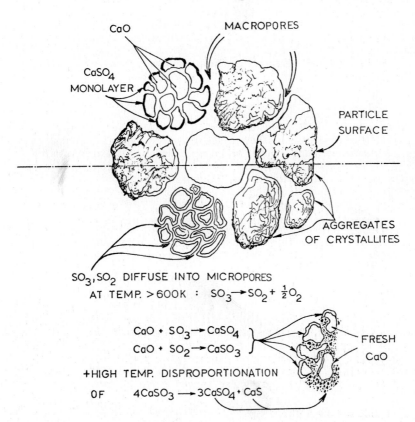

Figure 4. Model of CaO particles

TEMPERATURE-ADAPTABLE HOLLOW FIBERS CONTAINING
INORGANIC AND ORGANIC PHASE CHANGE MATERIALS

Tyrone L. Vigo and C. E. Frost
USDA, ARS, Textiles and Clothing Laboratory
1303 W. Cumberland Avenue
Knoxville, Tennessee 37916

INTRODUCTION

The use of inorganic phase change materials (PCM's) for thermal storage in residences (1) and in spacecraft (2) has been described. However, not until recently have these PCM's been incorporated into hollow textile fibers to achieve the same thermal effects on a micro scale (3, 4). This was accomplished in our laboratory by increasing the total thermal energy of rayon fibers when they were filled with aqueous solutions of $CaCl_2 \cdot 6H_2O$ (3, 4). Temperature-adaptable fibers were first prepared by a different concept--decrease of solubility of a gas in a liquid inside the hollow fiber when the liquid solidified (5). However, this method is useful only in cold weather applications and loss of any gas from the fiber would dramatically diminish the useful thermal effects.

Although our initial studies demonstrated that $CaCl_2 \cdot 6H_2O$ caused desired endotherms and exotherms on heating and cooling in rayon fibers, this system was not useful in hydrophobic fibers and also had the disadvantage that it was only partly congruent in its melting. Thus, congruent melting inorganic ($LiNO_3 \cdot 3H_2O$ and $Zn(NO_3)_2 \cdot 6H_2O$) and organic (polyethylene glycol 600) PCM's were evaluated for their ability to produce desirable thermal changes in hydrophilic (rayon) and hydrophobic (polypropylene) hollow fibers that showed little decay after an extensive number of heating and cooling cycles. The results of this investigation are described in this paper.

EXPERIMENTAL

Materials. Filament polypropylene (0.66 mg/m) with a single round cavity cross section--furnished by Hercules*--and staple rayon (0.19 mg/m) with a similar geometric cross section--furnished by Courtaulds*--were the hollow fibers used in this study. Phase-change materials (PCM's) included: $LiNO_3 \cdot 3H_2O$ (certified), $Zn(NO_3)_2 \cdot 6H_2O$ (reagent), and polyethylene glycol 600 (Carbowax 600)*.

Filling Hollow Fibers. Hollow fibers were filled with the PCM's conditioned prior to thermal analysis as previously described (3). This consisted of aspirating aqueous or neat solutions of the PCM's through fiber bundles (rayon--38mm; polypropylene--135 mm) tightly aligned inside an O-ring for about 30 mins. or until visual observation indicated that the fibers were completely filled. Concentration ranges (w/w) of PCM's aspirated through the fibers were: (a) 12.5, 25, 50% aq. and 100% $LiNO_3 \cdot 3H_2O$, (b) 89.7% aq. $Zn(NO_3)_2 \cdot 6H_2O$, and (c) 8, 16.2, 32.4, & 64.7% aq. PEG 600. After aspiration, the filled fibers were placed horizontally and cooled at

258°K for 1 hr., then dried over anhydrous $CaSO_4$ for 24 hrs. These samples (with unsealed ends) were subsequently conditioned in a dessicator at 293°K and 45% RH (latter achieved with a saturated solution of KNO_2) prior to thermal analysis. Saturated salt solutions of KNO_2 were prepared as previously described (3). Unmodified rayon and polypropylene fibers and various concentrations of the PCM's were dried and conditioned in the same manner to serve as controls for comparison to the modified hollow fibers. Whenever possible, any extraneous PCM was removed from the fibers prior to thermal analysis.

DSC Scans. Modified and control fibers (rayon--38mm; polypropylene--66mm) were placed in non-volatile aluminum pans and the variation of thermal energy with time (dH/dt) measured on a Perkin-Elmer DSC-2 at temperature intervals of 233 to 333°K. The instrument was temperature-calibrated with distilled water and calorifically-calibrated with 0.0282 & 0.1294 g sapphire discs. Instrument settings were: heating and cooling rates of 10°K/min, chart speed of 40mm/min, DSC ranges of 5-10 mcal/sec, recorder ranges of 10-50 mv, positive pressure (137.9 Kpa) of dry N_2 through the sample holder block, and auto cool mode. Fibers and PCM's were first run on heat mode until the temperature reached 333°K, run on auto cool until the temperature reached 233°K, then recycled each time after baseline stabilization for the desired number of cycles. ΔH (total thermal energy at different temperature intervals) was determined by comparison with a sapphire standard. Aluminum pans of equal weight were selected to make the calculations more precise. A typical calculation is as follows:

$$\Delta H_{Sample} = \Delta H_{Sapphire} \times \frac{\text{Weight sample}}{\text{Weight sapphire}} \times \frac{\text{Area sample}}{\text{Area sapphire}} \times \frac{\text{Sensitivity Range Sample}}{\text{Sensitivity Range Sapphire}} \times \frac{\text{Chart Speed Sapphire}}{\text{Chart Speed Sample}}.$$

Microscopy. Photomicrographs of filled and unfilled fibers were made by using a Leitz Polarizing Microscope* equipped with a Polaroid camera* at a magnification of 700X.

RESULTS AND DISCUSSION

Thermodynamic Properties. Specific heat, heat of fusion, and other ΔH measurements reported in the literature for hollow fibers and phase-change materials used in this study are listed in Table I. These values will be discussed in conjunction with the thermal results obtained with modified hollow fibers containing various PCM's.

---

*Mention of a trademark or proprietary product does not constitute a guarantee or warranty of the product by the U. S. Department of Agriculture and does not imply its approval to the exclusion of other products that may also be available.

Table I. Selected Thermodynamic Properties of Hollow Fibers and PCM's Used in This Study[a]

System	$C_p(298°K)$ cal/g-°K	$\Delta H_{fusion}(298°K)$ cal/g	Other ΔH cal/g
Hollow rayon fiber	0.45[3]	-	-
Hollow polypropylene fiber	0.46[3]	-	-
$Zn(NO_3)_2 \cdot 6H_2O$	b	32.0[1]	$\Delta H_{soln.}(298°K)$ = +18.8 [6]
$LiNO_3 \cdot 3H_2O$	0.39[7]	70.7[2]	$\Delta H_{form.}(296.5°K)$ $LiNO_{3(s)} + 3H_2O(l)$ = -60.8 [8]
PEG 600	0.54[2]	35.0[2]	-

[a]References listed in brackets next to C_p or ΔH values. [b]Not reported or not readily available from the literature.

DSC Scans. The thermal behavior of three PCM's--$Zn(NO_3)_2 \cdot 6H_2O$, $LiNO_3 \cdot 3H_2O$, and PEG 600, and their behavior in both rayon and polypropylene hollow fibers, is shown in Table II after several heating and cooling cycles. Although various concentrations of these PCM's in the hollow fibers increased their ΔH values, the best thermal performance of the filled fibers, particularly after cycling, was achieved with the concentrations of the PCM's listed in Table II: 89.7% aq. $Zn(NO_3)_2 \cdot 6H_2O$, neat $LiNO_3 \cdot 3H_2O$, and 64.7% aq. PEG 600. As previously noted (3), the only thermal changes in untreated rayon and untreated polypropylene hollow fibers that occurred in this temperature range (233-333°K) were due to their specific heat.

Since $Zn(NO_3)_2 \cdot 6H_2O$ is one of the few salts that undergo congruent melting and crystallization in the temperature range of interest, it was investigated for its ability to increase the ΔH values of hollow fibers. The salt itself exhibited somewhat erratic thermal behavior, since its initial endotherm at 310°K diminished greatly at 5 heating cycles, and its initial exotherm at 266°K disappeared at 5 cooling cycles. These decreases could be attributed to the evaporation of water to make the salt more concentrated, and thus cause enthalpy due to only fusion rather than fusion and heat of solution. Although polypropylene fiber filled with 89.7% aq. solution of this salt behaved less erratically than the salt alone, it did not show an exotherm at 5 cooling cycles. Rayon fiber filled with the same salt concentration, however, exhibited higher enthalpy values than the salt alone or the salt in the polypropylene fiber. As previously noted in our earlier studies with $CaCl_2 \cdot 6H_2O$ in hollow rayon fibers, this may be due to the ability of the hydrophilic fiber to promote or prolong congruent melting of inorganic salt hydrates or facilitate other processes having high ΔH values. Even though the rayon/$Zn(NO_3) \cdot 6H_2O$ system was promising, its thermal fluctuation within 5 heating and cooling cycles (Table II) precluded it from further investigation.

Table II. Thermal Properties of Filled Hollow Fibers and PCM's as Determined by DSC

Fiber and/or PCM	Endotherm[a] (°K)	ΔH[b] cal/g	Exotherm[c] (°K)	ΔH[d] cal/g	No. of heating & cooling cycles
$Zn(NO_3)_2 \cdot 6H_2O$ neat	310	86.7	266	10.2	1
" "	314	15.9	none	-	5
89.7% aq. $Zn(NO_3)_2 \cdot 6H_2O$ in PP[e]	312	21.4	291	12.3	1
" "	319	15.2	none	-	5
89.7% aq. $Zn(NO_3)_2 \cdot 6H_2O$ in rayon	300	26.2	280	18.5	1
" "	306	38.3	280	15.4	3
" "	306	32.8	274	15.0	5
$LiNO_3 \cdot 3H_2O$ neat	308	105.8	284	96.0	1
" "	304	18.3	263	3.1	10
$LiNO_3 \cdot 3H_2O$ neat in PP[e]	306	65.1	280	7.5	1
" "	303	40.9	none	-	10
" "	303	24.0	none	-	50
$LiNO_3 \cdot 3H_2O$ neat in rayon	308	125.1	284	103.2	1
" "	308	129.7	274	68.3	10
" "	304	57.2	256	8.5	50
PEG 600 neat	288	89.3	270	57.5	1
" "	294	87.4	278	66.8	10
64.7% aq. PEG 600 in PP[e]	290	36.1	278	38.8	1
" "	297	37.8	282	46.7	10
" "	298	42.8	282	38.3	75
64.7% aq. PEG 600 in rayon	289	71.8	270	59.2	1
" "	298	77.9	280	64.5	50
" "	296	74.8	280	62.7	75
" "	298	73.6	280	66.9	150

[a] On heating. [b] Total thermal energy on heating. [c] On cooling. [d] Total thermal energy on cooling. [e] Polypropylene.

Lithium nitrate trihydrate, another inorganic salt hydrate that melts congruently in the temperature range of interest, was also investigated for its effects in both hollow fibers. The salt itself (Table II) underwent extensive decay from 1 to 10 heating and cooling cycles (ΔH values decreased 90-95% of their initial value) and also showed some supercooling (shift of exotherm to a lower temperature) on cycling. Since it has been reported that dehydration of the trihydrate to either anhydrous $LiNO_3$ or $LiNO_3.1/2H_2O$ also occurs in the same temperature range as melting of the trihydrate (8), the trihydrate available for melting after extensive cycling could diminish, and thus provide less thermal energy. When the neat salt was incorporated into hollow polypropylene fibers (Fig. 1), the initial endotherm also decreased markedly from 1 to 10 heating cycles and the exotherm disappeared at 10 cooling cycles. In contrast to this, rayon hollow fibers containing $LiNO_3.3H_2O$ (Fig. 2) still had a significant endotherm ($308°K$) at 10 heating cycles (ΔH = 129.7 cal/g) and a significant exotherm ($274°K$) at 10 cooling cycles (ΔH = 68.3 cal/g). On prolonged cycling however, thermal decay was noted in both heating and cooling (50 cycles) and pronounced supercooling occurred (exotherm shifted to $256°K$). If supercooling can be overcome or minimized (no attempt was made to do so in this study), the rayon $LiNO_3.3H_2O$ system may be useful, since the total thermal energy on heating (ΔH = 57.2 cal/g) is still substantial. It appears that the water in the rayon facilitates congruent melting with this system by reducing conversion to anhydrous or lower hydrates of $LiNO_3$.

Polyethylene glycol 600, an organic PCM, was also evaluated for its thermal effects in both hollow fibers. Although the PEG 600 itself did not exhibit the initial intensity that the inorganic PCM's exhibited on melting and crystallization, it underwent little thermal decay from 1 to 10 heating and cooling cycles or supercooling during cycling (ΔH of 87-89 cal/g on heating and of 57-67 cal/g on cooling). When a 64.7% aq. solution of the PEG 600 was incorporated into hollow polypropylene fibers (Fig. 3), similar thermal behavior was observed, even after 75 heating and cooling cycles. Enthalpy values associated with the endotherm at $298°K$ are 42.8 cal/g and those associated with the exotherm at $282°K$ are 38.3 cal/g (Table II). When the same concentration of PEG 600 was incorporated into hollow rayon fibers (Fig. 4), they also exhibited stable and thermally reproducible effects at 75 heating and cooling cycles, although the magnitude of the enthalpy values was about twice that of the modified polypropylene fibers. This is probably due to the difference in the diameters of the cavity of the fibers and their ability to retain the organic PCM. The modified rayon fiber was still thermally stable at 150 heating and cooling cycles, and had ΔH values on heating of 73.6 cal/g and on cooling of 66.9 cal/g. Since the enthalpy of the organic PCM is due only to heat of fusion and not thermally dependent on hydrate formation, this system appears to be suitable for increasing the total thermal energy of representative hydrophilic (rayon) and hydrophobic (polypropylene) hollow fibers (a) after extensive thermal cycling, (b) with little supercooling, and (c) at temperature ranges useful for indoor textile applications (endotherm at 298°K or 77°F and exotherm at 280°K or 45°F).

Microscopy. Preliminary results indicate that major portion of $LiNO_3 \cdot 3H_2O$ and of the PEG 600 resides inside the hollow cavity of both the polypropylene [Fig. 5 (b)] and rayon [Figs. 6 (b) and (c)] fibers, although there are some discontinuities of the solid PCM inside the cavity. A photomicrograph of the PEG 600 inside the polypropylene fiber is not shown, since there was not enough dark field contrast to show the presence of this PCM with black-and-white photographs. Additional photomicrographs are planned (using scanning electron microscopy) to further characterize the distribution of the PCM's inside the hollow fibers.

CONCLUSIONS

Hydrophilic (rayon) and hydrophobic (polypropylene) hollow fibers containing an organic phase-change material (polyethylene glycol with a molecular weight of 600) exhibited changes in their enthalpy (due to ΔH_f) at desirable ambient temperature ranges that were of much greater magnitude than that of the heat capacity exhibited by untreated hollow fibers. Endotherms (on heating) and exotherms (on cooling) showed little supercooling or thermal decay after extensive cycling (75-150 cycles).

Of the two inorganic PCM's evaluated in the hollow fibers (zinc nitrate hexahydrate and lithium nitrate trihydrate), only the latter in rayon showed any promise. However, this system underwent extensive supercooling at 75 cycles (exotherm shift), and would only be useful if supercooling could be avoided. It appears that the water in the rayon fiber facilitates congruent melting and crystallization of inorganic salt hydrates, and thus is more suitable with these systems than are hydrophobic fibers.

ACKNOWLEDGMENTS

The authors are grateful for the photomicrographs and preparation of the fibers for microscopy by Miss Dong-Hwa Shin, a graduate student in the Dept. of Textiles, Merchandising and Design at the University of Tennessee, Knoxville.

REFERENCES

1. B. Carlsson, H. Stymme and G. Wettermark, "Storage of Low-Temperature Heat in Salt-Hydrate Melts--Calcium Chloride Hexahydrate," Swedish Council for Building Research, Document D12: 1978.
2. D. V. Hale, M. J. Hoover and M. J. O'Neill, "Phase Change Materials Handbook," NASA Contractor Report CR-61363, September, 1971.
3. T. L. Vigo and C. E. Frost, Textile Res. J., accepted for publication.
4. Chem. & Engr. News, 59(36), 67 (1981).
5. R. H. Hansen, "Temperature Adaptable Fabrics," U.S. Pat. 3,607,591, Sept. 21, 1971.
6. W. W. Ewing, J. D. Brandner and W. R. F. Guyer, J. Am. Chem. Soc. 61, 260 (1939).

7. H. Koshi, Suom. Kemistilehti B, 45 (4), 135 (1972).
8. L. Nedeljkovic, Z. Anorg. Allg. Chem. 357 (1-2), 103 (1968).

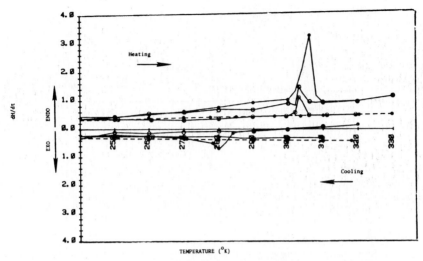

Fig. 1. dH/dt vs temperature for hollow polypropylene fibers (a) unmodified [●----●----● 1 heating and cooling cycle]; (b) filled with 100% $LiNO_3 \cdot 3H_2O$ [●———●———●-1, ○———○———○-10, and ●———●———● 50 heating and cooling cycle(s)].

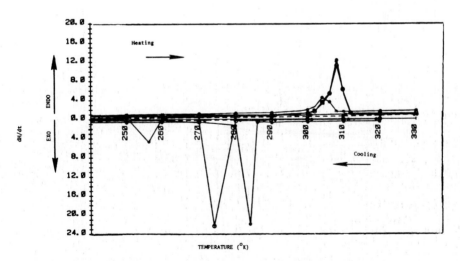

Fig. 2. dH/dt vs temperature for hollow rayon fibers (a) unmodified [●----●----● 1 heating and cooling cycle)]; (b) filled with 100% $LiNO_3 \cdot 3H_2O$ [●———●———●-1, ●———○———● 10, and ●———●———● 50 heating and cooling cycle(s)].

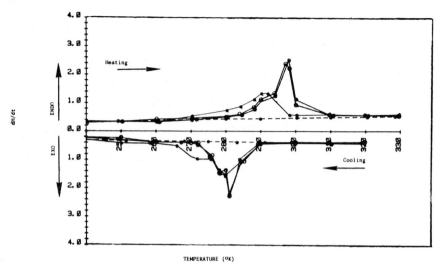

Fig. 3. dH/dt vs temperature for hollow polypropylene fibers (a) unmodified [●──●──●──●1 heating and cooling cycle]; (b) filled with 64.7% PEG-600 (●────●────●────●-1, ○────○────○-10, and ●────● -75 heating and cooling cycle(s)].

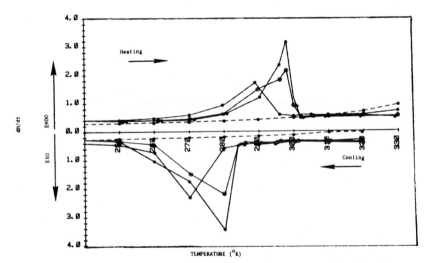

Fig. 4. dH/dt vs temperature for hollow rayon fibers (a) unmodified [●───●───●; (b) filled with 64.7% PEG-600 [●────●-1, ○────○-50, and ●────●-150 heating and cooling cycle(s)].

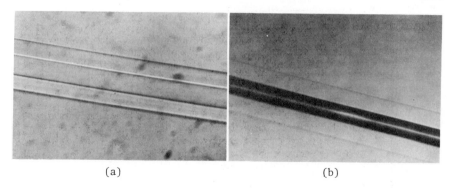

Figure 5. Photomicrographs, 700X, showing polypropylene (a) untreated (b) filled with 100% $LiNO_3 \cdot 3H_2O$.

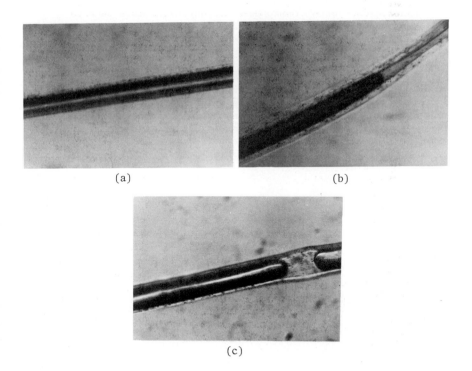

Figure 6. Photomicrographs, 700X, showing rayon (a) untreated (b) filled with 100% $LiNO_3 \cdot 3H_2O$ (c) filled with 64.7% PEG-600.

INVESTIGATION OF THE ROLE OF CHEMICAL ADMIXTURES
IN CEMENTS - A DIFFERENTIAL THERMAL APPROACH

V.S. Ramachandran, Head
Building Materials Section
Division of Building Research
National Research Council of Canada
Ottawa, Canada K1A 0R6

INTRODUCTION

Most concrete used in North America contains at least one admixture, which may be defined as any material other than water, aggregate or hydraulic cement used as an ingredient of concrete or mortar and added to the batch immediately before or during mixing [1]. Several types of admixture are used, conferring one or more beneficial effects such as better workability, early strength development, acceleration or retardation of setting, improved frost resistance, lower water requirements, increased resistance to sulfate attack, and higher long-term strength.

The technique of differential thermal analysis (DTA) has been applied extensively in studying the reactions occurring in organic and inorganic systems. Its application to cement chemistry is well recognized [2]. It has been found recently that DTA can also be used to advantage in studying the role of various admixtures in concrete [3]. The technique yields quick, reliable and sometimes new information not readily obtained by other techniques. This paper describes the types of information that can be derived by applying DTA to a study of the role of accelerators, retarders, water reducers and superplasticizers in hydrating portland cement and its minerals.

EXPERIMENTAL

Tricalcium silicate (C_3S), dicalcium silicate (C_2S), and tricalcium aluminate (C_3A) were synthesized from pure compounds. Portland cement, Type I, was a commercial sample. All the admixtures were chemically pure materials. The sugar-free lignosulfonates were obtained by a fractional method that has been described [4]. The pastes were prepared by mixing the constituents on rollers [4-7]. The DTA, XRD and adsorption-desorption experiments were carried out according to procedures described previously [6,8,9].

RESULTS AND DISCUSSION

Accelerators

Evaluation. Calcium chloride acts as a very efficient accelerator and, since it is relatively cheaper than other chemicals, is the most widely used accelerating admixture. It has deleterious effects on concrete, however, and attempts to find an

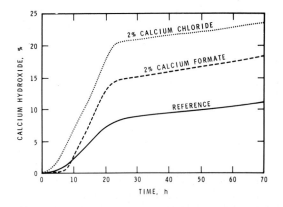

Figure 1. Effect of accelerators on the formation of calcium hydroxide in hydrated 3 CaO·SiO$_2$

alternative continue. Calcium formate is one such non-chloride alternative admixture used in many formulations.

The DTA technique may be applied to compare the relative accelerating effects of potential admixtures suggested as replacement for calcium chloride. Figure 1 gives the rate of hydration of C$_3$S, measured in terms of Ca(OH)$_2$ formed at different times, and that of 2% Ca-chloride and Ca-formate as well. Although Ca-formate accelerates hydration of C$_3$S, it is not so effective as Ca-chloride.

Kinetics of Hydration. A systematic study of the hydration of C$_3$S in the presence of 0, 1 and 4% CaCl$_2$ has shown that the estimation of Ca(OH)$_2$ cannot be used to compare the degree of hydration accurately because the CaO/SiO$_2$ of the calcium silicate hydrate formed in the presence of 4% CaCl$_2$ is higher than that formed in the presence of 0 or 1% CaCl$_2$ [5].

In terms of Ca(OH)$_2$ estimation the degree of hydration of C$_3$S at 30 days would be: C$_3$S + 1% CaCl$_2$ > C$_3$S + 0% CaCl$_2$ > C$_3$S + 4% CaCl$_2$. The relative degrees of hydration, however, in terms of the amount of C$_3$S reacted (by XRD) were found to be C$_3$S + 4% CaCl$_2$ > C$_3$S + 1% CaCl$_2$ > C$_3$S + 0% CaCl$_2$.

Possible States of Chloride in Hydrating C$_3$S. Thermal investigations of C$_3$S hydrated in the presence of CaCl$_2$ indicate the emergence of new peaks that are not registered by the hydrated products containing no admixture [3,6]. DTA results in conjunction with the estimation of chloride in leached and unleached samples can be interpreted to imply that chloride exists in four or five states in hydrating C$_3$S. During the early periods of hydration chloride exists in a free state. In the dormant or induction period (when hydration of C$_3$S is negligible) chloride is adsorbed on the C$_3$S surface. After the acceleration stage has been reached the chloride is chemisorbed on the C-S-H surface, with some existing in the interlayers of the hydrate. At later periods some 20% is incorporated strongly into the lattice of the C-S-H phase.

Effect of Water:Cement Ratio (w/c). A cement paste was hydrated with 3½% CaCl$_2$ at water:cement ratios of 0.25 and 0.4; heat of hydration was used to measure the degree of hydration. At 24 h the total heat of hydration measured by conduction calorimetry was compared with the amount of Ca(OH)$_2$ determined by DTA:

the ratio $\dfrac{\text{heat developed at w/c} = 0.4}{\text{heat developed at w/c} = 0.25} = 1.27$

and the ratio $\dfrac{\text{Ca(OH)}_2 \text{ estimated at w/c} = 0.4}{\text{Ca(OH)}_2 \text{ estimated at w/c} = 0.25} = 1.50$

Comparison of these ratios suggests that less $Ca(OH)_2$ is formed at a w/c ratio of 0.25. Thus there is a possibility that the excess lime is combined in the C-S-H phase, yielding a higher C/S ratio for the product prepared at a w/c = 0.25.

Role of Triethanolamine. Figures 2 and 3 compare the thermal behaviour of C_3S hydrated with 0 and 0.5% TEA [8,9]. It is evident from the thermograms that the endothermal peak in the range 480 to 500°C caused by $Ca(OH)_2$ decomposition, appearing at 3 h and later, is not evident in samples containing 0.5% TEA until about 10 h. Similar results, indicating the formation of low amounts of $Ca(OH)_2$ in TEA-treated samples, were obtained by TEA and chemical methods of analysis. Conduction calorimetric work confirmed the retarding effect of TEA. The exothermic peak observed at about 400°C in the sample containing TEA can be attributed to a complex of TEA formed on the surface of the hydrating C_3S. It may play a role in the retardation effect.

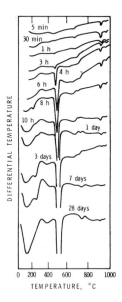

Figure 2. Thermograms of 3 CaO·SiO₂ hydrated to different periods

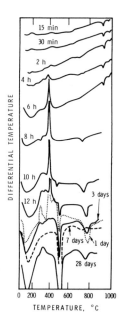

Figure 3. Thermograms of 3 CaO·SiO₂ hydrated to different periods in presence of 0.5% TEA

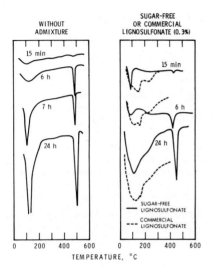

Figure 4. DTA of cement hydrated in the presence of lignosulfonates.

Triethanolamine is also a retarder of hydration of C_2S [8]. In the presence of TEA two endothermal effects are registered at 450 to 500°C. The higher temperature peak may be ascribed to crystalline $Ca(OH)_2$ and the other to the formation of non-crystalline $Ca(OH)_2$. Although TEA retards the hydration of C_3S, it seems to accelerate the reaction between C_3A and gypsum [10].

Retarders

Hydration. Differential thermal technique was used to follow the hydration of C_3S, C_3A and cement in the presence of different amounts of commercial lignosulfonate, sugar-free Na-lignosulfonate and sugar-free Ca-lignosulfonate [4]. In Fig. 4 the reference cement without admixture exhibits an endothermal effect at 450 to 500°C caused by the dehydration of $Ca(OH)_2$; its intensity indicates the extent of hydration of the C_3S component of cement. Cement hydrated in the presence of lignosulfonates is retarded, as evidenced by the relatively lower peak areas for $Ca(OH)_2$ decomposition. These results suggest that both commercial and sugar-free lignosulfonates act as retarding agents. Thermograms of C_3S and C_3A hydrated in the presence of sugar-free lignosulfonates also show the retarding effects of lignosulfonates.

Water Reducers

Interaction with Calcium Aluminate Hydrates. The C_3A

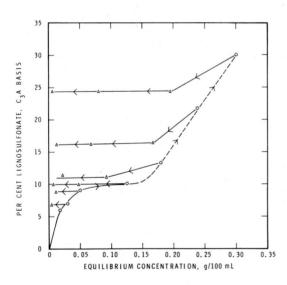

Figure 5. Adsorption-desorption isotherms of calcium lignosulfonate on the hexagonal phases.

component present in small amounts in cement dominates early setting and dispersion characteristics. Initially, hydration of C_3A produces a mixture of hexagonal phases (C_4AH_{13} and C_2AH_8) that ultimately converts to the cubic phase (C_3AH_6). A study was carried out to determine the effect of calcium lignosulfonate on the C_3A, $C_4AH_{13}-C_2AH_8$ and C_3AH_6 phases [11]. The adsorption-desorption isotherm of lignosulfonate on the hexagonal phases is shown in Fig. 5. The amount of adsorption increases as the concentration of lignosulfonate increases. Irreversibility also indicates that substantial amounts are strongly adsorbed. The XRD results show peaks at 7.9 Å for C_4AH_{13} and 10.5 and 5.23 Å for C_2AH_8. The hexagonal phases containing lignosulfonate exhibit a decrease in intensity of the line at 7.9 Å and enhancement in intensity of the peak at 10.5 Å. This may imply that lignosulfonate enters the interlayer of the C_4AH_{13} phase and increases the C-axis spacing.

Late Addition of Calcium Lignosulfonate. In portland cement it has been observed that hydration is retarded to a greater extent if calcium lignosulfonate is added to paste that has been prehydrated for 5 min than if it is added with the mixing water [12]. In portland cement C_3A hydrates as soon as it comes into contact with water, and this phase may partly explain the effect of delayed addition. The hydration kinetics of C_3S were followed by adding different amounts of C_3A and lignosulfonate. Estimate by DTA of the amount of $Ca(OH)_2$ formed during hydration indicated the extent to which C_3S hydration was retarded.

Hydration of C_3S was completely inhibited by the addition of 0.5 to 0.8% lignosulfonate. Addition of 5% C_3A was sufficient to promote hydration of C_3S, indicating that calcium lignosulfonate, which retards the hydration of C_3S, is preferentially adsorbed by C_3A and its hydration products. The inhibitive action of lignosulfonate on C_3S hydration is more pronounced when it is added to a mixture of C_3S and C_3A prehydrated for a few minutes. The prehydrated C_3A contains hexagonal and cubic phases that adsorb less lignosulfonate than C_3A, so that more lignosulfonate is left in the solution to retard hydration of C_3S.

Interaction of Lignosulfonate with Hydrating C_3S. Adsorption-desorption data of lignosulfonate on the C_3S-H_2O system provide information about the mechanism of hydration of C_3S in the presence of lignosulfonate [13]. Thermograms can be used to explain the adsorption of lignosulfonate in the C_3S-H_2O system exposed to different concentrations of lignosulfonate (Fig. 6). A small amount of adsorption at higher concentrations is due to dispersion effect and at low concentrations largely irreversible adsorption results due to hydration.

Superplasticizers

Hydration Characteristics. Some of the effects of superplasticizers seem to depend on the composition of cement in concrete. It is therefore necessary to appreciate the influence of superplasticizers on the hydration characteristics of individual cement components such as C_3A, C_3A + gypsum, C_3S, C_2S and C_4AF.

Addition of SMF in small amounts influences the kinetics of hydration of cement and individual cement minerals [14]. DTA results

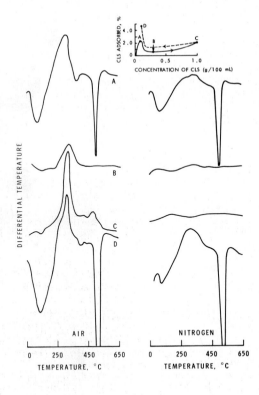

Figure 6. Thermograms of C_3S treated with CLS in an aqueous medium (curves A, B, C, D correspond to the samples taken at different points on the adsorption-desorption curve drawn).

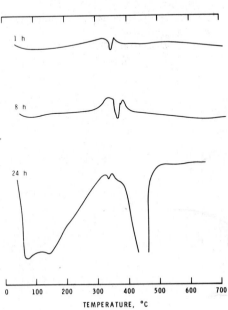

Figure 7. DSC curves for C_3S hydrated in the presence of SMF.

show that SMF retards the hydration of C_3S and C_3A but accelerates the formation of ettringite in the $C_3A + CaSO_4 \cdot 2H_2O + H_2O$ system. There is evidence also that the C/S ratio of the C-S-H product during hydration of C_3S is affected. Figure 7 shows thermograms of C_3S hydrated in the presence of SMF for 1, 8 and 24 h. There is no indication of $Ca(OH)_2$ in samples hydrated for up to 8 h. C_3S hydrated without superplasticizers, however, shows the existence of $Ca(OH)_2$ (not shown in the figure), indicating that hydration of C_3S is retarded by SMF.

ACKNOWLEDGEMENTS

The author wishes to thank G.M. Polomark for the experimental assistance. This paper is a contribution from the Division of Building Research, National Research Council of Canada, and is published with the approval of the Director of the Division.

REFERENCES

1. Annual Book ASTM Standards, Philadelphia, Part 14, p. 124, 1980.
2. V.S. Ramachandran, Applications of Differential Thermal Analysis in Cement Chemistry, Chemical Publishing Co., USA, 1969, 308 p.
3. V.S. Ramachandran, Thermochimica Acta, 1972, 3, 343-366.
4. V.S. Ramachandran, Zement Kalk Gips, 1978, 31, 206-210.
5. V.S. Ramachandran, Thermochimica Acta, 1971, 2, 41-55.
6. V.S. Ramachandran, Materiaux et Constr., 1971, 4, 3-12.
7. V.S. Ramachandran and R.F. Feldman, Il Cemento, 1978, 75, 311-322.
8. V.S. Ramachandran, J. Appl. Chem. Biotech., 1972, 22, 1125-1138.
9. V.S. Ramachandran, Cem. Concr. Res., 1976, 6, 623-631.
10. V.S. Ramachandran, Cem. Concr. Res., 1973, 3, 41-54.
11. V.S. Ramachandran and R.F. Feldman, Materiaux et Constr., 1972, 5, 67-76.
12. G.M. Bruere, Effect of Mixing Sequence on Mortar Consistencies when Using Water Reducing Admixtures, Highway Research Board, Special Report 90, 26-35, 1966.
13. V.S. Ramachandran, Cem. Concr. Res., 1972, 2, 179-194.
14. V.S. Ramachandran, Influence of Superplasticizers on the Hydration of Cement, III International Congress on Polymers in Concrete, Koriyama, Japan, Vol. II, 1071-1081, 1981.

APPLICATION OF DTA AND TGA TO TROUBLE-SHOOTING
OF THE PORTLAND CEMENT BURNING PROCESS

Hung Chen, Central and Research Laboratory,
Canada Cement Lafarge Ltd., P.O. Box 398,
Belleville, Ontario, Canada, K8N 5A5

INTRODUCTION

In the manufacture of Portland cement clinker, a finely pulverized raw mix of calcareous and argillaceous materials is subjected to dehydration, decarbonation, and then sintering at about 1450°C, in a continuous process within a rotary kiln system.
To obtain optimum operation of the sintering process and the resulting clinker and cement properties, the composition of CaO, SiO_2, SiO_2, Al_2O_3, and Fe_2O_3 in the raw mix must be accurately controlled. Unfortunately, there are always other minor constituents in the raw materials and fuels which are difficult to control. They have however, great effects on the burning process as well as the product characteristics. The subject was most recently reviewed by Bucchi [1].
The minerals containing the minor constituents K, Na, S, Cl, and F are the most troublesome. These elements tend to form low melting and volatile salts, such as CaF_2, KCl, K_2SO_4, Na_2SO_4, and $CaSO_4$. These salts volatilize in the hottest zone of a rotary kiln, being carried back to the cooler areas of the kiln system by flue gas. They condense on the refractory surfaces and entrained dust and raw mix particles. Volatile recirculation had caused many plugging problems in suspension preheaters during the fifties and sixties [2]. In the seventies the damaging effects had spread to conventional rotary kiln systems because of the recycling of electrostatic precipitator dusts in compliance with more stringent environmental laws.
DTA and TGA have been widely used to study cement hydration and the thermal reactivities of raw materials or raw mixes [3-11]. However, they have rarely been used in trouble-shooting of the above mentioned process problems [12-14]. The special emphasis of this paper is on the analysis of spurrite, i.e. $Ca_5(SiO_4)_2CO_3$ or $2C_2S \cdot CaCO_3$ in cement chemists' notations, and the determination of the melting/freezing points and the volatility of the problem compounds. This information is essential in understanding the operational problems caused by volatile recirculation.

EXPERIMENTAL

The thermal analysis system used is a non-simultaneous DTA (Tracor Model-202) - TGA (CSI/Stone Model 1050). The instrumental conditions used for most of the experimentation are shown in Table 1. Unless otherwise specified, these apply to all DTA and TGA scans. To ensure good representation of the small sample tested, considerable precautions were necessary. It was reduced to a fineness of less than 20 microns in a Spex shatterbox to ensure uniformity.

The qualitative analysis of a sample by DTA or TGA is often difficult or uncertain. In these circumstances, other techniques, such as XRD, microscopy, and chemical analysis were used to substantiate the results of thermal analysis.

In the determination of spurrite concentration, the extraction procedure using either 20% salicylic acid or 20% maleic acid solution in methanol was adopted [15,16]. A sample was subjected to TGA scans before and after the extraction.

TABLE 1 - THE INSTRUMENTAL CONDITIONS FOR DTA AND TGA SCANS

	DTA	TGA
Sample weight	50 mg	50-100 mg
Reference material	Al_2O_3, 50 mg	
Gas	air or N_2	air or N_2
Gas flow rate	0.05 SCFH	0.04 SCFH
Heating/cooling rate	20°C/min	20°C/min.
ΔT or ΔWt. sensitivity	40 μν/F.S.	4-20 mg/F.S.

RESULTS AND DISCUSSION

Spurrite Determination.

Spurrite which may co-exist with calcite in some geological materials [17] was also identified in kiln materials [18-20]. Steuerwald et al [21] and Glasser's [22] studies indicated that, at a given CO_2 partial pressure, spurrite decomposed at temperatures which were 25-50°C higher than for calcite. Gross [23] on the other hand observed that, for natural spurrite and calcite, these two decarbonation peaks were more than 100°C apart. He was able to determine the concentration of both minerals by DTA and TGA. Based on many of our observations, these two decarbonation peaks were rarely well separated in kiln samples. Shown in Fig.1 are DTA curves of kiln charge samples taken along a

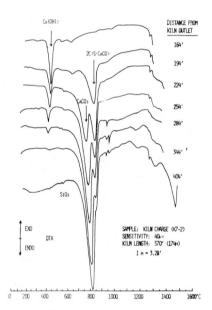

Fig.1 DTA Curves of the Samples of a Kiln Charge

570' dry-process kiln. The distances shown are those from the clinker discharge of the kiln. It is apparent that with the exception of the 194' - sample where only spurrite exists, quantitative analysis is difficult due to peak overlap. Qualitatively, one can see that as the charge moves along the kiln, more and more spurrite forms at the expense of calcite and, less apparently, quartz. At 284', the calcite simultaneously decomposes and transforms into spurrite. Near 224', spurrite appears to have reached the maximum and also starts to decarbonate.

A TGA scan of the 254'-sample is shown as trace B in Fig.2. The weight loss between 750° and 1020°C is due to the decarbonation of both calcite and spurrite. The two reactions are not distinct, and the amount of CO_2 so obtained can only be used to calculate the total $CaCO_3$. The same sample was subjected to an acid extraction [15,16] in order to dissolve spurrite, along with KCl, Ca_2SiO_4, CaO, and $Ca(OH)_2$. The spurrite was completely dissolved without precipitating any $CaCO_3$. The calcite was left intact during the extraction. Scan C in Fig.2 is the TGA curve of the residue after extraction. This residue

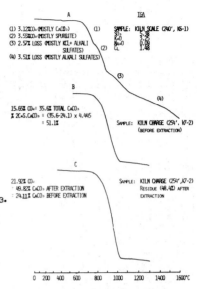

Fig.2 TGA Scans of a Kiln Scale Sample and a Kiln Charge Sample

contained 50% calcite, or equivalent to 24% on a weight basis before extraction. The difference between this value and that determined from scan B multiplied by 4.445 is 51% of the spurrite concentration.

The above method permits an accurate determination of calcite and spurrite in any proportion, if precautions are taken to thoroughly purge the TGA apparatus with air or nitrogen before each run, to prevent the in-situ carbonation of any CaO and $Ca(OH)_2$ in the sample. The accuracy of the spurrite value is not as good as that of calcite because any error is multiplied by a factor of 4.445. However, it can be expected to be within 1%.

In the event that a sample contains considerable amounts of volatile compounds, particularly KCl, the volatilization of these may introduce a substantial error in the analysis. Curve A in Fig. 2 is a TGA scan of a coating sample from another dry-process kiln. This sample contains 5.38% SO_3, 3.25% K_2O, 0.09% Na_2O, and 1.48% Cl. The seemingly distinct four stages of weight loss due to calcite, spurrite, KCl, and alkali sulfates actually overlap. Since the volatile salts are readily soluble in water, the reliable determination of calcite and spurrite still can be carried out by the "water extraction-TGA-acid extraction-TGA" procedure.

<u>Spurrite Profiles of Kiln Charges</u>. The spurrite profiles of kiln charges taken from five of our dry-process rotary kilns are plotted in Fig. 3 for comparison. Representative samples were taken along each kiln 20' or 30' apart after the kiln was cooled down. The positions of the profiles shown are thus closer to the kiln discharge than they would be when the kilns are running. This is because a rotary kiln must be kept rotating during the cooling off period to prevent the shell from warping.

Three kilns which exhibit maximum spurrite contents in the range of 50 to 60% had experienced operational problems.

Kiln K7-2, which was previously mentioned in relation to Fig. 1, shows that there was more than 30% spurrite in the kiln charge even in the chain zone near 400'. A very hard ring build up in the chain zone was found to contain up to 60% spurrite. The kiln charge and ring material in this zone contained 5.5 and 7.0% volatile compounds, respectively. The formation of certain

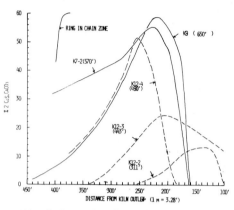

Fig.3 Spurrite Profiles of Kiln Charges of Dry-Process

types of rings and coatings is directly related to the volatile laden dust entrained by the combustion gases. After recondensation and before solidification of the volatiles, the dust particles will have a tendency to agglomerate and adhere to solid objects such as chain systems. This build-up phenomena may be intensified by the presence of spurrite whose formation is catalized by the liquid volatile phases. It was concluded that in this case the chain system was too extensive. Temperatures in the chains promoted liquid formation of the volatile compounds. Some chains were subsequently removed, which improved the operation. Fig. 3 also illustrates that as the kiln charge moved downward in the calcining zone, the spurrite gradually built up to about 50%, with a corresponding increase of liquid volatile compounds to 7.0%, and to 10.5% when approaching the burning zone [13] . The consequence was an unstable kiln operation, which adversely affected the production and product qualities. The volatiles in this kiln were mainly K_2O and SO_3 from the fuel. Much less important was Cl. These recirculated in the forms of $2CaSO_4 \cdot K_2SO_4$ (calcium langbenite), $2C_2S \cdot CaSO_4$ (sulphospurrite), KCl, and $2Ca_5 [(Si,S)O_4]_3 (OH,Cl,F)$ (ellestadite) [12,24] .

Kiln K12-4 is a 490' dry-process kiln. The chain section is located 388'-464' from the discharge end. As indicated in Fig.3, there was a very low spurrite content in the charge of this zone. However, it increased rapidly to 51% at 260', in the second half of the calcining zone. This peak of spurrite coincides with the maximum Cl content (1.5%) and close to the peaks of K_2O (4.5%) and SO_3 (4.0%). The liquid volatile content in this zone was in the range of 9 to 10%. The flow of kiln charge was impaired by spurrite and liquid, which caused cyclic patterns in such parameters as secondary air temperature, % O_2 in kiln exhaust gas, kiln drive amperage, chain material temperature, chain gas temperature, and feed end draft. The unstable kiln operation shortened the refractory lining service life and became a production bottleneck.

The two smaller dry-process kilns, K12-3 and K12-2, which used the same raw mix as that of K12-4, had no operational problems. The spurrite profiles of these two kilns are also shown in Fig. 3. The maximum contents were only 24% and 14%. The volatile liquid contents were also much lower, being 5% and 4%. These data suggest that volatile recirculation is greatly affected by equipment design. For example, flame impingement on the kiln charge creates a localized reducing condition. This greatly increases the volatilization of the minor constituents creating a concentration build-up.

Kiln K9 is a 650' dry-process kiln equipped with a one-stage preheater. The spurrite profile of its charge appears to be similar to that of kiln K12-4, with a maximum of 60%. Again this peak coincides with a maximum volatile liquid content of 11%. This high volatile content, creates an abnormally large amount of liquid formation in the kiln charge at the burning zone. The result is inconsistent mobility of the material through the kiln and unsteady operations. The solution in this case was a gas by-pass to reduce the potassium and chloride levels.

<u>Melting and Volatilization of a Sulfate and Chloride Eutectic Mixture</u>. Since the melting of this eutectic mixture promotes the formation of spurrite and impairs the flow of kiln charge, it is essential to know its melting temperature and to some extent its degree of volatilization. DTA and TGA can be used for such purposes. Curve A in Fig. 4 is a DTA scan of a kiln riser duct deposit from Kiln K9. This sample contains nearly 60% volatile compounds. The onset of melting is at 550°C and peaks at 630°C. This is much lower than the melting temperatures of the individual compounds. Curve B is a DTA scan of a chain deposit from the same kiln.

The onset and peak temperature of melting is shifted upward to 610° and 642°C, respectively, because of the lower KCl content (21% as opposed to 29% in the former case). On cooling, the crystallization starts at 640°C and peaks at 610°C. The super-cooling results in a sharper twin-peak, indicating the first precipitation of the double-salt of alkali sulfates and the subsequent appearance of

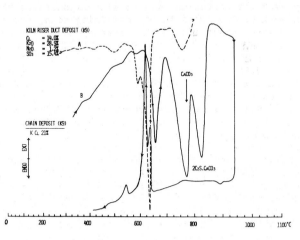

Fig.4 DTA Curves of a Kiln Riser Duct Deposit and a Chain Deposit

potassium chloride. The crystallization peaks can not be observed if the sample is heated beyond 950°C, where volatilization occurs.

In Fig.5, Curves B and D again demonstrate the sensitivity of the melting point of a eutectic mixture to composition. Scan B reveals melting peaks at 660° and 770°C, which are due to potassium

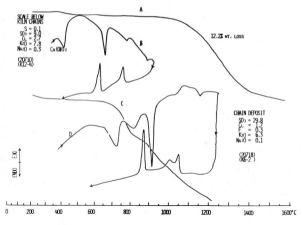

Fig.5 DTA and TGA Scans of Chain Deposits

chloride and probably calcium langbenite. Scan D exhibits two melting peaks at 920°C and 1050°C, because of a much higher SO_3 and lower Cl. The assignment of these peaks is difficult, but XRD indicates the presence of relatively high contents of calcium langbenite and ellestadite, and some potassium chloride. The variable melting and freezing temperatures of the eutectic mixtures are related to where the rings, deposits, scale, and balls exist in a kiln system, and where the restriction of material flows occur.

It is impossible to simulate in a laboratory the volatilization of sulfates and chloride that occurs in a rotary kiln. However, TGA can reveal the temperature ranges at which different salts volatilize. Scan A in Fig.5 shows the volatilization that takes place between 950° and 1550°C. The first step in trace C is due to calcite and spurrite, which is followed by potassium chloride and alkali sulfate volatilization in that order. Similarly, in Fig. 2, scan A, it is evident that potassium chloride volatilizes rapidly between 900° and 1150°C, and alkali surfates volatilize more slowly from 1000°C to 1600°C. Such an observation seems to be consistent with the volatile profiles of kiln charges we have studied [13] .

CONCLUSION

The spurrite content, and melting and volatilization behaviours of kiln charges, coatings, scale, dust, and balls can be determined by DTA and TGA. This information combined with that obtained by XRD, chemical analysis, and microscopy is useful in understanding the mechanism and causes of problems in Portland clinker burning. Such investigations very often lead to practical solutions.

ACKNOWLEDGMENT

This author would like to thank Mr. R. W. Suderman, Manager of Research Laboratory and Quality Control, for his contribution through discussion, and the staff of the Central and Research Laboratory of Canada Cement Lafarge Ltd. for performing the experiments.

REFERENCES

1. R. Bucchi, World Cement Technology, 1981, 12, 210; ibid., 1981, 12, 258.
2. C. Howlett and H. Garrett, 1981, May 10-12, IEEE Cement Industry Technical Conference, Lancaster, Pa., U.S.A.
3. V. S. Ramachandran, R. F. Feldman, and P. J. Sereda, Highway Research Record, Highway Research Board, No. 62, Publication 1246, (1964), p.40.
4. V. C. Farmer, "Chemistry of Cement," ed. H. F.W. Taylor, Academic Press, New York, (1964), Vol. 2, Chapter 22, p.22.
5. P. Longuet, Proc. 5th Intern. Symp. Chem. Cement, 1968, Tokyo, Japan, Vol. 1, p. 239.
6. V. S. Ramachandran, "Application of Differential Thermal Analysis in Cement Chemistry," Chemical Publishing Co., Inc., New York, (1969).
7. R. Barta, "Differential Thermal Analysis," ed. R.C. Mackenzie, Academic Press, New York, (1972), Vol. 2, Chapter 33, p.207.
8. G. L. Kalousek, and K. T. Greene, Transp. Res. Circ., 1976.
9. J. Bensted, Il Cemento, 1980, 77, 169.
10. J. Bensted, Ibid., 1980, 77, 237.
11. G. R. Gouda, Proc. 16th Intern. Cement Seminar, (1981), p.51.
12. P. Hawkins and R.J. Wilson, 30th Pacific Coast Regional Meeting of Amer. Ceram. Soc., Los Angeles, Ca., Oct., 1977.
13. H. Chen and P. S. Grindrod, 80th Annual Meeting of Amer. Ceram. Soc., Detroit, Mich., May, 1978.
14. M. S. Y. Bhatty, Proc. 16th Intern. Cement Seminar, (1981), p. 110.
15. S. Takashima, Ann. Rept. Japan Cement Eng. Assoc., 1958, 12, 49.
16. A. A. Tabikh and R.J. Weht, Cem. Concr. Res., 1971, 1, 317.
17. Y. K. Bentor, S. Gross, and L. Heffler, Amer. Mineral., 1963, 48, 924.
18. B. Courtault, CERILH Technical Pub. 140, (1963).
19. B. Courtault, Rev. Mater. Constr., 1963, p.110-24, 143-56, 190-203.
20. F. Becker and W. Schräml, Cement and Lime Manuf., 1969, 42, 91.
21. F. Steuerwald, P. Hackenberg, and H. Scholze, Zem. Kalk Gips, 1970, 23, 579.
22. F. P. Glasser, Cem. Concr. Res., 1973, 3, 23.
23. S. Gross, Israel J. Chem., 1971, 9, 601.
24. A. E. Moore, World Cement Technology, 1976, 7, 85; ibid., 1976, 7, 134.

MECHANISM OF HYDRATION AND THE ROLE OF AN ADMIXTURE
IN CEMENT - A THERMAL ANALYTICAL APPROACH

A.A. RAHMAN*

Department of Metallurgy and Science of Materials,
Oxford University, Parks Road, Oxford, OX1 3PH, U.K.

ABSTRACT

Thermal methods have been used to calculate the degree of hydration of hydrated cement pastes. A series of degree of hydration data at different times under hydrostatic pressure was compared with a corresponding set of control samples with no pressure. Mercury porosimetric studies have been used to supplement the hydration results from thermal analytical methods.

A semiquantitative relative thermal method (combination of TGA and DTA has been used to investigate the role of calcium chloride as an admixture in cement hydration. Differences in thermal parameters demonstrate the effects of the amounts of the admixture on the production of $Ca(OH)_2$ and microstructure of the paste.

This paper illustrates the importance of thermal methods in the study of different aspects of chemistry and material properties of hydrated cement pastes.

1. INTRODUCTION

Thermal analytical techniques $e.g.$ TG, DTG, DTA and DSC are widely employed in studies of the chemistry and physics and materials properties of cements. Ordinary Portland Cement (OPC) is a multi-phased, heterogeneous inorganic substance with a typical weight percent chemical or oxide composition of (1) CaO - 63, SiO_2 - 22, Al_2O_3 - 6, Fe_2O_3 - 2.5, MgO - 2.6, SO_3 - 2.0, K_2O and Na_2O - 0.9.

The hydration products of cements are very complex but the two major components are the calcium silicate hydrate (C-S-H) gel and calcium hydroxide (C-H). Reaction with CO_2 in the atmosphere often results in the formation of $CaCO_3$. Many factors, $viz.$ water-cement ratio (w/c), composition of the cement, admixtures used, temperature, pressure etc., greatly affect the hydration products. Thermal analytical techniques can be used to follow the hydration process and also help in the qualitative and quantitative determination of the hydration products in a hydrated or hardened cement paste (HCP).

DTA studies of different hydrate phases have been reviewed (2). Recently, thermal methods for measuring the degree of hydration have been described, $e.g.$ semi-isothermal differential thermogravimetry (3) and DTA (4). Sabri and Illston (5) have reported on the complexity and distribution of evaporable water from HCP which is assumed to be removed at around 105^0C. Midgley (6) has utilized DTA and TG to record the thermal changes and weight changes involved

---

* Present address: The Department of Chemistry, University of Aberdeen, Meston Walk, Old Aberdeen, AB9 2UE, Scotland, U.K.

in driving off the water (combined water) in the course of
decomposition of calcium hydroxide at around 500^0C and concluded
that thermal techniques give most reliable results of the total
calcium hydroxide content in a cement paste.

Bensted (7) and Ben-Dor (8) reported that thermal removal of
calcium hydroxide could give double peaks in DTA thermograms. Odler
et al. (9) studied the hydration of tricalcium silicate, the major
component of OPC, using TG, DTG and DTA and found evidence that the
thermal behaviour of the calcium silicate hydrate (C-S-H) changed
with hydration.

Ramachandran (10) has reviewed the thermal analysis literature
on cements modified by $CaCl_2$ admixture. He favoured (11)
differential thermal method as an useful alternative to chemical
methods for the estimation of $Ca(OH)_2$ in HCP.

The object of this paper is to report two different experimental
studies using thermal analytical techniques, e.g. TG, DTG, and DTA
with two different hydrated cement pastes (Water/cement ratio 1.0
and 0.35 respectively). Attempts are made to inter-relate the
thermal results with those obtained from other independent techniques
e.g., Mercury Intrusion Porosimetry (MIP), Scanning Electron
Microscopy (SEM) and other results (e.g., mechanical strength,
surface areas etc) involving comparable hydrated cement systems
reported in the literature.

2. EXPERIMENTAL METHODS, RESULTS AND DISCUSSIONS

2.1 Hydration Under Hydrostatic Pressure :

The effect of applied hydrostatic pressure on the rate of
hydration of cement paste (water/cement ratio = 1.0) was investigated.
The paste was contained in a stainless steel "bomb" and a pressure of
75 MPa was applied by means of a sliding piston   (Fig. 1).
Experimental details have been
reported in an earlier paper (12).
The pressure was maintained for a
number of hours with the bomb held
at constant temperature,
immediately on release samples were
immersed in acetone to stop
further hydration.

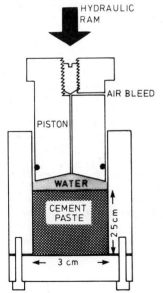

Figure 1

Diagram in cross section of
pressure "bomb".

2.1.1. "Freezing" of Hydration by acetone :

Figure 2 shows that immersion into actone freezes the hydration process. A control HCP Sample 'A', where no pressure was applied, was removed after 60 hours of hydration from the excess bleed-water. A part was immersed in acetone and thus hydration process was frozen; while the other portion continued hydrating because of the presence of excess eavporable water held tenaciously either in the pore or by the hydrated cement paste. Identical experiments were carried out on a sample (C in Figure 2) which was subjected to 60 hours under pressure, and placed in acetone **after about** half an hour. Similar results were obtained in the latter case, but being under pressure there was a total upward shift of about 8% in the hydration curves 'A' and 'C' corresponds to points described in Figure 3. The freezing of hydration by acetone is likely to be due to the removal of water from the paste by the actone.

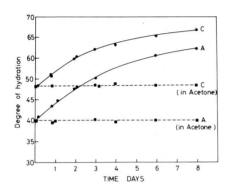

Figure 2

Degree of hydration as a function of time and "freezing" of hydration in acetone. A and C correspond to points in Figure 3.

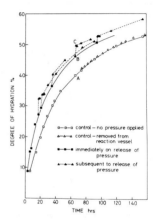

Figure 3

Degree of hydration as a function of time in experiments with and without pressure.

2.1.2. Degree of Hydration and Mechanism of Hydration :

The degree of hydration of the cement samples were measured by established thermogravimetric methods (3, 13). The samples were equilibrated at 105^0 and TGA was completed in a Stanton Redcroft TG 750 thermobalance. The method of calculating the degree of hydration has been reported (12) and was based on the method of Berger et al. (13).

The degrees of hydration is plotted against time of hydration (with and without pressure) in Figure 3. It can be seen that, relative to control samples the samples subjected to pressure and (tested immediately after release of pressure) gave consistently higher value (5 - 8%) of hydration. Further, when removed from excess bleed water, control samples with no pressure followed exactly the curve 'A' which was obtained with excess bleed water. When pressure is released the hydration is most pronounced within about half an hour following released from the pressure bomb. The mechanism of this "jump" has been explained (12) on the basis of an osmotic membrane model of cement hydration proposed by Double (14) and Birchall (15).

2.1.3. Hydration and Mercury Intrusion Porosimetry. (MIP):

Specimens corresponding to points A,B,C in Figure 3 were removed and immersed in acetone for 48 h to "freeze" the hydration process as described in Section 2.1.1. The pastes were then oven dried at $105^{0}C$ for 24 h, transferred to a desiccator for cooling and tested by MIP assuming a contact angle of 140^{0}. Figure 4 shows cumulative pore volumes against pore sizes for three pastes corresponding to the points in the hydration curves in Figure 3.

It has been demonstrated (16,17) that under similar chemical environment, an increase in the degree of hydration is associated with a decrease in the intruded pore volume. An increase in the hydration generates more hydration products which fill the pore space in a set cement paste. Successive decrease in pore volume from A to C in Figure 4 demonstrates this pore-filling. The effect of the "jump" of about 3% increase in degree of hydration in Figure 3 between B and C is reflected in the MIP curves which shows about 2% decrease in the intruded volume. However, the decrease in the pore volumes between A and B is significantly larger. Apart from the primary effect of hydration, the change in physical properties of the paste probably had secondary effects. Application of hydrostatic pressure on an open porous system such as a cement paste of water-cement ratio of 1.0 is likely to have an effect that will decrease the pore volume.

2.2 Effect of Calcium Chloride on the thermal properties and and microstructure.

2.2.1. Thermal Studies:

Calcium chloride is a major set accelerating admixture used in cement (10, 17, 18). In this study hydrated cement pastes (of water/cement ratio = 0.35) were prepared with 0.0, 0.5, 2.0 and 5.0 per cent $CaCl_2$. Combined TG, DTG and DTA studies were performed at 1, 7 and 14 days of hydration, with a Stanton Redcroft 781 Thermal Analyser. Typical combined thermal plots are shown in Figure 5.

The DTG and DTA gives characteristic endothermic peaks due to dissociation of the $Ca(OH)_2$ in the paste between 450 and $550^{0}C$. The area under the DTG curve had been used to calculate the weight loss (Δm) during this process. The area under the endothermic DTA peak (ΔE) had been considered as a measure of the energy required to bring about the mass change Δm.

Figures 6 and 7 represent the relative mass ($\Delta m/m$) and energy ($\Delta E/m$) changes respectively with various amounts of $CaCl_2$ admixture. 'm' is the initial mass of the paste at $105^{0}C$. Clear maxima were observed between 0.5% and 2% $CaCl_2$. But when these values were represented as energy change per unit mass change ($\Delta E/\Delta m$) as shown in Figure 8, minima are observed between the same 0.5 - 2% $CaCl_2$

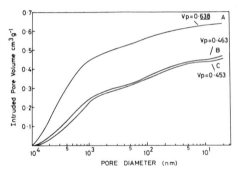

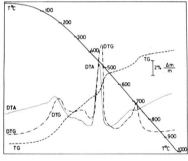

Figure 4

Cumulative Pore Volume curves corresponding to point A, B and C in Figure 3.

Figure 5

Typical experimental plots of TG, DTG and DTA responses with temperature.

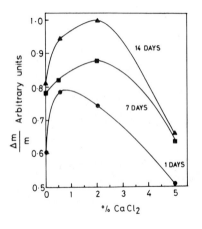

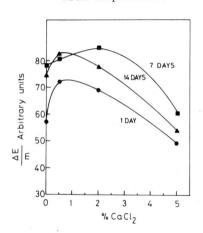

Figure 6

Variation of relative mass changes of calcium hydroxide (area under DTG peak) with calcium chloride concentration.

Figure 7

Variation of calcium hydroxide endothermic peak (DTA) with calcium chloride concentration.

range. This indicated that the $Ca(OH)_2$ within this range is thermally less stable and thus inherently different.

2.2.2. Microstructure

Collepardi and Marchese (19) have shown that tricalcium silicate paste with 2% $CaCl_2$ has a significantly higher surface area than that with 0% $CaCl_2$. Compressive strengths of pastes with $CaCl_2$ also show a maxima around 2% (10).

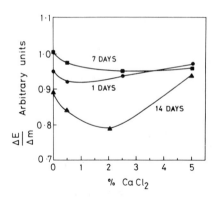

Figure 8

Variation of DTA endothermic response per unit mass change of calcium hydroxide with calcium chloride concentration.

Addition of $CaCl_2$ to the hydrating cement affects the microstructural features of the cement pastes (20). Microstructural features obtained by Scanning Electron Microscopy of cement pastes containing 0.0, 0.5, 2.0 and 5.0 percent $CaCl_2$ are shown in Figure 9. Similar studies of the various SEM features have been described (20). Microstructural features presented in Figure 9 are comparable to those reported by Ramachandran and Feldman (21).

Further, Mercury Intrusion Porosimetry studies also show microstructural changes between 0% and 0.5% $CaCl_2$ pastes. The addition of 0.5% $CaCl_2$ decreases the total intruded volume

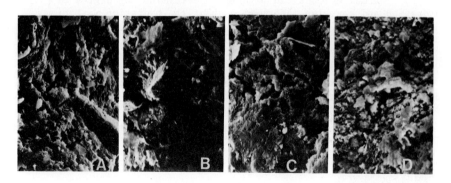

Figure 9

Figure 9, SEM Micrographs (X 2100) showing microstructural features of cement pastes at different calcium chloride concentrations :

(A) 0%, (B) 0.5%, (C) 2%, (D) 5%.

(Figure 10); while it enhances the porosity at low pore radius (Figure 11) because of an increase in hydration products as indicated by thermal studies.

Thus, the effect of $CaCl_2$ on the microstructure of cement pastes obtained from specific area, density, porosity, strength and SEM work is further supplemented by thermal analytical studies, particularly by this semi-quantitative relative method of combined DTG/DTA studies reported in this paper.

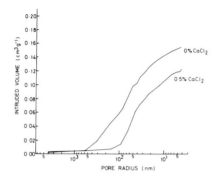

Figure 10.

Cumulative MIP curve of cement pastes.
(W/C = 0.35, 14 days hydration)

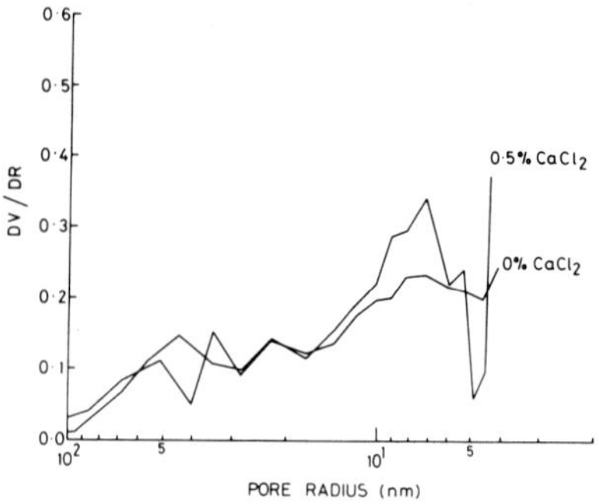

Figure 11.

Pore size distribution of cement pastes in the narrow pore range.

3. CONCLUSION :

1. Thermal methods can be used to determine the degree of hydration and can give information about the hydration mechanism.

2. The effect of admixtures can be studied by thermal methods.

3. Microstructural differences in cement pastes can be obtained by combining TGA and DTA.

4. Special care wil have to be taken about crucible packing, position of the crucible, identical pretreatment of the paste etc., before a semiquantitative approach can be used.

5. As many independent techniques as possible should be employed in order to understand the microstructural features in hydrated cement pastes.

ACKNOWLEDGEMENT :

Helpful discussions with Dr. D.D. Double and Dr. N. McN. Alford (Oxford University) and Dr. F.P. Glasser (Aberdeen University) are gratefully acknowledged.

REFERENCES :

1. S. Mindess and J.F. Young "Concrete" Pub. Prentice-Hall, Inc., New Jersey, (1981).
2. F.P. Glasser, "Differential Thermal Analysis" Vol. 1, Ed. R.C. Mackenzie, Academic Press, London (1970).
3. B. El-Jazairi and J.M. Illston, Cement and Concr. Res., 1980, $\underline{10}$ (3), 361.
4. H.W. Dorner and M.J. Setzer, Cement and Concr. Res., 1980, $\underline{10}$ (3), 403.
5. S. Sabri and J.M. Illston, "Proc. 7th. Int. Cement Congress," 1980, Vol. III, page VI, 52, Paris.
6. H.G. Midgley, Cement and Concr. Res., 1979, $\underline{9}$ (1), 77.

7. J. Bensted, Il Cemento, 1979, 76, 117.
8. L. Ben-Dor and Y. Rubinsztain, Thermochim Acta, 1979, 30, 9.
9. J. Odler and J. Schneppsthul and H. Dory, Thermochim Acta 1979, 29, 261.
10. V.S. Ramachandran, "Calcium Chloride in Concrete", Applied Science Pub., London 1976.
11. V.S. Ramachandran, Cement and Concr. Res., 1979, 9 (6), 677.
12. A.A. Rahman and D.D. Double, Cement and Concr. Res., 1982, 12 (1), 33 - 38
13. R.L. Berger, J.H. Kung and J.F. Young, J. Testing and Evaluation, 1976, 4 (1), p. 85.
14. D.D. Double, A. Hellawell and S.J. Perry, Proc. Roy. Soc. (Lond.), 1978, A 359, P 435.
15. J.D. Birchall, A.J. Howard and J.E. Bailey, Proc. Roy Soc. (Lond.), 1978, A360, P 445.
16. M. McN. Alford and A.A. Rahman, J. Materials Sci., 1981, 16, 3105.
17. J.F. Young, Powder Technology, 1974, 9, 173.
18. V.S. Ramachandran, Can. J. Civil Eng. 1978. 5, 213.
19. M. Collepardi and B. Marchese, Cement & Conc. Res., 1972, 2, 57.
20. P.J. Sereda, R.F. Feldman and V.S. Ramachandran. "Proc. 7th Int. Cement Cong." Paris 1980, Vol. I, p. VI-1-32.
21. V.S. Ramachandran and R.F. Feldman. Il-Cemento, 1978, 75, 311.

SOME ASPECTS OF THE APPLICATION OF TMA IN THE SELECTION
OF BINDING AND CONSTRUCTION MATERIALS

H.G. Wiedemann, Mettler Instruments AG, CH-8606 Greifensee
M. Roessler, Institut für Steine und Erden, TU Clausthal,
D-3392 Clausthal-Zellerfeld

To show the possibilities of the TMA (Thermo Mechanical Analysis) combined with other thermoanalytical methods in the section of binding- and construction -materials, some classical examples out of this section have been studied in this paper.

INVESTIGATIONS IN THE SYSTEM $CaSO_4-H_2O$

The hemihydrate of calcium-sulfate is one of five phases [1] of this system. Hemihydrate (HH) appears in two different forms α and β, both of which have almost rhomboedric lattice structure [2,3]. The typical characteristics of the α-HH are large, well-grown and compact crystals, whereas β-HH forms cryptocrystalline and flocculent particles full of fissures.

The knowledge of the exact α/β-percentage of hemihydrates is of great technological interest, because then there is a possibility to regulate the chemical and physical properties (e.g. setting-conditions, strength, porosity, etc.). Already during the production process (until today hemihydrates are mainly produced by cooking or burning natural calcium-sulfate-dihydrate) a possibility to change the α/β-ratio and to optimize the above mentioned and the later properties of the raw materials is given [4].

A distinction by x-ray diffraction is very sophisticated. The differences between the interlattice plane distances and so the angular positions are very small. There are only a few but remarkable intensity differences of some interferences [3-6], even if the samples are specially treated [7]. Moreover the XRD-diagrams of the hemihydrates and another phase of the system (anhydrite III; A III) are very similar.

The difficulties of the DTA of plaster of Paris and of HH have been studied extensively, but TMA has hardly been used for research of this matter [8]. Compressed powder beds of pure α- and β-HH (α = 97.5 %; β = 99 % phase content) with the same granulometric properties were studied in a special TMA-cell (Fig.1) and combined DSC and TGA were carried out. The results of that are reported in Fig.2. The TG-curves correspond with the chemism of the hemihydrates: first step to approx. 100°C evaporation of physically adsorbed water (possibly partial rehydration to dihydrate/α = 1.3 %; β = 0.54 %); second step to 200°C stoichiometric evaporation of crystal water (α=β= 6.21 %); third step from 600° to 700°C decalcination of $CaCO_3$, impurity of the raw product. The contents of $CaCO_3$ are in α = 3.7 % and in β = 4.2 %.

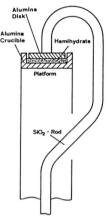

Fig.1: TMA-cell
Mettler TA 3000

A distinction of the α- and β-form of HH with TGA is not possible.

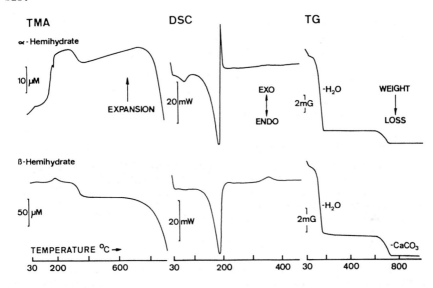

Fig.2: TMA-, DSC- and TGA-curves of α- and β-hemihydrate

A comparison of the DSC-curves shows obvious differences between the α- and β-HH. But unfortunately these are not easily suited for quantitative analysis of α- and β-forms in mixtures [4-8]. — By integration of the first peaks one gets by using the hydration heat of the conversion from dihydrate to hemihydrate a dihydrate content of 2.8 % for α and 0.58 % for β. In this case the calculation seems to be only correct for β-HH as here - because of the huge specific surface - the adsorbed water has been quantitatively combined as dihydrate water. In the case of α-HH the main part of the water is most probably adsorbed at the surface only. — The second peak renders a hydration heat of 211.00 J/g for α and 178.44 J/g for β. This shows a very good coincidence with [9] and also with [1], considering that there a mixture of 50 % α- and 50 % β-HH has been used (194.8 J/g in [1] to 194.7 J/g in this paper). — The exothermic third peaks [conversion from A III (hexagonal-trapezoedric) to A II (orthorhombic - dipyramidal)] render 7.199 J/g for α (at 200°C) and 10.7 J/g for β (at 350°C). No literature references are known for comparison. For the exothermic peak for α-HH at 200°C, which - depending on the gypsum- does not always occur, there are many interpretations from overshot to lamination of α-A III [8].

Fig. 3: Schematic graph of the formation of hydrate phases with cement hydration. A - D SEM-microphotos of cement pastes after different times of hydration (CSH = Calciumsilicatehydrate, AFm-phase = Fe_2O_3 - containing Tetracalciumaluminahydrate). See next page.

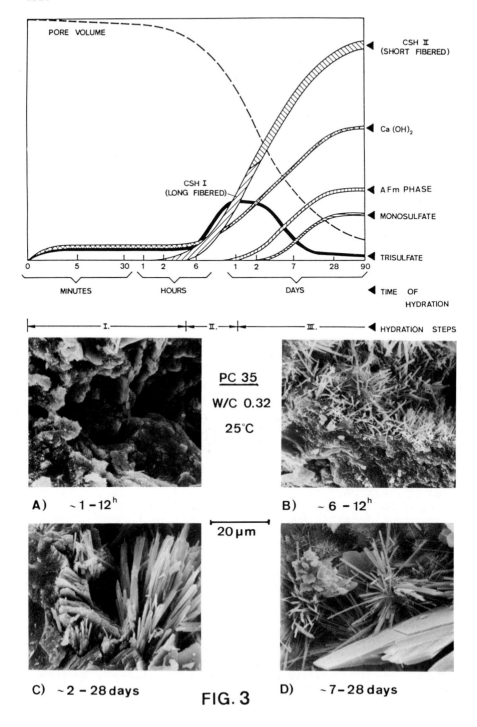

FIG. 3

The TMA curves show a distinct difference between α- and β-HH especially from 30° to 200°C up to about 750°C. The β-HH experiences a little change in length except in the region of dehydration to A III and at the point of conversion from A III to A II. The α-HH shows a rather great expansion between about 80 and 180°C; at the range of conversion from A III to A II (370°C) the shrinkage isn't as distinct as in the case of β-HH. Additionally the α-HH expands slightly in the range between 400 and 750°C. The shrinkage from 750°C upwards is smaller than with β-HH.

Further TMA are necessary to show if the expansion of α-HH between 80 and 180°C has to be traced back to a swelling of the powderbeds by water vapour and if this effect gives perhaps the possibility of a quantitative distinction of α- and β-HH.

INVESTIGATIONS IN THE FIELDS OF CEMENTS

In the fields of cements the different methods of TA offer a good possibility to pursue the hydration progress. As the cement hydration is a very complex process, which cannot be exhaustively treated in a few words and as it still isn't completely researched, it is shown schematically in Fig.3 [10]. According to Fig.3 the hydration of a normal portland cement is mainly subdivided into steps, which however blend in their time sequence:

First step: stiffening of the still viscous cement paste by the formation of longer trisulfate needles starting from the surface of the cement particles.

Second step: development of strength by originally long fibered CSH (Fig.3 B,C), which link the spaces between the cement particles filled with water.

Third step: filling of the interspaces by fine crystalline CSH (Fig.3 A-D), which can be long fibered (Fig.3 B-D) depending on the seize of the interspaces (pores) - at the same time formation of $Ca(OH)_2$; partially very large hexagonal crystals (Fig.3 D).

The results of the TGA with the aim to pursue the hydration progress are shown in Fig.4. The best conditions as given in the picture yield the shown 5 thermogravimetric curves. They show the hydration progress especially by the increase of $Ca(OH)_2$-content, but also by the characteristic changes of hydration products. To show smaller peaks better the thermogravimetric curves -due to the big mass loss below 200°C- are figured from 200°C upwards. - At the shorter hydration times 1 h and 6 h the decomposition of ettringite (6-calcium aluminate trisulfate-32-hydrate) at about 130°C is anyhow completely superimposed by the dehydration of the CSH-phases.

The percentage mass losses at the different steps (automatic step analysis) are additionally listed for better presentation in table 1. As the test samples have not been predried, one can assume that at low hydration times (especially after 1 h and 6 h) a few percent H_2O are adsorbed at the surface. So one gets an increase of the CSH-phases (decomposition between 30 - 420°C; not considering here exotic phases like brucite, etc. [11]). An increase of $Ca(OH)_2$ and $CaCO_3$ (carbonation of free lime) can also be observed. The degree of hydration of a pc can be calculated from the content of $Ca(OH)_2$ [12-14]. The mass losses between 790 and 950°C can possibly be explained by alkali-volatilization [11], which decreases with increasing hydration time as

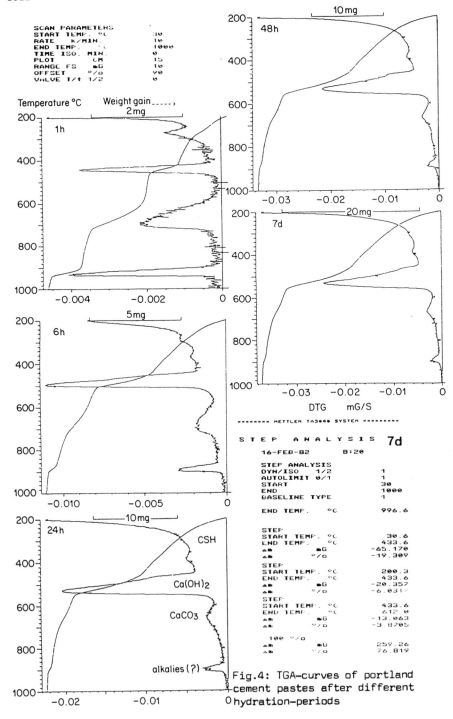

Fig.4: TGA-curves of portland cement pastes after different hydration-periods

TABLE 1

Results of the TGA with PC-samples after different hydration times

hydration time	30 - 420°C [CSH-phases]	mass loss (%) 420 - 566°C [Ca(OH)$_2$]	566 - 787°C [CaCO$_3$]	787 - 950°C [alkalies]
1 h	22.6 *	1.1	1.05	0.3
6 h	20.5	4.4	1.11	0.28
24 h	17.7	7.3	1.4	0.2
48 h	17.2	11.7	1.7	0.16
7 d	19.3	14.3	2.1	0.07

* 0.4 % mass loss at 270°C possibly decomposition of syngenite [K$_2$Ca(SO$_4$)$_2$H$_2$O] [11].

the alkalies are bound stronger.

INVESTIGATIONS OF CEMENT MORTAR

A completely hydrated pc-paste shows a more or less big shrinkage in a TMA from 130°C up to 1000°C due to the decomposition of the above mentioned constituents. This behaviour changes completely with a cement mortar. Fig.5 shows TMA-curves of a pc, a cement mortar and a quartz. The influence of the two parts of pc in the mortar is almost neglectable. The (quartz-)sand superimposes the shrinkage of the pc

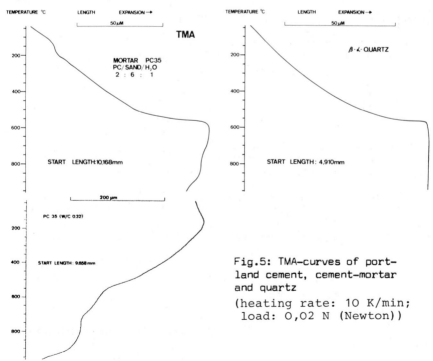

Fig.5: TMA-curves of portland cement, cement-mortar and quartz
(heating rate: 10 K/min; load: 0,02 N (Newton))

nearly completely; only with the decomposition of the CSH-phases and above the quartz transition in the region of the decalcination of $CaCO_3$ a small influence can be observed.

EXPERIMENTAL

The thermoanalytical measurements of this paper have been carried out with a measuring and evaluation system for DSC, TMA and TGA - the Mettler TA 3000 system. The TA-processor TC10 is the basic unit for all configurations of the TA-system. Control and regulation of the respectively used analyzers are brought about by the TA-processor. The evaluation with the basic programs are realised by the setvalue memory. The test data, coming up while analysing are stored in the working storage and can be recalled for further evaluations at any time. The printed graphs are automatically provided with scales.

CONCLUSION

In the section of binding- and construction-materials the TA offers -by the usage of combined methods including TMA- a good possibility to analyse the raw materials, to determine their purity and to predict certain properties of the hydrated materials. As further studies will show, the TMA offers an excellent possibility to predict the properties of construction materials especially in their aplication at higher temperature (load capacity, fire protection, etc.), too.

REFERENCES

1. Ullmanns Encyklopädie der technichen Chemie, Vol.
2. Der Baustoff Gips, VEB Verlag für Bauwesen, Berlin, 1978
3. H. Lehmann, K. Rieke, Sprechsaal 106, (1973), p.924 - 930
4. M.J. Ridge, J. Beretka, Rev. Pure and Appl. Chem. 19, (1969), p.313 - 316
5. W. Krönert, P. Haubert, ZKG 25, (1972), p.546 - 552
6. H.-G.Wiedemann, G. Bayer, Z. Anal. Chem. 276, (1975), p.21 - 31
7. A. Abdul-Aziz Khalil, Thermochim. Acta 53, (1982), p.59 - 66
8. K. Rieke, Diss. TU Clausthal, 1974, p.23, 27
9. V. Schlichenmaier, Thermochim. Acta 11, (1975), p.335 - 338
10. F.W. Locher, W. Richartz, VDZ-Zement-Taschenbuch 1979/80, Bauverlag Wiesbaden - Berlin
11. J. Bensted, il cemento 3/1979, p.117 - 126; 3/1980, p.169 - 182
12. F.W. Locher, W. Richartz, S. Sprung, ZKG 12, (1976), p. 294
13. H. Lehmann, F.W. Locher, D. Prussog, TIZ 94, (1970), p.230 - 235
14. H. Dörr, Diss. TU Clausthal, 1979

Test conditions of Fig. 2:
heating rate: 10 K/min
sample length TMA: α-HH 3,433 mm; β-HH 3,534 mm
sample weights: DSC α-HH 35,068 mg; β-HH 18,681 mg
               TGA α-HH 146,35 mg; β-HH 107,03 mg

FLAME RETARDANCY EFFECTS ON THE THERMAL DEGRADATION OF POLY(ETHYLENE TEREPHTHALATE) FABRICS AS STUDIED BY THERMOGRAVIMETRIC ANALYSIS*

J.D. Cooney, M. Day and D.M. Wiles
Division of Chemistry
National Research Council of Canada
Ottawa, Canada, K1A 0R9

INTRODUCTION

The derivation of kinetic data from thermogravimetric analysis (TGA) is a subject of immense activity, and several comprehensive reviews have been written on the subject [1-6]. The actual significance of the numbers obtained by the many interpretive methods for the rate coefficient k, the pre-exponential factor A, activation energy E_a and order of reaction n have to be interpreted with caution [6-11]. This arises due to difficulties in establishing the significance of a particular kinetic or mathematical model for data obtained from non-isothermal experiments when both the temperature and weight loss are changing simultaneously. The complexity of the situation with respect to the thermal degradation of poly(ethylene terephthalate) (PET) is illustrated by the non-isothermal data of Birladeanu et al. [12] obtained using the Coats-Redfern method of analysis [13]. They found that the apparent kinetic parameters determined in air and nitrogen were dependant upon heating rate, mean molecular weight and degree of conversion, reporting values for E_a between 72.4 and 272.2 kJ/mol. Interestingly, with the exception of those by Granzow [14] and by Bechev [15] no studies have reported the effect of flame retardant chemicals on the kinetic parameters for the thermal degradation of PET. Most of the studies in the literature present TGA curves for the treated and untreated PET and then include general statements such as "the flame retardants were responsible for an increase in the initial weight loss, a decrease in the temperature of maximum decomposition and an increase in the char residue".

In this paper we report the application of various experimental techniques and analytical procedures to derive kinetic parameters for the thermal decomposition of two samples of PET fabric. Based upon the results of these preliminary evaluations, experiments and analyses will be performed on flame retarded PET samples in an attempt to elucidate the role of the retardants in the kinetic processes occurring during the thermal degradation of PET. Specifically, the multiple heat rate method of Ozawa [16, 17] was employed for part of the kinetic study and the results were compared with those obtained using the temperature jump method of Flynn and Dickens [18]. In addition the multiple heat rate data have been analysed according to the method of Kissinger [19].

EXPERIMENTAL

Materials. Two fabrics were used in the TGA analysis. One was a 100% spun Dacron Type 54 polyester yarn woven fabric (127 g/m^2) obtained from Testfabrics Inc., Middlesex, N.J. (Style 767). The second fabric was also 100% polyester (132 g/m^2) obtained from a local supplier and referred to as "untreated PET".

Apparatus. Thermogravimetric analysis was performed on a DuPont 951 TGA/1090 Thermal Analyzer. Rectangular fabric samples (20 ± 3 mg) were stacked on top of one another in a platinum sample pan and heated in either air or nitrogen gas at a flow rate of 50 ml/min. The Chromel-Alumel sample

* Issued as NRCC #20079

thermocouple was positioned near the side of the platinum pan approximately in line with the bottom of the attaching triangular ring. This sample thermocouple was calibrated at room temperature.

For the heating rate method, an ICTA #761 nickel Curie Point Standard (T_2 = 353°C [20]) was used as a single point temperature calibration of the actual temperature experienced by the sample in the bottom of the platinum pan.

The Curie Point temperature of the #761 nickel standard was determined in air at each heating rate using a 17.7 mg sample placed in the platinum pan used in each experiment. The appropriate temperature correction at each heating rate was obtained from a comparison of the machine calculated derivative peak, T_2 with the recommended T_2 value (353°C).

Temperature Jump Method. Only a preliminary experiment with the temperature jump method is reported. The temperature was maintained at each temperature for 7 minutes and continually cycled between the three temperatures: 351, 360 and 368°C in a flowing nitrogen atmosphere.

$$351°C \underset{\longleftarrow}{\overset{\longrightarrow 360°C \longrightarrow}{}} 368°C$$

The average rate of weight loss (ie. the rate of weight loss at the mid-point of each step) was plotted against the fraction weight loss [18] as shown in Figure 1.

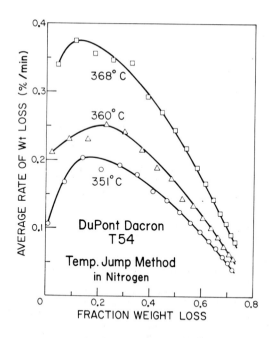

Figure 1. Variation of average rate of weight loss as a function of fraction weight loss for DuPont Dacron T54 in nitrogen at temperatures of 351°(O), 360°(Δ), and 368°C (□).

Heat Rate Method. Each fabric was heated at eight heating rates, "β", of 0.1, 1, 2, 5, 10, 20, 30, and 50 deg/min from 30 → 900°C with β converted to deg/sec for construction of Figures 2 - 4. TGA curves were analyzed at fraction weight loss "α" values of 0.01, 0.05, 0.10, 0.20, 0.35, 0.50, 0.65, 0.80, 0.85, 0.90, 0.95 and 0.99, and depicted by the letters A → L, respectively, in Figures 2 and 3. All the heat rate method experiments were

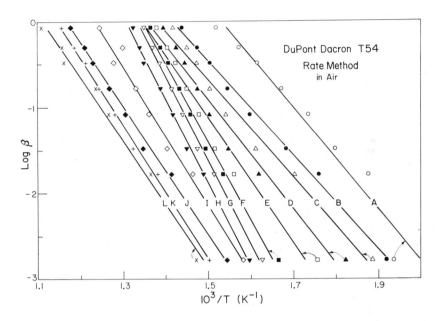

Figure 2. Ozawa plot of log heating rate versus reciprocal absolute temperature for DuPont Dacron T54 in air at the following fractional weight losses: A: 0.01 (O); B: 0.05 (●); C: 0.1 (Δ); D: 0.2 (▲); E: 0.35 (□); F: 0.5 (■); G: 0.65 (∇); H: 0.8 (▼); I: 0.85 (◇); J: 0.9 (◆); K: 0.95 (+); and L: 0.99 (X).

conducted in flowing air. A proportional band heater board was installed for these experiments and operated at a setting of 9 for all experiments except those at heating rates of 20, 30, and 50 deg/min where a setting of 16 was used. The sample pan was zeroed with the system assembled and the air flowing at 50 ml/min. The system was opened to load the sample then reassembled prior to weighing the sample.

The data obtained from the heat rate method experiments was also subjected to analysis by the Kissinger method [19]. This involved obtaining the values of the temperature (Tmax) at the maxima of the first derivative weight loss curve. The plot of $\ln[\beta/(T_{max})^2]$ vs $10^3/T_{max}$ for the three identified peaks is shown in Figure 4. This method of analysis, which gives $-E_a/R$ from the slope, only applies to a first order reaction but allows comparison with the activation energies calculated by the Ozawa method from the same experimental data.

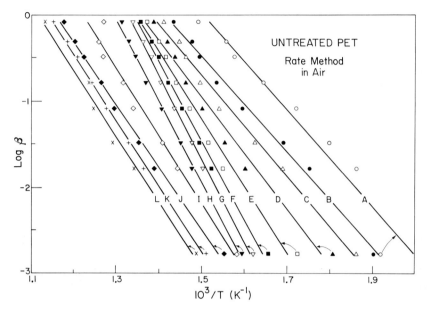

Figure 3. Ozawa plot of log heating rate versus reciprocal absolute temperature for untreated PET in air at the following fractional weight losses: A: 0.01 (O); B: 0.05 (●); C: 0.1 (Δ); D: 0.2 (▲); E: 0.35 (□); F: 0.5 (■); G: 0.65 (▽); H: 0.8 (▼); I: 0.85 (◇); J: 0.9 (◆); K: 0.95 (+); and L: 0.99 (X).

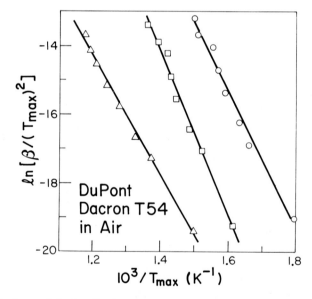

Figure 4. Kissinger plots for the three main stages in the thermal decomposition of DuPont Dacron T54. First stage (O); Second stage (□); Third stage (Δ).

TABLE I. ACTIVATION ENERGIES FOR DUPONT DACRON T54 IN NITROGEN VIA THE TEMPERATURE JUMP METHOD

Fraction Weight Loss α	Activation Energy, E_a (kJ/mol)	Correlation Coefficient r
0.05	154.7	-0.993
0.10	131.1	-0.971
0.20	111.5	-0.977
0.30	110.8	-0.986
0.40	111.5	-0.986
0.50	114.2	-0.964
0.60	122.2	-0.989
0.70	139.8	-0.994

RESULTS

Temperature Jump Method. The activation energies determined from the Temperature Jump Method (Fig. 1) for DuPont Dacron T54 fabric are listed in Table I. This was a preliminary experiment in nitrogen without temperature correction. Since only three data points are utilized in this technique, the "Student's t" statistical analysis (Probability = 0.1-0.2) indicates "no significance" in the results. The trend in the E_a values is decreasing, then increasing as the conversion level (α) increases. This experiment will be repeated in air at slightly lower temperatures to improve the data for α = 0.1-0.3 and hopefully will generate more accurate and meaningful E_a values.

Heat Rate Method. Similar Flynn-Wall-Ozawa isoconversional plots were obtained for two different samples of PET fabric as illustrated in Figs. 2 and 3. The best fit straight lines drawn in these Figures were obtained by least-squares linear regression and the correlation coefficients quoted in Table 2 indicated a highly significant correlation (0.1% level) as tested by the "Student's t". Both fabrics gave similar "global" E_a values at each iso-conversional level listed in Table 2. Examination of the conversion levels 0.01 to 0.99 (Figs. 2 & 3, Table 2) indicates three distinct stages in the PET thermal degradation in air as listed in Table 3. There is also good agreement between the average E_a values obtained for the two PET fabrics (Table 3). The three stages of PET degradation have average "global" E_a values of 107, 176 and 138 kJ/mol. The general trend is increasing, then decreasing E_a's as "α" increases with a maximum E_a of 180 kJ/mol occurring at 75% weight loss.

In addition to the three main stages in PET oxidative thermal degradation, a pre-stage was indicated during the slowest heating rate (0.1 deg/min). This pre-stage occurs at low conversion, α = 0.01.

Application of the Kissinger method to the heat rate data for DuPont Dacron T54 gave the plot shown in Fig. 4 and activation energies for three stages in the thermal degradation of PET in air (Table 4). The pre-stage and stage 1 were only visible in the weight loss first derivative curve during the slowest heating rate (0.1°C/min) with the stages coalescing at more rapid heating rates. At higher heating rates, stage 1 was a shoulder on the larger derivative curve of stage 2 creating difficulty in determining Tmax and generating error in E_a. The Tmax for stages 2 and 3 were more reliable and generated E_a values of 201 and 145 kJ/mol, respectively, which were similar to those obtained from the isoconversional plots.

TABLE 2. ACTIVATION ENERGIES FOR PET IN AIR VIA THE HEAT RATE METHOD

	DUPONT DACRON T54		UNTREATED PET	
Fraction Weight Loss α	Activation Energy, E_a (kJ/mol)	Correlation Coefficient r	Activation Energy, E_a (kJ/mol)	Correlation Coefficient r
0.01	107.7	-0.975	102.7	-0.973
0.05	99.94	-0.998	101.1	-0.996
0.10	103.4	-0.991	107.3	-0.995
0.20	115.9	-0.982	119.2	-0.987
0.35	133.0	-0.980	142.9	-0.988
0.50	167.5	-0.995	168.4	-0.996
0.65	179.7	-0.998	177.6	-0.998
0.75	181.9	-0.998	177.9	-0.999
0.80	178.5	-0.998	173.1	-0.998
0.85	146.3	-0.992	143.8	-0.988
0.90	134.2	-0.998	132.4	-0.997
0.95	142.2	-0.999	139.8	-0.999
0.98	141.8	-0.996	142.4	-0.998
0.99	134.1	-0.993	139.2	-0.994

TABLE 3. SUMMARY OF "GLOBAL" ACTIVATION ENERGIES FOR PET IN AIR

Region	Graph Designation Figs. 2 & 3	Fraction Weight Loss α	Average E_a (kJ/mol)	
Stage 1	A → D	0.01 - 0.20	106.7*	107.6‡
Cross-over	E	0.35	133.0	142.9
Stage 2	F → H	0.50 - 0.80	176.9	174.3
Cross-over	I	0.85	146.3	143.8
Stage 3	J → L	0.90 - 0.99	138.1	138.5

* values for DuPont Dacron T54
‡ values for Untreated PET

TABLE 4. CALCULATED FIRST ORDER ACTIVATION ENERGIES BY KISSINGER'S METHOD

Region	Fraction Weight Loss* α	Activation Energy, E_a (kJ/mol)	Correlation Coefficient r
Pre-stage	0.00 - 0.09	---	---
Stage 1	0.09 - 0.36	167.8	-0.990
Stage 2	0.36 - 0.92	201.1	-0.994
Stage 3	0.92 - 1.00	144.7	-0.997

* values determined from the 0.1°C/min experiment.

DISCUSSION

Initial emphasis has been placed on TGA in air since it relates to the burning process and the effect of flame retardants on PET. A pre-stage and three main stages were observed in the thermal degradation of PET in air when examined by the heat rate method. The three stages were found to have "global" average E_a values of 107, 176 and 138 kJ/mol. These values of E_a are in close agreement with the values of 122-185 kJ/mol reported by Zimmermann and Kim [21] for isothermal studies between 280 - 300°C, and similar to values of 155 and 126 kJ/mol reported by Granzow et al. [14] at temperatures reportedly consistent with the surface burning of PET (ie. 400 - 500°C).

At the symposium, the results reported so far, will be compared with those obtained on flame retardant PET.

REFERENCES

1. J.H. Flynn and L.A. Wall, J. Res. Nat. Bur. Stand., 1966, 70A, 487.
2. L. Reich and D.W. Levi, Macromol. Rev., 1967, 1, 173.
3. W.W. Wendlandt, "Thermal Methods of Analysis", 2nd ed. Wiley-Interscience New York, 1974.
4. R.R. Baker, Thermochim. Acta., 1978, 23, 201.
5. J.H. Flynn, "Analysis of Kinetics of Thermogravimetry: Overcoming Complications in Thermal History" in Thermal Analysis in Polymer Characterization, Ed. E. Turi, Heyden, Philadelphia (in press).
6. B. Dickens and J.H. Flynn in Advances in Chemistry Series, Ed. C.D. Craver, ACS Publication (in press).
7. J.R. MacCallum, Brit. Polym. J., 1979, 11, 120.
8. T.B. Tang and M.M. Chaudri, J. Therm. Anal., 1980, 18, 247.
9. G.G. Cameron and A. Rudin, J. Polym. Sci., Polym. Phys. Ed., 1981, 19, 1799.
10. M. Arnold, G.E. Veress, J. Paulik and F. Paulik, "Thermal Analysis", Proc. 6th ICTA, 1980, Bayreuth, H.G. Wiedemann, Ed., Birkhaeuser Verlag, Boston, 1, p. 69.
11. T.A. Schneider, "Thermal Analysis", Proc. 6th ICTA, 1980, Bayreuth, W. Hemminger, Ed., Birkhaeuser Verlag, Boston, 2, p. 387.
12. C. Birladeanu, C. Vasile and I.A. Schneider, Makromol. Chem. 1976, 177 121.
13. A.W. Coats and J.T. Redfern, Nature (London) 1964, 201, 68.
14. A. Granzow, R.G. Ferrillo and A. Wilson, J. Appl. Polym. Sci., 1977, 21, 1687.
15. K. Bechev, R. Lazarova, L. Dimova and K. Dimov, Khim. Ind. (Sofia) 1980, 8, 341.
16. T. Ozawa, Bull. Chem. Soc. Japan, 1965, 38, 1881.
17. J.H. Flynn and L.A. Wall, Polym. Lett. 1966, 4, 323.
18. J.H. Flynn and B. Dickens, Thermochim. Acta., 1976, 15, 1.
19. H.E. Kissinger, Anal. Chem., 1957, 21, 1702.
20. P.D. Garn, O. Menis and H.G. Wiedemann, "Thermal Analysis", Proc. 6th ICTA, 1980, Bayreuth, H.G. Wiedemann, Ed., Birkhaeuser Verlag, Boston, 1, p. 201.
21. H. Zimmermann and N.T. Kim, Polym. Eng. Sci., 1980, 20, 680.

THERMAL CHARACTERIZATION AND STABILITY STUDIES OF A POLYESTER-BASED POLYURETHANE LAMINATE

J. T. Stapler and F. H. Bissett
Chemistry Branch, Materials Application Division
U.S. Army Natick Research and Development Laboratories Natick MA

ABSTRACT

Thermal characterization and stability studies of a textile laminate consisting of a polyester based polyurethane foam sheet having an adhesive bonded layer of charcoal on one side, and a flame-laminated nylon tricot on the other have been performed using thermogravimetric analysis (TGA) and differential scanning calorimetry (DSC) in nitrogen and air. Comparison of the thermograms from the individual components and the laminate itself provided a unique means of characterizing laminates for quality control purposes.

INTRODUCTION

This study was undertaken to develop a reliable, thermal quality control method for a laminated material prepared and furnished by contractors. The flexible textile laminate studied consisted of a polyester-based polyurethane foam sheet having an adhesive-bonded layer of charcoal on one side and a flame-laminated nylon tricot on the other. The tricot material was 6-nylon while the precise polyester-urethane was not identified.

Thermal analysis by thermogravimetric and differential thermogravimetric analyses (TGA-DTG) and differential scanning calorimetry (DSC) on materials similar to the laminate components have been reported: 6-nylon [1], polyurethane foams [2], latex rubber [3] and charcoal [4]. Although flame-bonded laminates were developed more than a decade ago [5] very little is reported about their behavior under thermal stress. The nearest parallel to our investigation involved thermal analysis on fiber and powder blends [6].

In our study, we concentrated our attention on thermal responses of the laminate as it relates to those of the separate components. With quality control as the main objective, the data would fingerprint and identify components as they appear in the thermogram of the laminate and detect missing or improperly substituted materials. Other material properties would be ascertained such as physicochemical changes, temperature degradation reactions and thermal stability.

EXPERIMENTAL PROCEDURES

INSTRUMENTATION

a. DuPont 1090 Thermal Analyzer having a digital temperature programmer, printer/plotter, visual display, disk memory and data analyzer capability; Modules: 950 Thermogravimetric Analyzer, and 910 Differential Scanning Calormeter calibrated as directed by operational procedures.

b. Revolving Leather Punch, C.S. Osborne and Co., Harrison, NJ.

MATERIALS

a. Dry nitrogen, Matheson Co., was bubbled through a basic solution of paragallol to insure oxygen removal and dried by passage through a silica gel trap.

b. Dry air was passed through a silica gel trap prior to use.

MATERIALS INVESTIGATED

Material	Designation	Source
Tricot knit nylon, dyed black knitted on 28 gauge machine, 40 denier	NYL	Continental Knitting Co., NY
Unpigmented polyester-based polyurethane foam sheet, 90 mil	PEPU	William T. Burnett Co., MD
Charcoal-latex rubber binder slurry containing Barneby-Cheney XZ charcoal (325 mesh), latex, water, unidentified emulsifier or antifoaming agent	CR	Grant Chemical Co., NJ
Laminate: 6-nylon/polyester-based 90 mil polyurethane foam/charcoal-latex	NYL/PEPU/CR	Peacedale Processing Co., RI
Laminate, same as above, 50 mil, without flame retardant	NYL/PEPU/CR WOFR	Same as above
Laminate, same as above, 50 mil, with flame retardant	NYL/PEPU/CR WFR	Same as above

SAMPLE PREPARATION

For TGA-DTG runs, material samples of uniforms diameter were prepared using Osborne punch #7 (13/64" dia) and #2 (3/32" dia) for DSC. The carbon slurry was dried to form brittle film and broken to fit sample pans. All samples were pre-dried under vacuum at 125°C for 3 hours and stored in screw-capped vials. In runs involving laminated samples, disks were placed in sample pans with nylon sides down.

PROCEDURES

Samples for TGA were prepared as noted above and placed in the 950 Analyzer which was continuously flushed with nitrogen or air (50 cc/min). Runs were then temperature programmed from ambient to 700°C at a rate of 10°C/min. The TGA-DTG curves obtained were analyzed using the 1090 TGA Data Analysis Program. This program calculated the weight loss for each transition and reported the extrapolated onset and endpoint temperatures as well as the inflection point.

Samples for DSC were run under the same conditions using the 910 Calorimeter except the maximum temperature was 600°C. The curves obtained were analyzed using the 1090 DSC Interactive Data Analysis Program which calculated peak areas and onset and inflection temperatures.

RESULTS AND DISCUSSIONS

The laminate (NYL/PEPU/CR) and all components (NYL, PEPU and CR) were each run six times in nitrogen or dry air in TGA-DTG and DSC modes. The reproducibility in nitrogen was excellent. By data analysis, thermal transition peak temperature varied $\pm 1.5^\circ$C. In air, general curve traces were good and peak temperatures varied $\pm 4^\circ$C. However, some traces were marked by a few non-reproducible eruptions during oxidative, exothermic reactions.

Typical reproducible TGA-DTG curves (fingerprints) from nitrogen and air runs are shown in Figures 1 and 2 for the laminated material NYL/PEPU/CR. DSC curves are illustrated in Figures 3 and 4. Use of the DuPont 1090 data analysis mode for TGA gave DTG curves as well as a print out of extrapolated temperature transition values in terms of weight loss and residue. DSC data was obtained by the aforementioned "interactive" means. DSC data (Tables 1 and 2) reflect energies involved in physical and chemical changes such as degassing, melting, oxidation and degradation. Certain DSC transitions could be correlated to TGA-DTG transitions. The influence of some exothermic reactions was reflected in reproducible, but distorted curves (Figures 2 and 6). This distortion is caused by a rapid rise in temperature from the reaction which exceeds the oven temperature. Once the reaction is complete the temperature falls back to the programmed level.

Figures 5 and 6 are DTG curves in N_2 and air of the components which make up the laminate and the laminate itself. The laminate transition which occurs at 400°C (inflection temp.) in Figure 5 corresponds to the largest weight loss and arises from decomposition of foam (PEPU) and charcoal (C).

The peak at 463°C is due to nylon decomposition. The PEPU transition at 300°C either does not occur in the laminate or is reduced and contributes to the shoulder at 340°C. It has been reported that some polyurethane decomposition occurs during the flame-lamination process [7]. A comparison of a delaminated PEPU sample from a nylon/foam laminate (NYL/PEPU) and a pure PEPU specimen showed no detectable differences by DTG or DSC. The almost identical N_2 and air thermograms for PEPU (Figures 5 and 6) suggest a non-oxidative degradation mechanism. Both NYL and CR and the laminate exhibit large oxidative changes (Figure 6).

The DSC thermograms of the components (Figure 7) show a large endotherm at 59°C for CR degassing, an endotherm at 220°C for nylon melting and at 298°C for PEPU melting. These same endotherms are evident in the laminate except the 298°C peak is broader and is shifted slightly (287°C). The DSC thermogram of the laminate in air is much less definitive.

An example of the ability of this procedure to detect changes in the laminated material is shown in Figure 9. The material used for thermogram B contains the same components as the control A except the PEPU is only half as thick. Thermogram C contains the same amount of components as the control A but has been treated with a flame retardant which accounted for the observed major DTG inflection shift from 450°C to 389°C.

CONCLUSION

It has been demonstrated that currently available thermal analysis equipment provides dependable, reproducible data for the characterization of textile laminates and are suitable for product quality control. The data also reflect the thermal stability of the laminate material in relation to the step by step addition of components. The reinforcement, reduction or disappearance of transitions in the final product represent the net effect of all component transitions which occur at or near the same temperature. While unresolved transitions are minimal, they are essential to the laminate "fingerprint" process.

ACKNOWLEDGEMENT

The authors are grateful to George P. Dateo (for advice and equipment use), Armando C. Delasanta and Gil M. Dias (for all samples used in this study), Materials Application Division, Individual Protection Laboratory.

DISCLAIMER

The use of any product or equipment in the conduct of this study and listed by trade name or manufacturers does not constitute an official indorsement by the U.S. Government.

REFERENCES

1. S. L. Madorsky, "Thermal Degradation of Organic Polymers," Interscience Publishers, NY, 1964, p. 262.
2. a. K. C. Frisch and S. L. Reegen, "Advances in Urethane Science and Technology, 1978, 6, p. 173.
   b. J. H. Engel, Jr., S. L. Reegen and P. Weiss, J. Applied Polym. Sci., 1963, 7, 1979.
   c. J. H. Saunders and J. K. Bachus, Rubber Chem. and Technology, 1966, 39, 461.
   d. N. Grassie and M. Zulfiquar, J. Polym. Sci., 1978, 16, 1563.
3. Reference #2 above, p. 213.
4. L. Hemphill, "Thermal Regeneration of Carbon", U.S. Department of Commerce, PB-284 065, 1978, p. 83.
5. J. A. Hart, Canada 878,560, 1971.
6. a. P. Neumeyer, et al, Thermochimica Acta, 1976, 16, p. 133.
   b. R. F. Schwenker, Jr., et al, Amer. Dyestuff Reporter, 1964, 53, p. 817.
7. W. Koenig, Z. Gesante Text. Ind., 1965, 67, p. 636.

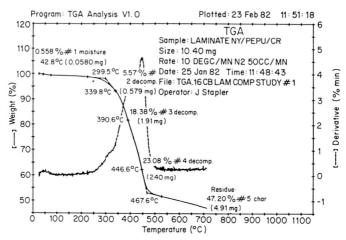

Fig. 1. TGA-DTG: NYL/PEPU/CR in N_2

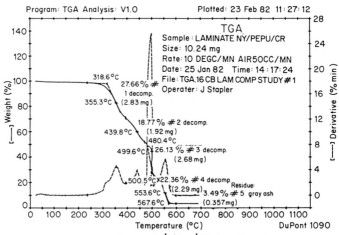

Fig. 2. TGA-DTG: NYL/PEPU/CR in air

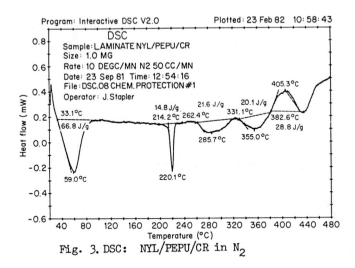

Fig. 3. DSC: NYL/PEPU/CR in N_2

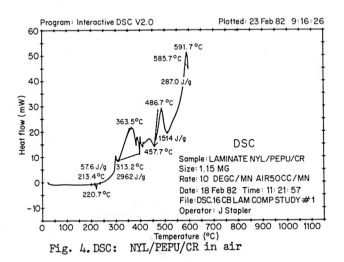

Fig. 4. DSC: NYL/PEPU/CR in air

Table 1. TGA-DTG Transitions of Components and Laminate

COMPONENT	IN NITROGEN				IN AIR			
	ONSET (°C)	INFLECTION (°C)	COMPLETION (°C)	WT.LOSS (%)	ONSET (°C)	INFLECTION (°C)	COMPLETION (°C)	WT.LOSS (%)
NYL		55		1		50		3
	436	471	492	92	432	456		87
				7 (Residue)		492	471	7
							518	3(Residue)
PEPU		50		4		50		1
	273	300		27	270	296		28
		321		4		403	427	61
		392		65		521	560	8
								2(Residue)
CR		50		3		50		2
	328	381		10		310	337	8
		681		7			510	85
				80 (Residue)				5(Residue)
NYL/PEPU/CR		43		6	31.6	44		1
	300	340		5		324		6
		400		23		355		20
		463		19		440		21
				47 (Residue)	481	498	500	26
						553	568	22
								4(Residue)

Table 2. DSC Transitions of Components and Laminate

COMPONENT	IN NITROGEN				IN AIR			
	THERMIC TRANSITION	ONSET (°C)	INFLECTION (°C)	PEAK AREA (J/g)	THERMIC TRANSITION	ONSET (°C)	INFLECTION (°C)	PEAK AREA (J/g)
NYL	ENDO	25	54	52	ENDO	20	48	42
	ENDO	213	221	86	ENDO	211	216	76
	ENDO	325	451	324	EXO	321	351	309
					EXO	375	445	5277
					EXO	470	511	2295
PEPU					ENDO	253	291	235
	ENDO	268	298	261	EXO	310	390	>1300
	EXO	368	398	412				
	EXO	412	501	>104				
CR	ENDO	18	47	58	ENDO	47	80	105
	ENDO	139	200	48	EXO	285	342	146
	EXO	306	358	33	EXO	375	581	≥2500
NYL/PEPU/CR	ENDO	21	59	33	ENDO	213	220	60
	ENDO	214	220	15	EXO	295	300	84
	ENDO	262	285	22	EXO	316	364	2760
	ENDO	331	355	20	EXO	409	437	422
	EXO	434	550	>84	EXO	585	592	>287

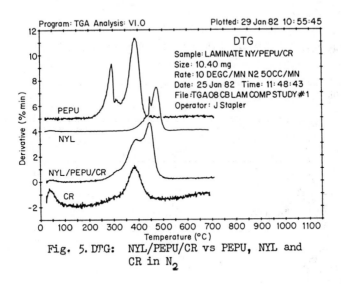

Fig. 5. DTG: NYL/PEPU/CR vs PEPU, NYL and CR in N_2

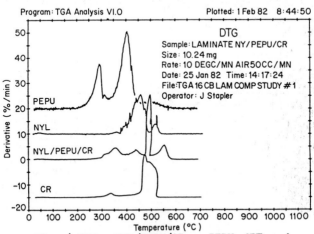

Fig. 6. DTG: NYL/PEPU/CR vs PEPU, NYL and CR in air

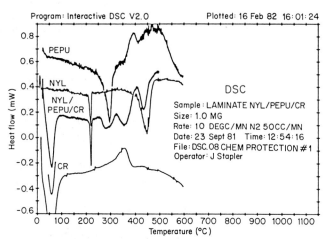

Fig. 7. DSC: NYL/PEPU/CR vs PEPU, NYL and CR in N_2

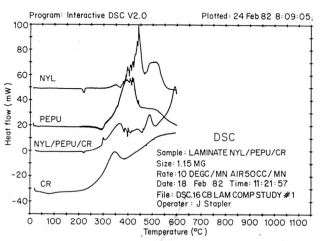

Fig. 8. DSC: NYL/PEPU/CR vs PEPU, NYL and CR in air

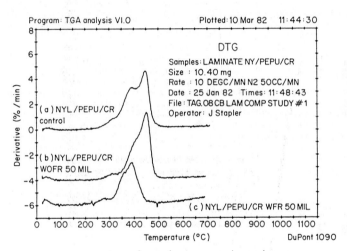

Fig. 9. DTG: NYL/PEPU/CR vs NYL/PEPU/CR WOFR and NYL/PEPU/CR WFR

A THERMAL ANALYTICAL TECHNIQUE FOR EVALUATING

BROMINE CONTAINING FLAME RETARDANTS*

M. Day and D.M. Wiles
Textile Chemistry Section
Division of Chemistry
National Research Council of Canada
Ottawa, Canada K1A 0R9

INTRODUCTION

The liberation of hydrogen bromide or bromine during the thermal decomposition of bromine containing polymers has long been recognised as one of the principal methods of reducing the combustability of polymeric materials. Although the liberation of HBr has been shown to effect the condensed phase thermal breakdown of certain polymers [1-3], the generally accepted mechanism of flame retardation is the release into the gas phase of HBr and/or Br radicals which suppress the free radical chain branching reactions in the flame. This mechanism is based on studies of hydrogen bromide inhibited premixed flames [4,5]. For a bromine containing flame retardant to be effective, therefore, it should not only be capable of producing HBr or Br but these species must be released at the right time if they are to interact with the pyrolysis products and inhibit the flaming combustion.

In this paper we report results obtained with a novel thermogravimetric technique for monitoring the evolution of Br_2 and HBr from materials under rapid heating conditions. The technique has been applied to the study of flame retarded poly(ethylene terephthalate) (PET) and the screening of several organic bromides as potential flame retardants for PET. This approach of monitoring HBr release in mechanistic studies of brominated flame retardants is not new, but previous experiments employed [6,7] TGA equipment at conventional heating rates. The use of our equipment at the higher heating rates reflects more realistically the condition encountered during the burning process.

EXPERIMENTAL

Materials. Two commercial flame retarded polyester fabrics were evaluated. One material was treated with tris (2,3-dibromopropyl) phosphate (TRIS) and had a bromine content of 6.6% (w/w). The other fabric was a sample of Dupont's Dacron 900F polyester (900F) in which the flame retardant is incorporated into the polymer by replacing some of the glycol with an ethoxylated tetrabromo bisphenol A. The concentration of bromine in this fabric was 4.3% (w/w). In addition to the commercially treated TRIS fabric, three samples of laboratory treated PET fabric were examined in which the concentrations of TRIS were 5.35%, 9.97% and 20.09% to give bromine contents of 3.68%, 6.86% and 13.8% respectively.

Several organic bromides were also examined and are listed in Table II. These chemicals were used as received without further purification and the calculation for bromine release are based on 100% purity.

* Issued as NRCC #20064

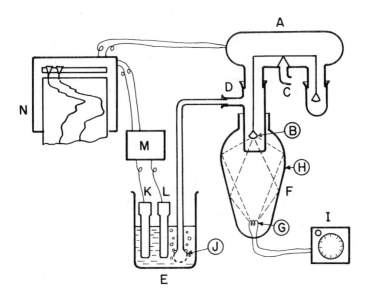

Figure I. Schematic of the experimental test arrangement

Equipment. Figure I provides a schematic of the general experimental layout. Essentially the equipment centres around a Cahn R.G. Electrobalance (A) which is used to monitor the weight loss of the sample (B) contained in a spherical quartz dish approximately 10 mm in diameter. The balance and sample are enclosed in a closed system through which air or nitrogen can be passed entering at (C) and exiting at (D). This gas serves two purposes, one it prevents contaminants reaching the delicate balance mechanism and two, it transports the gaseous products to the absorption solution contained in the 250 ml beaker (E).

Heat for the thermal decomposition is provided by a high intensity infrared spot heater (F) (Model 4085) obtainable from Research Inc. This heater provides radiant energy from a tungsten filament iodine "quartzline" lamp (G) which is focused into a spot about 5 mm in diameter at the sample location by a specular aluminum ellipsoidal reflector (H). The rate of heating can be adjusted by varying the lamp voltage with the control (I). The effluent gases were passed into the adsorption solution via a 8 hole bubbler (J) and the bromide ion concentration detected with an Orion model 94-35A selective bromide electrode (K) used in conjunction with a standard reference electrode (L). The electrode output was measured using a Beckmann model 3550 digital pH meter (M) which was directly recorded on a two pen strip chart recorder (N) which simultaneously also monitored the weight loss of the sample under test. All the results were obtained using air at a flow rate of 8 litres/min. and a voltage to the heater of 40 volts unless otherwise stated. In order to ensure uniform and rapid detection of the bromide ions the solution was rapidly stirred throughout the experiments.

Procedure. The adsorption solution comprising 100 ml of 1×10^{-6}M NaBr, and 2 ml of 5M $NaNO_3$ (ionic strength adjustor I.S.A.) was pipetted into the 250 ml beaker. A 15-20 mg sample of fabric or chemical under test was then loaded into the sample pan and weighed accurately. After the appropriate connections had been made the air supply was turned on and the system allowed to equilibriate for 2 minutes before the heater was switched on. The signals from the balance and bromide ion electrode were monitored for at least 10 minutes or until a 10% weight loss was obtained. From the output of the recorder and appropriate calibration graphs the bromide ion concentration and percentage weight losses as a function of time were calculated.

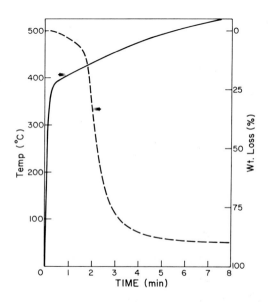

Figure 2. Sample Temperature and weight loss of standard untreated PET during a typical experiment.

RESULTS AND DISCUSSION

The weight loss curve for untreated PET is shown in Figure 2 along with the temperture recorded by a thermocouple in the sample position. It will be noted that this heating technique gives a rapid rate of heating (of the order of 15°/sec) up to about 400°C after which there is a more gradual temperature increase associated with secondary radiative effects. In some respects, it can be seen to represent the types of temperatures and heating rates encountered in flame spread studies. The delay in apparent onset of weight loss for the PET can be associated with the thermal capacity of the sample holder and sample in being brought up to the decomposition temperature.

It goes without saying that in order for a chemical to be an effective combustion inhibitor it must be present when required. Thus in the case of PET [whose weight loss curve is given in Figure 2] it is essential for the bromine entity to decompose between the 1.5 min. and 4 min. mark (in our decomposition setup) at a rate such that the concentration of HB or Br in the gas phase is sufficient to inhibit the combustion of the flammable mixture of fuel vapour produced from the polymer which undergoes rapid thermal decomposition in this range.

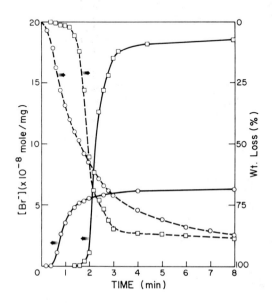

Figure 3. Bromide ion concentration [—] and percentage weight loss [---] of TRIS (O) and 900F (□) flame retardant PET system as a function of heating time.

Examination of the data obtained from the two commercially flame retarded polyesters [see Figure 3] do in effect reveal that these two systems release HBr or Br in the region of maximum weight loss and hence are capable of being in the right place at the right time to inhibit combustion reactions. The slight delay in the observed Br^- ion concentration curve can be attributed to the time delay in the transfer of the liberated bromine or hydrogen bromide from the sample to the absorption solution due to the volume of the sample tube and connecting tubing.

With the TRIS treated sample it is assumed that the detected bromide ion originates from the generation of HBr formed as a result of dehydrobromination via the classical four centred transition mechanism. (8)

$$RCHX-CH_2Br \rightarrow R-CX\overset{H---Br}{\underset{}{=\!=\!=}}CH_2 \rightarrow RCX=CH_2 + HBr$$

With the 900F material, however, because of the aromatic nature of the bromine substitution the above mechanism cannot take place and an alternative route for the Br^- ion generation must be proposed. The liberation of $Br\cdot$ radicals by straight chemical cleavage seems the most likely followed by hydrogen abstraction to give HBr.

$$RH + Br\cdot \rightarrow HBr + R\cdot$$

Alternatively, it can react by combination with other active radicals ie. $Br\cdot + H\cdot \rightarrow HBr$ or $Br\cdot + Br \rightarrow Br_2$
Even the formation of Br_2 will be detected since it is sufficiently soluble in water that the following reaction can take place in the absorption solution.

$$3Br_2 + 6OH^- \rightarrow BrO_3^- + 5Br^-$$

Whilst both the commercially treated PET samples liberate bromine species capable of inhibiting chain radical processes (i.e. HBr and $Br\cdot$) their actual quantitative yields are low. Only 7.9% of the available bromine is monitored in the case of the TRIS sample and 34.9% with the 900F sample.

In order to determine the influence of the decomposing atmosphere on the bromide ion formation the experiments were repeated in nitrogen and the results (Table I) indicate little change due to the effect of the atmosphere.

TABLE I

BROMIDE ION RESPONSE FROM TREATED PET FABRICS

Samples	Expts in Air		Expts in N_2	
	Moles/mg $\times 10^{-8}$	% Br Available	Moles/mg $\times 10^{-8}$	% Br Available
Mich TRIS	6.50	7.9	7.07	8.5
900F	18.80	34.9	18.23	33.9
Lab Treated TRIS at				
5.35% Add on	3.19	6.9	3.03	6.6
9.97% Add on	7.17	8.4	6.85	8.0
20.09% Add on	14.90	8.6	13.01	7.5

The TRIS chemical was also evaluated on PET at several chemical treatment levels in both air and nitrogen [Table I]. Whilst the evolved bromine was observed to be a function of treatment level in all cases, only 6.6 - 8.6% of the available bromine was monitored. Interestingly, the pure chemical itself only gave 8.3% of the available bromine in a measurable form. [Table II].

The results obtained with other organo bromides are summarised in Table II in which, not only is the amount of monitored Br^- ion concentration given but also the times to achieve weight losses of 10, 20, 40, 60, 80 and 90%. Theoretically, for a flame retardant to be an effective inhibitor of the gas phase combuston of PET it must release HBr or $Br\cdot$ into the gas phase between the 1.5 and 4 min. mark.

It will be noted that with the exception of TRIS the only aliphatic bromide to produce any detectable bromide ions was 2,3 dibromopropanol. This initially appears a little surprising since the majority of these compounds are known to liberate HBr as a result of dehydrobromination under these conditions [6]. However, it will be noted that the weight losses for all of these species were rapid and

TABLE II

% WEIGHT LOSSES AND BROMINE RELEASE FROM SEVERAL ORGANO BROMIDES

Test Sample	Time to reach the following %Wt losses (min.)						$[Br^-]$ monitored	
	10%	20%	40%	60%	80%	90%	$\times 10^{-8}$ moles/mg	% of Br available
Std. PET	1.7	1.9	2.1	2.4	3.2	8.0		
Mich Treated PET	0.5	0.7	1.2	2.1	4.6	12.0	6.50	7.9
TRIS Liquid	1.3	2.1	3.2	4.0	4.7	4.9	71.34	8.3
1,2 Dibromo ethane	0.1	0.1	0.1	0.1	0.2	0.2	0	0
1,1 Dibromo ethane	0.1	0.1	0.1	0.1	0.1	0.1	0	0
1,1,2,2 Tetrabromo ethane	0.4	0.5	0.6	0.7	0.8	0.9	0	0
1,1,2,2 Tetrabromo ethylene	0.6	0.7	0.8	0.9	1.0	1.0	0	0
1,2 Dibromo propane	0.1	0.1	0.1	0.1	0.2	0.2	0	0
1,3 Dibromo propane	0.2	0.2	0.2	0.3	0.3	0.3	0	0
2,3 Dibromo propanol	0.2	0.2	0.3	0.3	0.4	0.4	10.12	1.11
1,2,3,4 Tetrabromo butane	0.3	0.4	0.6	0.7	0.8	0.9	0	0
Pentaerythrityl Tetrabromide	0.4	0.5	0.7	0.8	0.9	1.0	0	0
Bromo benzene	0.1	0.1	0.2	0.2	0.2	0.2	0	0
o-Dibromo benzene	0.2	0.3	0.4	0.5	0.7	0.7	0	0
m-Dibromo benzene	0.4	0.5	0.7	0.9	1.0	1.0	trace	trace
p-Dibromo benzene	0.2	0.3	0.4	0.4	0.5	0.5	0	0
1,2,4 Tribromo benzene	0.4	0.5	0.7	0.9	1.0	1.1	0	0
1,2,4,5 Tetrabromo benzene	0.6	0.8	1.2	1.5	1.7	1.9	0	0
Hexabromo benzene	4.2	5.3	6.9	8.1	9.3	10.1	trace	trace
Hexabromo biphenyl	2.4	3.2	4.5	5.6	7.0	7.5	0	0
Decabromo diphenyl oxide	13.2	18.0	33.5	---	---	---	3.04	0.29
Dibromo triphenyl phosphorane	0.0	0.2	1.7	3.6	5.6	7.0	17000	250

and hence the failure to monitor Br⁻ ions in the volatiles may be ascribed to the rapid volatilisation of the undecomposed compound prior to degradation. Since a flow system is being employed this will transfer the compound quickly from the reaction zone. Thus since these species fail to yield HBr in our system it is possible to speculate that in actual burning conditions similar non-degrading volatilisation can occur.

The aromatic bromides examined were, on the whole, less volatile than the aliphatic ones. However, only three, namely 1,2,4,5 tetrabromo benzene, hexabromo biphenyl and dibromo triphenyl phosphorane gave weight loss values within the acceptable time limit restraints. The former two species, unfortunately did not yield detectable bromine ions and hence are unlikely contenders as gas phase combustion inhibitors. The dibromo triphenyl phosphorane, however, gave a massive response on the detection system (≥ 100%) which suggests that some other ionic species are being generated to which the selective bromide electrode responds. It should be pointed out, however, that this compound is sensitive to moisture and readily undergoes hydrolysis when it comes into contact with water [9].

$$Ph_3PBr_2 + H_2O \rightarrow Ph_3PO + 2HBr$$

Hence even if no thermal decomposition occurred the volatilisation of undecomposed compound would yield 100% HBr when it comes into contact with the aqueous absorption solution.

The only other compound to give measurable bromide ions was decabromodiphenyl oxide, which at 50% weight loss after 50 minutes pyrolysis only gave 0.29% of the available bromine. The weight loss curve for this compound would suggest that as a flame retardant for PET it would not be very effective when used alone, in the absence of antimony oxide, due to its thermal stability.

CONCLUSIONS

A technique to thermally decompose polymeric materials under conditions similar to those encountered during the burning process has been developed, and applied to a study of flame retarded PET. By monitoring the liberation of Br• and HBr from flame retarded PET samples, the feasibility of the technique to evaluate chemicals as potential gas phase combustion inhibitors has been demonstrated. The system has been employed to screen a whole range of organo bromides but of the chemicals examined so far none have demonstrated the effectiveness attributable to tris (2,3 dibromopropyl) phosphate.

REFERENCES

1. M.D. Carabine, C.F. Cullis and I.J. Groome, Proc. Roy. Soc. A, 1968, 306, 41.
2. H. Feilchenfeld, Z.E. Jolles and D. Meisel, Combust. Flame, 1970, 15, 247.
3. S.K. Brauman, J. Fire Retardant Chem., 1977 4, 38.
4. C.K. Westbrook, Comb. Sci. and Tech. 1980 23, 191.
5. M.C. Drake and J.W. Hastie, Combust. Flame, 1981 40 201.
6. W.C. McNeill, M.J. Drews and R.H. Barker, J. Fire Retd. Chem., 1977, 4, 222.
7. A. Mey-Marom and D. Behar, Thermochimica Acta, 1979, 30, 381.
8. A. Maccoll, Chem. Rev., 1969, 69, 33.
9. A.D. Beveridge, G.S. Harris and F. Inglis, J. Chem. Soc., (A) 1966, 598.

THERMAL ANALYSIS OF BROMINATED FIRE RETARDANTS
USING PYROLYSIS-MASS SPECTROMETRY

R. M. Lum, R. P. Jones and X. Quan
Bell Laboratories
Murray Hill, New Jersey 07974

ABSTRACT

The pyrolysis chemistry of the brominated aromatic fire retardants tetrabromobisphenol-A (TBBPA) and decabromodiphenoxy ethane (DBDPE) was investigated using pyrolysis-mass spectrometry techniques. The composition of the thermal degradation products was determined for a fire-retarded polyethylene formulation, the isolated fire retardant compounds, and the reaction products formed through pyrolysis of these compounds in the presence of Sb_2O_3. Only very low levels of HBr were detected during pyrolysis of the polyethylene compound containing DBDPE. Results from the isolated fire retardant compounds indicated that HBr yields were not dependent upon the initial bromine content of the materials. Our results indicate that no single reaction scheme adequately characterizes the pyrolysis of both compounds. While the data support a gas phase mechanism as the primary fire-retardant mode for TBBPA, the underlying chemistry governing high temperature interactions of DBDPE is unsettled, and a significant condensed phase component to its overall mode of operation cannot be ruled out. Finally, the data presented here provide the first evidence for the production of $SbBr_3$ through direct reaction of $SbCl_3$ with DBDPE without the intermediate formation of HBr.

TG/MS OF SOME PHOSPHATE SYSTEMS

H. G. Langer, J. D. Fellmann & C. D. Wood
Dow Chemical USA
Central Research New England Laboratory
Wayland, MA 01778

Our previous studies indicated that gaseous decomposition products from fire retardant additives may have an effect on flame propagation.* Thermogravimetry was used to relate flammability of wood with reactions of added phosphates.[1] Subsequently evolved gases from these systems were analyzed by direct-probe mass spectrometry.[2]

For this report, three different methods of thermal analysis by mass spectrometry have been compared in an effort to evaluate phosphate fire retardants. They are:

1) DIP-MS: The sample is heated in the ion source of an HP 5985 Quadrupole Mass Spectrometer from room temperature to 350°C at 30°/min and continuously scanned.
2) GC-MS: The sample is injected at 250°C and chromatographed through a capillary column, heated from 50 to 240° at 20°/min before entering the mass spectrometer.
3) TG-MS: The sample is heated in a du Pont 900 Thermobalance at 50°/min from room temperature to 600°C. Evolved gases enter the mass spectrometer through a 70 micron fused silica capillary, approximately 12 inches long. Using helium as a carrier gas, the capillary is sufficient to reduce the pressure to about 3×10^{-6} torr.

The TG/MS system was first tested with calcium oxalate monohydrate (Fig. 1). In addition to demonstrating the method, the previously detected oxidation of CO to CO_2[3] was verified.

Of the various phosphates investigated, four samples were chosen for this report.

1. A commercial ammonium phosphate labelled: $(NH_4)_3PO_4 \cdot 3H_2O$ (I)
2. A sample of (impure) dibutoxyethyl phosphoric acid

$$\begin{array}{c} O \\ \| \\ HO-P-O-CH_2-CH_2-O-bu \\ | \\ O-CH_2-CH_2-O-bu \end{array} \quad (II)$$

3. A mixture of mono- and di-butoxyethyl phosphoric acid (III) and
4. Bis(butoxyethyl)diphosphoric acid

$$\begin{array}{c} O \quad O \\ \| \quad \| \\ bu-O-CH_2-CH_2-O-P-O-P-O-CH_2-CH_2-O-bu \\ | \quad | \\ OH \quad OH \end{array}$$

*This is not intended to reflect on the hazards of these or any other materials under actual fire conditions.

At least 5 regions of weight loss were observed for I (Fig. 2) which are strongly affected by the flow rate of the carrier gas. DIP-MS (Fig. 3) confirms an overlapping release of ammonia, water, CO_2, phosphoric acid and P_4O_{10}.

For TG-MS the carrier gas flow rate becomes even more important than for TG alone. At 10 ml/min only broad bumps were recorded, which sharpen up considerably at 25 ml/min (Fig. 4).

The sequence of evolved species seems to be changed by the transport through the capillary. Phosphoric acid is barely noticeable and P_4O_{10} could not be detected even when the sensitivity of the multiplier was increased after the temperature had reached 600°C.

Obviously, GC-MS could not be done on this salt.

Decomposition of dibutoxyethyl phosphoric acid (II) occurs around 100°C and is essentially complete at 300°C. High temperature release of H_3PO_4 or P_4O_{10} seems to be absent. Apart from an impurity introduced during the synthesis, only butoxyethanol and an oligomer of butyl-vinyl ether could be identified with GC-MS. On the other hand, from DIP-MS very little of the alcohol evolved, with butene and the oligomers of the ether dominating the spectrum (Fig. 5). For TG-MS, butene, the alcohol and lesser amounts of ether were observed simultaneously (Fig. 6).

The mixture of mono- and diesters (Fig. 7) shows the pattern of the diester at lower temperatures, followed by evolution of products attributed to the mono ester. The amount of alcohol in this case increased significantly.

While the weight-loss curve for the diphosphate ester (IV), is remarkably simple, its GC-MS trace is very complicated (Fig. 8). Both DIP-MS (Fig. 9) and TG-MS (Fig. 10) confirm butene as the major product.

From these data a number of conclusions can be reached:
1. The phosphate esters decompose by two mechanisms:
   a) pyrolytic, with formation of unsaturated compounds,
   b) hydrolytic, with formation of alcohols.
2. The decomposition is affected by molecular structure as well as by external conditions, such as temperature and pressure.
3. The three mass spectrometric methods provide partial and differing results, which are caused by discriminating transport phenomena through a capillary, and also by secondary reactions.
4. Identification of thermal decomposition products can lead - at least in part - to an evaluation of phosphate-containing fire retardant agents.

REFERENCES

1. T. P. Brady and H. G. Langer in "Thermal Analysis", W. Hemminger, ed., Vol. 2, Birkhauser Verlag, 1980, p. 443.
2. H. G. Langer and J. D. Fellmann, presented at the 11th NATAS Conf., New Orleans, 1981.
3. Wiedemann, H. G., in "Thermal Analysis", Vol. 1, R. F. Schwenker and P. D. Garn, eds., Academic Press, New York, 1969, p. 229.

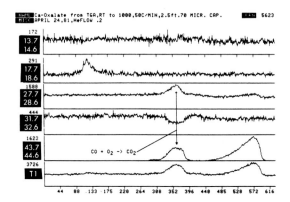

Fig. 1. TG-MS of Calcium Oxalate, RT to 1000°C.

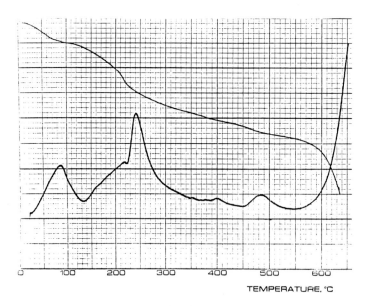

Fig 2. TGA of $(NH_4)_3PO_4 \cdot 3H_2O$

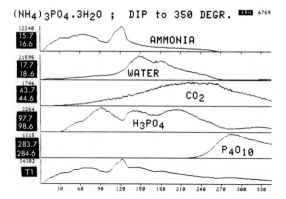

Fig. 3. DIP-MS of $(NH_4)_3PO_4 \cdot 3H_2O$

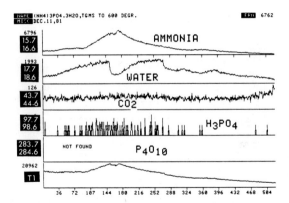

Fig. 4. TG-MS of $(NH_4)_3PO_4 \cdot 3H_2O$

DI-BUTOXYETHYL PHOSPHORIC ACID

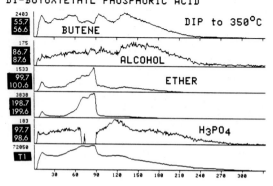

Fig. 5. DIP-MS of Di-butoxyethyl Phosphoric Acid

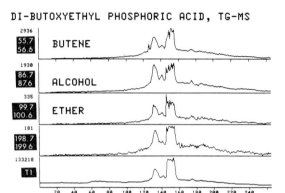

Fig. 6.   TG-MS of Di-butoxyethyl Phosphoric Acid

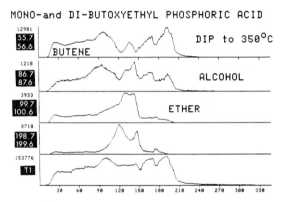

Fig. 7.   DIP-MS of Mono- and Di-butoxyethyl Phosphoric Acid

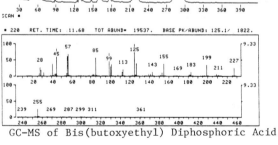

Fig. 8.   GC-MS of Bis(butoxyethyl) Diphosphoric Acid

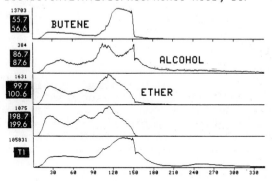

Fig. 9. DIP-MS of Bis(butoxyethyl) Diphosphoric Acid

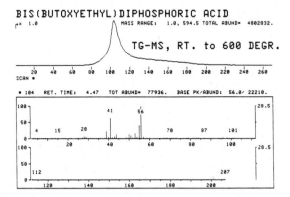

Fig. 10. TG-MS of Bis(butoxyethyl) Diphosphoric Acid

THERMAL INVESTIGATION OF ELECTRIC INSULATING MATERIALS POLYOLEFIN TYPES

G.LIPTAY[x], M.LACZKÓ[xx], L.LIGETHY[xx], E.PETRIK-BRANDT[x]

[x]Department of Inorganic Chemistry, Technical University, Budapest, H-1521 Budapest, Hungary

[xx]Hungarian Cable Works, H-1117 Budapest, Hungary

INTRODUCTION

Polyolefines gained wide-spread applications in the electric industry by their excellent properties. The electric insulating materials are stressed in service conditions both electrically and thermally. The authors have published some thermoanalytical investigations of polyethylenes applied as insulating materials in high voltage cables [1], and some results about the service properties and changing of silicone rubbers used as insulating materials in cable accessories and terminals [2]. We have found both electric field and temperature effects on ageing of insulating materials. It was emphasized the thermogravimetric methods being usefully to evaluate the effects of both stresses separately [3].

In this paper there are discussed the thermoanalytical properties of polyethylene (PE), voltage stabilized polyethylene (VSPE), cross-linked polyethylene (XLPE), and in the practice newer more wide spread applicated polypropylene-polyethylene copolymers (PP-PE). From results of our investigations we found a relative stability sequence of this materials. This information gained a direct industrial utility.

EXPERIMENTAL PROCEDURE

The investigations were carried out partly by dinamic, partly by static thermoanalytical methods. The dinamic thermoanalytical investigations were made by MOM Derivatograph Hungarian made. Every tests were carried out air atmosphere, in platinum crucible with a heating rate 3 °C/minute, and sample weight of 500 mg.

Isotherm thermogravimetry was applied as static methods. An constant temperature the weight change of sample was recorded in the function of time. The tests were carried out by an analytical balance applicated for this special purposes. The sample in crucible took place on balance arm, being protruded into the oven. First the oven without sample was heated up, then for a short time removed, during sample was put quickly on balance-arm, carefully prepared befor, and then oven placed back. In consequence of great heat capacity of oven, the temperature of sample was reacked very rapid (in some minutes). Constant temperatures of the oven were controlled by PROGRAMIK temperature controller (made by Industrial Research Institute, Budapest), which could maintaine the temperature within 0.2 ^{o}C-range. For this investigation we chose 1 gr weight in, and weight change in air atmosphere was recorded. The samples for both dinamic and static methods were cut in 1x1x2 mm prisms.

The investigations were carried out in air and neither in oxygen, nor in inert atmosphere because we would like to investigate the thermooxidative processes existing in real service situations to make use industrial utility of our data.

We investigated some polyolefine granulates and cable insulation-samples:

1/ processing stabilised polyethylene with very high molecular weight and a narrow molecule-weight distribution,

2/ polymer as above, but containing some percent p-phenylene-diamine additives (as voltage stabilizer),

3/ peroxyd containing crosslinkable PE,

4/ PP-PE copolymer with bigger amount of PP.

RESULTS AND DISCUSSION

After an informative investigation carried out by dinamic methods, the samples were tested by isotherm thermogravimetry. We found PE-samples showed the caracteristic curve shape according to figure 1., where IWL = initial weight loss, WI = weight increase, EWL = extent of weight loss during measurement, WLO = weight loss from oxidation maximum. Generally there were caracteristic the time to IWL and WI too.

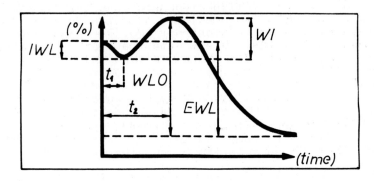

Fig. 1.

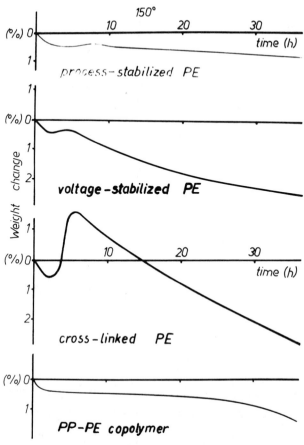

Fig. 2.

The thermooxidative processes of polyethylenes consist of consecutive reactions, in which some reactions running paralelly and contrary too. For practical reason, to compare and evaluate the results got from investigating samples in same situations, we devided the curves: beginning weight loss, oxidation with weight increase, and degradation with weight loss, but it means no reason only one type of process running in the investigated route of curves. The recorded curves shows only the resultant of separate processes.

On figure 2. some isothermal curves are shown about several polyolefine granulates measured on 150 °C.

a/ In case of process-stabilized PE after a beginning weight loss in 7 hours a weight increase in

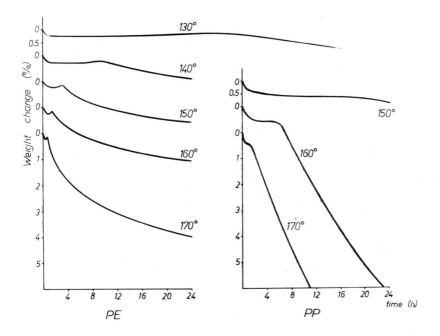

Fig. 3.

amount of 0.1 % to beseen. The measure of weight loss is relative small during the testing time.

b/ Voltage stabilized PE (enhanced electrical properties) showed weight increase after 4 hours, and the rate of degradation rather increased later during the investigation.

c/ After a beginning weight loss, in case of cross--linked polyethylene we found weight increase some 2 %, followed with greater rate of weight loss (degradation), greater than in case of other samples.

By this result it can be seen the increasing of mechanical and electrical properties causes a loss in thermal stability.

d/ Also an isothermal curve of PP measured 150 °C are shown and can be seen no weight loss process to be mentioned.

The shape and form of isothermal thermogravimetric curves depends very much on temperature measured. The rate of decomposition reactions changed and we gained data applicable directly in the industrial utility of PE types.

On fig. 3. isothermic curves of a PE and PP-PE are shown measured on several temperature range. From isothermal curves of voltage stabilized polyethylene measured on 130 °C, 140 °C, 150 °C, 160 °C, 170 °C can be evaluated the time shortening to the beginning of the oxidation process with weight increase by enhanceing of temperature representing the acceleration of degradation processes. In case of PP-PE no weight increase (WI) at 150 °C. At 160 °C after a relative short induction time began the degradation process.

By comparison of above curves the relative thermal stability sequences of samples can be evaluated and this data serves informations about their industrial applicability.

Dinamic methods are applied at the investigation of the thermal exposure effect of samples. We would like to investigate what changes in oxidation, degradation and morphologycal properties can be followed. Fig. 4. contains the curves of thermal aged (on different temperatures) PP-PE, but only the DTA-curves, in comparison with DTA-curves of the original unaged samples.

Derivatogramm of samples aged on 150 °C is quite similar to those of unaged samples. The DTA peak temperatur appeares at 160 °C, which caracterizes the melting process of polymer. The grade of thermal effect decreses it with increasing stressing temperatures. The reason of it is, that after ageing the polymer didnot recover to its original cristalline state after heat exposure, so this smaller recovery effect didnot cause as high enthalpy effect as earlier.

The evaluations of curves showed, the thermal ageing of PP-PE above 150 °C for more than 24 hours had a mentionable effect on thermal properties of PP-PE too.

REFERENCES

[1] G.Liptay, L.Ligethy and E.Petrik-Brandt, 6th ICTA 1980, Thermal Analysis, Birkhäuser Verlag, Basel 1980, Vol.I. p.427-434.
[2] G.Liptay, L.Ligethy and J.Nagy, 6th ICTA 1980, Thermal Analysis, Birkhäuser Verlag, Basel 1980, Vol. I. p.477-482.
[3] G.Liptay, L.Ligethy and E.Petrik-Brandt, Hungarian Symposium on Thermal Analysis, Budapest,1981.

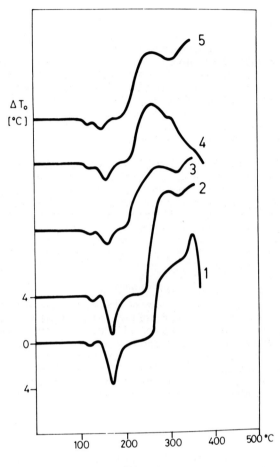

Fig.4.

DTA-curves of thermally aged PP-PE copolymers:
1. original, 2. after 24 hours heat treatment at 150 °C,
3. after 24 hours heat treatment at 160 °C, 4. after
24 hours heat treatment at 170 °C, 5. after 24 hours
heat treatment at 180 °C.

THERMAL-OXIDATIVE DEGRADATION OF POLY(PHENYLENE SULFIDE)

C. L. Markert*
H. E. Bair
P. G. Kelleher

Bell Laboratories
Murray Hill, NJ 07974

INTRODUCTION

Poly(phenylene sulfide) (PPS) is a commercially available heat-resistant thermoplastic used in the Bell System for the 946 and 947 connectors in 1A Processors. It is formed by the reaction of p-dichlorobenzene and sodium sulfide in a polar solvent.[1] The resulting sodium chloride is removed in aqueous solution. Various formulations incorporate glass fiber reinforcements, a vinyl silane coupling agent, a lithium carbonate acid scavenger, and a zinc stearate or ethylene bis stearamide mold releasing agent into the resin.

We are interested in the thermal stability of PPS because it is heated at several stages in the manufacture and assembly of molded parts. During processing, the PPS resin is injection molded at cylinder temperatures of 300-325°C. In certain applications the resin is also exposed in thermocompression bonding to a die at 750°C for about 15 seconds, during which time it heats up to approximately 225-250°C. Although thermogravimetric studies show that polymer decomposition does not begin until around 400°C, evolved gas measurements indicate that outgassing occurs at much lower temperatures (~200°C.)[2] The outgassing is thought to be absorbed or otherwise easily evolved materials.

The purpose of this study is to explore PPS weight loss both dynamically and isothermally and to determine the activation energy for polymer decomposition.

EXPERIMENTAL

Apparatus

Dynamic and isothermal studies were conducted in a Perkin-Elmer Thermogravimetric System TGS-2 interfaced with a Perkin-Elmer 56 two-channel recorder.

Materials

The compositions of the various formulations are shown in Table 1. The vinyl silane is a coupling agent incorporated on the surface of the glass fiber reinforcement. Sample C contains Li_2CO_3 as an acid scavenger, and a zinc stearate mold release agent. An experimental low molecular weight formulation designed to improve flow properties for encapsulation, uses ethylene bis stearamide as the mold release agent.

Previous thermal characterization of the formulations has been carried out by differential scanning calorimetry.[3] Samples A, B and D had glass transition temperatures of 95°, 93° and 76°C respectively. In addition, T_g of the virgin resin, E was 78°C and that of the final cured resin before compounding, was 94°C.

---

* Summer Program Employee. Current address is Lafayette College, Easton, Pa. 18042.

Experimental Results

Of all the known organic polymers either a fluorinated polymer like Teflon or various polymeric structures containing benzene and or heterocyclic rings such as Kevlar, Kapton or Vespel appear to be among the thermally most stable plastics. A comparative scan of the volatilization behavior of these polymers against sample A in nitrogen is plotted in Fig. 1. The latter resin's resistance to short term, high temperature degradation is comparable to these other thermally stable materials.

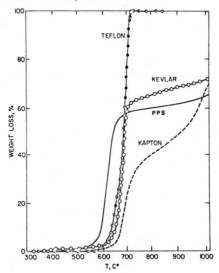

1. Comparison of volatilization behavior of PPS (A) (R6) with several high temperature polymers in nitrogen (heating rate, 40°/min).

Typical examples of weight loss as a function of temperature at a heating rate of 450°/min in N_2 and O_2 atmospheres are shown in Figure 2A and B. For comparison, the values of weight loss for samples B, C and D are based on the resin only. Polymer degradation as indicated by weight loss begins at about 400°C and is essentially complete at 750°C. As noted by Lum,[2] the pyrolysis of PPS in O_2 is characterized by two stage decomposition.

Isothermal studies were carried out in nitrogen at 300, 350, 400, 450, and 500°C. Figure 3 is a typical plot showing weight loss as a function of time at 400°C. Again, all of the values are based on the percent resin. The 400°C series was also run in O_2 to compare rates of degradation (Fig. 4). Note, in each of these experiments a single pellet was used for a sample. In addition, the fine powder collected from the filing of a sample A pellet was run versus the pellet and compared with the volatilization behavior of sample E which was received in a powdered form (Figs. 5, 6). Gases from the former evolved about four times slower in nitrogen and 7 times more slowly in oxygen from PPS pellets as compared to PPS in the form of fine powders.

Rate constants were determined from the first linear portion of the weight loss versus time curves, following the evolution of the adsorbed species. The losses were initially linear with time, and then began to decrease slightly as degradation progressed. The rate constants are listed in Table 2. A comparison of PPS's rate of degradation in nitrogen and oxygen yielded activation energies of 37 and 30 kcal/mole respectively.

DISCUSSION

Figures 2A and 2B show the degradation of PPS with increasing temperature in N_2 and O_2 atmospheres. Because it has a lower molecular weight as determined by melt flow and T_g measurements, sample D began volatilization at a lower temperature in N_2 and proceeds more quickly initially, ending with a greater weight loss. The stabilized sample C appears to degrade slightly faster in N_2 than the sample without the stabilizer; this difference, though, is small and may be explained in part by the nonhomogeneity of the dispersion due to the twin-screw extrusion method used for mixing. The actual amounts of resin, glass filler and Li_2CO_3 vary from pellet to pellet. In O_2, the stabilizing effect of the Li_2CO_3 is readily observable; degradation occurs at a significantly higher temperature in sample C than in the other formulations. Note that in N_2, there is only a 50-60% weight loss at 800°C. In this inert atmosphere the bulk of the residue was stable at temperatures higher than 1000°C in an inert atmosphere. Typically after cooling to room temperature the residue was found in the shape of a thin walled bubble whose diameter was many times greater than that of the original pellet. In O_2, there was essentially complete combustion.

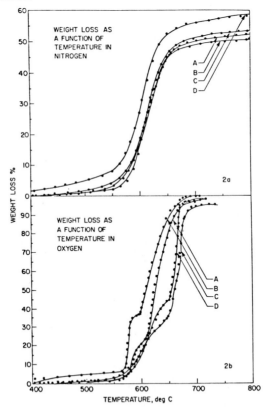

2. Weight loss curves for several commercial PPS formulations in nitrogen (2A) and in oxygen (2B) (Heating rate, 20°/min).

The isothermal runs in N_2 support the conclusions drawn from the dynamic runs. In each case, sample D loses 2-5% of the resin during the initial heating period, probably representing the lowest molecular weight fragments. Its rate of weight loss then slows to approximately that of the other PPS formulations. Again, there is a seeming disparity between the observed and expected behavior in the formulations with and without the Li_2CO_3 stabilizer. At 300° and 500°C, the sample with the additive appears to degrade slightly faster. This is again explained partly by the fact that the method of mixing does not insure constant or homogeneous dispersion. The different initial losses are due to varying amounts of adsorbed and/or easily evolved materials.

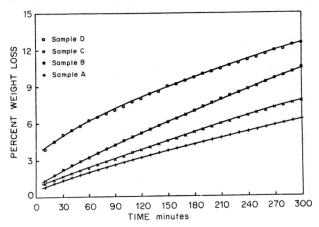

3. Isothermal weight loss curves of PPS formulations at 400°C in nitrogen.

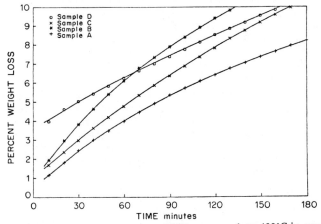

4. Isothermal volatilization curves of PPS compounds at 400°C in oxygen.

Arrhenius plots give an activation energy of 35 ± 2 kcal/mole for the various formulations in nitrogen. Although all of the activation energies are close, sample D, which showed the greatest tendency to degrade in both the dynamic and isothermal runs, had the lowest activation energy. Samples A and B had similar activation energies, and they behaved nearly identically in the dynamic and isothermal runs.

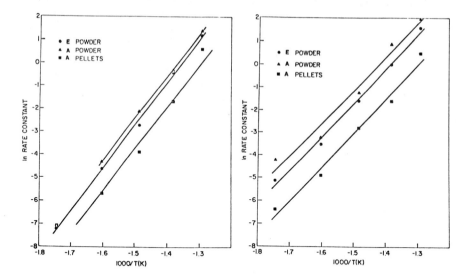

5. Arrhenius plots of PPS degradation in nitrogen.

6. Arrhenius plots of PPS degradation in oxygen.

The size of the PPS sample had no effect upon a sample's activation energy for degradation in either oxygen or nitrogen. However a sample's volatilization rate was about four times slower in nitrogen and 7 times more slow in oxygen from a pellet as compared to a fine powder. We attribute these differences in rate to the fact that the pellet retains its basic shape when exposed to temperatures as high as 500°C and thus provides a relatively thick barrier to the outward diffusion of degraded species or the inward movement of oxygen.

The sample E powder had a slightly lower volatilization rate in either O_2 or N_2 than sample A powder. The former resin is the low molecular precursor of the latter. Apparently the complex curing reaction which yields higher molecular weight resin also produces some branched or lower molecular fragments which easily volatilized at processing temperatures or higher. The latter products are presumably formed and trapped in sample D during compounding and lead to its relatively marked low temperature volatilization behavior.

CONCLUSION

In this work we characterized the degradation of PPS as a function of temperature in nitrogen and in oxygen. In oxygen, degradation was a two step process that began at about 400°C and was essentially complete with no residue at 750°C. The stabilizing effect of a Li_2CO_3 acid acceptor caused the weight loss-temperature curve of the stabilized formulation in oxygen to shift about 50°C above that for the unstabilized formulation. In nitrogen, the compounds behaved similarly, beginning degradation at about 400°C and in a single step process losing 50-60% of the resin by 800°C. There was no apparent stabilization due to the acid acceptor. The low molecular weight formulation began volatilization at a slightly lower temperature and proceeded initially more rapidly.

Isothermal degradation rates were obtained at temperatures between 300° and 500°C in nitrogen and oxygen and yielded activation energies of 37 and 30 kcal/mole, respectively. Volatilization rates were found to be dependent on sample size. Gases evolved about four times slower in nitrogen and 7 times more slowly in oxygen from PPS pellets as compared to PPS in the form of a fine powder.

An acid acceptor additive, Li_2CO_3, reduced the rate of degradation of PPS in oxygen to about two thirds that of the unstabilized polymer. The same additive had no apparent effect of PPS's volatilizations behavior in nitrogen.

REFERENCES

[1] J. N. Short and H. W. Hill, Jr., Chem. Technol. 2, 481 (1972).

[2] R. M. Lum, ACS Polym. Preprints 18(1), 761 (1977).

[3] J. E. Bennett, T. M. Paskowski and H. E. Bair,

Table 1

Composition of PPS*

	Sample A	Sample B	Sample C	Sample D	Sample E
Resin	100[b]	60	58.9	34.8	100[a]
Glass Fiber	-	40	40	15	-
Fused Silica	-	-	-	50	-
Vinyl Silane	-	<1	<1	<1	-
Zinc Stearate	0.1	-	-	-	-
Lithium Carbonate	-	-	1	-	-
Acrawax-C	-	-	-	0.2	-

Table 2

Rate Constants For Polyphenylene Sulfide Degradation*

T,°C		Sample A wt. %/min	Sample B wt. %/min	Sample C wt. %/min	Sample D wt. %/min
300		7.9×10^{-4}	4.4×10^{-4}	8.9×10^{-4}	1.3×10^{-3}
350		3.2×10^{-3}	5.2×10^{-3}	4.1×10^{-3}	4.2×10^{-3}
400	N_2	2.0×10^{-2}	3.9×10^{-2}	2.7×10^{-2}	4.4×10^{-2}
450		.19	.26	.20	.21
500		1.8	1.0	1.3	1.7
400°C, O_2		5.8×10^{-2}	9.3×10^{-2}	5.8×10^{-2}	3.9×10^{-2}
E_a^{**}, N_2		34.1	34.3	32.4	32.1

OXIDATIVE STABILITY APPLICATIONS FOR THE DU PONT 1090 THERMAL ANALYSIS PROGRAM

T. A. Blazer and R. L. Blaine,
E. I. Du Pont de Nemours & Co., (Inc.)
Analytical Instruments Division,
Concord Plaza, Wilmington, Del 19898

A common goal of the manufacturers of plastics, lubricating oils and greases, and food and drug products, is to increase the lifetime of their products. These products are subject to oxidation leading to the loss of desirable properties. Additives or stabilizers are often added to such products to increase their useful lifetime.

Over the years, several tests have been developed to determine the effects on the product lifetime due to the addition of these stabilizers. These tests, in general, suffer from being either too time consuming or inaccurate.

In 1959, Baum[1] published the results of unstabilized low density polyethylene exposed to oxygen, in which he noted the exotherm denoting the oxidation of the polyethylene. In 1961, Rudin and coworkers[2] published on what is probably the first use of Differential Thermal Analysis (DTA) in oxidation studies. Within the last few years, DTA has been supplemented by Differential Scanning Calorimetry (DSC) in work on oxidative stability. As an empirical technique, thermal data has found common acceptance as a measure of oxidative stability and are used in such test methods as ASTM Test Methods D3350 and D3895; Western Electric Specification MS-17000; REA Specification PE-39; British Post Office Specification M131C and M133B; and British Gas Standard BGC/PS/PL2.

Du Pont's line of Thermal Analysis equipment, particularly DSC and Pressure DSC (PDSC), has found widespread use as a research and quality control tool to estimate oxidative stability. The Du Pont 1090 controller is particularly useful with its capability for automatic operation and computer reduction of the data is well suited to the use.

A recent addition to the 1090 family of data reduction programs is the Oxidative Stability program. This program permits automatic operation and data processing while still allowing the operator maximum flexibility in the selection of parameters. This capability for operator interaction is especially useful for those samples where competing thermal processes may be at work.

There are two principal techniques employed in DSC oxidative stability studies. The first technique involves heating the sample at a programmed rate and measuring the temperature at the onset of oxidation as indicated by the exotherm (Figure 1). This temperature is called the oxidation induction temperature. The second technique involves holding the sample at a given temperature and measuring the time to the onset of oxidation (Figure 2). This time is called the oxidation induction time (OIT). The choice of which of these two approaches to use depends upon the material under test and the use to which the resultant data is to be put.

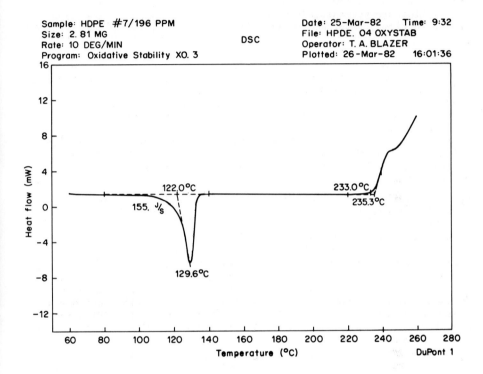

FIGURE 1

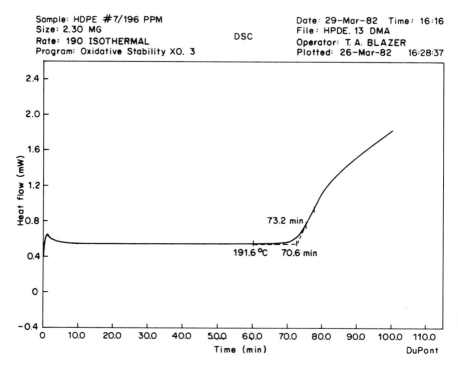

The extrapolated onset of oxidation is defined as the intersection of the baseline with a line that is tangent to the oxidation exotherm. The 1090 Data Analysis program permits the operator to fix the tangent line to the oxidation exotherm at either (1) the tangent to a particular point (stop point) or (2) the tangent to the point of maximum slope between the start and stop limits. The first option is illustrated in Figure 3 where a double break in the oxidation exotherm is shown and the operator has selected a stop point for the exotherm tangent line. The second option is illustrated in Figure 4 where the operator permitted the 1090 to choose the point of maximum slope and draw the line tangent to that point. The baseline is always the tangent line at the analysis start point.

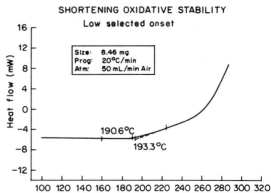

FIGURE 3

Another operator selectable parameter is the threshold value which marks the "first deviation from the baseline". The default value for this "first deviation" is set at 0.2mW. The temperature or time at this "first deviation" can be especially useful for those samples which do not give a well-defined oxidation exotherm from which an extrapolated onset value can be obtained.

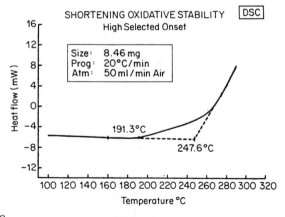

FIGURE 4

Figure 5 shows a sample of polypropylene run isothermally at 195°C with a purge gas flow of 50 ml air/min. This sample shows a first deviation of 19.2 min and an extrapolated onset time of 23.8 min. These values together with the 195°C isothermal temperature valure, are automatically plotted on the thermogram by the 1090.

For those samples where the oxidative stability is so great that very long times or very high temperatures are required to evelute it,

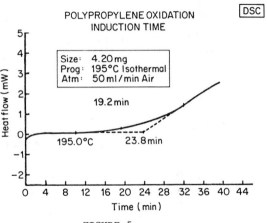

FIGURE 5

Pressure DSC is a very useful technique giving faster and often more reliable results. Since the Oxidative Stability program is independent of the analytical technique, it applies equally well to PDSC as well as standard DSC runs.

Figure 6 also demonstrates the use of differing atmospheres. For many samples, air will give oxidative stability results of moderate time or temperature values. For the most stable samples, the use of an oxygen atmosphere may be required. In other cases, the operator may choose to use special gas mixtures.

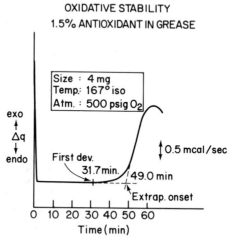

FIGURE 6

SUMMARY

The Oxidative Stability Program for Du Pont 1090 Thermal Analysis systems provides rapid and reliable oxidation stability results. It functions in an automatic fashion while still permitting the operator maximum flexibility in setting key analysis parameters. Lastly, the program processes data obtained by DSC or PDSC using any desired purge gas.

BIBLIOGRAPHY

1. Baum, B. J. Appl. Polymer Sci., 2, 281, (1959).
2. Rudin, A. Schreiber, H.P. and Waldman, M.H., Ind. Eng. Chem., 53, 137, (1961).
3. DuPont Thermal Analysis Applications Brief 41.
4. DuPont Thermal Analysis Applications Brief 48.

THERMOGRAVIMETRIC PROFILE OF PHOTOCURED ACRYLATE SYSTEMS

D. M. La Perriere and J. A. Ors
Engineering Research Center
Western Electric Company
P. O. Box 900
Princeton, NJ 08540

F. R. Wight
Bell Telephone Laboratories
Whippany, NJ 07981

INTRODUCTION

The printed circuit board (PCB) industry has become increasingly dependent on photopolymers for many of its technological and manufacturing advances. Photopolymers are currently used in a variety of applications including photoresists, conformal coatings, etc. Several classes of materials, generally varying in backbone composition, are currently being used. Some of the more common resins employed are the acrylates, cinnamates, epoxies, and urethanes.

Certain advantages can be derived by coupling a thermo-cured resin (e.g. epoxy) with a photoreactive system (e.g. cinnamate) in a dual cure fashion. The result is a photodefinable-thermally curable system. An example of such a polymer network is the Probimer system from Ciba-Geigy. Some recent developments in the photocure of epoxy systems via photocationic initiators [1] widen the range of application of epoxy resins for PCBs. These systems, however, appear to lack the imaging speed encountered with acrylate systems. This definition problem is partly being addressed by the addition of multifunctional acrylates (MFAs) to the epoxy resins. These systems are also cured via the photocationic initiator, albeit through different mechanisms [2].

Acrylate systems in general are economical, easily applied, fast room temperature curing materials that yield desirable physical and electrical properties for use in PCB manufacturing. As technology is driven to higher circuit density; hence fine line requirements, the demand for improved materials is increasing. In order to fully evaluate the commercially available materials and/or design systems to meet these technological needs, a thorough understanding of the behavior of these materials during processing is necessary. Materials are designed to yield specific properties, but the fully cured material can oftentimes differ from the formulated one. This difference is a function of the process preceeding final cure. An overview of the various steps is given in Scheme 1.

SCHEME 1

COATING ⟶ IMAGING ⟶ DEVELOPING ⟶ HARD CURE

This study is concerned with the curing profile of some acrylated-epoxy resin based mixtures when processed according to the above scheme. The profile has been developed by means of thermogravimetric analysis (TGA) data coupled with gas chromatographic mass spectrometric (GC/MS) analysis. These techniques allow the monitoring of changes in composition of the material during processing and enable modification of the formulation and the process to obtain the desired composition of the end material.

MATERIALS

All materials used in this study are commercially available and were used as received (Table 1). Mixtures were prepared according to weight percent as reported in Table 2. All mixtures are solvent free.

EXPERIMENTAL

Substrates were coated using a Jack Film applicator (40 thread/in) which yielded a film thickness of approximately 3.5 mils. The films were imaged using an Oriel system model 8103. The lamp flux was 9.9 mW/cm^2 as measured with an IL-745 UV curing radiometer. The final film irradiation was carried out in a CoLight UV curing system, model UVC24, equipped with three 200 W medium-pressure Hg-vapor lamps. All cure experiments were done using one lamp on 'high' power. The lamp flux, 75 mW/cm^2, was measured using an International Light 440 Photoresist Radiometer.

The TGA data were obtained with a DuPont 1090 thermal analyzer coupled to a 951 thermogravimetric analyzer or a Perkin Elmer TGS-2 Thermogravimetric analyzer, and are reported as percent weight loss.

A Hewlett-Packard 5985B GC/MS system was used in conjunction with various sample introduction procedures, pyrolysis gas chromatography and direct insertion probe, for acquisition of mass spectrometric data.

TABLE 1
MATERIALS USED IN CURING PROFILE STUDY

Name	Code	Source
Celrad 3700	3700	Celanese
Isobornyl acrylate	IBOA	Rohm & Haas
2-Ethylhexyl acrylate	EHA	Polysciences
2,2-dimethoxy-2-phenyl acetophenone	DMPA	Ciba-Geigy
magenta pigment	MP	Penn-Color

TABLE 2
MIXTURE COMPOSITION BASED ON WEIGHT PERCENT

Mixture	Component %				
	Celrad 3700	IBOA	EHA	DMPA	MP
I	60	37		1	2
II	60		37	2	1

RESULTS AND DISCUSSION

The nature of thermogravimetry lends itself to empirical use in monitoring the changes in composition occurring in a polymer film as a function of time and temperature. It is in this fashion that TGA has been utilized here, thus offering a contrast between the compositions of the formulated and the final (polymerized) material. Figures 1 and 2 show the TGA traces of mixtures I and II, respectively, before and after final cure. In each case the percentages derived from the TGA trace for the uncured mixture correlate with the composition in Table 2 and are consistent with GC results [3]. Post-irradiation traces show the disappearance of the monomer, presumably due to incorporation into the polymer network. In the uncured samples the monomers are volatilized between 100°-250°C (transition 'A'). In the cured sample, the IBOA containing mixture (I) shows an early onset of decomposition leading to an approximate 28% weight loss within the range of 250°C to 350°C (transition 'B', Figure 1). In contrast, the EHA containing mixture (II) does not exhibit the 'B' transition implying that the lower temperature (250-350°C) decomposition occurs due to the presence of polymerized IBOA. In order to substantiate this a film of poly-IBOA was prepared; and its TGA curve is shown in Figure 3. The temperature range for the largest transition corresponds to the 'B' transition of mixture I (see

Figure 1). The curve also shows about 20% of the film decomposing above 350°C (transition 'C'). GC/MS data of the products obtained during transition 'B' revealed the presence of one major component for both poly-IBOA and mixture I. This component has been identified as camphene. A possible mechanism for its generation is given in Scheme 2.

SCHEME 2

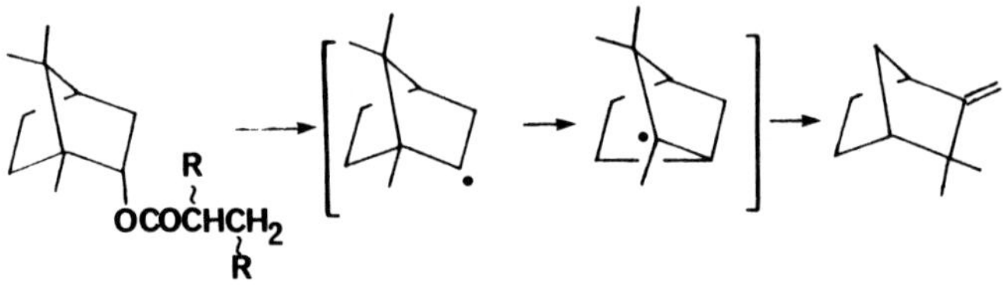

Similar TGA decomposition profiles have been reported for polymers containing acrylates of secondary alcohols (i.e. isopropyl acrylate) yielding substituted olefins (i.e. propene) [4].

Due to the decomposition mode of the incorporated IBOA, transition 'B' can then be used to detail the changes in material composition throughout the curing process. In the general process, as outlined in Scheme 1, the first occasion for a possible change in composition is during the imaging step where the material is photodefined. Upon irradiation of the sample, a percentage of the monomer will remain unreacted due to the O_2 inhibition. The degree of inhibition is dependent on: a) the reactivity of the monomer(s), b) the viscosity of the mixture, c) the concentration of photoinitiator and d) the light intensity and length of exposure. In this study all process parameters were fixed excluding variations in the soft exposure (vide infra). The quantity of unreacted monomer can be used as a qualitative measure of the degree of cure for any given system. Table 3 shows the TGA data for mixture I at various stages of the process. After soft exposure (ii) approximately 25% of the total mixture (i.e. 64% of the formulated monomer content) is comprised of unreacted IBOA. The balance of the IBOA has reacted, as is evident from the presence of transition 'B'. Unreacted monomers can easily be removed from the film by a subsequent bake step (iii) resulting in a monomer depleted film.

The O_2 inhibition takes place predominantly at the surface of the film (Figure 4). This is due primarily to the O_2 diffusivity in the viscous material, thus resulting in an inhibition gradient. TGA of a sample, (iv), physically removed from the surface after soft exposure yielded a composition similar to that of the uncured formulation (i).

The development step is designed to give the desired image by removing the unexposed and/or inhibited material from the film. Table 3 shows that after the soft/development step, (v), a large amount of low boiling materials are present. These volatile materials can be attributed to inclusion of development solvent along with some unreacted monomer remaining in the bulk [5]. These materials are eliminated from the film by the post-development bake. This bake is done to improve adhesion of the coating to the substrate as well as to eliminate any residual solvent and/or water which may get trapped in the partially cured film causing possible cracking of the film upon final irradiation. The final irradiation completes the cure of the film but should not alter its composition with respect to the monomer.

Changes in the ratio of 'B' and 'C' depends on the degree of cure of both resin and monomer, which in turn is dependent on the initial imaging step. This is shown in Table 4 where increases in soft exposure yield direct increases in the incorporation of monomer.

TABLE 3
THERMOGRAVIMETRIC ANALYSIS OF MIXTURE I DURING VARIOUS PROCESS STEPS

Process Step	% Weight Loss		
	A	B	C
i) Uncured	39.2		55.1
ii) Soft[a]	25.5	12.0	59.9
iii) Soft/Bake	-	13.4	72.1
iv) Inhibited Surface Sample (Soft)	36.3	-	63.7
v) Soft/Develop[b]	24.1	13.2	56.0
vi) Soft/Develop/Bake[c]	-	18.6	80.4
vii) Soft/Develop/Bake/Hard[d]	-	18.4	74.3

a - Soft exposure corresponds to 30 secs irradiation in the Oriel System.
b - Development consisted of 2.5 min in butyl carbitol followed by 2 min. H_2O rinse.
c - Bake: 30 min. @ 100°C
d - Hard: Corresponds to 11 secs exposure in CoLight

TABLE 4
CHANGES IN FILM COMPOSITION WITH VARIATIONS IN SOFT EXPOSURE OF MIXTURE I[a]

Irradiation time (sec)	TGA (% weight loss)		
	A	B	C
20	–	13.9	78.6
30	–	18.4	74.3
45	–	21.0	73.8
60	–	21.3	72.7

a – All samples were subjected to the entire process (Scheme 1)

SUMMARY

This study has shown the importance of monitoring the complete process that any photopolymer undergoes during PCB manufacture. This monitoring allows the necessary formulation and process design insight to achieve the desired end properties of a polymer film. Use of IBOA in this study has facilitated the derivation of a curing profile due to the nature of its decomposition. GC/MS data have shown that decomposition of poly-IBOA takes place via generation of camphene. The same decomposition product is found for IBOA polymerized into a resin system although other products exist.

REFERENCES

1. J. V. Crivello, Chem. Tec., 524 (1980) and references therein; S. P. Pappas and C. W. Lam, J. of Rad. Cur., 2 (1980).
2. W. C. Perkins, J. of Rad. Cur., 8, 16 (1981) and references therein.
3. GC analysis of mixture I shows a concentration of IBOA of 36.5%. This is in agreement with the proposed formulation and TGA results.
4. N. Grassie and J. G. Speakman, J. Polym. Sci. (A1), 9, 919 (1971).
5. J. A. Ors, unpublished results. The percentage of monomer that remains in the film after development is dependent on the film thickness.

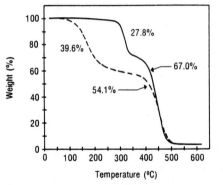

Figure 1. TGA of Mixture I
(---) Uncured Mixture; (–) Fully Cured Film

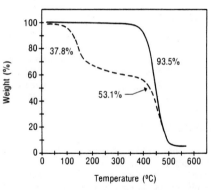

Figure 2. TGA of Mixture II
(---) Uncured Mixture; (–) Fully Cured Film

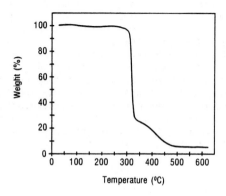

Figure 3. TGA of Poly –IBOA

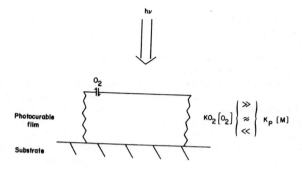

Figure 4. O_2 inhibition scheme of a polymer film

USE OF THERMAL ANALYSIS TO STUDY THE DEGREE OF CURE OF EPOXY RESINS
T.R. Manley and G. Scurr
Faculty of Engineering
Newcastle upon Tyne Polytechnic
Ellison Place
Newcastle upon Tyne
NE1 8ST, UK.

INTRODUCTION

The first patent on epoxide resins is claimed by W.H. Moss (1) in December 1937 but the first commercial development was founded on the work of P. Castan (2). These resins, crosslinked with phthalic anhydride were originally intended for dental applications but provided a new tool that revolutionized the design of high voltage electrical equipment (3), and founded a new species of adhesives. The use of epoxide resins as coatings, where amine hardeners replaced the caustic sensitive ester linkages is traced back to S.O. Greenlee (4). Castan's resins were licensed to Ciba and the coating resins were developed by Shell. In 1957 the cycloaliphatic resins were introduced initially by Union Carbide.

Shortly after the introduction of fluidized beds for coatings epoxy resins began to be applied in this way and powder coatings are now widely used. The solvent free coatings may be applied to give thick (50 mil) coatings very rapidly, covering complex shapes and providing good mechanical strength with resistance to moisture and chemicals.

The commonest epoxy resin is the diglycidyl ether of bisphenol A or DGEBA,

$$CH_2 - \overset{O}{CH} - CH_2 - O - Ph - \underset{Me}{\overset{Me}{C}} - Ph - O - CH_2 - \overset{O}{CH} - CH_2$$

where Me is a methyl group and Ph is a phenyl residue. This is obtained by reacting epichlorhydrin with bisphenol A in the presence of caustic soda. Polyglycols are co-reacted to give flexible resins and phenol formaldehyde novolac resins are co-reacted to improve the mechanical and thermal properties.

Cycloaliphatic resins are made by reaction of a diene with peracetic acid and consequently the range of resins available is limitless. Having no aromatic groups they are more resistant to UV radiation than the bisphenol resins. The mechanical properties of epoxy resins are dependent on the degree of cure. In modern applications of epoxy resins such as matrices for composites in aerospace applications or as coatings for gas pipelines the ability to determine the exact degree of cure is essential and there is therefore interest in techniques of establishing the degree of cure.

The commercial introduction of epoxide resins was contemporaneous with the discovery of thermal analysis by polymer chemists and studies on the cure of epoxies were amongst the first topics studied (5),(6). The use of DTA and TG to identify epoxides and to prognosticate their service life was fully studied (7); later the technique of DSC was used to identify changes in Tg associated with increase in cure (8). TBA, TMA and DMA have also been used. The following factors were taken into consideration in choosing the technique to be used.

DTA. DTA is well established as a method of determining polymerization reactions in solution. In measuring degree of cure, however, where most of the reaction has already occurred a very high degree of sensitivity is required. In addition classical DTA is less suitable for quantitative measurements of solids and powders. The technique can detect undercure and also Tg.

Thermogravimetry. In early work on anhydride cured resins TGA was successfully used to determine serious undercure (7). Where greater precision is required and especially since epoxy resins do not produce condensation products thermogravimetry is unlikely to be sufficiently sensitive.

Torsional braid analysis TBA. This requires a support for the epoxy resin. In the case of fibre reinforced materials this is no hardship but for coatings a further complication is introduced. TBA can produce a wealth of fundamental information but is less suited to a quality control laboratory.

Thermomechanical analysis. Thermomechanical analysis whether in the expansion or penetration mode is an excellent method for determination of Tg but cannot determine the degree of cure directly.

Dynamic Mechanical Analysis. Like TBA this requires a support and is excellent for obtaining fundamental information but is somewhat expensive and complicated for routine quality control.

Differential scanning calorimetry. DSC techniques whether heat flux or power compensation (enthalpy) tend to be more suitable than DTA for quantitative work. DSC will measure residual heat of cure or Tg (8). It was decided therefore to use DSC to determine the Tg of epoxy coatings.

EXPERIMENTAL

A DuPont 990/910 heat flux DSC cell was used. The pretreatment involved heating to $150°C$ at maximum power, holding at $150°C$ for one minute then quenched with liquid nitrogen. Each Tg determination was made by heating to $280°C$ and the degree of undercure was estimated from the increase (if any) in Tg on the repeat determination. Films of a grey epoxy powder coating were prepared by applying to a PTFE coated steel plate at a temperature of $240°C$ using a Gema 710 electrostatic gun. On striking the hot substrate, the powder particles coalesce and form a film. The residual heat of the substrate partially aids the crosslinking reaction which was completed by returning to an oven at a temperature of $240°C$ for three or six minutes.

After cooling, the films were removed from the plate and stored in a dessicator over silica gel at ambient temperature. Samples of the film were made in three ways, viz:

(i) Discs were cut from the film to give a close fit in the DSC sample pans by the use of an appropriate cork borer.

(ii) Sections of film were milled in a Glen Creston micro hammer mill to a particle size between 0.1mm and 2.0mm.

(iii) These coarse particles were then sieved through a 150 micron aperture brass sieve to produce a fine particle size.

As a first stage, disc samples were used to determine the effects of the heating rate and equipment sensitivity on the value of Tg determined from the DSC trace using film cured at three and six minutes at $240^\circ C$ without further treatment.

At a heating rate of $1^\circ C$ per minute, the movement of the pen was so slow that migration of the ink gave an illegible trace. At heating rates of $50^\circ C$ per minute, there were practical difficulties of confining the trace to the chart. Heating rates of 5, 10 and 20 degrees per minute produced satisfactory traces. The sensitivity setting on the instrument had a varying degree of influence on the values of Tg obtained and differences of up to $4^\circ C$ could be found. The lower sensitivity settings gave obviously less base line shaft making the Tg somewhat difficult to determine and this was exacerbated at low heating rates.

Almost all of the traces showed more than one inflection or considerable base line drift which could affect Tg determination. By heating the samples to $150^\circ C$ (9) to remove effects of thermal history very much smoother traces were obtained as may be seen in Figure 1 showing samples stoved for six minutes at $240^\circ C$ at heating rates of $20^\circ C/min$ and sensitivity of 2mv/cm. It was therefore decided to make a more detailed study of the following factors:

(1) Particle size (disc, coarse particles, fine particles).
(2) Curing at 240° for 3 or 6 minutes.
(3) Heating rates of 5, 10, 20 and 50 degrees per minute.
(4) Pretreatment as in experimental section.

The results are shown in Table 1. The significance of the various factors was analysed by a Students t-test since it was clear that interactions were occurring amongst the parameters. The results of the t-test are given in Table 2. It is clear that the pretreatment causes a significant rise in Tg [$p < .1\%$] as does the increase in heating rate above 10 degrees per minute. The effect of particle size and of the extra minutes of cure are significant at the 5% level which in the present experiment is comparatively insignificant, therefore the effect of particle size will not be considered further.

The effect of the extra cure is surprising at first sight but it should be remembered that three minutes is considered to give an adequate cure.

CONCLUSIONS

The most convenient technique at present for determining the degree of cure in epoxy coatings is by heat flux DSC. The heating rate must be standardized and 20 degrees/min is recommended. The T_g can be determined much more easily if a short heat pretreatment (9) is used. As this does affect the T_g however we propose to investigate alternative pretreatments.

TABLE 1 - T_g OF EPOXY POWDER COATINGS

Sample		Heat Rate	Sens	T_g Pretreatment	
Form	Mins Cure			None	Yes
D	3	5	2	82	89.5
	6	5	2	85	90
	3	10	2	86	92
	6	10	2	88.5	89
	3	20	2	88	94.5
	6	20	2	94	93.5
	3	50	5	91.5	99.5
	6	50	5	94.5	97.5
CP	3	5	2	86	94
	6	5	2	87	91
	3	10	2	86	92
	6	10	2	88.5	94
	3	20	2	92	96
	6	20	2	90	97.5
	3	50	5	98.5	98
	6	50	5	92.5	96.5
FP	3	5	2	92	93
	6	5	2	88	87
	3	10	2	86	92
	6	10	2	88.5	97.5
	3	20	2	93.5	95
	6	20	2	90.5	95
	3	50	5	97.5	97.5
	6	50	5	93.5	96

TABLE 2 - COMPARISON OF EFFECTS BY T-TEST

Effect	t	Dof	Significance level %		Remarks
Pretreatment	3.81	43	.04	VS	
Hearing rate	0.89	21	38.3	NS	$5° \text{ v } 10°$/mm
do	3.53	20	0.2	VS	$5° \text{ v } 20°$
do	5.88	19	0.01	VS	$5° \text{ v } 50°$
do	2.51	20	2.08	S	$10° \text{ v } 20°$
do	4.82	19	0.01	VS	10 v 50
do	2.64	21	1.52	S	20 v 50
Particle size	1.00	29	32	NS	Disc v CP
do	1.18	28	25	NS	Disc V FP
do	.13	29	90	NS	F v CP
Curing	.26	44	80	NS	3' v 6'

REFERENCES

1. Brit. Pat. 506, 999.
2. Swiss Pat. 211, 116.
3. T.R. Manley, K. Rothwell and W. Gray, Proc. Inst. Elect. Eng., 1960, 107A, 213.
4. U.S.P. 2, 521, 911.
5. C.H. Klute, W. Viehmann, J. Applied Poly Sci., 1961, 5, 86.
6. T.R. Manley, "Techniques of Polymer Science", Monograph 13 Soc. Chem. Industry, London 1963.
7. T.R. Manley, "High Voltage applications of epoxy resins", M. Billings Ed. U of Manchester Press, 1967.
8. J.M. Barton, p.125, T.R. Manley, p.53 in "Polymer Characterization by Thermal Analysis", J. Chiu, Ed. M. Dekker N.Y. 1974.
9. British Gas ERS standard for coating steel pipe PS/CW6 Pt. II July 1980.

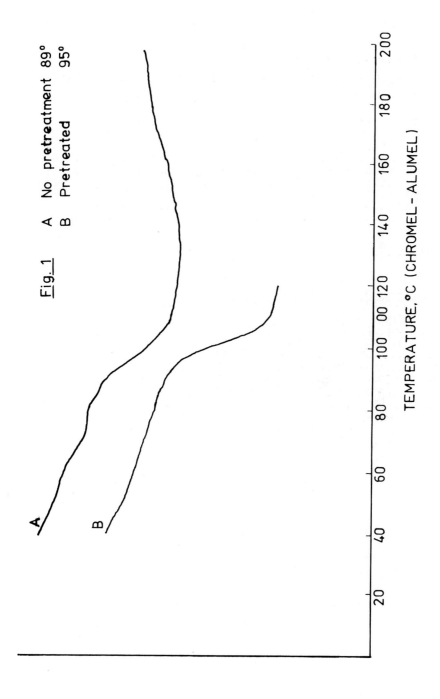

Fig. 1　A　No pretreatment　89°
　　　　B　Pretreated　　　　95°

THE STUDY OF THE CURE OF EPOXY COMPOUNDS BY THERMOELECTROMETRY

Lin Fu

Harbin Institute of Electrical Engineering

INTRODUCTION

This paper reports the results of measurement on the electrical conductivity variation curves of some epoxy conpounds during their cure through the use of a self-constructed thermoelectrometric analyser of autoregistration. The parameters used to evaluate the rate and the degree of cures are obtained. The dynamic parameters of curing reactions are also measured. The test results of this paper are valuable to the industry for the proper selection of technological parameters.

TEST RULES OF THE CURE STUDY BY THERMOELECTROMETRY

Fig. 1 shows the typical thermoelectric test curve of epoxy compound during constant temperature raising. For the ease of interpretation, two extreme cases are first studied.

Thermoelectric tests of constant temperature cure. The conductivity-time curves of cure under constant temperature are shown in Fig. 2 and Fig. 3. During cure, the electrical conductivity of epoxy cempound decreases.

As shown from Fig. 2, under the same temperature, $\lg \gamma$ changes with time, their reproducibility is fine. This shows there exists a relation between $\lg \gamma$ and concentration of group of curing reaction C:

$$C = f_1(\lg \gamma) \quad \text{or} \quad \lg \gamma = f_2(C) \tag{1}$$

after transformation

$$d\lg \gamma/dt = df_2(C)/dt = f_3(dc/dt) \tag{2}$$

where C is the concentration of group of curing reaction, γ is the electrical conductivity, t is the time, and f_1, f_2, f_3 are functions. From (2), it is possible to evaluate curing rate by $d\lg \gamma/dt$. After the cure reached a certain time, where $d\lg \gamma/dt=0$, this means the cure is completed. This time is called as cure time (tc).

Since $\lg \gamma$ is not only the function of time but also the function of temperature, in order to evaluate the

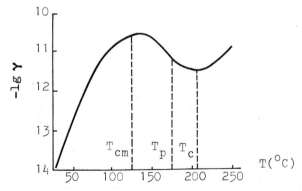

Fig. 1 Thermoelectric test curve of epoxy compound during constant temperature raising.

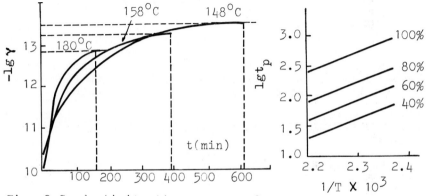

Fig. 2 Conductivity-time curves of cure under constant temperature.

Fig. 4 Arrhenius plot of cure.

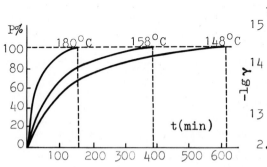

Fig. 3 Variation of P with time.

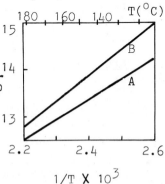

Fig.5 $\lg \gamma$ - $1/T$ plot of completely cured samples.

curing rate at different temperature, we introduce a conductivity correlation factor P, whose definition is

$$P = \frac{lg\gamma_t - lg\gamma_b}{lg\gamma_c - lg\gamma_b} \qquad (3)$$

where γ_b is the conductivity before cure, γ_t is the conductivity at curing time t, and γ_c is the conductivity after the completion of cure.

Take the logarithm of corresponding time t_P of same P, and plot it against the reciprocal of absolute temperature, we obtain a set of parallel straight lines, as shown in Fig. 4. The relation between lg t_P and 1/T holds Arrhenius relationship:

$$lgt_P = A + \frac{E}{2.303RT} \qquad (4)$$

where t_P is the reaction time up to P, E is the activation energy of curing reaction, A is a constant, R is the gas constant, and T is the absolute temperature in K.

We can use dP/dt to evaluate curing reaction rate, use P to evaluate the degree of cure, and use (4) to calculate the activation energy of curing reaction.

Thermoelectric tests of completely cured samples. Fig. 5 shows the conductivity-temperature curves of completely cured samples. Owing to the fact that cure is already completed, in the measurement process no reaction takes place within the sample. Conductivity increases with temperature, the relation between lg γ and 1/T is

$$lg\gamma = lg\gamma_0 + \frac{U}{2.303RT} \qquad (5)$$

where $lg\gamma_0$ is a constant, and U is the energy barrier of transition for ionic motion. There exists a linear relation between lg γ and 1/T, that means that the mechanism of conduction of completely cured epoxy compound is ionic.

Thermoelectric tests of constant rate of temperature raising. The thermoelectric test curves of constant rate of temperature raising is the consequence of temperature raising which increases the conductivity and curing which decreases the conductivity.

Fig. 6 is the derivative curve of Fig. 1 and Fig. 7 is the curve obtained by the plotting of lgγ with 1/T on the low temperature values of Fig. 1. From these curves, several values are characteristic:

(1) T_i: This is the temperature when cure begins. As shown in Fig. 7, curing reaction in the low temperature region is very slow, its influence on conductivity can be neglected, hence there is a linear relation between lg γ

and 1/T. As the temperature increases further, the curing reaction rate accelerates. Their influence on conductivity is no longer small and the relation between $\lg \gamma$ and 1/T deviates from linear. The temperature where deviation begings is called cure beginning temperature T_i.

(2) T_{cm} : This is the temperature of maximum conductivity (see Fig. 1). In this temperature, the influences to conductivity from temperature raising and from curing reaction are equal.

(3) T_p: This is the temperature of maximum curing rate. T_p is the temperature which corresponds to the negative maximum value of Fig. 6, which means the curing rate is maximum.

(4) T_f: This is the temperature at the end of cure, which corresponds to the minimum value of Fig. 1. In this temperature range, the effect of cure to conductivity turns out to be a minor factor, which means the curing reaction is already finished.

According to the measured results of T_p, which correspond to the maximum curing rate of epoxy compound at various rates of temperature raising (β), the relation between T_p and β obeys Kissinger equation, which is

$$\ln\beta/T_p^2 = A - E/RT_p \qquad (6)$$

(6) can also be used to calculate the activation energy E of the curing reaction.

EXPERIMENTAL EXAMPLES

Evaluation of dynamic parameters of curing reaction by thermoelectrometry. The results of thermoelectric test for three epoxy compounds are shown in Table 1.

As shown from above test results, the activation energy of curing reaction obtained by two methods are close.

The effect of ionic impurities to curing. Two samples of epoxy compound are studied. The epoxy resins used in these samples are different, the one (B) uses purified resin, while the other (A) uses unpurified one. As shown in Table 1, the ionic impurities in unpurified resin (sample A) accelerate the curing reaction, and decrease the activation energy of curing reaction. Ionic impurities also decrease the electrical properties of insulation compounds greatly (see Fig. 5).

Comparison of results between thermoelectrometry and differential thermal analysis. At present, the study of cure is mostly by differential thermal analysis.

Table 1. The activation energy of curing reactions of three epoxy compounds.

materials	No.	Constant temp.			Constant rate of temp. raising.		
		cure temp. (°C)	cure time (min)	E ($\frac{kcal}{mole}$)	β (°C/min)	T_p (°C)	E ($\frac{kcal}{mole}$)
Epoxy compound A	1 2 3 4	170 160 150 132	370 480 790 1560	13.8	0.462 0.567 1.05 1.65	149 159 169 180	14.6
Epoxy compound B	1 2 3 4	170 160 150 137	300 440 780 1310	16.4	0.383 0.567 1.05 1.65	158 167 179 190	16.6
Epoxy compound C	1 2 3 4	205 190 176 158	100 220 500 1300	22.5	0.420 0.567 1.05 1.65	185 193 204 210	22.1

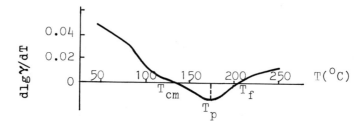

Fig. 6. Derivative curve of Fig. 1.

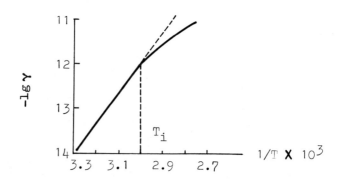

Fig.7. lgγ-1/T curve of lower temperature part of Fig.1.

Through the use of a OD-102 Type Differential Thermal Analyzer, the author studies three samples of epoxy compounds. From the exothermic peak of cure, three characteristic temperatures T_i, T_p, and T_f are obtained.

To study cure by thermoelectrometry and differential thermal analysis, different methods base upon different properties change, yet their reaction processes are the same, hence there must be some internal connections between them.

Table 2 shows the results of characteristic temperatures of three samples of epoxy compounds, which are measured by thermoelectrometry and differential thermal analysis respectively.

Table 2. The Characteristic Temperatures of Three Samples of Epoxy Compound by Thermoelectrometry and Differential Thermal Analysis.

Materials	Character. temp. (°C) methods	T_i	T_{cm}	T_p	T_e
Epoxy compound A	T E M	60	125	169	210
	D T A	100		162	212
Epoxy compound B	T E M	63	120	179	215
	D T A	103		178	220
Epoxy compound C	T E M	/*	/*	204	225
	D T A	/*		198	/*

* Because of the occurrence of other process, we can't obtain this value.

As shown in Table 2, T_i determined by thermoelectrometry is much lower than those from differential thermal analysis. The sensitivity of measurement of low temperature cure of thermoelectrometry is higher. The values of T_p and T_e are very close and this means the characteristic temperatures obtained by thermoelectrometry are accurate.

CONCLUSIONS

(1) The advantages for the use of thermoelectrometry to study the cure of epoxy compounds are: high sensitivity and good reproducibility. It can obtain the data of cure time and dynamic parameters. This method is used in industry to determine the optimum techological conditions of cure and to monitor the curing process.

(2) Since the mica tapes used in electric machines contain sufficient quantities of epoxy compounds, thermoelectrometry is an effective means to judge the quality of mica tapes.

APPLICATION OF THERMAL X-RAY DIFFRACTION IN
ELECTRONIC MATERIALS RESEARCH
D. D. L. Chung, Center for the Joining of
Materials and Department of Metallurgical
Engineering and Materials Science, Carnegie-
Mellon University, Pittsburgh, Pennsylvania
15213

ABSTRACT

Thermal x-ray diffraction is found to be valuable for characterizing structural changes, such as alloying, reaction, recrystallization and phase transformation, which occur upon heating electronic materials. This paper introduces thermal x-ray diffraction as a major thermoanalytical technique in materials research. The technique is illustrated by a study of the interfacial processes between metals and gallium arsenide.

INTRODUCTION

Thermal analysis has recently become an important tool in the electronics industry because of the growing demands of reliability and performance of miniaturized devices in integrated circuits and solar cells.[1] Alloying, reactions, recrystallization and phase transformations are some of the phenomena which occur upon heating in materials constituting various devices. To characterize such structural changes, x-ray diffraction is one of the most relevant experimental techniques. Although room temperature x-ray diffraction is widely used, thermal x-ray diffraction is not. Thermal x-ray diffraction refers to x-ray diffraction performed at various temperatures. This technique allows in situ observation of the structural changes during heating, during cooling or isothermally. The in situ measurement is valuable for process characterization and kinetics study.

As an illustration of the use of thermal x-ray diffraction in electronic materials research, this paper describes a study of the processes which occur at the interface between a metal thin film and the compound semiconductor GaAs upon heating. The metals chosen for the study are gold and palladium, which are used in the Ohmic contact technology for GaAs. The processes involved are reaction (compound formation) and phase transformation (melting and solidification).

THERMAL X-RAY DIFFRACTION TECHNIQUE

A thermal x-ray diffraction set-up is made up of an x-ray diffraction system, a high temperature attachment and a temperature controller. The x-ray diffraction system can be a diffractometer or a camera. There are two heating methods. Method I involves the use of a sample chamber which contains a heater surrounding the sample. The sample chamber can be evacuated or purged. The sample does not need to be sealed in glass. Method II involves no sample chamber and heating is achieved by using a heater or a hot gas stream. The sample is usually sealed in glass to avoid being heated in the presence of air. Method I is superior because of usually better thermal contact between the heater and the sample, more accurate measurement of the sample temperature, and the possibility of controlling the vacuum or gaseous environment of the sample. It is often used when an x-ray diffractometer is involved. Method II is often used when an x-ray camera is involved. However, Method I is experimentally more difficult because of the necessity of aligning the sample chamber on the diffractometer. Moreover, Method I is more expensive due to the cost of the sample chamber. In this work, Method I is used.

High temperature attachments are commercially available. Those using Method I and designed for diffractometers are available from Rigaku, which manufacturers a 1500°C attachment (Cat. No. 2311B1) and a 2500°C attachment (Cat. No. 2315). Available from Blake Industries, Inc., are a 900°C attachment using Method I (Cat. No. D1600NH) and a 1500°C attachment using Method II (Cat. No. D4231). Available from Enraf Nonius is a 1000K attachment using Method II (Cat. No. FR559).

Because the technological use of thermal x-ray diffraction is usually at temperatures above room temperature, this paper is concerned with this high temperature range. However, low temperature x-ray diffraction is useful for studying phase transformation and other processes which occur at low temperatures. A low temperature x-ray diffraction attachment using Method I is a cryostat which houses the sample. It is often equipped with a heater so that the temperature range extends to $\sim$200°C. The lowest temperature is close to liquid nitrogen temperature or liquid helium temperature, depending on the coolant used. A low temperature attachment using Method II involves directing a cold gas stream on to the sample. To prevent freezing of water vapor from the air to the specimen, the cold gas stream is often surrounded by a warm gas stream and the entire sample is usually enclosed in a plastic sheet such that the interior of the plastic enclosure is kept dry. Method I is superior because of the capability of a wider temperature range, the smaller possibility of water condensation on the sample, and the capability of controlling the vacuum or gaseous environment of the sample. Low temperature attachments using Method I are commercially available from Air Products, Rigaku (Cat. No. 2351B1), Oxford Instruments (Cat. No. CF100), and Enraf Nonius (Cat. No. FR537). Low temperature attachments using Method II are available from Enraf Nonius (Cat. No. FR524).

In this work, thermal x-ray diffraction above room temperature was performed by using the 1500°C high temperature attachment manufactured by Rigaku. A Rigaku D/MAX II powder x-ray diffractometer system was used, with a fine-focus Mo x-ray tube and detection by a scintillation counter. The high temperature attachment allows the sample to be heated up to 1500°C in vacuum and to 1400°C in air or inert gas. The airtight window of the airtight chamber is made of Al foil to reduce the amount of x-ray absorption. The heater is made of platinum and permits direct contact between the thermocouple and the sample. A Pt - Pt Rh (13%) thermocouple is used.

APPLICATION IN REACTION STUDIES

Identification of Reaction Products. Identification of the reaction products is one of the most important aspects of a reaction study. In many cases, the reaction products depend on the reaction temperature. Thermal x-ray diffraction can be used for phase identification at temperature. An illustration of this use of thermal x-ray diffraction is the identification of the products of the interfacial reaction between a palladium thin film and gallium arsenide.[2]

The (100) GaAs wafers were n-type, Te-doped, with a dopant concentration of $10^{17} cm^{-3}$. The palladium film was deposited by boat evaporation ~3 Å/s and was ~1000 Å thick.

The compound formation at Pd/GaAs was observed in situ by using x-ray diffraction performed at various constant temperatures from 150°C to 550°C as a function of time. Shown in Fig. 1 are in situ x-ray diffraction patterns obtained before heating and after various times and temperatures of heat treatment in argon (1 atm). No compound formation was observed below ~250°C, so that the diffraction patterns at these temperatures consisted of only Pd and GaAs peaks. Isothermal heating at 250°C resulted in the appearance of PdGa and a small quantity of Pd_2Ga after 30 - 40 min of heating. Both phases grew as heating proceeded further, with PdGa remaining as the dominant compound phase. The $PdAs_2$ phase, observed ex situ by Olowolafe et al.[3] after 60 min at 250°C, was not observed even after 3 hr at 250°C. Note from Fig. 1 that the Pd peaks became smaller and eventually vanished as heating proceeded. Isothermal heating at 350°C resulted in the appearance of Pd_2Ga, PdGa and $PdAs_2$ after 15 - 20 min of heating. On the other hand, isothermal heating at 500°C resulted in the appearance of PdGa alone after ~5 min. The PdGa

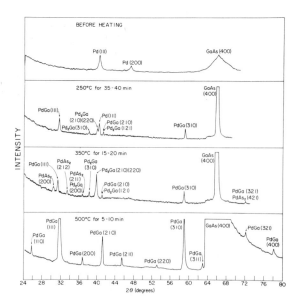

Fig. 1 Thermal x-ray diffraction patterns of Pd/GaAs heated in 1 atm argon.

phase remained as the only compound even upon further heating at 500°C, in agreement with the ex situ result of Olowolafe et al.[3]

Kinetics. Kinetics information, such as the activation energy, is difficult to obtain by ex situ measurement. The in situ measurement in thermal x-ray diffraction enables such information to be obtained on structural changes more precisely and more conveniently. Experimentally, this can be done by measuring the integrated intensity of a diffraction peak of the reaction product as a function of time isothermally at various temperatures. Because the same peak at the same 2Θ angle is followed, no correction is necessary in comparing the intensities at various times. The reaction is considered complete when the peaks of the reaction product stop growing with time. Hence, this time corresponds to 100% transformed. From the plot of percent transformed vs. time, the time for 50% transformed can be obtained. Let the time for 50% transformed be τ. The reaction rate is proportional to $1/\tau$. Hence,

$$\frac{1}{\tau} \propto e^{-\Delta Q/RT},$$

where ΔQ is the activation energy for the reaction. The slope of the plot of log τ against $1/T$ is then $\Delta Q/(2.3R)$. This yields the value of ΔQ.

An illustration of the use of the above method of activation energy determination is the determination of the activation energy of the solid state interfacial reaction between a gold thin film and GaAs[4].

The (100) GaAs wafers were n-type, with a dopant concentration of $10^{16}-10^{17} cm^{-3}$. The gold film was deposited by electron beam

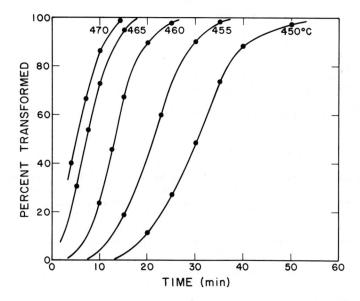

Fig. 2 Plot of percent AuGa formed vs. time for various constant temperatures.

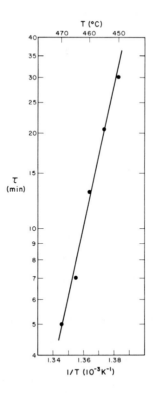

Fig. 3 Plot of log τ vs. 1/T, where τ is the time for 50 % AuGa formed.

evaporation and was ∼1000 Å thick. The reaction product has been tentatively identified as the orthorhombic AuGa compound (50 at.% Ga).

To obtain quantitative information on the kinetics of the compound formation, isothermal in situ x-ray diffraction was performed at 450, 455, 460, and 470°C. The plot of percentage transformed as a function of time is shown in Fig. 2 for these constant temperatures. Compound formation at temperatures below 450°C was found to be too slow for the experiment to be completed in a reasonable amount of time, so the temperature range used for activation energy determination was limited. The percentage was obtained from the integrated intensity of the (211) AuGa diffraction peak. Figure 3 shows a plot of log τ against 1/T. The slope of this plot yields an activation energy of 106±5 kcal/mol for the AuGa compound formation.

Effect of the Atmosphere Around the Sample. Because the sample chamber can be evacuated or purged, thermal x-ray diffraction can be performed in vacuum or in a gaseous environment such as nitrogen, argon or air. Due to the importance of a controlled atmosphere during the processing of electronic materials, this capability is valuable for electronic materials research.

An illustration of the effect of the atmosphere around the sample is the difference in the minimum reaction temperature and in the reaction products between heating in argon (1 atm) and heating in vacuum (10^{-6} torr) for the case of a boat evaporated Pd thin film (1000 Å) on (100) GaAs.

Figure 1 shows thermal x-ray diffraction patterns of Pd/(100) GaAs heated in argon (1 atm.). At this pressure, the compound formation was observable only at temperatures above ∼250°C. Similar thermal x-ray diffraction performed in vacuum (10^{-6} torr) showed that the compound formation was observable at temperatures above ∼200°C. The appearance of PdGa and a small amount of Pd_2Ga was observed after ∼15 min of heating at 200°C. Hence the required reaction temperature was lowered by decreasing the pressure. Moreover, $PdAs_2$ was not observed at any temperature, in contrast to its appearance during heating in 1 atm argon. This is attributed to the enchanced arsenic vapor evolution in vacuum[5]

APPLICATION IN PHASE TRANSFORMATION STUDIES

Other than reaction, phase transformation is one of the most common phenomena that occur in electronic materials at elevated temperatures. The phase transformations include melting, solidification and solid-solid transformations. Thermal x-ray diffraction enables in situ observations of the phase transformation during heating or cooling, so that the transformation temperatures during heating and cooling can be determined and the nature of the transformation can be studied. Furthermore, the effect of the cooling rate on the solidification product can be investigated. This effect is particularly relevant in the study of mestastable phase formation.

<u>Melting and Solidification</u>. Melting is perhaps the most common phase transformation which occurs in solders, polymers, certain reaction products and various electronic materials. Given here is an illustration of the use of thermal x-ray diffraction in studying the melting and solidification of the interfacial reaction product at Au/GaAs(100).[4]

Shown in Fig. 4 are x-ray diffraction patterns of Au/GaAs obtained in a flowing argon ambient (1 atm) at different temperatures during the first, second and third melting-solidification cycles. The temperature changes during all three cycles are shown in the plot of sample temperature vs. time in Fig. 5. The AuGa compound formed during the first heating was observed to melt around 458°C, as shown by the absence of the AuGa peaks in Fig. 4 (a). Upon subsequent cooling, solidification was observed at around 277°C, resulting in the appearance of the β phase[6-8] as shown in Fig. 4(b). On second heating, second melting was observed at around 415°C, as shown by the absence of the β peaks in Fig. 4(c). Second cooling resulted in second solidification at around 385°C and the reappearance of β, as shown in Fig. 4(d). Third melting and third solidification were observed at ∼413° and ∼397°C, respectively. Fig. 4(g) shows the diffraction pattern at 150°C after the third solidification. Excess Au coexisted with the Au-Ga alloy during all the heating and cooling cycles, as shown by

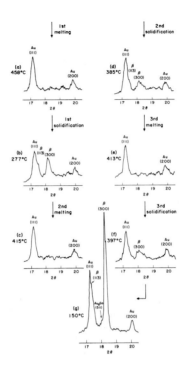

Fig. 4 Thermal x-ray diffraction patterns of Au/GaAs in argon during the first three melting-solidification cycles.

the presence of Au peaks in Fig. 4(a) - (g).

By in situ x-ray diffraction observation during heating and cooling to various temperatures, the melting and solidification temperatures were found to fall in the ranges shown in Table 1. The melting and solidification temperature for second and subsequent cycles are essentially the same. The first melting temperature in Table 1 agrees with the known melting temperature of 461.3°C for AuGa. The second melting and second solidification temperatures in Table 1 agree with the known melting temperature of 409.8°C for β. The first solidification temperature decreased considerably with increasing heating rate, so that its values fall in a relatively large temperature range.

TABLE 1 MELTING AND SOLIDIFICATION TEMPERATURES IN Au/GaAs

	Melting temperature (°C)	Solidification temperature (°C)
1st cycle	456-500	277-390
2nd cycle	410-415	390-398

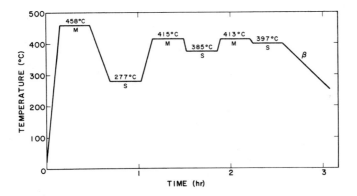

Fig. 5 Temperature vs. time during the melting-solidification cycles of Fig. 4.

Effect of Cooling Rate. The cooling rate in thermal x-ray diffraction can be controlled by the temperature controller and the temperature of the gas purging the sample chamber. By using a cold gas purge, a maximum in situ cooling rate of ~40°C/min was obtained with the set-up used in this work. Given here is an illustration of the use of thermal x-ray diffraction to study the effect of the cooling rate on the solidification product at the Au/GaAs (100) interface[4].

The effect of cooling rate on the solid phases formed on solidification was also investigated. Shown in Fig. 6(a) - (c) are x-ray diffraction patterns obtained at (a) 265, (b) 238, (c) 255°C after melting and subsequent solidification at three successively higher cooling rates, respectively. The respective cooling curves, labeled a, b and c, are shown in Fig. 7. The three cooling curves mainly differ in the cooling rate in the temperature range from 465°C to 350°C. The cooling rates for curves a, b and c around this temperature range are ~2.5, ~5.2 and ~40°C/min., respectively. The arrows in Fig. 7 indicate the temperatures for the representative x-ray patterns of Fig. 6. Note that these temperatures are quite close together for ease of comparison. At the lowest cooling rate (a), β was observed, as shown in Fig. 6 (a). At the intermediate cooling rate (b), additional peaks were observed. These peaks grew as the cooling rate increased. At the highest cooling rate (c), the new peaks dominated the β peaks. The new peaks have been <u>tentatively</u> identified as being due to the Au_2Ga compound, which is a metastable phase absent in the Au-Ga phase diagram. Fig. 7 shows that the cooling rate in the temperature range 465° - 350°C determine the relative amounts of Au_2Ga and β to form. A high cooling rate (40°C/min.) in this temperature range is necessary for Au_2Ga to dominate. This is reasonable since the β phase field in the Au-Ga phase diagram extends from 282°C to 409.8°C. Note the absence of the AuGa compound in Fig. 6 (a) - (c).

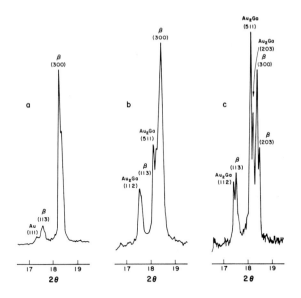

Fig. 6 Thermal x-ray diffraction patterns after melting and subsequent solidification at three successively higher cooling rates.

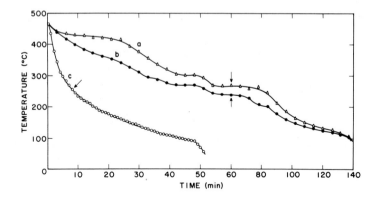

Fig. 7 Cooling curves which gave the diffraction patterns in Fig. 6.

DISCUSSION

Because of the growing importance of structural characterization of electronic materials at temperatures above room temperature, thermal x-ray diffraction is found to be a valuable technique in electronic materials research. In this paper, this experimental technique is introduced and its capabilities are illustrated by means of one of the forefront problems in current electronic materials research, namely the interfacial processes between a metal thin film and a semiconductor. The capabilities include phase identification, in situ process characterization, kinetic parameter determination, pressure and atmosphere control and cooling rate control for metastable material formation. Specialized uses such as stress measurement[9] and texture measurement are not considered in this paper.

Akin to thermal x-ray diffraction is thermal electron diffraction, which is typically performed by using a transmission electron microscope (TEM). The main advantages of thermal x-ray diffraction over thermal electron diffraction are (i) better control of the sample temperature, (ii) intensity measurement capability for crystallographic studies, (iii) capability of control of the atmosphere around the sample. On the other hand, thermal electron diffraction enables microdiffraction, and the use of the imaging capability of TEM allows correlation of crystallographic and microstructural information.

It is time for thermal x-ray diffraction to be used as one of the main thermoanalytical techniques in materials research. The structural information obtained by thermal x-ray diffraction is essential for interpretation of the results of DSC, TGA and other common thermoanalytical techniques.

REFERENCES

1. Proceedings of the 10th Annual Conference of North American Thermal Analysis Society, 1980.
2. Xian-Fu Zeng and D. D. L. Chung, J. Vac. Sci. Tech., July/Aug. 1982.
3. J. O. Olowolafe, P. S. Ho, H. J. Hovel, J. E. Lewis and J. M. Woodall, J. Appl. Phys. 50, 955 (1979).
4. Xian-Fu Zeng and D. D. L. Chung, J. Appl. Phys. (1982).
5. S. Leung, L. K. Wong, D. D. L. Chung and A. G. Milnes, J. Electrochem. Soc. (1982).
6. T. J. Magee and J. Peng, Physica Status Solidi A32, 695 (1975).
7. K. Kumar, Japanese J. Appl. Phys. 18, 713 (1979).
8. J. M. Vanderberg and E. Kinsbron, Thin Solid Films 65, 259 (1980).
9. J. Angilello, J. Baglin, F. d'Heurle, S. Petersson and A. Segmuller, IBM Research Report, RC 8106 (#35182), Solid State Physics, 2/6/80.

ACKNOWLEDGEMENT

The x-ray equipment grant from the Division of Materials Research of the National Science Foundation under Grant No. DMR-8005380 was essential for this work. Support from the Materials Research Laboratory Section, Division of Materials Research, National Science Foundation under Grant No. DMR 76-81561 A01 is also acknowledged.

INDUSTRIAL APPLICATIONS OF CURIE
TEMPERATURE DETERMINATION OF ALLOYS

Bengt O. Haglund
Sandvik AB, Hard Materials
Coromant Research Center,
P.O. Box 42 056
S-126 12 Stockholm, Sweden

1. INTRODUCTION

Only a few elements of the periodic table are known to show ferromagnetism. The most well-known of these are the three elements in the iron-group, viz. iron, cobalt and nickel. But at very low temperatures a number of elements in the rare-earth group also become ferromagnetic, viz. gadolinium, terbium, dysprosium, holmium, erbium and thulium.

2. DEFINITIONS AND SOME BASIC RELATIONS

The ferromagnetic materials are characterized by a spontaneous magnetization and by the magnetization reaching a very high level even when a relatively small external field is applied. When the external field is very strong, e.g. in the neighbourhood of 1 MA/m, also relatively large pieces of the ferromagnetic metals become saturated and comprise only a single domain. The saturation magnetization M_s decreases when the temperature is increased. When the material is heated further M_s falls rapidly to zero at a temperature named the Curie temperature, T_C. Above this temperature the material is paramagnetic.

If the spontaneous magnetization is normalized as a "reduced magnetization", M_s/M_o, and studied as a function of the "reduced temperature", T/T_C, the curves for the ferromagnetic elements Fe, Co(FCC), and Ni are very similar and almost coincide [1]. It is thus to be expected that there exists a close relation between T_C and M_s.

According to Weiss this relation can be written:

$$T_C = \frac{J+1}{3J} \cdot \frac{\gamma}{k} \cdot \mu_m \cdot M_s$$

where J   = total orbital impulse moment
        γ = the molecular field constant
        k   = Boltzmann's constant
        μ_m = the magnetic moment of a dipole

3. CURIE TEMPERATURES OF PURE ELEMENTS.

The magnetic transformation temperatures or Curie temperatures of pure, ferromagnetic metals are given in Table 1. There is an appreciable difference in the literature data, but those given in the table seem to be the most preferred ones for high purity metals. Cobalt exists in two ferromagnetic allotropic modifications, the low-temperature HCP phase and the high-temperature FCC phase. The phase transition temperature is reported to be between 388°C and 450°C [2]. By extrapolation of low-temperature data of cobalt alloys the Curie temperature of the HCP phase of pure cobalt has been estimated.

TABLE 1. CURIE TEMPERATURES OF FERROMAGNETIC METALS

Metal	Allotropic modification	Curie temperature K	°C	Ref
Fe	BCC	1043	770	[2]
Co	FCC	1394	1121	[2]
Co	HCP	1070	797	[2]
"	"	1150	877	[2]
Ni	FCC	631	358	[2]
Gd	HCP	289	16	[2]

4. CURIE TEMPERATURE OF ALLOYS

In binary phase diagrams the solubility curves of several metallic solutes in iron, cobalt or nickel have a similar curvature. This curvature is generally approximated by an Arrhenius equation, in which the logarithm of the solubility is a linear function of the reciprocal absolute temperature.

However, more thorough studies by, among others, Hillert, Wada and Wada [3], Harvig, Kirchner and Hillert [4] and recently by Takayama, Wey and Nishizawa [5] have revealed anomalies of this solvus curve in the neigbourhood of the Curie temperature. The work of Hillert, Wada and Wada [3] was an experimental test i.a. in the systems Fe-Mn and Fe-Ni of a model proposed by Zener [6]. According to this there is an effect of Mn and Ni on the Curie temperature of BCC-Fe and on the standard free energy of reaction of the solutes when transferred from the BCC-Fe to the FCC-Fe. According to Zener the free energy of pure BCC-Fe can be split into a magnetic term and a non-magnetic term. When a small amount Δx of another element is alloyed in the iron the Curie temperature is displaced a certain amount ΔT_C and there is also a change in the free energy. Hillert, Wada and Wada have described this relation according to the equation:

$$\left(\Delta°G_\gamma^{\alpha\to\gamma Fe}\right)_{Mag} = \frac{1}{\Delta x}\Delta T_C \cdot \frac{\delta(°G_{Fe}^\alpha)_{Mag}}{\delta T} = -\frac{dT_C}{dx}\left(°S_{Fe}^\alpha\right)_{Mag}$$

Thus there is an intrinsic slope dT_C/dx of the Curie temperature vs concentration curve which is specific for the solute element in question.

5. MEASUREMENT

5.1 Thermogravimetric methods. There are several methods available for measurements of the transition from the ferromagnetic to the paramagnetic state (or vice versa) based upon the attraction force on the sample from a magnetic field. When this determination is made to study the influence of the temperature it is often called "thermomagnetometry".

A determination of the Curie temperature of a cemented carbide binder phase sample by means of an automatic recording magnetic balance was presented already in 1958 by Nishiyama and Ishida [7]. During the latest decades there have been several publications on the use of thermobalances for thermomagnetometry [8, 9, 10]. The basic design has generally been the same, i.e. the sample in the furnace is subjected to a magnetic field gradient, generally from a permanent magnet. The necessary magnetic field strength is of the order of magnitude of 10-100 kA/m for Curie temperature measurements.

5.2 Inductive methods. There are a number of methods described in the literature which make use of the change in permeability which occurs when a material is heated through its Curie temperature [11]. One example is based on the direct measurement of the permeability of the sample by measuring the impedance of a coil surrounding the sample [12]. The field strength used is of the order of magnitude 0.1-4 kA/m, depending on the sensitivity of the amplifier.

5.3 The Foner magnetometer. For determination of the saturation magnetization and of the Curie temperature the so-called Foner magnetometer can also be used [13]. In this instrument the sample has been fastened to a vertical rod in the gap of the magnet. The rod vibrates at a low frequency. When the sample vibrates an alternative voltage is induced in two detector coils which are situated between the sample and the pole pieces of the magnet.

5.4 Calorimetric methods. The transition from the ferromagnetic to the paramagnetic state is accompanied by a sharp peak in the specific heat capacity [14]. Differential thermal analysis (DTA) or differential scanning calorimetry (DSC) are methods which are very well suited to study this change in specific heat capacity at rising or descending temperatures.

6. APPLICATIONS

6.1 Control of dissolved elements in ferromagnetic alloys. Most solute elements lower the Curie temperature of the alloys continuously to the point where the Curie temperature curve intersects the solubility curve. In alloys with a composition within the two-phase field the Curie temperature remains constant. Thus, the Curie temperature can be used to determine the concentration of the solute element in a binary system up to the solubility limit.

6.2 Phase transformations. Curie temperature measurements can be used for studying the presence of magnetic phases in alloys at different stages during thermal treatment. Thus thermomagnetometry can provide complimentary information to other methods, e.g. X-ray diffraction, electron microprobe analysis combined with microstructural studies at different magnifications and differential scanning calorimetry. As thermomagnetometry is rapidly performed and also often can be made to reproduce an industrial heat treatment cycle it can give information which is difficult or even impossible to obtain by other methods.

6.3 Cemented carbides. Fukatsu [15] studied already in 1961 the structure and properties of binder phase alloys of compositions corresponding to those of WC-Co cemented carbide. He studied i.a. the magnetization curves vs. the temperature and he was thus able to determine the Curie temperature as a function of the tungsten carbide content in solid solution in cobalt. He found the relation

$$\theta_x = \theta_o - 17 x_{WC}$$

where θ_o = the Curie temperature of the pure cobalt

Θ_x = the Curie temperature of the alloyed cobalt binder phase

x_{WC} = the concentration of WC(wt %) in solid solution.

Nishiyama and Ishida [7] investigated the cobalt binder phase of several types of cemented carbide grades containing not only Co and WC, but also TiC and (Ta,Nb)C. They found that the cobalt binder phase of cemented carbide always has a lower Curie temperature than that of cobalt saturated with carbon. This indicates that the binder phase always contains more or less tungsten and also (according to Nishiyama and Ishida) fairly large amounts of various other elements in solid solution at room temperature.

Later studies [16] have, however, not confirmed that any appreciable amounts of titanium, tantalum or niobium are present in the binder phase. It is thus suggested that the Curie temperature of normal cobalt-based cemented carbide alloys is, approximately, a function of only the carbon and tungsten contents in the binder phase.

6.4 Use of the Curie temperature for calibration purposes.

In most thermobalances the temperature of the sample is measured at some distance from the sample (or reference material). As thermal analysis generally is performed at a certain rate of heating or cooling and as there always exists a temperature gradient inside the furnace, the measured temperature generally differs more or less from the true temperature of the sample. To enable corrections for this difference the ICTA Committee on Standardization recommends the use of Curie temperature measurement on a series of Certified Reference Materials with recognized Curie temperatures [17].

6.5 Artifacts in thermogravimetry introduced by ferromagnetic samples.

Gallagher et al [18] have reviewed and reinvestigated work of other authors which claim to have found a pronounced effect of an external magnetic field on the reaction rate of materials below their magnetic transition temperature. This effect was found in thermogravimetric studies of the reduction rates of some oxides of iron and cobalt [19].

Gallagher et al used a complimentary technique to thermogravimetry, namely evolved gas analysis (EGA). This technique is not sensitive to any influence of external magnetic fields on the sample and Gallagher et al was also unable to reproduce the claimed influence of the magnetic field on the reduction rate. The earlier results which imply such an influence were thus suggested to be artifacts brought on by magnetic forces on the samples. Similar magnetic effects on thermogravimetric measurements can also result from the interaction between a ferromagnetic sample and strong magnetic fields e.g. from the furnace windings, see for instance Moskalewicz [20].

7. SUMMARY

The basic relations between magnetic saturation and Curie temperature versus the composition are discussed for ferromagnetic metals and alloys of technical importance. There exist a number of methods for the determination of the transition from ferromagnetic to paramagnetic behaviour, i.a. thermogravimetry (thermomagnetometry), inductive methods and calorimetric methods.

As most non-magnetic solute elements lower the Curie temperature of ferromagnetic metals the Curie temperature can be used to determine the concentration of the solute element in binary systems up to the solubility limit.

In ternary and higher systems there is a combined effect of different elements on the change of Curie temperature. Thus it is not generally possible to use the Curie temperature as a measure of the solute element concentration in such alloys. Under certain circumstances, e.g. when the composition of a multiphase material is close to the equilibrium composition, it is still possible to use the Curie temperature to characterize the composition of the ferromagnetic phase. Such a system is exemplified by normal cobalt-based cemented carbide.

The Curie temperature can also be used as a means for temperature calibration, e.g. of thermobalances. Certified Reference Materials have been selected for this purpose by the ICTA Committee on Standardization.

8. ACKNOWLEDGEMENT

The author wishes to thank Sandvik AB for the permission to publish the manuscript. He is also indebted to Professor B. Aronsson, Drs L.J. Aschan, B. Uhrenius and U. Smith for valuable criticism and to Mrs Mariana for typing the manuscript.

9. REFERENCES

1. D H Martin: Magnetism in Solids.
   Iliffe Books Ltd, London (1967), p.9
2. Cobalt Monograph
   Centre d'Information du Cobalt, Brussels (1960)
3. M Hillert, T Wada, H Wada
   J Iron and Steel Institute 1967, 205, 539-546
4. H Harvig, G Kirchner, M Hillert
   Met. Trans. 1972, 2, 329
5. T Takayama, M Y Wey, T Nishizawa
   Trans. Japan Inst.Metals 1981, 22, 315-325
6. C Zener
   J. Metals, 1955, 7, 619
7. A Nishiyama, R Ishida
   Trans. Japan Inst. Metals 1962, 3, 185-190
8. W R Ott, M G McLaren
   Proc. Second Int. Conf. on Thermal Analysis. Acad Press, New York, London (1969), p 1439-1451
9. U Bäckman, B O Haglund, B Bolin
   Proc. Third Int. Conf. on Thermal Analysis.
   Birkhäuser Verlag, Basel, Boston, Stuttgart (1972), p 759
10. P K Gallagher, E M Gyorgy
    Thermochim Acta 1979, 31, 380
11. A de Sa
    J. Phys. E: Sci. Instrum.1968, 1, 1136-1137
12. V I Tumanov, E A Shchetilina, A A Cheredinov, O I Serebrova, E A Korchakova, V N Elizarov, S M Elmanov
    Soviet J. Nondestructive Testing 1975, p 641-646
13. S Foner
    Rev. Sci. Instrum. 1959, 30, 548
14. M Braun, R Kohlhaas
    Phys. Status Solidi 1965, 12, 429
15. T Fukatsu
    J. Japan Soc. Powder Metallurgy 1961, 8, 183-194
16. A. Henjered, M. Hellsing, H. Nordén, H.O. Andrén
    Int.Conf.Science of Hard Materials,
    Jackson Lake Lodge, Wyoming, (1981)
17. P D Garn
    Thermochim. Acta 1980, 42, 125-134
18. P K Gallagher, E M Gyorgy, W R Jones
    J. Chem. Phys 1981, 75, 3847-3849
19. M W Rowe, R Fanick, D Jewett, J D Rowe
    Nature 1976, 263, 756
20. R Moskalewicz
    Proc. Fourth Int.Conf. on Thermal Analysis, Heyden, London, New York, Rheine (1975):3, p 873-880

EFFECTS OF POWDER PROCESSING UPON THE

THERMAL BEHAVIOR OF THE TiC-Ni-Mo$_2$C SYSTEM

By

Masood A. Tindyala, Inland Steel Company, Research
Department, East Chicago, Indiana 46312; and
Ronald A. McCauley, Department of Ceramics
College of Engineering, Rutgers - The State
University, Piscataway, NJ 08854

INTRODUCTION

In spite of the extended use and high state of practical development of the cemented titanium carbides, consistency in industrial production is still a matter of chance. The question put forward by Norton [1] some twenty years ago is still not fully answered. "How does one exercise the kind of control necessary to obtain the desired microstructure?"

The mechanical properties of the components of the system TiC-Ni, and the ideal microstructure consistent with the properties of the components have been well established [2,3]. Blumental and Silverman [4] attempted to establish the effects of processing upon the microstructure of titanium carbide but the introduction of multiple variables weakened their conclusions. However, these studies pointed to the fact that processing of carbide-metal powders had a very significant influence on the microstructure.

An understanding of the effects of powder processing upon the thermal behavior of the TiC-Ni-Mo$_2$C system is essential to the understanding of the control necessary to obtain the desired microstructure. The thermal processes such as decarburization in the system TiC-Ni-Mo$_2$C can be studied by investigating the weight loss as a function of temperature (TG), the rate of weight loss (DTG) and gas analysis of the evolved gas (EGA) during heating of the carbide-metal powder. The DTA study will further enhance the understanding of the reactions leading to decarburization.

EXPERIMENTAL

Titanium carbide was produced by the Menstruum process in large units that were then dry crushed by jaw crushers to pass a 100 mesh sieve. The Mo$_2$C powder was produced by the dry carburization process. The resulting Mo$_2$C powder was milled in hexane in tungsten carbide lined mills with tungsten carbide mill balls. The nickel powder was produced by the carbonyl process.

A Mettler Recording Vacuum Thermoanalyzer was used to monitor the temperature, weight change (TG), rate of weight change (DTG) and the differential temperature (DTA) of specimen powders during heating. Specimen powders were heated in 16mm alumina crucibles to 1500°C. Vacuum runs were made at 2×10^{-5} torr. In the case of helium, a flow rate of 10 l/hr of ultra high purity helium was maintained. At least two runs were made for each powder examined. For

CO and CO_2 analysis a Gastec Precision Gas Detector System was employed. The sampling was done at 50°C intervals and at the temperatures of DTG peaks.

Surface area determinations of selected powders were made by the BET method using a Model 2205 High Speed Surface Analyzer. The specific surface areas reported represent the averages of two analyses having a standard deviation of less than 0.05 m^2/gm. The total and free carbon contents of selected powders were determined prior to the thermal analysis. A gravimetric combustion technique was employed whereby a known amount of sample was heated in a stream of flowing oxygen. The accuracy of the measurement was $\pm$ 0.015% at 6% carbon. The analysis of chemically uncombined (free) carbon was accomplished by the dissolution of a known quantity of the sample in a mixture of concentrated acids.

Oxygen contents were determined with a Leco Model 760-400 RO-16 Oxygen Determination Instrument. The accuracy of the oxygen determination was $\pm$ 3%.

DISCUSSION

Oxygen Content and Surface Area. A summary of chemical and physical properties of the experimental powders is given in Table I. The most significant aspect of these data was the relative oxygen contents and the specific surface areas of the experimental powders. The increase in the surface area of the TiC powder during milling clearly indicated that the oxygen was primarily adsorbed at the newly created surfaces. In contrast the adsorption of oxygen may have taken place both on the existing as well as newly formed surfaces in the Mo_2C. For the ternary composition 70TiC-12Ni-18Mo_2C a logarithmic relationship between the surface area and milling time, similar to that observed for WC-Co compositions [5], was exhibited. The oxygen analyses represent an average concentration over all phases present in these powders. It was evident from the relative amounts of the free carbon and oxygen, however, that much of the oxygen present in these powders was associated with carbon surfaces.

TGA of Virgin Powders. The percent weight loss in vacuum as a function of temperature for the virgin powders is shown in Figure 1. The magnitude of weight loss for Mo_2C powder as compared to other two powders is noticeable. The composite curve is also shown.

TGA of Milled Individual Powders. Weight loss curves for individually milled TiC and Mo_2C powders, heated in helium at 2°C/min., are shown in Figure 2.

In both TiC and Mo_2C powders the adsorption of oxygen may have taken place on old as well as new surfaces. However, these two powders showed two different types of behaviors as shown in Figure 3. In this figure oxygen contents are shown as a function of the specific area of the experimental powders. A three-fold increase in the specific area of Mo_2C powder yielded a five-fold increase in the oxygen content. For TiC powder it took an increase of 44 times to yield a seven-fold increase in oxygen content. The manufacturing and processing differences not withstanding, Mo_2C powder has a much higher affinity to adsorb oxygen than does TiC powder.

TGA of Milled Ternary Composition. The percent weight loss in vacuum as a function of temperature for experimental powder composition 70TiC-12Ni-18Mo$_2$C is shown in Figure 4. This figure shows the weight loss curves for powders milled for 24, 48, and 72 hrs. All three powders showed a rapid weight loss in the range 100 to 350°C. The slope of the weight loss curve became steeper for the powders with finer particle size. The three powders showed a fairly uniform weight loss for the range 350 to about 800°C. There was an increase in weight loss rate at about 800°C and again a uniform weight loss to about 1300°C where the weight loss became rapid due to nickel melting and vaporization. The percent weight loss at a given temperature was higher for the longer milled powders.

The weight loss curves for the experimental composition 70TiC-12Ni-13Mo$_2$C milled for 48 hrs. and heated in vacuum at different heating rates were very similar in form to the 48 hrs. curve shown in Figure 4. Higher weight loss was shown with slower heating rates. Weight loss curves for the same 48 hrs. milled powder heated in helium at different heating rates, again similar in shape, showed that the percent weight loss decreased with slower heating rates. There was no rapid weight loss following the nickel melting.

DTG Curves. The titanium carbide DTG curves showed increasing rate of weight loss with increasing milling time, at both low and high temperature ranges. It also showed appearance of two maxima at low temperature, at 200 and 230°C with increasing milling time, and one at high temperature, at 1100°C.

The virgin Mo$_2$C powder was different from the virgin TiC powder in that it was milled in heptane by the supplier and had very fine particle size, in addition to the crystallographic and manufacturing process differences. The DTG curves for the Mo$_2$C powders (Fig. 5) showed dramatic increases in the weight loss maxima with increasing milling time. Additionally, a very large new peak at 980°C appeared as a result of the milling operation.

DTA Curves. The DTA curves obtained did not give any information for the desorption of CO and CO$_2$. The only reliable peak shown by the DTA curve was an endotherm at approximately 1352°C for the nickel melting in the ternary composition 70TiC-12Ni-18Mo$_2$C. This nickel melting temperature varied a few degrees with heating rates. It was higher for the faster heating rate and lower with a slower heating rate.

Carbon Monoxide and Carbon Dioxide Spectra. Oxygen is chemisorbed on carbon surfaces more readily than most other elements and the resulting covalent bonds do not allow desorption of adsorbate atoms without simultaneous removal of carbon atoms as well. The presence of impurities on the carbon and carbide surfaces promotes adsorption and chemisorption of impurity gases such as oxygen. Carbon monoxide and carbon dioxide desorption spectra did not correspond to the total weight loss since water, some nitrogen and hydrocarbons were also evolved. These gases were not identified in the present study.

Figure 6 shows the CO and CO$_2$ desorption spectra for the virgin TiC powder and indicated that some amount of gases existed on the surface of unprocessed TiC but was insignificant. The reason for the low amount of adsorbed oxygen during storage lies in the coarse nature of the virgin TiC particle size.

Figure 7 shows the effect of milling operation on the character of CO and CO_2 spectra. These desorption spectra are shown for TiC powder milled for 96 hrs. The amount of the sample used in this run was 1/5th of the virgin TiC shown in Figure 6. The major CO peak observed in virgin TiC showed up as a shoulder on the low temperature side of the 1155°C peak for the 96 hrs. milled powder. The CO_2 was evolved primarily below 1100°C having maxima at 230, 400 and 925°C. This very large amount of evolved CO and CO_2 gases correspond to the seven-fold increase in the oxygen contents of the TiC powder milled for 96 hrs.

The CO and CO_2 desorption spectra of the ternary composition 70TiC-12No-18Mo_2C milled for 72 hrs. are shown in Figure 8. A large increase in evolution of both CO and CO_2 gases is apparent. A comparison of these spectra and the DTG curves for Mo_2C powders indicated the influence of Mo_2C presence in the ternary composition powder. The shift in CO maximum from 1200 to 1080°C was attributed to the presence of nickel, which destabilized the adsorbed oxygen on the carbide surface.

The desorption spectra shown in Figures 6-8 indicated that the composition of the product gas was most affected by Mo_2C and nickel additions to the titanium carbide powder. In TiC powders the adsorption of oxygen took place at the unsaturated sites during initial exposure to the atmosphere. Due to the coarse nature of the TiC particles the small number of these active sites soon approached saturation and the adsorption process apparently ceased. Figure 6 suggests the existence of at least two strongly bonded adsorbed phases decomposing to CO and a single adsorbed state giving CO_2 at the surface of the unmilled powder. The higher desorption temperature for these states suggested that the amount of oxygen adsorbed in storage was found at both carbide and free carbon surfaces.

Several distinct states for oxygen are formed during milling operation. Some of the adsorbed phases so formed during milling are strongly bonded and contribute a lot to the total quantities of CO and CO_2 desorbed during heating. This may have resulted in extensive decarburization of the milled powders. The concentrated acids solution of the powders, heated for CO and CO_2 studies, showed no presence of the free carbon, confirming its complete consumption. The one to nine ratio of free carbon and oxygen contents indicated that the major portion of the oxygen was adsorbed on the carbide surfaces during milling and unless sufficient free carbon was provided in the carbide powders the resultant carbide phase would be deficient in carbon.

ACKNOWLEDGMENT

The financial assistance of the V. R. Wesson Company is deeply appreciated.

TABLE 1 CHARACTERIZATION OF THE EXPERIMENTAL POWDERS

POWDER	CARBON CONTENTS (%) TOTAL	CARBON CONTENTS (%) FREE	OXYGEN CONTENTS (PPM)	AVE. PARTICLE SIZE (ESD, μm)	SURFACE AREA (M²/g)	SULPHUR (PPM)
TiC	19.74	0.09	1376	30	0.23	276
Mo$_2$C	6.08	0.12	4521	1.6	1.49	107
Ni	—	—	—	5	0.36	63
TiC milled for 24 hrs.	—	—	3775	1.58	2.80	276
TiC milled for 96 hrs.	—	—	9411	0.69	10.11	276
Mo$_2$C milled for 24 hrs.	—	—	12989	1.00	2.16	107
Mo$_2$C milled for 96 hrs.	—	—	21702	0.76	4.82	107
70TiC-12Ni-18Mo$_2$C	—	—	5257	1.76	2.13	—
70TiC-12Ni-18Mo$_2$C milled for 48 hrs.	—	—	6457	0.91	3.94	—
70TiC-12Ni-18Mo$_2$C milled for 72 hrs.	—	—	7343	0.58	4.81	—

REFERENCES

1. J. T. Norton in *High-Temperature Materials*, page 119, ed., R. F. Hehmann, John Wiley & Son, New York (1959).
2. G. H. Price, C. J. Smithells and S. V. Williams, J. Instt. Metals, 1938, *62*, 239.
3. J. Gurland and J. T. Norton, J. Metals 4, Trans. AIME, 1952, *194*, 1051.
4. H. Blumenthal and R. Silverman, J. Metals 7, Trans. AIME, 1955, *203*, 317.
5. H. F. Fishmeister and H. E. Exner, Planseeber Pulvermet., 1965, *13*, 178.

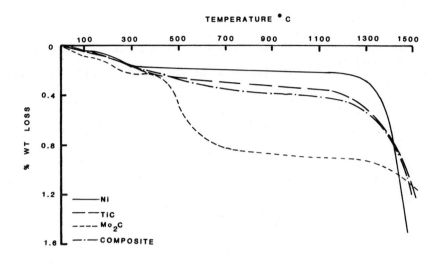

FIGURE 1 TG CURVES FOR THE VIRGIN POWDERS HEATED AT 2°C/MIN. IN VACUUM.

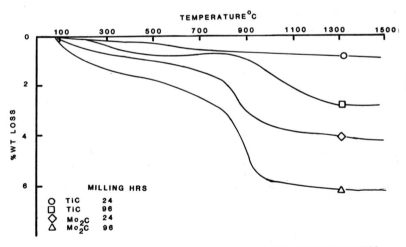

FIGURE 2 TG CURVES FOR THE INDIVIDUALLY MILLED POWDERS HEATED AT 2°C/MIN. IN HELIUM.

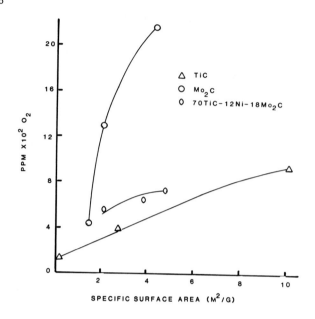

FIGURE 3 THE OXYGEN CONTENTS AS A FUNCTION OF SPECIFIC SURFACE AREA

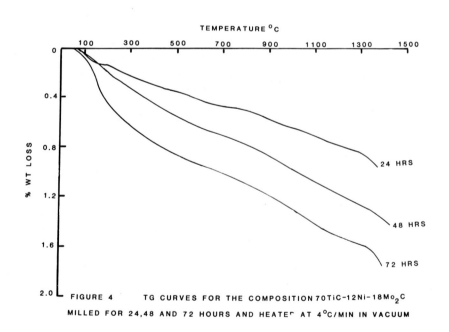

FIGURE 4 TG CURVES FOR THE COMPOSITION 70TiC-12Ni-18Mo$_2$C MILLED FOR 24, 48 AND 72 HOURS AND HEATED AT 4°C/MIN IN VACUUM

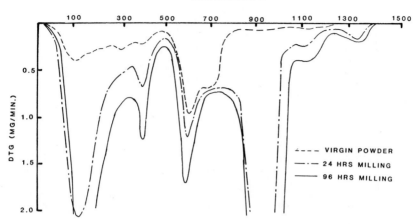

FIGURE 5   DTG CURVES FOR THE Mo$_2$C POWDERS HEATED AT 10°C/MIN. IN HELIUM.

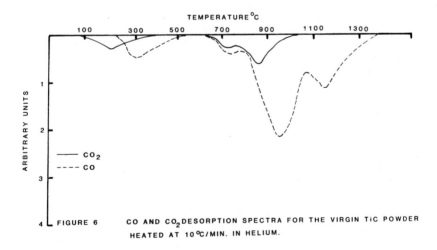

FIGURE 6   CO AND CO$_2$ DESORPTION SPECTRA FOR THE VIRGIN TiC POWDER HEATED AT 10°C/MIN. IN HELIUM.

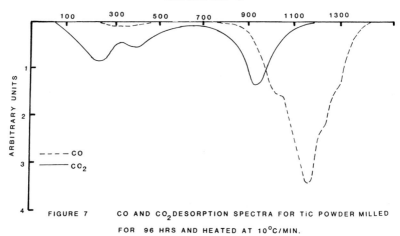

FIGURE 7   CO AND CO$_2$ DESORPTION SPECTRA FOR TiC POWDER MILLED FOR 96 HRS AND HEATED AT 10°C/MIN.

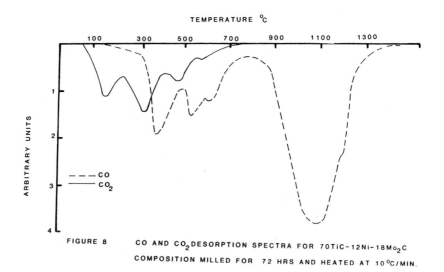

FIGURE 8   CO AND CO$_2$ DESORPTION SPECTRA FOR 70TiC-12Ni-18Mo$_2$C COMPOSITION MILLED FOR 72 HRS AND HEATED AT 10°C/MIN.

THERMAL ANALYSIS IN MANGANESE METALLURGY AND THE PROSPECTS OF ITS DEVELOPMENT IN COMBINATION WITH OTHER INVESTIGATION METHODS

D. V. Mosia, L. K. Svanidze, T. N. Zagu, T. I. Sigua
The Institute of Metallurgy under the Georgian SSR
Academy of Sciences
380042, 15 Pavlov St.
Tbilisi, USSR

Under thermal treatment, manganese concentrates undergo complex transformations, the nature of which is, to a large degree, determined by chemical, mineralogic and granulometric composition and also by other properties of ores.

The diversity and complexity of energy-consuming physico-chemical transformations in the thermal treatment of manganese ores, concentrates and their agglomeration products can be successfully registered by the method of differential thermal analysis.

The processes of thermal dissociation and of the reduction of oxides and manganese carbonates were studied by us on pure reagents, manganese ores and concentrates obtained from them and also on various ore-fuel mixtures.

The tests were conducted on the "MOM" derivatograph (Hungary) used in the following operational mode: heating rate: 1° per minute, thermocouple plate: of platinum and rhodium, sample: 1-3 g, crucible of corundum, standard: alumina calcinated up to 1000°C, heating time: 100 min.

Owing to the limited length of this report, data on only the most typical material investigated by us are listed here.

Peroxide (Fig. 1). The concentrate of oxide-bearing ore containing 89 per cent of manganese dioxide and 0.24 per cent of manganese monoxide. 180°C-loss of adsorbed water; 700°C-dissociation of pyrolusite with the formation of β-kurnakite; 990°C-conversion of β-kurnakite into β-gaussmanite. Endothermal bends at 360° and 760°C point out to a small admixture of psilomelane.

In argon flow the decomposition temperature of manganese dioxide is 30° and of β-kurnakite 40° lower than in air. The loss of weight is 3.9 per cent higher.

In oxygen flow the same effects manifest themselves at temperatures 15° higher and with weight loss 1.2 per cent lower than in air, which indicates the slowing down of dissociation.

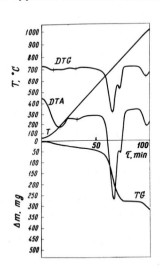

Fig. 1 A derivatogram of peroxide. Weight 2000 mg.

Manganese Concentrate of the 1st Grade (Fig. 2). 160°C-removal of adsorbed water; 370°C-decomposition of manganite with the formation of α-Mn_2O_3 (α-kurnakite) and possibly, dissociation of manganese carbonate; 510°C -oxidation of α-kurnakite with the formation of solid solution of oxygen in it; 670°C-total effect of the decomposition of the solid solution and of the pyrolusite dissociation; 750°C -conversion of tetragonal modification of Kurnakite (α)-into cubic one (β); 810°C-decomposition of permanganite (an admixture of psilomelane). At the same temperature the decomposition of calcium carbonate occurs. 990°C dissociation of β-kurnakite. In argon flow the above-mentioned processes occur at temperatures 20-40° lower; except for the decomposition of α-Kurnakite solid solution which lags behind by 50°C.

In oxygen flow the decomposition temperature of manganite remains the same, the formation of α-kurnakite solid solution occurs, naturally, at a lower temperature-480°C, the dissociation reaction proceeds 30° lower than in air.

Carbonate Concentrate (Fig. 3). The prevalent effects of the derivatogram are: dissociation of manganese carbonate-600°C and of calcium carbonate-850°C, 160°C the loss of water. Small waves at 420° and 680°C-a slight admixture of vernadite-hydrate of manganese dioxide which is the result of the oxidation of carbonate ore.

In argon flow the dissociation temperatures of carbonates do not change. The loss of weight increases largely (by 3.1 per cent) which is brought about by a more complete removal of reaction products. In oxygen flow the weight loss is 2.4 per cent less than in air.

The thermoanalytic curve obtained in thermal analysis reflects, besides the thermodynamics of the conversion process, the kinetics of the process in nonisothermic conditions. Horowitz and Metzger[1] showed the possibility of computing from these curves the energy of activation, the order of the reaction and preexponential factor for comparatively simple processes of dissociation not accompanied by the formation of intermediate products. On the basis of thermogravimetric curves obtained by us there was calculated the energy of activation for the most common minerals of industrial type manganese ores: -pyrolusite and calcium rhodochrosite.

To obtain the apparent energy of activation the thermoanalytic curves given in Fig. 1 and 3 were treated by the modified Horowitz-Metzger method[2].

For the sake of comparison the same derivatograms were treated by a method suggested by G. R. Allakhverdov and B. D. Stepin[3].

The areas on the derivatograms were calculated by Y. V. Sementovsky's method[4]. The results of the calculations are given in Table 1.

A slight deviation in values of activation energy is accounted for principally by the influence of a number of experimental factors on the thermal decay process-such as heating speed, weight value, sensitivity in the circuit of the differential thermocouple, shape and material of the crucible, packing density of the substance in the crucible,atmosphere of the furnace space, degree of sample wear, etc. The properties of the substance affecting the character of the thermogram include its composition, structure and degree of its perfec-

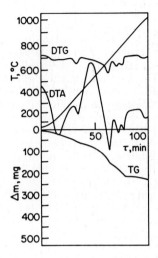

Fig. 2. A derivatogram of manganese concentrate, 1st grade. Weight 2110 mg.

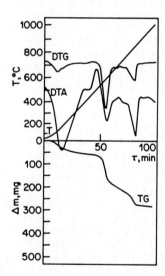

Fig. 3. A derivatogram of carbonate concentrate. Weight 1480 mg.

tion, crystallinity, dispersibility, isomorphic substitutions and also heat capacity and thermal conductivity.

Generally speaking, both methods of calculating activation energy are suitable for an approximate evaluation of the energy barrier for the process of mineral thermal decomposition.

On the basis of investigating a large number of carbonaceous materials there was shown the suitability of thermal analysis in the study of evaluating their reactive properties in relation to manganese oxides and industrial-type manganese ores.

The thermal analysis of peroxide-black mixture (Fig. 4) showed that a temperature of 670°C on the DTA curve of the derivatogram emerged an abrupt exothermic peak, corresponding to a violent reaction of MnO_2 reduction with a release of large amount of heat, the result of it being a jump-like rise of the furnace temperature by 50°C.

The tests conducted earlier, showed that the interaction of black with air oxygen started about 600°C. The second exothermal effect at 940°C, indicating the reduction of MnO_4 is considerably less pronounced owing to a comparatively small value of the reaction thermal effect at this stage. The intermediate effect of Mn_2O_3 reduction is not present at all, being absorbed by the explosive reduction of MnO_2.

For comparison under the same conditions, we obtained the derivatogram of manganese dioxide reagent (86 per cent MnO_2) on the DTA curve (a dotted line in Fig. 4) the step-like character of the reaction of manganese oxide reduction is very sharply pronounced. In this reaction the peak dimensions of MnO_2 and Mn_2O_3 reduction correspond to the relationship between the values of their thermal effects. As to the third peak, Mn_3O_4 reduction it is evident that at the temperature of its manifestation the simultaneous processes of direct and indirect reduction are taking place, and this peak is the total of them.

For ore-fuel mixtures (manganese concentrate of the 1st grade + black) up to 500°C the derivatogram is similar to the one obtained while heating the sample in air without a reducer. The exothermal peaks at temperatures of 600° and 640°C correspond to the reduction of MnO_2 and Mn_2O_3. After the endothermal effect of psilomelane decomposition at 820°C there follows a slight exothermal bend corresponding to the reduction of MnO_2 and Mn_2O_3. As the content of manganese in the concentrate decreases, the derivatograms of ore-fuel mixtures show thermal effects corresponding to manganite decomposition (350°C), manganese carbonate dissociation (570°C), permanganite decomposition (830°C) and calcium carbonate dissociation. Consequently, the kinetic features of reducing manganese pure oxides by solid carbon can hardly be applied to manganese ores and concentrates having rather a complex structure and chemical composition. To determine the optimum technologic parameters at the stage of preliminary reduction and burden preparation for smelting, it is necessary to investigate the kinetic features of the reduction by solid carbon used in the production of manganese ores and concentrates.

The study of the interaction process of manganese ores with bituminous coals and coke breeze is of interest in connection with the developing trend of using previously reduced agglomerated materials in the burden of ferroalloy furnaces. By way of illustration Fig. 5

Table 1. APPARENT ACTIVATION ENERGY OF DISSOCIATION REACTION (KJmol^{-1})

Investigated material (n=1)	Horowitz-Metzger's method		Alakhverdov-Stepin method	Literary Sources [6]
	in air	in argon flow		
Peroxide	150,3	90,9	152,0	178,4 160,4
Rhodochrosite	162,5	116,0	178,0	—

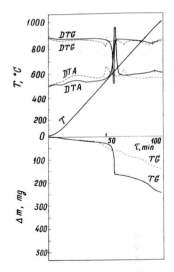

Fig. 4. A derivatogram of peroxide-black mixture Weight 930 mg

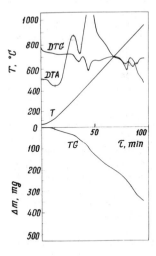

Fig. 5. A derivatogram of ore-coal mixture. Weight 1730 mg.

shows a derivatogram of ore-coal mixture (85 per cent of manganese concentrate, 1st grade and 15 per cent of Tkibuli coal concentrate). On the DTA curve a high endothermal effect at 360°C is caused by manganite decomposition. In a temperature range of 360-650°C a complex process occurs including the dissociation of manganese dioxide accompanied by air intensive oxidation of the coal. The endothermal effect at 820°C is caused by the decomposition of psilomelane and the consequent bends of the DTA curve by the reduction of Mn_3O_4 to manganese monoxide and by the gasification of carbon.

Replacement in the mixture of Tkibuli coal concentrate by Tkvarcheli concentrate caused the shift to higher conversion temperatures (by 10-15°). The comparison of the ore-coke derivatogram with ore-coal ones confirms the higher reactive properties of the coals (manganese higher oxides up to Mn_3O_4 are reduced by coke breeze at 610°C).

The process of physico-chemical conversions in the interaction of carbonaceous reducers with ore components of the burden during its thermal preparation for electric smelting are closely interconnected, extremely complex and difficult to interpret.

According to numerous investigations [5,6] and practical industrial experience, this interaction already starts at low temperatures, but it becomes especially intensive within a range of 300-600°C. At this time there occurs strong catalytic action of manganese oxide on coal pyrolysis and simultaneous reduction of manganese higher oxides. The cause of the process, to a large degree, depends on the type of a reducer, its amount, composition and the nature of manganese ore stock. The interpretation of effects of complex physico-chemical conversions obtained by the method of thermoanalysis in many cases presents considerable difficulties. To overcome these difficulties thermoanalysis should be combined with various physical and physicochemical methods: X-ray phase analysis; IR-spectroscopy; simultaneous registration of changes in electric conductivity, dielectric conductivity, dielectric sensitivity, volumetry, gas chromatographic analysis, etc.

The complex of these data will make it possible to find with high accuracy and reliability various physical and chemical processes going on in the thermal treatment of ore stock, to establish the nature of the observed process.

In particular, the continuous measuring of the amount of gaseous reaction products with the discrete determination of their composition by gas chromatography will enable us to determine not only the total amount of volatile products escaping in the thermal treatment of manganese ore burdens, but also the dynamics of their escape and quality at separate stages of the thermal treatment.

It also helps to observe experimentally that the reduction of MnO_2 in the field of lower temperatures 570-600°C is carried out through the stage of its dissociation, which is confirmed by the presence of free oxygen in the system. In the presence of coal the reaction is intensified as the escaping O_2 is consumed by carbon oxidation, reducing thereby the partial pressure of oxygen and accelerating the dissociation of MnO_2. The exothermicity of the carbon oxidation reaction is also conducive to it.

The obtained results to a certain degree, make it possible to estimate the behavior of ores and the products of their dressing in thermal treatment in various gas atmospheres and help to choose the optimum parameters for the thermal treatment of manganese ore burdens prior to smelting.

REFERENCES

1. H. H. Horowitz, G. Metzger, Anal. chem., 10, 1963.
2. A. K. Shkarin, N. D. Toper, G. M. Zhabrova, J. "Phisicheskays Khimia", 11, XII, 1968.
3. G. K. Allakhverdov, B. D. Stepin, J. "Phisichaskaya Khimia", 9, 1969.
4. Y. V. Samentovsky. Plant Laboratory, 22, 122, 1956.
5. V. B. Matveev. Complex Thermoanalysis of Manganese-Containing Ore-Coal Mixtures. Novosibirsk, "Nauka" 1975.
6. L. K. Svanidze, T. N. Zagu, D. V. Mosia, T. I. Sigua. The Thermal Analysis of Manganese-Containing Materials, Tbilisi, ed. "Metsniereba", 1979.

THERMOCHEMISTRY OF MIXED EXPLOSIVES

J. L. Janney and R. N. Rogers
Los Alamos National Laboratory
MS C920
Los Alamos, NM 87545

INTRODUCTION

In order to predict thermal hazards of high-energy materials, accurate kinetics constants must be determined [1]. Predictions of thermal hazards for mixtures of high-energy materials require measurements on the mixtures, because interactions among components are common.

A differential-scanning calorimeter (DSC) can be used to observe rate processes directly, and isothermal methods enable detection of mechanism changes [2,3]. Rate-controlling processes will change as components of a mixture are depleted [3,4], and the correct depletion function must be identified for each specific stage of a complex process.

A method for kinetics measurements on mixed explosives can be demonstrated with Composition B [5,6]. Comp B is an approximately 60/40 mixture of RDX (1,3,5-trinitro-1,3,5-triazacyclohexane) and TNT (2,4,6-trinitrotoluene), and it is an important military explosive.

Kinetics results indicate that the major process is the decomposition of RDX in solution in TNT with a perturbation caused by interaction between the two components. The kinetics constants measured for the global decomposition process have provided an excellent predictive model for the thermal initiation of Comp B.

EXPERIMENTAL

Apparatus. All measurements were made with a Perkin-Elmer Model DSC-1B differential-scanning calorimeter. Samples were sealed in aluminum cells, Perkin-Elmer Part Number 219-0062, and the cells were perforated with a single hole approximately 0.15 mm in diameter.

Sample. Composition B, Grade A, was used. It has a nominal composition of 59.5% RDX, 39.5% TNT, and 1% wax, and it is produced according to military specification MIL-C-401C, dated 15 May 1968.

Procedure. Temperature accuracy is extemely important in kinetics measurements; therefore, both the average and differential temperatures of the DSC must be calibrated with care. Calibrations must be made as close as possible to the temperature of measurement. The differential-temperature calibration can be made and the cell support can be checked for quality by "emission balancing" [7].

Both sample and reference cells must be perforated to maintain constant pressure. Unperforated sealed cells bulge at higher temperatures, causing baseline discontinuities. Uniform holes are easily made by adjusting the conical angle of the tip of a needle and punching through the sample-cell cover and a sheet of plastic that is 0.05-0.10-mm thick. The effect of lid perforation on apparent rate constant can be observed by making replicate runs at the same temperature with different numbers of holes of different sizes. If significant sample is lost by volatilization, the apparent rate constant will be seen to increase with hole size. The sample must not be allowed to spatter or boil out, as significant emittance changes will be made by contamination on the outside of the cell.

Mixtures present a sampling problem. Since samples of only 1 to 5 mg are usually used, extreme care must be taken to ensure as much homogeneity as possible without altering the sample.

After calibration with the appropriate temperature standard, the DSC is set at the desired test temperature, and the differential temperature control is adjusted again by emission balancing. The empty cell is removed from the sample support, the desired range is selected, the recorder is started (and the automatic data acquisition system, if one is used), and the sample is dropped quickly onto the sample support. A sharp break will appear on the record as the sample touches the support, but thermal equilibrium will normally be reestablished within one minute. If the recorder pen does not reappear within 75 s, an endothermic process is probably involved. If the pen recovers quickly but no additional signal is obtained, either the reaction was completed during the warmup time or a very

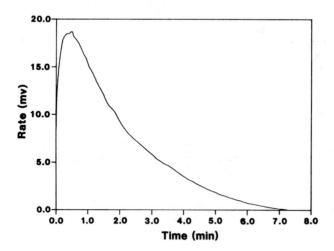

Figure 1. Isothermal DSC rate curve or Comp B at 500 K, normalized to 1 mg, range 1.

long induction time is involved. This can be determined by opening the cell and observing the condition of the sample. When isothermal runs of highenergy materials are made within a suitable temperature range, rate curves similar to Figure 1 will be obtained. Each run should be allowed to continue long enough to identify the infinite time baseline. When automatic data-acquisition and processing equipment are not available, rate curves can be evaluated with a ruler by extending the infinite-time baseline back to zero time and measuring the deflection from this baseline at as many equally spaced times as are practicable. The rate curve can be integrated by using Simpson's Rule, and the fraction reacted at any time, α, is obtained by dividing the fractional area to that time by the total area. These simple measurements of the deflection above the infinite-time baseline, the fraction reacted, and time are the only input data required for all further calculations.

As the first step in the evaluation of the kinetics data obtained from mixtures of high-energy materials, it is necessary to determine whether there is a significant interaction among the components during the primary self-heating reaction. The simplest observation of complexity and/or measurement of reaction order and reaction stoichiometry can be obtained from an "order plot" [4,8]. It is dangerous to use kinetics evaluations made from assumed rate laws to make thermal-hazards predictions for high energy materials. The correct reaction order function, or, in the case of complex reactions, the correct depletion function must be used for the evaluation of rate data. We strongly recommend that order plots be produced early in the evaluation of any rate data.

An order plot is a graph of the log of the deflection (in any consistent units) versus the log of the residual fraction $(1 - \alpha)$. The reaction order for a specific extent of reaction is shown by the slope of a straight line. Negative slopes and curved lines indicate complex processes such as melting with decomposition, autocatalysis, and solid-state reactions. When complexity is observed, autocatalytic or nucleation-growth-type functions can often be used for data evaluation [2,3,9].

RESULTS AND DISCUSSION

An order plot for the decomposition of Comp B is shown in Figure 2. The order of the global reaction is exactly 1.0, so Comp B rate data can be evaluated according to a first-order rate law, $d\alpha/dt = k(1 - \alpha)$. Rate constants can be determined very simply from the isothermal DSC data [10]. Relative recorder deflections, b, and time, t, are the only data required, as follows:

$$\ln b = C - kt, \qquad (1)$$

where k is the rate constant (s^{-1}), and C is a constant.

A total of 16 independent rate measurements were made with Comp B over a temperature range of 30 K (from 495 to 525 K). The resulting Arrhenius plot (Figure 3) is linear, and it gives an activation energy (slope = -E/R, where R is the gas constant) of 180.2 kJ mole^{-1} and a pre-exponential (intercept) of 4.62×10^{16} s^{-1}.

Pure RDX gives first-order kinetics constants through much of its decomposition in the pure liquid phase, with E = 197.1 kJ mole^{-1} and $Z = 2.02 \times 10^{18}$ s^{-1}. We believe that the currently "best" kinetics constants for pure TNT are E = 143.9 kJ mole^{-1} and $Z = 2.51 \times 10^{11}$ s^{-1} [1]. It must be emphasized that these kinetics constants apply primarily to predictions of critical temperatures: they are not intended to be used for all predictive models. A simple comparison of activation energies is not especially informative with regard to hazard analysis.

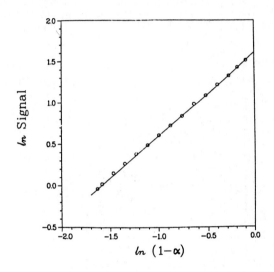

Figure 2. Order plot for the decomposition of Comp B. The segment shown was calculated from the part of the rate curve that showed a positive order.

The most important fact to establish with regard to thermal hazards is whether a high-energy material will self heat catastrophically. Consequently, we use the kinetics constants determined for any system to produce a predictive model for the critical temperature, T_c, of that system. The critical temperature is defined as the lowest constant surface temperature at which catastrophic self heating can occur in a high-energy material of a specific size and shape. A predictive model for critical temperature can often be obtained from the following expression [11]:

$$\frac{E}{T_c} = R \ln\left[\frac{a^2 \rho Q Z E}{T_c^2 \lambda \delta R}\right], \qquad (2)$$

where R is the gas constant (8.314 J mole^{-1} K^{-1}), a is the radius of a sphere or cylinder or the half-thickness of a slab, ρ is the density, Q is the heat of reaction during the self-heating process (not expected to be the same as the heat of detonation or combustion), Z is the pre-exponential and E the activation energy from the Arrhenius expression, λ is the thermal conductivity, and δ is a shape factor (0.88 for infinite slabs, 2.00 for infinite cylinders, and 3.32 for spheres). E and Z are the largest numbers in the equation, and their accuracy determines the accuracy of the predictions made from the model.

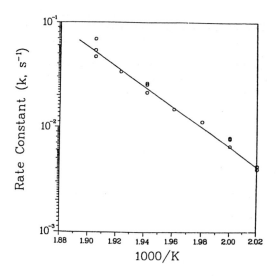

Figure 3. Arrhenius plot for Comp B.

We strongly believe that it is dangerous to rely on predictive models based only on chemical kinetics determinations, because of several types of errors that may occur without being detected. The most probable cause of error is that the values of E and Z may be based on the wrong part of the decomposition reaction. Identification of the mechanism that leads to catastrophic self-heating is particularly difficult with complex materials and mixtures. Also, errors can be made in the thermal conductivity measurement, and the Frank-Kamenetskii equation may not provide an accurate model for low-energy systems. Heat-of-reaction values may also be in error.

The heat of reaction for the decomposition of a material may be observed to vary with pressure and/or confinement, and high-energy materials differ in their responses to pressure and confinement. Therefore, it is important to test predictive models under different confinement than used for kinetics measurements. A system sensitive to confinement will be detected by a lack of agreement between the predicted T_c and the experimental T_c, but a lack of agreement does not always indicate a system sensitive to confinement. If a lack of agreement is observed, a "fail-safe" (but often unrealistic) predictive model may be produced by using the maximum possible heat of reaction (for example, the heat of detonation) for the system. The largest heat of reaction for Comp B that could be measured under any experimental conditions with the DSC was 2.85 kJ g^{-1}.

Because of the difficulties with measurements of kinetics constants, thermal conductivity, and heat of reaction, we strongly recommend independent methods for testing the accuracy of T_c predictions [1,12]. We do this on a laboratory scale with a time-to-explosion test. This involves weighing samples into aluminum blasting caps, confining them, measuring their thicknesses, and immersing them in a liquid metal bath at a known temperature. Times to explosion are measured for different temperatures, and the lowest temperature at which the sample will explode is the critical temperature for that size and shape.

Our experimental determination of T_c [1] was run with a measured average sample thickness of 0.795 mm for a 40 mg sample at a calculated density of 1.577 g cm^{-3} (91% TMD). Times to explosion were measured as a function of temperature until the lowest

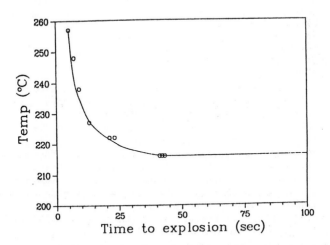

Figure 4. Time-to-explosion plot for 40-mg confined samples of Comp B.

temperature giving an explosion was identified, as shown in Figure 4. The experimental T_c for our particular size and shape was determined to be 489 K (216°C).

The predicted value for T_c with the size, shape, and density used in the experimental determination is 489 K. The values used in equation (2) were the following: a = 0.040 cm, ρ = 1.577 g cm^{-3}, Q = 2.85 kJ g^{-1}, Z = 4.62 X 10^{16} s^{-1}, E = 180.2 kJ mole^{-1}, λ = 1.98 X 10^{-3} J cm^{-1} s^{-1} K^{-1}, and δ = 0.88 (for a slab). The perfect check between the experimental and predicted values is unusual, but we normally hope to obtain agreement within ±5 degrees.

CONCLUSIONS

A combination of chemical kinetics and experimental self-heating procedures provides a good approach to the production of predictive models for thermal hazards of high-energy materials. Systems involving more than one energy-contributing component can be studied. Invalid and dangerous predictive models can be detected by a failure of agreement between prediction and experiment at a specific size, shape, and density.

Rates of thermal decomposition for Composition B appear to be modeled adequately for critical-temperature predictions with the following kinetics constants: E = 180.2 kJ mole^{-1} and Z = 4.62 X 10^{16} s^{-1}.

REFERENCES

1. R.N. Rogers, Thermochimica Acta, 1975, 11, 131.
2. R.N. Rogers and J.L. Janney, Proceedings of the Eleventh North American Thermal Analysis Society Conference, Vol. II, John P. Schelz, Editor, Johnson & Johnson Products, Inc., New Brunswick, NJ, 1981, 643.
3. J.L. Janney and R.N. Rogers, ibid, 651.
4. R.N. Rogers, Fraunhofer-Institut für Trei- bund Explosivstoffe, International Meeting, "Testing Methods for Propellants and Explosives," Karlsruhe, Germany, 1980, 59.
5. T.R. Gibbs and A. Popolato, "LASL Explosive Property Data," University of California Press, Berkeley, (1980), p.11.
6. AMCP 706-177, "Explosives Series, Properties of Explosives of Military Interest," 1971, 46. Foreign requests for this document must be submitted through the Washington, D.C., embassy to: Assistant Chief of Staff for Intelligence, Foreign Liaison Office, Department of the Army, Washington, D.C. 20310, USA.
7. L.W. Ortiz and R.N. Rogers, Thermochimica Acta, 1972, 3, 383.
8. R.N. Rogers, Thermochimica Acta, 1972, 3, 437.
9. T.B. Tang and M.M. Chaudhri, J. Thermal Anal., 1979, 17, 359.
10. R.N. Rogers, Anal. Chem., 1972, 44, 1336.
11. D.A. Frank-Kamenetskii, Acta Physicochem. URSS, 1939, 10, 365.
12. D.L. Jaeger, LA-8332, University of California, Los Alamos Scientific Laboratory, Box 1663, Los Alamos, NM 87545, USA, 1980.

The work reported was performed with support from the U. S. Army Armament Research and Development Command, Dover, N.J., Joseph Hershkowitz, Program Manager.

THERMOCHEMICAL EVALUATION OF ZERO-ORDER
PROCESSES INVOLVING EXPLOSIVES

R. N. Rogers and J. L. Janney
Los Alamos National Laboratory
Box 1663, MS C920
Los Alamos, NM 87545

INTRODUCTION

An accidental explosion occurred at a government laboratory in March 1960 during a pressing operation involving a mixture that contained both lead and HMX (1,3,5,7-tetranitro-1,3,5,7-tetrazacyclooctane). The incompatibility between Pb and HMX had not been observed with the vacuum stability test, but it can easily be observed with a DTA test. RDX (1,3,5-trinitro-1,3,5-triazacyclohexane) is the major impurity in HMX, and the solidus temperature of the RDX/HMX system appears to be near 180°C. A sharp exotherm is observed in the DTA record at about 180°C when Pb is added to impure HMX. Explosions are often obtained in the DTA apparatus at higher temperatures with HMX/Pb mixtures.

Both RDX and HMX decompose in the liquid phase to produce largely formaldehyde and N_2O. Observations on the HMX/Pb system indicate that one of the major contributions made by Pb to the overall reaction is a secondary reaction loop involving N_2O and formaldehyde. N_2O oxidizes Pb, and lead oxide oxidizes formaldehyde. The redox cycle increases the overall energy evolution rate during thermal decomposition. The heterogeneous reaction should be zero order, the rate depending on the surface area of the catalyst. The (impure) HMX/Pb reaction is extremely complex, but a zero-order reaction can be observed.

Several heavy metals other than Pb have been found to be dangerously incompatible with RDX and HMX, and it has become important to develop a general method for making quantitative predictions of the thermal hazards of such systems. All heavy metals with close-lying oxidation states appear to be incompatible with the nitramines. The metals, oxides, and salts of these elements all appear to be equally incompatible with HMX and RDX.

Although it has long been possible to detect incompatibilities between metals and explosives, it has not been possible to produce quantitative predictive models for hazards. We report here methods for identifying zero-order reactions and assessing their hazard potential. The general method proposed involves the following steps: (1) production of isothermal DSC rate curves, (2) identification of the parts of the reactions that obey a zero-order rate law by producing "order plots" of the rate data, (3) measurement of zero-order rates at different temperatures, (4) normalization of the rates to identical surface areas of catalyst, (5) determination of kinetics constants from an Arrhenius plot of the zero-order rate data, (6) measurement of critical temperatures for known mixtures with a time-to-explosion test, and (7) production of a mathematical predictive model for the critical temperature of any size, shape, and composition of mixture.

EXPERIMENTAL

All of the rate measurements were made with a Perkin-Elmer Model DSC-1B differential-scanning calorimeter. General methods for the production and evaluation of isothermal DSC rate curves are described in an accompanying paper by Janney and in reference [1]. We found that an 8-mg sample of 1/1-HMX/Pb was a convenient size for measurements on range 8 of the DSC at 535 K. We filled the excess free volume in the sample cell with aluminum-disc spacers.

A description of the time-to-explosion test apparatus and the general method for the measurement of a critical temperature (T_c) can be found in reference [2] and the accompanying paper by Janney. The critical temperature is the lowest constant surface temperature at which a material of a specific size, shape, and composition can self heat to explosion.

Samples. The HMX sample was obtained from the Holston Defense Corporation, Kingsport, TN, and it was identified as Lot 920-32. It is relatively pure (< 0.5% RDX). The lead powder is nominally spherical and shows a measured surface area of 0.11 $m^2 g^{-1}$.

RESULTS AND DISCUSSION

Figure 1 shows a comparison between an isothermal DSC rate curve obtained from the HMX sample alone and a rate curve obtained from the same amount of HMX with an equal volume of Pb powder added. The HMX sample shows an "autocatalytic-type" rate curve. Its initial rate is relatively low, but the rate increases with time. The HMX/Pb rate curve shows a significant initial rate, and the rate does not change greatly for approximately 300 s. The nearly constant rate indicates a significant contribution from a zero-order reaction.

Figure 2 shows the order plots obtained from the rate curves of Figure 1. The fraction decomposed at any time, α, is determined by integration of the rate curves. The slope through any range of reactant depletion is the reaction order for that reaction regime. Reaction regimes showing negative slopes or curved lines on the order plot involve complexity. Causes for complexity can be chemical autocatalysis, melting with decomposition, and/or solid-state reactions. Suitable autocatalytic or nucleation-growth rate laws must be used for the evaluation of reaction regimes showing complexity.

Figure 2 shows that the first 15% of the energy from the 1/1-HMX/Pb system is evolved according to a zero-order rate law [up to $\ln(1 - \alpha) = -0.16$]. Experiments with HMX samples of different purity indicate that the zero-order process is a result of reactions involving the RDX/HMX mixed melt. The rate is a function of the surface area of the catalyst, and the extent is a function of the amount of impurity (RDX) present.

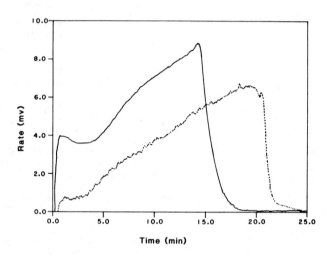

Figure 1. Isothermal DSC rate curves for 1.272 mg HMX (dashed) and 7.761 mg 1/1-HMX/Pb, containing 1.242 mg HMX, (solid) at 535 K. Curves normalized to 1 mg HMX at DSC Range 1.

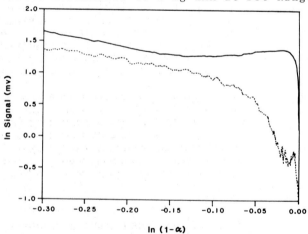

Figure 2. Order plots of rate data from Figure 1. HMX (dashed) and 1/1-HMX/Pb (solid) to 26% depletion [$\ln(1 - \alpha) = -0.3$].

We found that we could not make suitably homogeneous mixtures of HMX and Pb; composition varied from sample to sample. Consequently, considerable scatter was observed in the absolute rate data obtained from the different samples, but the overall characteristics of the rate curves remained the same. As a result of the data scatter, we cannot present accurate values for the kinetics constants for the zero-order reaction; however, the values obtained are in the RDX and HMX region. Our preliminary numbers for the zero-order reaction are the following: $E = 52$ kcal mole^{-1} and $Z = 1.2 \times 10^{18}$ s^{-1}. Our critical temperature predictive models for RDX and HMX are based on the following values: RDX, $E = 47.1$ kcal mole^{-1} and $Z = 2.02 \times 10^{18}$ s^{-1}; HMX, $E = 52.7$ mole^{-1} and $Z = 3 \times 10^{19}$ s^{-1}. The important observation is that Pb does not catalyze the elementary decomposition reaction during the zero-order process: there is no significant reduction in E. The only significant effect must be the increased energy-evolution rate that results from the heterogeneous redox cycle.

RDX and HMX are partially miscible in the solid state, and they show an apparent eutectic composition near 62.5 wt.% RDX. The heat of formation of RDX is 14.7 kcal mole^{-1}; that for HMX is 17.9 kcal mole^{-1}. The heat of reaction of the mixed melt should be 1075 cal g^{-1} when the products are $H_2O(g)$, N_2, and CO. The decomposition of RDX in the mixed melt should be the predominant reaction; therefore, RDX kinetics should best model the rate at moderate temperatures.

The thermal conductivities of HMX/Pb systems have not been measured; however, measurements have been made on RDX systems containing granular Fe and Cu. The thermal conductivity of metal loaded explosives increases slightly as the metal content is increased, depending on particle size and shape. Assuming the same slope for the thermal conductivity versus volume-percent-metal functions for RDX and HMX systems, we have used the following expression to estimate thermal conductivities for HMX/Pb mixtures:

$$\lambda = 9 \times 10^{-4} V + 1 \times 10^{-3} \tag{1}$$

where λ is the thermal conductivity in cal cm^{-1} s^{-1} K^{-1} and V is the volume fraction of metal.

Since the presence of Pb appears to convert an RDX/HMX melt nearly quantitatively to H_2O, CO, and N_2 without making an appreciable change in the rate of the elementary reaction, we can estimate a heat of reaction and kinetics constants for the zero-order reaction. Given those values, thermal conductivity, and density, we can produce a predictive model for catastrophic self heating according to the Frank-Kamenetskii equation [3],

$$\frac{E}{T_c} = R \ln \left[\frac{a^2 \rho Q Z E}{T_c^2 \delta \lambda R} \right] \tag{2}$$

where E and Z are the Arrhenius activation energy and pre-exponential, R is the gas constant, T_c is the critical temperature, a is a dimension (radius of a sphere or infinite cylinder or half thickness of an infinite slab), ρ is the mass of explosive per unit volume of Q is the heat of reaction, λ is the thermal conductivity, and δ is a shape factor (0.88 for infinite slabs, 2.0 for infinite cylinders, and 3.32 for spheres).

If predictions made from the model agree with experimental critical temperatures for HMX/Pb systems of known size and shape, it can be assumed that the increased energy-evolution rate during the zero-order process is responsible for the observed hazard. It should also be possible to use the predictive model for predicting critical temperatures for other sizes and shapes of HMX/Pb charges under similar confinement. A comparison between experimental and predicted values is shown in Table 1.

TABLE 1. CRITICAL TEMPERATURE AS A FUNCTION OF LEAD CONTENT FOR HMX/Pb SYSTEMS.

Sample	a (cm)	Expt. T_c (°C)	Calc. T_c (°C)
HMX	0.048	257	---
95/5-HMX/Pb	0.041	222	222
75/25-HMX/Pb	0.046	226	227
50/50-HMX/Pb	0.045	239	239

Compositions are given in volume percent. The experimental time-to-explosion method [2] can be modeled as a slab ($\delta = 0.88$). Estimated thermal conductivities used in the predictive model are the following: for 95/5, 1 X 10^{-3}; for 75/25, 1.2 X 10^{-3}; and for 50/50, 1.5 X 10^{-3}. The crystal density for HMX is 1.91 g cm^{-3}, and all of the samples were pressed to about 90% maximum density. Therefore, the following density values were used in the predictive model: for

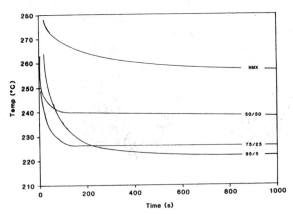

Figure 3. Time to explosion as a function of temperature and lead content for HMX/Pb systems.

95/5, 1.63 g cm^{-3}; for 75/25, 1.3 g cm^{-3}; and for 50/50, 0.86 g cm^{-3}. Heats of reaction were simply scaled according to the volume percent of HMX, as follows: for 95/5, 1021 cal g^{-1}; for 75/25, 806 cal g^{-1}; and for 50/50, 537.5 cal g^{-1}. The following kinetics constants for the decomposition of RDX in the liquid phase were used: $E = 47.1$ kcal mole^{-1} and $Z = 2.02 \times 10^{18}$ s^{-1}. As a result of the log function, predictions are not very sensitive to moderate errors in ρ, λ, and Q.

The experimental time-to-explosion curves are shown in Figure 3. It can be seen that the samples that contained lead had reduced times to explosion as well as lowered critical temperatures.

We normally determine kinetics constants in a self-generated atmosphere at ambient pressure (approximately 76 kPa in Los Alamos). The time-to-explosion test is run in a self generated atmosphere under significant confinement. A lack of agreement between predicted and experimental values is usually a result of pressure effects on secondary gas-phase and/or heterogeneous reactions. When agreement is achieved, it can be assumed that the system modeled is not very sensitive to pressure effects. In the case of the HMX/Pb system, the secondary reactions with Pb must be extremely fast; however, heavily confined systems may be significantly more hazardous than predicted by our model.

CONCLUSIONS

We believe that quantitative predictive models can be made for thermal hazards of explosives systems involving zero-order reactions. The zero-order reactions can be identified by making order plots of isothermal DSC rate data. Kinetics constants can be measured for the zero order reactions of complex processes (if homogeneous mixtures can be produced), and critical temperatures can be predicted. Predictions can be tested against experimental values.

Incompatibilities between explosives and materials that provide hazardous zero-order contributions to the self-heating process can easily be detected with a time-to-explosion test. However, because increasing the amounts of catalyst will increase the dilution as well as increasing catalyst surface area, tests at different compositions are required for greatest confidence. Predictions of hazards for sizes and shapes other than those tested require a kinetics-based predictive model.

REFERENCES

1. R.N. Rogers, Thermochim. Acta, 1972, 3, 437.
2. R.N. Rogers, Thermochim. Acta, 1975, 11, 131.
3. D.A. Frank-Kamenetskii, "Diffusion and Heat Transfer in Chemical Kinetics," Plenum Press, New York, (1969).

Much of the work reported was funded under contract to the United States Air Force. Time-to-explosion tests were run by M. H. Ebinger.

AN INVESTIGATION OF THE INFLUENCE OF ORGANIC
BINDERS ON THE REACTION OF PYROTECHNIC SYSTEMS
USING THERMAL ANALYSIS AND RELATED TECHNIQUES

E.L. Charsley & Jennifer A. Rumsey, Stanton
Redcroft Ltd., Copper Mill Lane, London SW 17
OBN, England

T.J. Barton & T. Griffiths, Royal Armament
Research & Development Establishment, Fort
Halstead, Sevenoaks, Kent, England.

INTRODUCTION

   Little information is available in the literature on the chemical role of organic binders in pyrotechnic systems and traditionally they have been considered to play a mainly physical part. However preliminary DTA studies had shown that for the titanium-strontium nitrate system, the presence of a chlorinated rubber, increased the observed exothermic reactions, thereby indicating direct participation of the binder in the pre-ignition reactions.
   A survey has therefore been undertaken to evaluate the effect of a wide range of binders on four metal-oxidant systems, based on the metals magnesium and titanium and the oxidants sodium nitrate and strontium nitrate. The range of binders was selected to include traditional materials such as beeswax and more modern ones such as poly (vinyl acetate) and are grouped as shown in Table 1.

TABLE 1. LIST OF BINDERS STUDIED

Constituent elements	Binders
C	Carbon black, graphite
C + H	Liquid paraffin, paraffin waxes, polyethylene
C + H + O	Poly (vinyl acetate), boiled linseed oil, beeswax, carnauba wax
C + H + Cl	Cereclors 51L - 70 (chlorinated straight chain paraffins), poly (vinyl chloride), Alloprene (chlorinated rubber)
C + Br + O	Saytex 102 (decabromodiphenyl oxide)
C + H + Cl + O	PVC (Breon) AS70/42 (vinyl chloride/vinyl acetate copolymer)

On the basis of practical experience the binders were incorporated at the 4% level in mixtures containing equal parts by weight of the metal and oxidant. This metal-oxidant ratio was chosen as being representative of that used in many pyrotechnic formulations and gives exothermicities close to the experimentally determined maxima. Carbon in the form of carbon black and graphite was included in the survey because of its possible role as a reaction intermediate.

The influence of the binder on the metal-oxidant reaction has been characterised by DTA (under both ignition and non-ignition conditions) and combustion calorimetry. Important pyrotechnic properties such as burning rate, light output and sensitivities have been measured and theoretical exothermicities have been derived using thermodynamically based computer models.

EXPERIMENTAL

DTA measurements were carried out using a Stanton Redcroft Model DTA 673. Sample weights of 20-50mg were used with heating rates in the range 5-20°C/min. Temperature of ignition measurements were made using the modified apparatus described previously (1) using 50mg samples at a heating rate of 50°C/min. Runs were carried out in quartz crucibles and an atmosphere of flowing argon was used to avoid oxidation of the metals. Reaction exothermicities were measured using a Gallenkamp Autobomb fitted with a Hewlett-Packard, Model 2801A quartz thermometer. Sample weights in the range 0.5-1.5g were used.

RESULTS AND DISCUSSION

DTA Measurements. The relative reactivity of the binary mixes as determined by temperature of ignition measurements is shown in Table 2. The results clearly show the greater reactivity of the magnesium based compositions, which gave ignition temperatures some 200°C lower than the titanium-sodium nitrate composition, the strontium nitrate composition failing to ignite.

TABLE 2. IGNITION TEMPERATURES OF THE BINARY METAL-NITRATE COMPOSITIONS

Composition	Temperature of Ignition (°C)
$Mg-NaNO_3$	575 ± 10
$Mg-Sr(NO_3)_2$	572 ± 10
$Ti-NaNO_3$	772 ± 4
$Ti-Sr(NO_3)_2$	Exo at 724 ± 1

Temperature of ignition measurements on ternary mixes showed that the addition of a binder could modify the behaviour of all four binary systems. This is illustrated in Table 3 for a range of titanium-sodium nitrate-binder compositions. It can be seen that the addition of 4% of boiled linseed oil has resulted in a drop in the ignition temperature of over 350°C, giving a value well below that for the magnesium binary mixes.

TABLE 3. IGNITION TEMPERATURES OF TITANIUM-SODIUM NITRATE-BINDER COMPOSITIONS

Binder	Ignition Temperature (°C)
Boiled linseed oil	407±9
Carbon black	533±4
Cereclor 65L	712±13
Beeswax	748±14
Saytex 102	749±5
Poly (vinyl acetate)	768±3
Binary mix	772±4

With the exception of the waxes and poly (vinyl acetate) the addition of a binder promoted ignition in the titanium-strontium nitrate system. The lowest reproducible ignition temperature was observed with carbon black, which gave a mean value of 506°C. Both magnesium ternary systems showed a lower range of ignition temperatures than was observed with the titanium compositions.

DTA curves for the ternary compositions showed that the presence of a binder could result in additional pre-ignition exotherms. The magnitude of these could be altered by increasing the concentration of the binder and this sometimes resulted in ignition taking place at a lower temperature exothermic reaction. This is illustrated in Fig. 1 for titanium-strontium nitrate compositions containing 0-10% Alloprene.

The ignition temperatures are given in Table 4. The results show that addition of as little as 1% of a binder can markedly alter the behaviour of a pyrotechnic mix. Where a multi-stage exothermic reaction is given, the ignition temperature can in some cases be reduced by increasing the sample weight or the heating rate (1).

TABLE 4. VARIATION IN IGNITION TEMPERATURE WITH % ALLOPRENE FOR TITANIUM-STRONTIUM-ALLOPRENE COMPOSITIONS

% Alloprene	0	1	2	5	10
Ignition Temperature (°C)	*	670	630	524	483

* Did not ignite, peak temperature 700°C.

Exothermicity Measurements. The effect of binders on the exothermicities of the binary compositions in 1 atm. of argon and of oxygen is shown in Table 5, where overall average values are given. The results show that the addition of binders at the 4% level, can significantly alter the exothermicity of a pyrotechnic system. In general the effect is to reduce the exothermicity in argon and to increase it in oxygen. The exothermicities of the magnesium mixes are higher than those of the titanium mixes and for both metals the higher exothermicities are given with sodium nitrate.

TABLE 5. OVERALL AVERAGE VALUES FOR THE REACTION EXOTHERMICITIES OF THE BINARY AND TERNARY NITRATE SYSTEMS

System	Exothermicities (k cal g^{-1})			
	Argon (1 atm)		Oxygen (1 atm)	
	Binary	Ternary	Binary	Ternary
Mg-NaNO$_3$	1.941	1.859	2.649	2.871
Mg-Sr(NO$_3$)$_2$	1.773	1.590	2.366	2.588
Ti-NaNO$_3$	1.522	1.413	1.830	1.860
Ti-Sr(NO$_3$)$_2$	1.376	1.286	1.707	1.754

Where the change in exothermicity produced by a binder is small, difficulties can arise in deciding if the change is real or due to batch to batch variation. To eliminate this problem, work is in progress on ternary systems selected on the basis of results at the 4% binder level, using mixes with binder concentration from 2-12%. To date a linear relationship between the exothermicity of reaction and the % binder has been found, the slope of the line giving a reliable indication of the effect of the binder. Typical plots are shown in Fig. 2.

Binder - Nitrate and Binder-Metal Reactions. In order to characterise the role of the binder reaction in the ternary system studies are being carried out to examine the binder-nitrate and the binder-metal reactions. To maximise reaction, mixes containing equal weights of the two components have been used. DTA experiments in argon showed that the majority of binders reacted exothermically with both nitrates. In some cases the reaction was extremely vigorous leading to ignition. This was observed with strontium nitrate mixes with Alloprene, Cereclor 70 and carbon black and with sodium nitrate mixes with Alloprene, Cereclor 51L, carnauba wax, boiled linseed oil and carbon black.

The binders in general reacted at lower temperatures with the nitrates than did the metals. Alloprene giving ignition temperatures with both nitrates below 300°C. This reaction is thought to be due to reaction between the nitrate and the hydrogen chloride from the binder. Mixes containing lower quantities of Alloprene also showed higher temperature exothermic reaction, attributed to the reaction of residual nitrate with the carbonaceous binder residue, since carbon black gave a highly exothermic reaction with both nitrates in the region of 500-600°C. The other chlorinated binders appear to react in a similar manner to Alloprene and a curve for the Cereclor 65L-

-strontium nitrate mix is shown in Fig. 3.

The reaction of the nitrates with carbon indicates that, even if the binder does not directly react with a nitrate, it can produce a significant effect in a ternary composition, if it decomposes to form a carbonaceous residue.

The DTA experiments showed that in all cases, except one, binder-nitrate mixes were more reactive than binder-metal mixes. The exception was Saytex 102 which gave ignition reactions with both metals, with ignition temperatures of 425°C and 405°C for magnesium and titanium respectively. A curve for the latter mix is also shown in Fig. 3.

The Effect of Binders on Pyrotechnic Performance. All pyrotechnic parameters measured have been profoundly altered by the presence of a binder. Thus in the titanium-strontium nitrate system the average light output changed by a factor of 17-fold over the range of binders examined. A correlation has been found to exist between the exothermicity in argon and the burning rate and has been explained on the basis of a simple model for flame propagation. The pyrotechnic measurements together with the thermodynamic calculations form the basis of a separate communication (2).

ACKNOWLEDGEMENTS

This work was carried out with the support of the Procurement Executive, Ministry of Defence.

REFERENCES

1. E.L. Charsley, C.T. Cox, M.R. Ottaway, T.J. Barton and J.M. Jenkins, Thermochimica Acta, 1982, $\underline{52}$, 321.
2. T.J. Barton, T. Griffiths, E.L. Charsley and Jennifer A. Rumsey, " Proceedings Eigth International Pyrotechnic Seminar ", Colorado, U.S.A., (1982), in press.

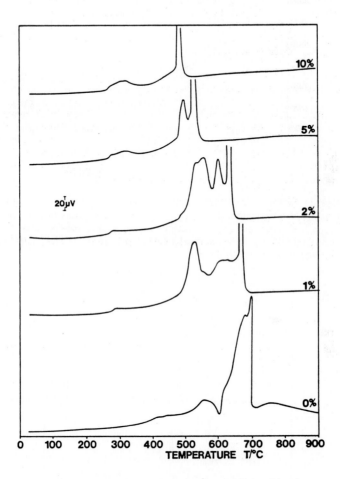

Fig. 1. DTA curves for titanium-strontium nitrate compositions with 0-10% Alloprene

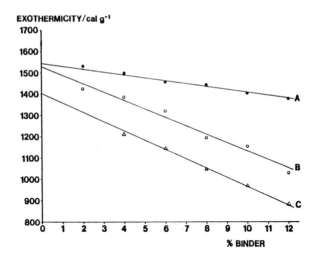

Fig. 2. Effect of binder concentration on exothermicities of reaction in argon for (A) titanium-sodium nitrate-Alloprene (B) titanium-sodium nitrate-boiled linseed oil (C) titanium-strontium nitrate-boiled linseed oil

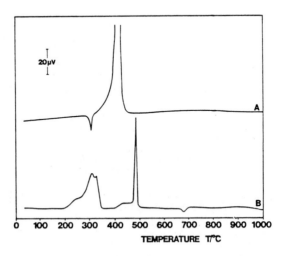

Fig. 3. DTA curves for (A) 50% Saytex 102-50% titanium and (B) 50% Cereclor 65L-50% strontium nitrate

INTEGRATED TESTING FOR THE
EVALUATION OF THERMAL HAZARDS

T. F. Hoppe & E. D. Weir
Ciba-Geigy Corporation
Toms River, NJ   USA

1. INTRODUCTION

Recently within the United States Chemical Processing Industry (CPI) there has been a trend to assess certain aspects of processing risks through the use of thermal analytical methods. The initial result of this trend has been the development of methods for the determination of thermal hazards through the use of global-kinetics. These methods are both rapid and easy to use but normally can only be applied to simple chemical systems. Unfortunately, a good portion of thermal safety problems facing CPI involve the analysis of condensed phase decompositional reactions of complex chemical systems. Seldom can an analysis of such complex systems be handled adequately in a quick and simplistic manner.

In terms of safety analysis, it is not often that one encounters problems of greater complexity than with the processing or storage of chemical systems that exhibit autocatalytic behavior. In order to address problems of this nature properly, a number of the following testing methods can be integrated:
(1) Isothermal - Constant Temperature Stability (CTS)
(2) Adiabatic Calorimetry   (3) Differential Thermal Analysis (DTA)   (4) Heat Accumulation   (5) Classical Chemical Analysis (GC, LC) in combination with isothermal aging techniques.

This integrated approach allows for the most efficient development of data and at the same time provides the researcher with an opportunity to internally cross-check this data. This cross-check method insures that the analysis is being performed in light of intrinsic material properties rather than in light of values that are strongly influenced by the measurement systems themselves.

Initially, an outline of the methods and the pitfalls involved in doing such a complex investigation will be discussed. Finally, in support of the theoretical section of the paper, two short examples will be presented where this type of integrated analysis has been used.

2. ASSESSMENT OF A THERMAL HAZARD

In order to properly define the thermal hazards potential of a particular chemical system, basically three considerations must be taken into account: (1) The Heat Production as a Function of Temperature and Time ($\dot{Q}$) of both the Wanted and Unwanted Reactions (2) The Engineering Aspects of the Unit or Heat Removal Capacity ($-\dot{Q}$) (3) The Operational Procedure or Time/Temperature Parameters.

The data necessary for the determination of $\dot{Q}$ (wanted reaction) can, in most cases, be developed in the laboratory using a heat flow calorimeter [1]. In many cases, corresponding data for decompositional type reactions can also be developed in the laboratory. This information can be obtained through the use of thermal analytical methods, however, a number of problems can be encountered with an investigation of this nature. These problems stem basically from two areas:
(1) The complex nature of the mechanisms involved in the exothermic reactions under investigation. In many cases, these investigations involve the condensed phase decomposition of multi-component chemical systems. Under these circumstances, rarely does one encounter a system that can be described by a single elementary reaction (A+B→C) during the entire course of its decomposition. (2) The limitations of the measurement techniques available to the researcher. This problem can be considered from two aspects. First, one must understand the limitations imposed by the physical aspect of the instrument itself (sensitivity, open/closed system, thermal inertia). Secondly, one must also understand the limitations imposed by the method. For example, the use of a temperature programmed mode in a DTA analysis may alter the course of the resulting decomposition when referenced to an adiabatic or isothermal method (parallel pre-reactions).

3. INTEGRATION OF THERMAL METHODS

In order to reduce the difficulty associated with the analysis of complex systems, it is often advantageous to integrate a number of the previously mentioned testing techniques. This approach offers two distinct advantages. First, it allows the researcher to apply the best and perhaps, least complicated technique for the information he is seeking. Secondly, it allows the researcher to compare results, cross-check, on data developed by the individual instruments. This "Cross-Check" method minimizes the possibility that an instrument influenced value can be mistakenly used as an intrinsic material constant.

As an example of how to use this integrated approach, let us consider the thermal analysis of a complex system undergoing a high energy decomposition and, as a result, a number of reaction regimes are encountered during the course of the thermal spiral. Let us assume further that the system displays self-accelerating properties. The reaction sequence can be segmented into three phases; induction, acceleration, and decay, corresponding to the hypothetical model.

(1) $A+(B) \xrightarrow{k_1} P+Q$ (2) $A+Q+(B) \xrightarrow{k_2} P+2Q$ (3) $P \xrightarrow{k_3}$ decompositional products

   Induction Phase   Acceleration Phase   Decay Phase

By segmenting the model into phases the concept behind the analysis of the problem can be simplified.

In sections 3.1 - 3.4 the relationship between the analysis of a particular reaction phase, the required information, and the recommended testing techniques will be discussed.

3.1 <u>Potential Energy Release (Induction, Acceleration, Decay Phase Analysis)</u>. The initial consideration in the evaluation of a possible thermal hazard is the overall energy potential (ΔH_{dec}) released during the decomposition of the chemical system. This analysis acts as a screening point in determining if further investigation is necessary.

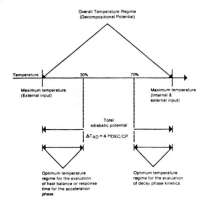

Figure 1 Temperature Regimes for Thermal Measurements

All segments of the hypothetical model need to be analyzed for this information. Two important aspects of this analysis should be considered in order to insure that the evaluation is properly done: (1) The possible temperature regime of the chemical system (top section, Figure 1). (2) The time restraints imposed by a test method relative to the thermal history a material will encounter as a result of real-life process conditions.

In order to accomplish these previously mentioned requirements the use of two methods of thermal analysis may be necessary to properly define the decompositional potential. Initially, a <u>temperature programmed method</u> is required. This technique insures that the material's full potential is observed by forcing the sample through all possible temperature regimes. Secondly, an <u>adiabatic method</u> is helpful in order to match the time trains associated with a long acceleration phase or long processing conditions (hold periods). In some cases, the comparison of the data obtained by these different methods is quite helpful. It has often been found, for example, that energy levels can, even within similar temperature regimes, differ considerably. In some cases, this difference can be attributed to instrument conditions such as sensitivity, open versus closed system (volatile components), container volume (vapor phase reactions), material of construction (metal <u>catalysis</u>) or thermal inertia (adiabatic systems).

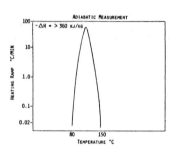
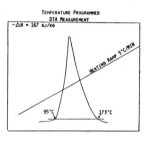

FIGURE 2
DECOMPOSITION AROMATIC NITRO COMP.

In other cases, however, it may be directly related to the time train or thermal history that the sample experiences as a function of the measurement technique (Figure 2).

3.2 Reaction Initiation (Induction Phase Analysis). It has been found that an induction period often precedes the decomposition of an organic system. The following classes of products have shown this dangerous self-accelerating behavior: aromatic nitro compounds; aliphatic nitroso compounds; thiophosphoric esters; acrylates; sulfones, esters and chlorides of sulfonic acids [2]. As such, thermal aging of the system may significantly alter its chemical reactivity. In order to properly define the hazard associated with this type of system, evaluation of the induction period is necessary. When evaluating this segment of the model, the most important information to obtain is temperature versus induction time (Figure 3). Based on this data, one can establish a maximum operational temperature in conjunction with acceptable time limitations which will prevent the initiation of the subsequent reactions. As a result, precise values of heat production versus temperature for the the acceleration phase, in some cases, need not necessarily be developed.

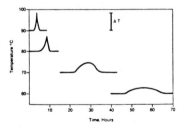

Figure 3 Isothermal Measurements of Induction Phase - Epoxy Resin

The best measurement technique for this type of data development is isothermal. A series of temperatures are chosen and a plot of the ln of induction time versus $1/T$ is made. If a straight line is obtained, one can assume consistency of mechanism within the experimental window and the extrapolation of the data may be possible. The slope of this line produces a temperature coefficient (activation energy). The value does not necessarily correspond to the temperature coefficient of the subsequent acceleration phase.

3.3 Control of the Potential (Acceleration Phase Analysis).

Assuming the temperature/time limitations for the induction period are such that the initiation of the subsequent high energy reaction series cannot with absolute assurance be avoided, analysis of the acceleration phase becomes of the utmost importance. Here two questions should be addressed: (1) Heat Balance (Steady State) (2) Response Time (The time available to react to a given situation).

Normally these questions are addressed in the lower portion of the overall possible temperature regime for the system under consideration (processing temperature + small % adiabatic potential) (See Figure 1).

In dealing with the question of heat balance, the necessary information to obtain is one of kinetics. The development of this data should be isolated within the lower temperature regime of the acceleration phase. With this information one can try to prevent the occurrence of a thermal runaway by establishing the proper limits for temperature versus heat transfer so that the likelihood of exceeding a steady state situation is minimized.

Since the energy release associated with the acceleration phase can be significant, evaluation of kinetics of this segment should be done by an <u>isothermal technique</u>. Using this technique, the problems of encountering a number of different reaction sequences, as might be the case with an adiabatic method, are reduced. Additionally, evaluation of the data is simplified by holding the temperature constant.

With the second question, response time, one generally assumes that conditions for a steady state have been exceeded and, on the large scale, an adiabatic system is the result. Questions pertaining to time versus the type of action to be taken (evacuation, dilution, etc.) are normally addressed. This time versus temperature concept for the adiabatic spiral is commonly referred to as the Time to Maximum Rate (TMR).

The natural extension of the previous assumption that near adiabatic conditions are experienced on a large scale dictates that, ideally, an <u>adiabatic measurement technique</u> should be used in the laboratory. In this way, the complicated relationship between the changing kinetics and energy potential versus temperature can be best simulated. One aspect of this type of measurement system, however, should be clearly understood. Since the vast majority of the experimental time associated with the evaluation of the acceleration phase is done in the lower conversion range (0-30%); it is also, in terms of the thermal spiral, correspondingly spent in the lower temperature regime and therefore lower heat production

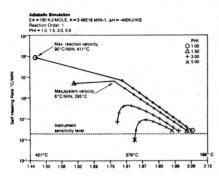

Figure 4 The Effect of Thermal Inertia

phase. As a result, the thermal inertia sometimes referred to as the PHI ($\bar{\Phi}$) Factor must be kept to a minimum. The effect of this thermal inertia can, in some cases, be quite dramatic and is often underestimated (Figure 4). In Figure 4, PHI = 1 simulates a pure adiabatic system.

3.4 <u>Release of Potential (Decay Phase Analysis)</u>. The final consideration in the evaluation of a thermal hazard is based on the assumption that its potential will be released and the effect of this release should be minimized, i.e. Emergency Relief Systems. At the present time, experience in the laboratory scale investigation of this type of problem is limited. One would assume to accomplish an analysis of this nature, however, the energy release rate associated with the thermal spiral around or at its maximum rate must be defined.* It has been demonstrated from simulation work with the simple n^{th} order and autocatalytic models of high energy reactions (400 KJ/kg), that under near adiabatic conditions this maximum rate normally occurs when > 70% of the total energy potential has been released [6]. One would further assume, therefore, that the temperature regime associated with this energy conversion range (70-100%) would be optimal for the evaluation of decay phase kinetics (Figure 1).

4. <u>INTEGRATION OF CLASSICAL ANALYTICAL METHODS AND SIMULATION TECHNIQUES WITH THERMAL ANALYTICAL DATA</u>

In the more complicated case of thermal hazard analysis the researcher is required to make predictions outside the experimental window (time/sensitivity) of the thermal analytical instruments. This is often the case when dealing with problems involving the bulk storage or transportation of materials. One is often required to extrapolate kinetic data obtained in a higher temperature region into a lower temperature region making the assumption that the reaction mechanism remains constant.

In some cases the use of classical analytical methods such as liquid and gas chromatography provide a means for verification of mechanism consistency. The increased sensitivity level of these instruments are such that the range of temperature extrapolation can, in some cases, be reduced.

Once the researcher has obtained the appropriate kinetic data to use for the extrapolation into the lower temperature regions, the question of heat balance is normally addressed. Classical calculations for steady-state solutions based on the Frank-Kaminetskii model can be performed. In the example to be demonstrated in the second case study, computer simulation calculations using finite differences equation techniques were applied. This simulation increased the accuracy of the steady-state predictions relative to using simpler analytical solutions.

*It is assumed here that the dp/dt_{max} coincides with the $°C/min_{max}$.

5. CASE STUDIES

5.1 Cold Storage - Formulated Thermal Set Resin.

After completion of the initial pilot plant run for this material, it was discharged from a Nauta mixer at 68°C and placed into polyethylene bags. These bags were placed in 16 cardboard cartons which were subsequently stacked tightly together in a cold storage room. Smoke was observed coming out of the area 4.5 hours later.

In light of both the information available on the incident and laboratory data, the following conclusions/correlations could be drawn (Table 1).

Table 1 Thermal Testing

Temperature Program Testing (DTA)

Onset Point °C	Heat of Reaction ΔH. KJ/KG	Remarks
109	213	—
304	16	—

Isothermal (Open System)

Temp °C	Exotherm	Induction Period Hours	Remarks
90	Yes	2.5	—
80	Yes	5.0	—
70	No	—	Test duration 8 hrs.
40	No	—	Test duration 24 hrs.

Heat Accumulation Method (Dewar Flask)

Temp °C	Exotherm	Induction Period Hours	Remarks
50	No	—	Test duration 90 hrs.
60	Yes	48	Exotherm quite
70	Yes	24	strong after initiation.

Adiabatic Testing

Onset Point °C	Sensitivity* °C/Hour	Runtime** Hours	Time to Max. Rate Hours	ΔH KJ/KG	Remarks
79	2.98	7.5	0.5	204	2 hrs. isothermal hold at 50°C.
81	8.8	6.6	1.0	204	

* Self Heating rate of the isolated chemical assuming a Cp = 2.09 KJ/KG °C.
** Accumulative experimental time at onset point.

DTA - Because of the potential energy release and packing conditions, internal temperatures of up to 170°C would be expected.

ISOTHERMAL - The material displayed autocatalytic behavior. Based on a discharge temperature of 68°C, however, an induction period of ~24 hours would be expected. Additional investigation revealed crusting had occurred on the plant thermocouple, therefore, the exact thermal history was unknown.

ADIABATIC TESTING - Finally, analysis of the adiabatic testing demonstrates the value of data cross comparisons. First, the onset points for both adiabatic experiments are relatively the same even though the sensitivity of the first experiment is approximately 3 times greater. This can best be explained in terms of an autocatalytic type mechanism which we have already observed with our isothermal testing. Secondly, both runs indicate the initiation of the reaction at ~80°C with an accumulative run time of 6.5-7.5 hours. This again shows good correlation with our isothermal experiment, where the induction time for the sample at 80°C was 5 hours. Thirdly, once the reaction was initiated, the time to maximum rate (TMR) was quite short (0.5-1.0 hours) especially in terms of the overall magnitude of the reaction (~50 kcal/kg RM). This again corresponds with the autocatalytic model. Finally, when one extrapolates the TMR versus temperature, only a modest 13°C down to 68°C, which corresponds with the observed discharge temperature, the TMR for this reaction is approximately 3.5 hours. This corresponds nicely to the data from the incident where the material was discharged at ~68°C and smoke was observed from the cold storage room approximately 4 hours later.

6. LONG TERM STORAGE - AGRO CHEMICAL

This study was undertaken because data from a previous study indicated that the thermal stability was questionable when

Table 2 Isothermal Aging Study

Temperature Programmed 4°C/min		Isothermal Aging				Analytical Data After Isothermal Aging (Closed glass container)		
				Residual ΔH Measurement		Liquid Chromatography Starting*		Material Balance
ΔH	Onset	Temp	Time	ΔH	Onset	Product	Material	Original Weight
KJ/KG	°C	°C	Days	KJ/KG	°C	%	%	%
-1128	190	100	2	-1262**	193	99.7	0.5	99.2
			4	-1172	190	96.5	0.8	99.1
			7	-1063	180	96.0	0.9	99.0
			9	-997	173	96.4	1.0	98.7
			11	-952	179	96.1	0.8	98.4
			13.4	-858	170	95.5	0.7	98.6
			15	-714	155	97.6	1.1	
			16	-506	80	96.6	1.0	
			17	-326	105	96.4	1.3	
			18	-260	115	92.5	1.5	
			21	—	—	94.2	1.8	
			22	—	—	92.4	3.0	
			25	-46	190	69.4	14.8	97.5
			30	—	—	45.8	24.4	96.5
			35	—	—	25.5	34.7	95.4

*One component of the decomposition was a starting material
**Initial peak not observed

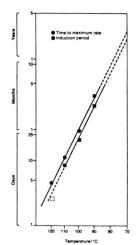

Figure 5 Induction Phase Analysis (Isothermal Aging)

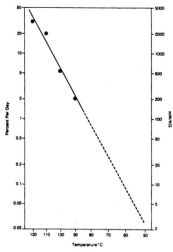

Figure 6 Acceleration/Decay Phase Analysis (Maximum Decay Rates)

tested in the 100°C range. Since the nature of the original DTA curve indicated that a multiplicity of reaction mechanisms were involved in the degradation, extrapolation into the lower temperature regions was not considered possible without additional experimentation. In this study, samples were aged isothermally in stainless steel, gold-plated DTA capsules (pressure tight to ~200 bar, 5 mg charge) and in closed glass containers (pressure tight to ~10 bar, 5 mg charge). The thermal characteristics of these aged samples were then investigated by comparing the ΔH (residual) against a ΔH value obtained on a sample without a previous thermal history. In conjunction with this evaluation the glass container contents were analyzed by LC (Table 2).

It was found during the experimentation that the residual heat measurements did not correspond well with the analytical data. For example, after 25 days at 100°C only 4% of the residual energy was measured by DTA analysis yet, 69.4% of the original material remained in the glass container experiment (Table 2). Based on the original energy potential of the starting material, a definite inconsistency was evident. The exact reason for the acceleration is not known, however, the question of material versus instrument constants is again brought into view.

A study of the degradation profiles obtained by these testing techniques indicated that a minimum of two distinct mechanisms were involved in the overall decomposition. Initially, an induction phase is observed followed by a decay phase. This type of profile was consistently observed over the entire isothermal testing range and was used as the basis for the determination of Activation Energy estimations (Figures 5 & 6).

Table 3 Critical Temperature for Storage and Shipment (Simulation Calculations)

Assumptions - Heat Production

EA$_{MEAN}$ 112 KJ/Mole
EA$_{MIN}$ 78 KJ/Mole } → Max rate / $\frac{1}{T}$

Zero order model after time to induction
Heat of decomposition = -1,250 KJ/KG

Assumptions - Heat Dissipation

Containers: Polyethylene 50 liter, 200 liter, 1,000 liter
Heat Transfer: Internal/Conduction
External/Convection

Results

Vessel Size	EA	Safe Temperature
50 liter	-112	70°C
	-78	60°C Δ 10°C (Max)
200 liter	-112	60°C
	-78	50°C Δ 7°C (Max)
1,000 liter	-112	50°C
	-78	<40°C

These estimated values were subsequently introduced into a simulation program for the determination of safe temperature/container limitations for the transportation and storage of the material in bulk quantities (Table 3). The computer method used in this determination is based on finite different equation techniques [5].

7. CONCLUSIONS

The analysis of thermal hazards of complex chemical systems through the use of thermal analytical methods, can be best accomplished through the use of an integrated testing method. The use of this integrated approach allows the researcher to simplify the problem by applying a systematic approach which is based on the relationship between reaction phase/temperature regime, information required, and the optimum test method. Additionally, the problem associated with measuring an instrument versus material constant is often reduced. This is a problem that is often underestimated, since consistency of data obtained within the regime of a particular measurement method is often regarded as sufficient evidence for the use of this information in scale-up.

8. REFERENCES

1. G. Giger and W. Regenass, NATAS Conference Proceedings, Vol. 2, p. 579, New Orleans (1981).
2. F. Brogli, P. Grimm, M. Meyer, H. Zubler, 3rd International Loss Prevention Symposium Proceedings, Vol. 2, p. 81665, Basel (1980).
3. F. Brogli, 3rd International Loss Prevention Symposium Proceedings, Vol. 2, p. 5/369, Basel (1980).
4. R. N. Rogers and J. L. Janney, NATAS Conference Proceedings, Vol. 2, p. 643, New Orleans (1981).
5. Computer Program for Simulation of Bulk Storage Problems, Central Safety Research Department, Ciba-Geigy, Basle, Switzerland.
6. Internal Computer Simulation Studies, Central Safety Research Department, Ciba-Geigy, Basle, Switzerland.

9. ACKNOWLEDGMENT

The authors extend their thanks to Dr. F. Brogli for his advice during the development of this paper.

THERMAL HAZARD EVALUATION OF STYRENE POLYMERIZATION BY
ACCELERATING RATE CALORIMETRY
L. F. Whiting and J. C. Tou
Analytical Laboratories
574 Building
Dow Chemical Company
Midland, MI 48640

INTRODUCTION

The evaluation of thermal and pressure hazards associated with the manufacture, transport, and storage of chemicals is an important area of research in the chemical industry. The engineering design of equipment to prevent, control or withstand runaway reactions which result in pressure increase is of great concern from a safety and loss point of view. In order to design a piece of equipment which will operate safely during an emergency situation, it is necessary to have data on the kinetics, thermodynamics, and physical properties of the potential runaway reaction.

One of the first commercially available instruments to be widely used in the evaluation of thermal runaway reactions is the Accelerating Rate Calorimeter (ARC). This instrument is designed to obtain time-temperature-pressure data on a small-scale runaway reaction. In this paper, ARC data on styrene monomer will be presented and applied to specific problems which might be of concern in the chemical industry.

EXPERIMENTAL

The Accelerating Rate Calorimeter (ARC) was developed by the Dow Chemical Company and was licensed to Columbia Scientific Industries of Austin, Texas, which currently markets the instrument under the trademark CSI-ARC^{TM}. The details of the design, operation, and performance of the ARC have been published elsewhere [1,2,3].

All ARC experiments presented in this paper utilized Aldrich styrene monomer, 98-99% pure, containing 10 to 15 ppm t-butyl catechol (TBC) as a polymerization inhibitor. Whether titanium, Hastelloy C or nickel, all sample containers, were one inch nominal internal diameter spheres with a one inch long by 1/8 inch outside diameter tube attached to the bomb for pressure measurement purposes.

RESULTS AND DISCUSSION

The thermal polymerization of styrene monomer inhibited with 10-15 ppm t-butyl catechol was examined using the Accelerating Rate Calorimeter. The data plots obtained from the first experiment are illustrated in Figure 1. Under the experimental conditions, the heat rate vs. temperature plot shows that the thermal polymerization of the monomer was detected at about 95°C and that the heat generated caused the temperature of the styrene and container to increase to 260°C where the reaction was apparently complete. The plot also indicates that the self-heating rate of the sample/container system reached a maximum of ~8.0°C/min at 210°C. Since pressure build-up in a closed vessel is an important parameter associated with runaway reactions, plots containing pressure data are also available from the ARC processor.

The adiabaticity of the sample during the ARC experiment is an important parameter that one must take into account before applying ARC data to a specific problem. The adiabaticity of the sample at any time is defined in Equation 1 as the portion of the heat of reaction being retained by the chemical, q, divided by the total amount of heat being generated by the sample, Q.

$$\alpha = q/Q \qquad (1)$$

For reactions where a thermal steady-state exists at all times between sample and container, i.e. temperature gradients are constant or small, the heat being produced by the sample will be partitioned according to the heat capacities of the sample and container as defined by Equation 2.

$$\phi = \frac{M_s\,C_{vs} + M_b\,C_{vb}}{M_s\,C_{vs}} \qquad (2)$$

where M_s and M_b are the masses of the sample and container or "bomb," respectively, and C_{vs} and C_{vb} are the specific heats of the sample and container at constant volume, respectively. ϕ, or thermal inertia, is thus a special case of sample adiabaticity and is related to α according to Equation 3 under conditions of thermal steady-state.

$$\phi = 1/\alpha \qquad (3)$$

Although under the stated assumption of thermal steady-state both α and ϕ are independent of time, they are still temperature dependent in most cases and C_{vs} is both temperature and composition dependent since in most reactions a change in heat capacity occurs as the reactant forms product. A more in-depth discussion of adiabaticity has been given elsewhere [4]. The data of Figure 1 can be applied to two hypothetical examples where the adiabaticities of the examples are chosen to be the same as that of the experimental data. This is done so that the ARC data can be applied directly to the example without further adjustments of the data. First, one must calculate the adiabaticity of the ARC experiment. The specific heat of the monomer/polymer mixture and the titanium container are 0.53 and 0.132 cal/g-°C, respectively [5]. Since the masses of the styrene, 4.74g, and the container, 9.47g, are known, the average adiabaticity can be calculated. For the experiment represented in Figure 1, $\alpha = 0.67$ and $\phi = 1.50$. This means that under these experimental conditions, on the average, the sample of styrene was 67% adiabatic with 33% of the heat generated from the reaction consumed in heating up the titanium container.

The first case involves a laboratory which routinely runs tests on the rates of polymerization of styrene in small bottles in an oven which can be set at various temperatures. The typical sample size is 10g (M_s) styrene of average specific heat 0.5 cal/g-°C (C_{vs} at 125°C) in a bottle weighing 12.5 g(M_b) of average specific heat 0.2 cal/g-°C (C_{vb} at 125°C). According to Equations 2 and 3, $\phi = 1.5$ for this example which is approximately the same as that of the data of Figure 1. Since the ARC under similar adiabatic (ϕ) conditions detected an exothermic reaction above 95°C, one might expect that if the laboratory runs its test near or above 95°C, (under conditions where little or no heat is lost from the bottle) an exothermic runaway reaction could occur as indicated in Figure 1 resulting in a pressure increase in the bottle. If the bottle was not constructed to withstand that pressure increase, the container would rupture.

Another example of how one might utilize ARC data would be a situation involving a distillation recovery system operating at 90 to 100°C in a vessel rated for 25 psig with ϕ = 1.5 for the vessel, the same as for Figure 1. The data indicate that loss of cooling to the condenser could result in a polymerization runaway in the vessel. In addition, if after loss of cooling the vessel were purged with nitrogen to minimize flammability of the styrene but the vessel was inadvertently valved off so that is was closed, these conditions would be similar to those of the ARC experiment and one can see from the ARC pressure data that the runaway reaction could generate internal pressures in the vicinity of 70 psia. The pressure is well above the rated pressure for the vessel.

In the examples presented thus far, an assumption has been made that the chemical and physial properties of the small ARC system accurately represent or simulate the large scale industrial system. This may not always be a valid assumption and under those circumstances the ARC data may not be conservative [5].

As another example of how ARC data can be used, suppose one would like to carry out a thermally initiated batch polymerization of styrene at 110°C. The rate of reaction and the rate of heat generation can be estimated at 110°C from the ARC data. If one can assume that the observed temperature rise is proportional to the conversion of monomer to polymer which assumes that the specific heat of the sample is independent of temperature, then the following equation can be written [5].

$$dx/dt = \frac{dT/dt}{\Delta Tab,s} \qquad (4)$$

Equation 4 relates the fraction reacted, x, at any temperature T to rate of temperature increase where $\Delta Tab,s$ is the observed experimental temperature rise.

The above equation does not account for the reaction rate dependence on concentration changes during the experiment or reaction order effects and, therefore, should only be applied under conditions of negligible concentration depletion near the initial temperature, T_o. At 110°C the styrene is approximately 8% polymerized for the reaction shown in Figure 1.

Since $\Delta Tab,s$ = 165°C and dT/dt = 0.13°C/min. at 110°C, the rate of reaction is approximately 4.7% per hour. This value is in close agreement with Platt's value of 4.5% per hour for initial rate of the polymerization of styrene at 110°C [6]. The initial rate of heat generation at 110°C, $\dot{q}_{110}$, can also be calculated from the ARC data according to Equation 5.

$$\dot{q} = Cvs \; \phi \; dT/dt \qquad (5)$$

Using Equations 2 and 5 along with the specific heats at 110°C, Cvs = .48 cal/g-°C and Cvb = 0.13 cal/g-°C, the estimated rate of heat generation at 110°C is 0.11 cal/g- min. It should be realized that these calculations of initial rates of polymerization and initial rates of heat evolution are only approximate and only apply to the early part of the reaction where percent conversion is low. This approach is not applicable to other portions of the ARC runaway curve except where the chemistry and kinetics of the reaction are known since the ARC data are dependent on degree of conversion. This technique is also limited to reactions that are not complicated by changes in mechanism early in the reaction such as autocatalytic processes.

The polymerization of styrene involves a combination of initiation, propagation, and termination of the polymer chains formed during the reaction. However, for any commercially available styrene, one must also consider the effect of inhibitor on the polymerization. The presence of inhibitor is readily seen in the self-heat rate plot of Figure 1. There is a very rapid rise in the self-heat rate when the polymerization reaction is first detected at 95°C. The inhibitor prevents the polymerization by consuming the free radicals formed in the styrene. The production of free radicals in the styrene is temperature dependent with more radicals being formed as the temperature is increased. As free radicals are formed they are consumed by the inhibitor until the inhibitor concentration becomes so low that it can no longer effectively terminate the free radically initiated polymerization process. The ARC can be operated in the isothermal mode to gain more information on the effectiveness of the inhibitor. Suppose one wished to distill styrene at 80°C in order to remove the inhibitor from the monomer prior to processing the material. How long would one have to correct an operating problem if a fresh batch of inhibited styrene was heated to 80°C but could not be distilled immediately? Figures 2 and 3 illustrate the runaway polymerization data acquired after isothermally aging the styrene in the ARC at 80°C under an air atmosphere for approximately 400 minutes. In terms of the stability of the inhibitor, one can now predict that under similar conditions the inhibited styrene can be held for 400 minutes at 80°C before the inhibitor is consumed and polymerization of the styrene is detected. As far as the distillation operation is concerned, one can assume that after 4 or 5 hours at 80°C that the risk of experiencing a runaway polymerization becomes quite high. In comparing the data from the isothermal age experiment to the data of Figure 1, one can readily see that the runaway appears to be less severe when the polymerization is allowed to begin at 80°C rather than at 95°C. The self-heat rates, the pressure rates, and the maximum pressure and temperature observed are all lower in the isothermal aging test. In order to see the effect of ϕ or adiabaticity on the ARC data, several experiments on styrene were carried out under varying conditions of ϕ. Figure 4 compares four heat rate curves for styrene where the calculated ϕ values varied from 1.12 for the nickel container to 4.17 for the heavy-weight Hastelloy C container. As one can see from Figure 4, a change in ϕ from 4.17 to 1.12, a factor of 3.7, results in a seventy-fold increase in the maximum self-heat rate, from 0.3°C/min. to 20°C/min., respectively. The increase in self-heat rate as ϕ decreases is dependent on the activation energy or the temperature sensitivity of the material of interest. For example, peroxides and high explosives typically have higher activation energies than that of styrene polymerization. Previous work on di-t-butyl peroxide showed that a factor of 2.8 decrease in ϕ resulted in a 300-fold increase in the maximum self-heat rate observed in the ARC [3]. The effects of varying ϕ on the pressures and pressure rates during a thermal runaway polymerization are shown in Figures 5 and 6. In this case the maximum pressure increased from 28 psia to 98 psia while the maximum pressure rate increased from 0.08 psia/min. to nearly 20 psia/min. when ϕ is varied from 4.17 to 1.12 respectively. All the above arguments pertaining to the effect of ϕ on runaway reaction data from the ARC also apply in general, but not necessarily quantitatively, to large scale equipment, and must, therefore, be carefully considered for accurate scale-up of processes.

REFERENCES

1. D. I. Townsend and J. C. Tou, Thermochim. Acta, 37 (1980), 1-30. 2. D. W. Smith, M. C. Taylor, R. Young, and T. Stevens, Amer. Lab., June, 1980.
3. J. C. Tou and L. F. Whiting, Thermochim. Acta., 48 (1981), 21-42.
4. J. E. Huff, "Determination of Emergency Venting Requirements from Runaway Tests in Closed Systems," to be presented at the Symposium on Venting of Runaway Chemical Reactions, AIChE Spring, 1982 Meeting, Anaheim, California, June, 1982.
5. L. F. Whiting and J. C. Tou, J. Therm. Anal., in press.
6. A. E. Platt, "Polymerization" in the Encyclopedia of Polymer Science and Technology, Volume 13, John Wiley and Sons, 1970, pp. 156-159.

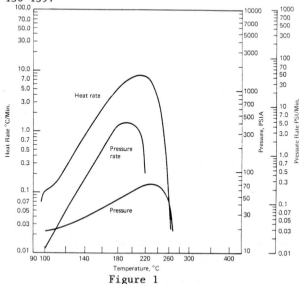

Figure 1

Data plots from ARC experiment on inhibited styrene monomer, 4.74 g styrene, 9.47 g titanium bomb, air atmosphere, 50°C start temperature, 5°C heat-step interval, 15 minutes wait time.

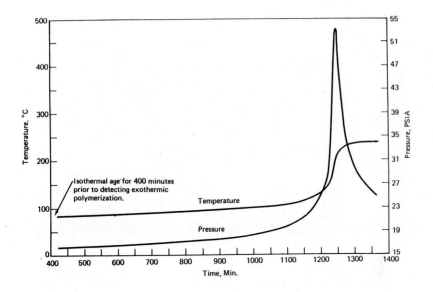

Figure 2

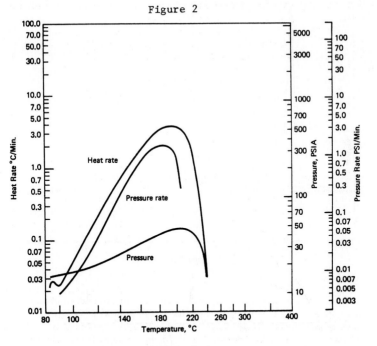

Figure 3

Data plots from 80°C isothermal age ARC experiment, 4.78 g styrene, 10.01 g titanium bomb, air atmosphere.

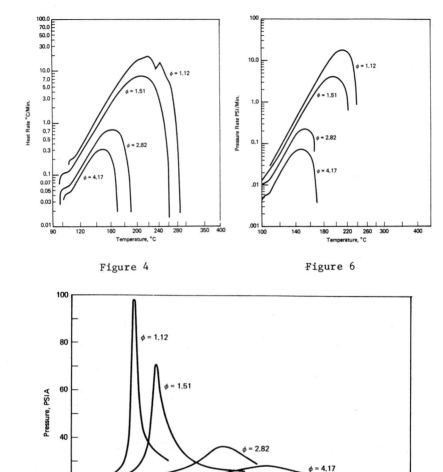

Figure 4

Figure 6

Figure 5

Comparison among ARC experiments on inhibited styrene run under different conditions of ϕ. Heavy-weight Hastelloy C-276 bomb with ϕ = 4.17. Heavy-weight titanium bomb with ϕ = 2.82. Light-weight titanium bomb with ϕ = 1.50. Light-weight nickel bomb with ϕ = 1.12. All tests were run under air atmosphere, 50°C start temperature, 5°C heat-step interval, 15 minute wait time.

GEOLOGICAL DATING BY THERMAL ANALYSIS

G. SZÖŐR

Department of Mineralogy and Geology
L.Kossuth University, H-4010 Debrecen
P.O.Box 4, Hungary

INTRODUCTION

Subsequent to a comparative examination of recent model material [1], [2] the author has recently completed the paleobiogeochemical evaluation of classical Quaternary and Pliocene inland fossils of vertebrata from Hungary [3], [4], [5].

The investigation has cleared up the formation, accumulation and diagenesis of sediments and the related fossilization process.

It has been established for the earliest, so-called syndiagenetic stage of embedding sediments that the changing geochemical effects of the microfacies and the climate-induced variations of microbiological decomposition may produce highly different /or even extreme/ stages of fossilization. However, in the case of continuous sedimentation, with the passage of time, these differences will become levelled for, as referred to the macrofacies, uniform.

At this stage the gradual and regular transformation, collagenic autohydrolysis, carbonation of apatite structure, impregnation of bone caverns with clay minerals can be influenced only by incidental drastic exogenous effects, such as the activity of thermal springs, effect of anthropogenic hearths, etc. Recognition of this regularity gave us the possibility to elaborate a complex thermoanalytical method /derivatography/, by means of which two thermoanalytical parameters can be determined, which are closely associated with the passage of geological time.

METHODS AND RESULTS

The underlying scientific basis for the selection of our particular sample material was provided by the comprehensive study of J.Chaline. This author, on summing up the results of international research reviewed the chronostratigraphy of the classic Eurasian vertebrate localities [6]. The essential basis of the biochronological classification, which was calibrated with the absolute chronological data, was provided by the material of finds revealed in the karstic region of Hungary.

In the course of our work this chronologically well-defined sample material was evaluated, i.e. we performed comparative studies on vertebral segments of

Ophidia indeterminata from the identical reddish-brown terra rossa of 14 karstic caves and crevices.

Since in our evaluation the total fine stratigraphic analysis of each locality was performed, data of hundreds of measurements were processed.

The derivatographic measurements were preceded by careful preparative work, the single particular steps being: clearance from the embedding sediment, grinding to identical particle size, removal of adsorptive water content in an desiccator.

After examining a number of methodological possibilities, the series of measurements were performed under the following circumstances. Temperature range 1000°, rate of heating 10°. minute an air current. Sample holder: platinum plate. Measured in: 200-300 mg.

The thermoanalytical curves reveal processes of thermal decomposition shown in Fig 1.
In the course of the first endothermic process the decomposition of the organic macro-molecules takes place and the water-content bound to the phosphate structure with weak forces is removed.
During the second, exothermic process, as a result of the catalytic effect of the platinum sample holder, all organic material burns out and is removed. Here the final fractionation, deamination, decarboxylisation of the collagen macromolecule takes place. Parallel with the removal of organic matter content the disappearance of water bound to the phosphate structure with stronger forces takes place continuously.
The third, endothermic process comes about only in the case of bone material older than 5000 years, and does not in recent and prefossil samples. Here the thermodissociation of inorganic carbonates secondarily built into the apatite structure takes place, and CO_2 is released. The disappearance of water bound to apatite stucture is continued in this temperature range, too.

The weight per cent values of the three well distinguishable material dissociations are called thermodecomposition products, and are denoted by A, B, C. When summing up the great number of measurement data we found that the B and /A+B/ values decreases as a function of embedding time. This trend is manifest in the relationship of the overlying layers of a given locality and of localities far away from one another. Parallel with this appears the increase in the C value. Thus, the organic matter content of the bone material gradually decreases with the passing of time, and parallel with this, the gradual dehydration and carbonation of the apatite phosphate system takes place.

This establishment was verified by supplementary research methods. In Fig 2 the fine stratigraphical evaluation of the bone material collected in the sediment of the karstic crevice called Rigó-lyuk is presented.

It can be seen that the trend of the time-dependent changes in the derivatographically determined B- value agrees with the temporal changes in nitrogen content determined by neutron activation analysis. This experiment justifies, on the one hand, our establishment, that the thermodecomposition products removed during the first two thermal reactions are in close connection with the organic matter content of the bone. On the other hand, the figure demonstrates the decrease in the organic material encountered parallel with embedding. The continuous trend is interrupted by two minima. This phenomenon is interpretable and very important. The evaluation of the rock and the embedded material of finds shows that in these time intervals the prevailing climate was warm and humid in these regions. As a result of enhanced microbiological decomposition the organic matter content of the bones is lower than the average. The minimum value of organic matter assigned to the two points of time appeared consistently in the fine stratigraphic evaluation of more than one locality, proving the informative usefulness of the derivatographical parameters in paleoclimatological reconstructions.

For a more exact interpretation of thermodecomposition, the derivatographic series of measurements were supplemented by gas titrimetric examinations.

The great number of data from derivatographic serial measurements were processed with computer programs. Our /A+B/ and Fk parameters were correlated with the absolute time values of Chaline's biochronological table. The exponential, logarithmic and power-form regression trend calculations were separately performed for the Holocene, Pleistocene and Pliocene periods and it was established that the regression coefficients of the latter two provide applicable numerical correlations /Table 1/.

Subsequently, the derivatographic time data of finds of unknown age were determined /Table 2/. The data were checked with C^{14} dating. The validity of our dating seems to be supported in the case of the other values too, since they are completely consistent with the stratigraphic determinations, with the relative chronological order based on the changes in vertebrate succession.

ACKNOWLEDGEMENTS

I wish to thank the co-workers of the Departement of Experimental Physics of L.Kossuth University for the neutron activation measurements and the Computer Center of the University for the work of calculation.

REFERENCES

1. G.Szöőr, Acta Mineralogica-Petrographica Szeged, 1971, 20, p. 149.
2. B.Mándi, M.Petkó, G.Szöőr and T.Glant, Acta Morphologica Acad. Sci. Hung., 1975, 23, p. 59.
3. G.Szöőr, Acta Mineralogica-Petrographica Szeged, 1975, 22, p. 61.
4. G.Szöőr, Ph.D.Thesis in the Hungarian Acad. of Sci. Library, 1979.
5. G.Szöőr, Journal of Thermal Analysis, 1982, 22 /In press/
6. J.Chaline, J.Michaux et P.Mein, Institut des Sci. de la Terre, Dijon, 1974.

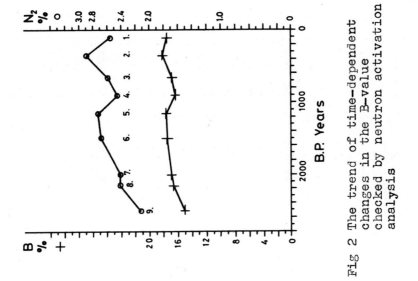

Fig 2 The trend of time-dependent changes in the B-value checked by neutron activation analysis

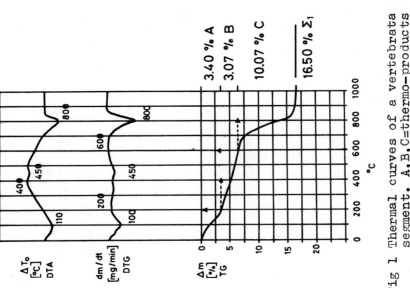

Fig 1 Thermal curves of a vertebrata segment. A,B,C=thermo-products

TABLE 1

THE CORRELATIONS FOR THE INDIVIDUAL GEOLOGICAL PERIODS

HOLOCENE	PLEISTOCENE	PLIOCENE
$Fk>4.0: /A+B/ >28.1\%$	$2.0<Fk<4.0: 11.0\%</A+B/<23.6\%$	$Fk<1.0: /A+B/<7.0\%$
$T_{abs}=10^2 \cdot e^{-\frac{/A+B/-29.71}{1.29}}$	$T_{abs}=10^3 \cdot \left(\frac{28.483}{/A+B/}\right)^{5.3124}$	$T_{abs}=10^3 \cdot \left(\frac{28.483}{/A+B/}\right)^{5.3124}$
$Fk>4.0: 23.6\%</A+B/<28.1\%$	$Fk>4.0: 11.0\%</A+B/<23.6\%$	
$T_{abs}=10^2 \cdot e^{-\frac{/A+B/-30.63}{1.78}}$	$T_{abs}=10^3 \cdot e^{-\frac{/A+B/-33.86}{2.60}}$	
$Fk>4.0: /A+B/<23.6\%$	$Fk<2.0: 7.0<\!/A+B/<11.0\%$	$T_{abs}=$ B. P. Years
$T_{abs}=10^2 \cdot e^{-\frac{/A+B/-33.86}{2.60}}$	$T_{abs}=10^3 \cdot \left(\frac{28.483}{/A+B/}\right)^{5.3124}$	Fk, /A+B/=derivato-graphical parameters

TABLE 2

THE NEW DERIVATOGRAPHIC TIME DATA

LOCALITIES /caves/	PERIODS	FAUNAL PHASES	DERIVATOGRAPHICAL AGES /B.P. years/	
Csontos	Holocene	Alföldi	15	/recent/
Békásmegyer		Bükki	3 897	/±250/
Baradla		Körösi	6 516	
Zöld		Körösi	7 314	
Balla	Pleistocene	Istállóskői	20 040	/±500/
Por-lyuk		Varbói	36 786	
Osztramos-4		Upponyi	259 981	/±5000/
Osztramos-5		Templomhegyi	347 200	
Osztramos-14		Betfiai	695 109	/±10 000/
Osztramos-12		Betfiai	979 800	
Osztramos-13	Pliocene	Estramontiai	3 077 942	/±100 000/

THERMOANALYTICAL MEASUREMENTS IN ARCHAEOMETRY

G. Bayer, Institute for Crystallography
and Petrography, ETH Zurich
CH-8092 Zurich, Switzerland

H.G. Wiedemann, Mettler Instrumente AG
CH-8606 Greifensee, Switzerland

INTRODUCTION

A variety of analytical methods - both chemical and physical - are applied in the area of archaeological investigations on ceramincs, glass, pigments, paper and other materials [1]. They all try to find the answers to questions such as: when and by whom was the material or object made and used; which were the raw materials and the technological processes; were there any environmental effects or changes? Basically, detailed characterization of the materials is required with respect to chemical and phase composition, microstructure and various physical properties. The choice of a specific technique is usually governed by its availability and cost, by its speed and accuracy, by the degree of damage to the specimen, by the type of material and the properties and elements to be determined. Most studies require the combination of several and quite different techniques for useful conclusions. The present paper demonstrates that also thermoanalytical methods combined with x-ray and mass spectrometry may be very useful in such archaeological investigations.

EXPERIMENTAL TECHNIQUE

For experimental investigations on ancient materials destructive methods can be tolerated only if they are sensitive and therefore require very small amounts of material. This is the case for x-ray and MS but also for the special thermoanalytical instrumentation (Mettler TA3000) which was used in the present studies. This new measuring and evaluation system is composed of a basic processor unit and several measuring cells for DSC, TG and TMA. The data acquired during these thermal analysis are held in a working storage which is accessible for the subsequent evaluation.

NABATEAN POTTERY

The typically painted Nabatean ceramic was manufactured within the first century B.C. and A.D., the period in which many of the famous rock tombs and monuments were created around Petra and Hegra [2]. A variety of fragments of this thin-walled ceramic is studied with respect to the phase composition and to the thermal behavior. Two of these samples with quite different chemical composition (table 1) will be briefly discussed. The x-ray patterns of these samples are shown in fig. 1. The Fe_2O_3-rich and CaO-poor sample 25 contains only quartz as crystalline phase. The typical red-brown color of this ceramic is due to the use of iron-rich clays for its manufacture. Heating this sample for 1 hour at

Fig. 1 X-ray photographs (CuK_α) of Nabataen pottery

900° C and at 1000° C did not cause any changes of the x-ray pattern or color. After heating at 1100° C, crystallization of mullite and cristobalite was observed with a decrease of the quartz content. The DSC, TG and TMA-analysis of this sample (fig. 2, left side) proved that exothermic reactions with weight loss (carbonaceous residues?) occur at around 300° C and 500° C, followed by the endothermic quartz transformation at 573° C. Strong sintering with corresponding deformation (see TMA) starts above 900° C. A completely different behavior was found for sample 26, which has a high concentration of CaO and also MgO, but much less SiO_2 and especially Fe_2O_3 in comparison to sample 25. The x-ray pattern (fig. 1) shows the presence of the following crystalline phases: quartz, augite (pyroxene), plagioclase and gehlenite. Again there was no change in phase composition after heating to 900° C and at 1000° C. Heating this cream-colored strong ceramic material at 1100° C caused the disappearance of gehlenite and a strong increase of the plagioclase concentration, whereas the pyroxene phase did not change significantly. Formation of mullite could not be observed. The DSC and TG-curves (fig. 2, right side) are somehow similar to those of sample 25, but the TMA-curve proves that this ceramic does not show any shrinkage due to secondary sintering up to at least 1000° C. These preliminary results suggest that a $CaCO_3$-rich clay was used in the manufacture of this material and that the firing temperature was higher than for the other material (sample 25) but probably not much above 1000° C.

Nabataen Ceramic	Loss on Ignition	SiO_2	Al_2O_3	Fe_2O_3	TiO_2	CaO	MgO	K_2O	Na_2O	TOTAL
NC 25	1,77	61,18	22,30	8,50	1,46	0,36	0,83	2,89	0,74	100,03%
NC 26	2,86	49,99	20,95	1,80	0,71	17,63	3,07	2,17	0,58	99,76%

Table 1

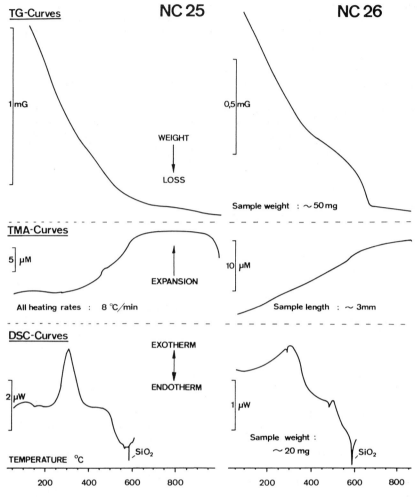

Fig. 2 DSC-, TG- and TMA-curves of Nabataen pottery

ANCIENT EGYPTIAN PIGMENTS

Many of such coloured materials were produced from natural minerals, however also a wide range of synthetic pigments was used mostly in form of glass frits. This is true especially for the blue pigments which were much in demand, since blue was the colour of the gods. During our research we studied a large number of ancient Egyptain pigments, especially blue ones, from various dynasties [3]. We also had the unique possibility to investigate the original pigments used for decoration of the bust of Nefertiti. These studies proved that the blue pigment of the bust is indeed the crystalline compound CaCu(Si_4O_{10}) - Egyptian blue - and not a blue glassy frit as assumed by Rathgen. In further investigations we also could find out more details about the experimental conditions for synthesis of this important blue pigment and its thermal stability. Egyptian Blue was in use already during the fourth dynasty in Egypt, i.e. at about 2600 B.C., and was produced in consistent quality for more than 2000 years. For the thermal synthesis of Egyptian Blue we used calcite, malachite (or azurite) and quartz (from Egyptian localities) as the raw materials.

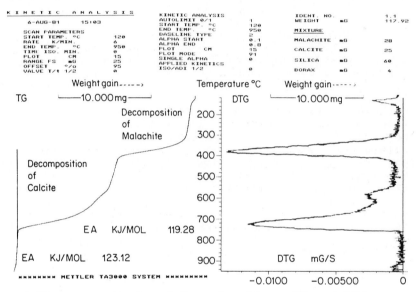

Fig. 3 Thermosynthesis of Egyptian blue, TG and DTG-curves

Four different fluxes were added to the above stoichiometric mixtures, namely borax, papyrus ash, salt or sodium sulfate. Fig. 3 shows the thermogravimetric analysis of the formation of CaCu(Si_4O_{10}) from the mixture limestone/malachite/sand/borax as an example. The TG curve proves that the decomposition of malachite to CuO occurs at 300 - 400° C, followed by the decomposition of limestone to CaO at 550 - 740° C. The maximum of the reaction rate

Fig. 4 X-ray photographs (CuK$_\alpha$) of Egyptian blue: A. Nefertiti (1355 B.C.), B) synthetic (1980 A.D.)

at 380° C and at 725° C respectively can be seen from the DTG. Above this temperature CaO and CuO react with SiO in the presence of the borax flux to form the compound CaCu(Si_4O_{10}). Fig. 4 shows the x-ray patterns of this synthetic Egyptian blue product and of the original blue pigment taken from the bust of Nefertiti. The agreement is obvious and unambiguous. These investigations on the thermal synthesis of Egyptian blue lead to certain conclusions with respect to the mixture ratio, the effect of fluxes, the heating temperature and atmosphere, all of which are important for the intensity of the colour and for the stability of this pigment.

WAX BINDERS IN ANCIENT PIGMENTS

Waxes like beeswax were frequently used as binding media for pigments. This is also true for the black carbon pigment which served for decorating the eyelashes of the bust of the Queen Nefertiti. A suitable method for investigating the type of wax used is DSC which allows to determine the melting point and the heat of fusion. In the previous investigation on the bust of the Queen Nefertiti F. Rathgen assumed that wax was used as a binding medium and quoted a melting range of 60 - 64° C. He did not try, however, to identify the type of wax which might have been used. Fig. 5 shows our DSC-curve of the original wax binder from the bust of Nefertiti. The melting range from 40 - 60° C is in agreement with the values given by Rathgen. The maximum of the melting peak was found at 64,4° C for the original wax binder and at 63,4° C for beeswax. The corresponding heats of fusion were 127.4 J/g and 207 J/g

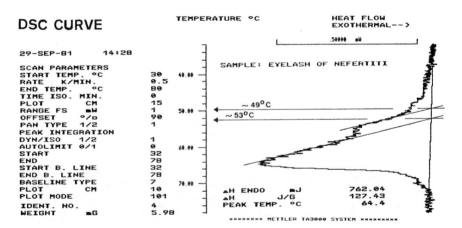

Fig. 5 DSC-curve of the ancient wax sample

respectively. This larger value for fresh beeswax is obviously caused by the high content of volatile components.

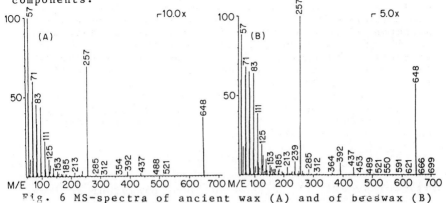

Fig. 6 MS-spectra of ancient wax (A) and of beeswax (B)

Further investigations on this sample have been carried out by means of a Finnigan 4510 quadrupol mass spectrometer equipped with a Carlo-Erba gas chromatograph. Two spectra from the same GC-fraction of the original sample (wax binder from the bus of the Queen Nefertiti) and of beeswax are shown in fig. 6. The similarity of the spectra is obvious, the mass numbers of the original sample correspond to the components typical for beeswax, e.g. higher paraffines, cerotin acid and especially esters of C16-C30 acids and C24-C36 alcohols.

ANCIENT PAPYRI

The application of thermoanalytical methods for differentiation between ancient papyri has been described previously [4]. A further problem encountered with such historic papyri is their overall state of preservation, since they are usually very fragile. Many papyri are in a bad shape mainly due to the effect of fungi. These fungi can be easily identified by means of electron microscopy but it is not possible to use DSC in such studies since there is an overlapping of the peaks of cellulose and chitine of fungi (fig. 7). In addition the swelling behavior of papyri was investi-

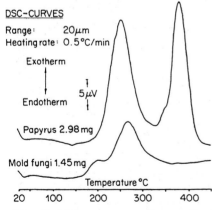

Fig. 7 DSC-curves of papyrus and of mold fungi

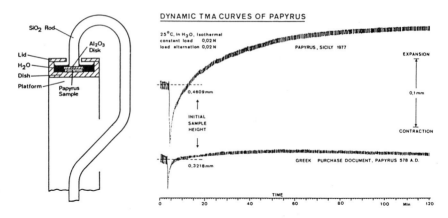

Fig. 8 TMA-cell METTLER 3000

Fig. 9 Swelling behaviour of fresh and ancient papyrus

gated by dynamic TMA measurements. Fig. 8 shows the TMA-cell with the punched disk-like papyrus sample and the posistion of the water container. The sample was placed between the vitreous silica platform and movable SiO_2 rod and subjected to a variable load. After equilibration ((3 - 5 minutes) water is added into the dish. The next figure (fig. 9) shows a comparison of the swelling behaviour of a fresh and an ancient papyrus sample. At first there is a sudden contraction due to the inhibition with water, followed by a parabolic swelling curve. If we take the fresh papyrus as reference (100 %), we can derive that the ancient papyrus shows a swelling in the order of 5 - 6 %. Further experiments with ancient papyri from different periods proved that papyri around 1000 B.C. show about 2 - 3 % swelling whereas the older papyri do not show such expansion behavior at all.

REFERENCES

1. J. Riederer, "Kunstwerke chemisch betrachtet", Springer Verlag, Berlin-Heidelberg-New York (1981).

2. K. Schmitt-Korte, "Die Nabatäer, Spuren einer arabischen Kultur der Antike", Veröffentlichungen der Deutsch-Jordanischen Gesellschaft, e.V., Hannover (1976).

3. G. Bayer and H.G. Wiedemann, Sandoz Bulletin 1976, 40, 19.

4. H.G. Wiedemann, Proceedings 4th International Conference on Surface and Colloid Chemistry, Israel, Jerusalem, July 5 - 10, 1981, p. 122.

THERMOLUMINESCENCE OF QUARTZ:
ITS USE IN ARCHAEOLOGY

Donald B. Nuzzio

Department of Chemistry,
Rutgers University
Newark, New Jersey 07102
U.S.A.

INTRODUCTION

Thermoluminescence is a process whereby light is emitted from a material upon heating. The recording of this light emission as a function of increasing temperature is known as a glow curve [1,2]. When a material such as quartz is exposed to ionizing radiation, electrons within the quartz become excited. When this excitation takes place, electrons are promoted from the valence band to the conduction band, leaving a hole behind (hole trap). The majority of excited electrons within the conduction band fall immediately back to the valence band emitting light. However, a few electrons become trapped in a state of higher energy known as electron traps [3,4]. These traps are created by imperfections within the crystal lattice and/or the presence of diffused impurities. The number of trapped electrons in these areas are proportional to the amount of ionizing radiation the material received (see Fig. 1a). When a material such as quartz has experienced ionizing radiation and is subsequently heated, electrons within the traps escape and fall back to the valence band emitting light. Either the electron trap is less stable and the hole trap becomes the emitting center (see Fig. 1b), or the hole trap is less stable and the electron trap becomes the emitting center (see Fig. 1c).

ARCHAEOLOGICAL APPLICATION

When a ceramic object is manufactured several basic criteria must be met. One of these criteria is that the clay used in the manufacture of pottery must be of a certain consistency. This consistency is affected by the amount of water and inert materials present within the clay fabric [5,6]. The presence of these inert materials, usually quartz, allows the ceramic object to "breathe" on subsequent firing and cooling. If the inert quartz does not exist at a critical percentage, the object will shrink and crack on firing. Thus by incorporating quartz sand into the clay fabric, the potter achieved his goal of producing an unflawed object and unknowingly incorporated a natural dosimeter. The use of inert materials such as quartz in ceramic objects to prevent cracking and shrinking is known as grog [7,8]. The quartz itself, used in ceramic manufacture, contains natural geological thermoluminescence. The geological thermoluminescence was erased when the ceramic object was fired. Thus any populated electron traps were emptied and subsequent dosing of the object would result in a repopulation of these traps. This subsequent dosing comes from the presence of ppm levels of uranium, thorium, and potassium-40 within the clay fabric and surrounding soils. Determining the amount of these radioactive substances is necessary in order to calculate an accurate annual dose rate [9,10]. See Fig. 2 for a summary of the thermoluminesence process of a manufactured ceramic artifact.

In firing the ceramic object, the quartz goes through at least three basic polymorphic states (see Fig. 3). When this occurs, impurities within the clay fabric can diffuse into the quartz grains present. Diffusion of these impurities can allow more electrons to become trapped, thus increasing thermoluminescence output. The basic firing temperatures of pottery are from 750°C to 1000°C [11,12].

Extraction of quartz grains from ceramic artifacts is normally done via magnetic separation [13] or liquid separation [14]. Once these grains are extracted and cleaned by a prescribed procedure [15], glow curves can be obtained. Knowing the annual dose rate as well as running standard irradiated quartz samples, the following equation can be used to determine age of the object under study.

$$\text{Age (yr)} = \frac{\text{Natural TL}}{(\text{TL/rad}) \times (\text{rads/yr})}$$

This equation is simple and is by no means complete. Correction factors and accurate dose measurements must be explored in order for an accurate date to be obtained.

Several methods of dating quartz grains are known [16,17,18]. Of these, the inclusion dating method (large grains), and the fine grain methods are the most popular.

INSTRUMENTATION

Thermoluminescence instrumentation consists of two major parts. The light measuring system and the sample heating system [19,20]. The use of such instrumentation consists of placing a sample on a heating planchette, positioning the light measuring system, purging the system with nitrogen, and then heating the sample at an extremely fast heating rate. In Fig. 4, a DuPont 990 thermal analyzer has been modified to record accurate glow curve data [21]. A typical glow curve of quartz obtained from the modified thermoluminescent instrument can be seen in Fig. 5. The sample weight was approximately 4mg and a heating rate of $50°\text{Cmin}^{-1}$ was used.

CONCLUSION

Dating of ceramic materials is a tedious and time consuming process. This paper only briefly discusses the use of quartz in this process. However, it is hoped that the thermal analyst has gained some insight into the use of thermoluminescence as a dating tool.

ENERGY LEVEL DIAGRAM OF INSULATING CRYSTAL EXHIBITING THERMOLUMINESCENCE

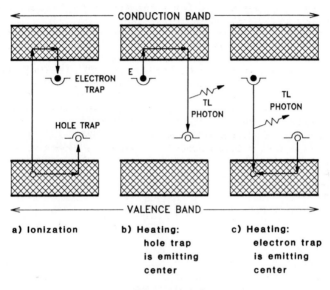

a) Ionization

b) Heating: hole trap is emitting center

c) Heating: electron trap is emitting center

Figure 1

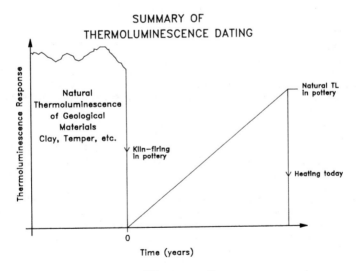

Figure 2

Polymorphic Forms of Quartz

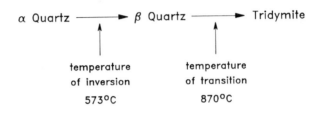

Figure 3

Block Diagram of Thermoluminescence Instrument

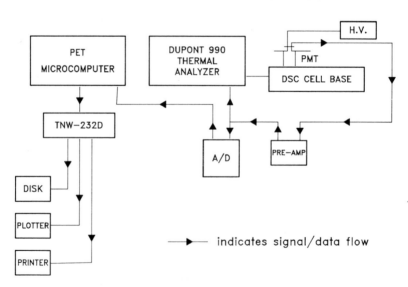

Figure 4

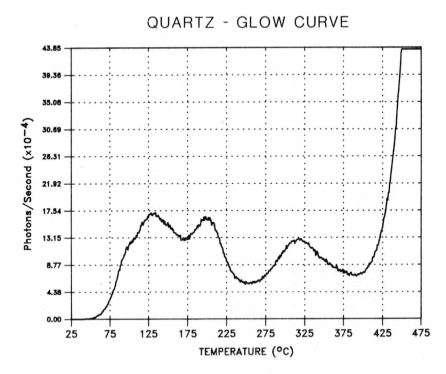

Figure 5

REFERENCES

1. J.T. Randall, and M.H.F. Wilkins, Proc. R. Soc., 1945, A184, 366.
2. A. Halperin, and A.A. Braner, Phys. Rev., 1960, 117, 408.
3. R.H. Babe, J. Phys. Chem., 1953, 57, 785.
4. M. Schlesinger, J. Phys. Chem Solids, 1965, 26, 1761.
5. W. Ryan, "Properties of Ceramic Raw Materials," Pergamon Press, Oxford, (1978), p.21.
6. R.E. Brim, "Clay Mineralogy," McGraw-Hill, New York, (1968), p.14.
7. A.O. Shepard, "Ceramics For The Archaeologist," Carnegie Inst. of Washington, Washington, (1956), p.25.
8. W. Ryan, "Properties of Ceramic Raw Materials," Pergamon Press, Oxford, (1978), p.23.
9. D.W. Zimmerman, Radiation Effects, 1972, 14, 81.
10. V. Mejdahl, Archaeometry, 1979, 21, 61.
11. M.J. Aitken, D.W. Zimmerman, S.J. Fleming, Nature, 1968, 219, 442.
12. H.H.M. Pike, Archaeometry, 1976, 18, 111.
13. H. Valladas Archaeometry, 1977, 19, 88.
14. S. Fleming, "Thermoluminescence Techniques In Archaeology," Clarendon Press, Oxford, (1979), p.40.
15. D.B. Nuzzio, to be published.
16. D.W. Zimmerman, Archaeometry, 1971, 13, 29.
17. M.J. Aitken, J. Huxtable, A.G. Wintle and S.G.E. Bowman, Fourth Int. Conf. on Luminescence Dosimetry, 1974, 1004.
18. S. Fleming,"Thermoluminescence Techniques In Archaeology," Clarendon Press, Oxford, (1979), p.40.
19. W.L Medlin, J. Phys. Chem., 1963, 38, 1132.
20. D.J. McDougall, "Thermoluminescence of Geological Materials," Academic Press, London, (1968), p.175.
21. D.B. Nuzzio, Thermochim. Acta, 1982, 52, 245.

INVESTIGATION OF THE THERMAL DEGRADATION
OF PAINTS USED IN THE BODY OF AUTOMOBILES.

D.Marjit
Physics Division
Forensic Science Laboratory
Government of West Bengal
Calcutta.

ABSTRACT
    In air atmosphere at different temperatures four typical paint samples used in the body of automobiles in India were aged and DTA and DTG curves of the questioned paint samples along with their unaged control were investigated. Having made a comparison among the results obtained from the thermoanalytical curves produced it was concluded that the condition of the paint samples could be characterised by the alteration of the quantity of heat evolved at the first exothermic DTA peak.

INTRODUCTION
    Paints may be considered to be composed of pigment with or without a dyestuff dissolved or suspended in a vehicle and thinned with a solvent. The pigments utilized in the automobiles paints are primarily combinations of Red Iron Oxide,Sienna,Carbon Black,Zinc Oxide,Zinc Yellow, Strontium Yellow,Barium White,Talc,Silicate Earths,Clay and driers of Pb,Co and Mn. The finish coats include nitrocellulose,lacquers,nitrocellulose alkyd resin-plasticized lacquers,acrylic resin lacquers,baking enamels formulated from alkyd-melamine resins and alkyd urea-soybean oil vehicle.
    In old paint the pigment remains more or less unchanged,the vehicle is present but may be oxidized or otherwise altered while that has evaporated due to its aging which leads to the rise of the operating temperature. It has an unfavourable influence on the aging characteristics of paints. To fulfil the requirements of the reliability it needs the application of thermal analysis technique in order to assess the identity or dissimilarity among the questioned paint samples from the stand point of forensic interest. The thermoanalytical methods have proved to be suitable as a differentiating technique for examination of paints [1,2].
    By the investigation of the paints we are interested in the alteration of the material's properties. But the reserves inherent in them have not been depleted yet.One of this reserves is the sensitivity of the DTA measurements which can be utilised by choosing a special material with respect to the special nature of the problem[3, 4].From this follows already that for the direction of aging processes it is suitable to select a method which

allows to make permanent comparison between the properties of the unaged paint and of the aged one. Furthermore this is advantageous for the DTA methodics also because in this case the thermal behaviours of the reference material approximate the thermal behaviours of the samples more closely than any common inert material. When the sample and reference are so matched that their thermal constants ie. thermal conductivity (L), specific heat (C) are more or less identical than the thermal diffusivity of the sample a_s will be equal to that of the reference a_r.

Where $a = L/d \cdot C$, d = density of the material.

To satisfy this condition, assuming equal volumes of the two specimens

$$m_s \cdot C_s / m_r \cdot C_r = L_s / L_r$$

Where m is the mass and the subscripts s and r refer to the questioned sample and reference material respectively. And the base line driff of the DTA peak will be absent if

$$m_s / m_r = C_r / C_s$$

but this condition will valid only if $L_s = L_r$.

FORENSIC INTEREST OF THE WORK

Different types of paints are frequently received by the Forensic Science Laboratories in connection with different types of offences including some very common accidental cases. Important factors in identification are shades of colour, chemical composition and the number of coats ie. layers of paint. The examination is carried out microscopically, chemically and spectroscopically. Paint which contain no metallic constituents are very difficult to identify, but under favourable conditions chemical as well as thermal analysis can be useful.

The paint samples received are usually in the form of smear, flake or chippings. While the first type may show smearing of different colours or tints, it is not possible in such samples to identify the paint structure as is possible with flake or chippings. A number of techniques are often utilized by the Forensic experts to prove that two paints are identical. Experience is, of course, essential to enable the Forensic scientist to evaluate the results of his tests in order to assess the specificity of his findings. For example, the fact that two paints are both pure white lead paints means that they are identical according to their composition, but it does not necessarily mean that they are from the same source. If both the paints have more than one identical additions or impurities present or if the composition is unusual the value of the evidence of identical origin rises very rapidly.

Four typical paints used in the body of automobiles in India were examined in this present work.

EXPERIMENTAL

All the four paints of this work were aged in air atmosphere at about 25C,30C,35C and 40C temperature levels. On this dinamic method the thermal deterioration of the paint sample suffered in its previous unaged condition. The particle size of the dried fresh and weathered aged paint samples were within the range of 150 to 200 mesh. The TG,DTG and DTA curves have been recorded photographically with a single sample in static air with a Paulik-Paulik-Erday type Derivatograph (MOM,Budapest, Hungery). The heating rate was 6C per minute and 200 mg of the samples were used in each programme.

The TG,DTG and DTA sensitivities were 100 mg, 1/10 and 1/10 respectively. Platinum crusible of medium size was used in each measurements.

RESULTS AND DISCUSSION

The derivatograms of the paint samples both fresh and aged were reproduced in Figures 1,2,3 and 4.

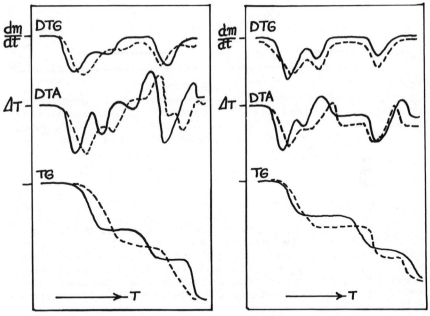

Figure 1
Red synthetic organic pigments(Thioindigo Maroon). Col. index No.73390,Pigment No.83 and Composition(5-chlor-O-tolylmorcapto)acetic acid cyclised and oxidized with chlorosulfonic acid.
Fresh (———) Aged (— — —)

Figure 2
Yellow synthetic orange pigments(Lithol:Fast Yellow NCR). Col.index No.12780,Pigment No.7 and Composition O-nitroaniline completed with 2,4-quinolinediol.
Fresh (———) Aged (— — —)

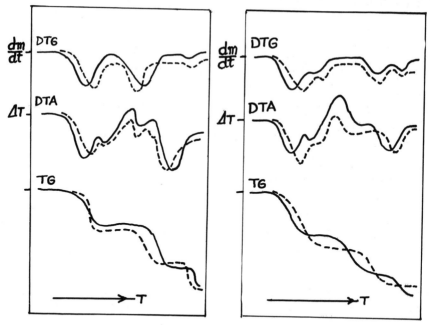

Figure 3
Black Inorganic Pigments(Carbon Black).Col.index No.77268 Pigment Black No.8 and Composition 80 to 95 per cent carbon,traces of sulphur and some organic matter.
Fresh (———) Aged (— — —)

Figure 4
Green synthetic organic pigments(Phthalocynine Green). Col.index No.74260,Pigment Green No.7 and Composition Chlorinated Cu,tetrabenzo-tetra-azoporphin.
Fresh (———) Aged (— — —)

A comparison of the results displayed in Tables 1 and 2 show that the shift of DTA peaks with aging is most prominent. The observed paint decomposition occured in four steps. The relative weight changes (TG) in the indisteps were nearly independent of whether aged or unaged paint samples were investigated. But the DTA peaks showed however significant difference between the samples. The area of the first exthermic DTA peak at 200C is in the case of unaged one (H_o).

For the purpose of direct comparison of aged and unaged paint samples DTA measurements were carried out using instead of the inert material the unaged paint sample itself (Figures 5 and 6). The heat transfer co-efficient (K) of the holders and the heat evolved by the reactions (dH) that took place through out the rise of furnace temperature of the aged and unaged samples may be shown as

$$dH = K \int_{t_1}^{t_2} dT \cdot dt.$$

Using the co-ordinates dT and t

$$S = \int_{t_1}^{t_2} dT \cdot dt.$$

dH= K.S., where S is the area of the peak on the DTA curve. The area of the exothmic peak $(H_o - H)$ is just proportional to the excess heat evolved from the aged sample.

Table 1

THERMAL PARAMETERS OF THE FRESH (UNAGED) PAINTS SAMPLES.

No	Samples Paint Unaged	Temperature Range $(T_i - T_f)$ C	Peak Temperatures (C) DTA	DTG
1.	Type 1	(95 - 120)	150/170	111/150
2.	Type 2	(98 - 125)	160/180	115/150
3.	Type 3	(95 - 130)	140/158	110/150
4.	Type 4	(98 - 120)	140/158	111/150

Table 2

THERMAL PARAMETERS FOR THE DEHYDRATION AND DECOMPOSITION OF THE AGED (WEATHERED) PAINT SAMPLES. (SAMPLING 180 DAYS)

No	Samples Paint Aged	Temperature Range $(T_i - T_f)$ C	Peak Temperatures (C) DTA	DTG
1.	Type 1	(100 - 140)	150/180	130/180
2.	Type 2	(115 - 195)	160/200	135/190
3.	Type 3	(105 - 155)	150/190	135/175
4.	Type 4	(110 - 135)	140/180	130/180

For the purpose of investigation of aging processes in the paint samples it seems to be suitable to use, such methods which compare the questioned aged sample with the unaged one by direct means. This technique is particularly valid for the DTA measurements of the samples. From the measurements of the exothermic peak areas it can be observed that not only the oxidation of the paint samples but also the greater heat evaluation by the oxidation of the questioned samples in comparison with the unaged paint can be characterised by the alteration of the quantity of the evolved heat indicated by the first exothermic peak of the DTA curve.

The oxidation suffered by the paint samples were also determined by the dielectric loss factor (f) measurements (f= 1Kcps freqency). The results of dielectric loss factor variation have the same tendency as it has observed during the investigation of the excess heat.

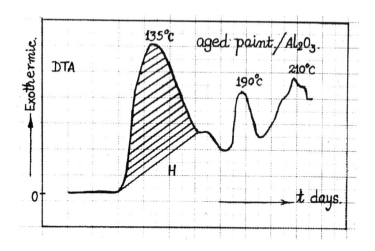

Figure 5
DTA curve of the aged (weathered) paint.
Reference material is Aluminium Oxide.

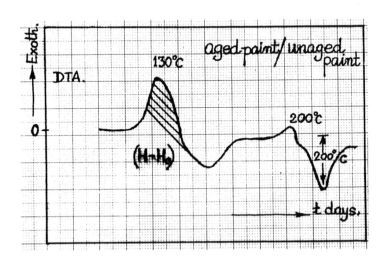

Fugure 6
DTA curve of the aged (weathered) paint.
Reference material is unaged fresh paint.

ACKNOWLEDGEMENTS

The author is grateful to the Director, Forensic Science Laboratory, Government of West Bengal, Calcutta, for his kind interest in this work. The author is also grateful to M/S Hindusthan Motor Co., Calcutta for supplying paint samples to this laboratory for present investigation.

REFERENCES

1. Pue Kolonits and G.Liptay, Hiradastech.Ipari Kut. Kozlemen, Budapest, 1963, 3, 63.
2. D.Marjit, Abstract cum Souvenir of the 2nd National Conference on Thermal Analysis, BARC, Bombay, 1980, 9.
3. N.H.Sze and G.T.Meaden, J.Therm.Anal., 1973, 5, 533.
4. E.A.Zelenyanszki and P.K.David, Thermal Analysis: Proceedings 4th ICTA, Budapest, 1974, 3, 329.

THERMOGRAVIMETRIC SIGNATURES OF COMPLEX SOLID
PHASE PYROLYSIS MECHANISMS AND KINETICS

Michael J. Antal, Jr.
Coral Professor of Renewable Energy Resources
Department of Mechanical Engineering
University of Hawaii
Honolulu, Hawaii 96822

INTRODUCTION

Pyrolysis plays a critical role in the thermochemical conversion of biomass materials such as wood, agricultural wastes, municipal solid wastes, and various animal wastes, into more useful fuels and chemicals. Pyrolysis also plays an important role in the gasification of coal, the retorting of oil shale, and most combustion processes. Consequently, there is some incentive to develop a better understanding of the detailed pyrolysis chemistry occurring in the thermal environment of conversion reactors, combustors, and related process equipment.

Much effort has been devoted to the study of simple pyrolysis mechanisms governed by the rate equation

$$\frac{dV}{dt} = k(V^*-V)^n \tag{1}$$

where $V = 1 - \frac{w(t)}{w_i}$, $V^* = 1 - \frac{w_f}{w_i}$, and $w(t)$ is the time dependent sample weight, w_i the initial sample weight and w_f the final sample weight, k is the Arrhenius rate factor and n the order of the reaction. Usually $k = A \exp(-E/RT)$ where A is the pre-exponential constant, E the activation energy, R the Universal Gas Constant, and T the sample temperature.

Thermogravimetry (TG) is often used to experimentally determine the values of A, E and n for a given material. The kinetic interpretation of thermogravimetry data involves certain mathematical complexities because the sample temperature T is usually chosen to be a linear function of time; thus $T = \alpha+\beta t$. Sestak et al [1] give an excellent review of methods for the kinetic interpretation of thermogravimetric data. Unfortunately, most of the techniques assume that pyrolysis can be characterized as a simple reaction mechanism governed by the rate equation (1). With the exception of the recent research of Flynn [2], more complex reaction mechanisms have been treated in only a cursory manner. The goal of research described in this paper is to develop a better understanding of complex pyrolysis mechanisms and kinetics through the study of selected model problems. The following sections describe the mathematical models studied to date, the methods of analysis, and early results.

MATHEMATICAL MODELS OF COMPLEX PYROLYSIS MECHANISMS

Table 1 lists various combinations of consecutive and competitive reaction mechanisms which may be used to describe the temperature dependent weight loss of a material undergoing pyrolysis. Although Table 1 only includes mechanisms with up to three steps, its extension to more complex mechanisms is straightforward.

For each of the mechanisms listed in Table 1, the appropriate rate equations describing weight loss as would be measured using thermogravimetry have been written down and solved numerically. For example, the rate equations governing the two reaction/competitive mechanism are

$$\frac{d\rho_A}{dt} = -k_1(\rho_A)^{n_1} - k_2(\rho_A)^{n_2} \quad (2)$$

$$\frac{d\rho_B}{dt} = s_1 \, k_1(\rho_A)^{n_1} \quad (3)$$

$$\frac{d\rho_C}{dt} = s_2 \, k_2(\rho_A)^{n_2} \quad (4)$$

where ρ_A, ρ_B, and ρ_C are the densities of solid reactant A and solid products B and C, s_1 and s_2 are stoichiometric coefficients giving the fractional weight of solid reactant A remaining as solid products B and C (respectively), and the k_i are rate factors as before. Here it is assumed that the volume of the sample remains constant during pyrolysis, so that the densities of A, B and C are directly related to the sample weight. Future research may modify this assumption by incorporating volume change into the rate equation. Solving Equations (2) - (4) numerically with an assumed linear heating rate produces a simulated TG curve for a material undergoing pyrolysis by the indicated mechanism. Values of the k_i, n_i and s_i used to generate the simulated weight loss curves are given in Table 2 and on Figures 1 - 5. The question of interest is to determine whether a set of simulated TG curves at various heating rates provides sufficient information to infer the mechanism and associated rate data which were used to construct the TG curves.

METHODS OF ANALYSIS

Although many methods are available for the kinetic analysis of TG curves, the author's experience with cellulose pyrolysis [3] lead to the emphasis of multiple heating rate methods in this research. These methods are attractive since they naturally synthesize the results of several (perhaps many) experiments. Because they do not assume that a single reaction dominates an entire TG curve, they appear to be especially well suited for dissecting the intricacies of complex pyrolysis mechanisms. As described in the following paragraph, two differential multiple heating rate methods have been used here.

The Friedman multiple heating rate method [4] assumes Equation (1) to be applicable to a set of TG curves spanning a narrow range of heating rates. If the values of ln(dV/dt) are plotted for various heating rates as a function of 1/T, where the values of dV/dt are each associated with the same degree of volatilization V(t) (i.e.

TABLE 1
REPRESENTATIVE REACTION MECHANISMS

Description	Pictorial Representation	Symbol
One Reaction/Simple	$A \xrightarrow{1} B$	AB
Two Reaction/Parallel	$A1 \xrightarrow{1} B1$ $A2 \xrightarrow{2} B2$	2AB
Two Reaction/Consecutive	$A \xrightarrow{1} B \xrightarrow{2} C$	ABC
Three Reaction/Consecutive-Competitive	$A \xrightarrow{1} B \begin{smallmatrix} \xrightarrow{2} C \\ \xrightarrow{3} D \end{smallmatrix}$	ABCBD
Three Reaction/Competitive-Consecutive	$A \begin{smallmatrix} \xrightarrow{1} B \\ \xrightarrow{2} C \xrightarrow{3} D \end{smallmatrix}$	ABACD

TABLE 2
VALUES OF THE RATE CONSTANTS USED TO GENERATE
THE SIMULATED WEIGHT LOSS CURVES

	E	A
k_1	35 kcal/gmol	5.42×10^7 s^{-1}
k_2	40	3.50×10^9
k_3	45	2.25×10^{11}

V(t) is a constant for a set of values (dV/dt)), then a straight line connecting the points has a slope of -E/R. Furthermore, if the values of ln(dV/dt) for various heating rates are plotted as a function of ln(V*-V(t)), where the values of dV/dt are each associated with the same value of T, then a straight line connecting the points has slope n. Thus a graph of E vs V(t) and n vs T are obtained using the multiple heating rate method of analysis from a set of TG curves spanning several heating rates. These graphs serve as useful signatures of complex reaction mechanisms since their shape varies according to the prevalent pyrolysis chemistry. Moreover, the values of E and n are often good first guesses for the values of the kinetic parameters associated with the various mechanisms. Research described here emphasizes the first method of analysis, which was found to be more useful.

RESULTS AND DISCUSSION

A Friedman analysis of simulated weight loss curves derived from the simple AB mechanism always results in a straight line parallel to the x axis for graphical displays of E vs 1-V and n vs T. Roundoff errors can lead to a loss of precision in the third significant figure of E. Only variations in excess of 1 kcal/gmol in the value of E should be considered significant.

Figures 1 - 3 display the signatures of the mechanisms 2AB, ABC and ABAC. The 2AB signature is characterized by two broad plateaus separated by an abrupt transition region. The lower temperature plateau is associated with the higher activation energy, as would be expected. The signature shifts to the right at higher heating rates, evidencing the rate limiting nature of the lower activation energy step. The signature of the ABC mechanism is characterized by many less broad plateaus with no abrupt transition regions. As before, lower temperatures are associated with higher values of E, and higher heating rates shift the signature to the right. By way of contrast, the competitive ABC mechanism shows a very smooth shift to higher activation energies at higher temperatures, and the usual shift to the right at higher heating rates. Of course, a clear manifestation of competitive mechanisms is a decreasing (or increasing) char residual at higher heating rates. In all three cases, values of the apparent E were intermediate to the actual values for most of the reaction sequence. However, towards the end of the reaction sequence the ABC mechanism customarily evidences rapidly increasing (or decreasing) values of E which exceed those actually used in the calculation.

Figures 4 and 5 display signatures of the more complex reaction sequences ABCBD and ABACD. The U shaped signature of ABCBD reflects the rate limiting high activation energy competitive steps at low temperatures shifting to the rate limiting low activation energy initiation step at higher temperatures, shifting back to the high activation energy steps following the consumption of reactant A. The signature shifts to the right at higher heating rates. The sequence ABACD evidences both the plateau structure typical of consecutive reactions, and the rapidly increasing values of E at the end of the reaction sequence typical of competitive reactions. Unlike earlier examples, this signature shifts downwards at higher heating rates.

CONCLUSIONS

The five mechanisms discussed in this paper exhibit relatively distinct signatures when analyzed by the Friedman multiple heating rate method. Recent research, which has identified the effects of variations in the parameters A, E, n, s and heat transfer on the signature, will be described in a forthcoming publication. The technique of signature identification has already shown its utility in a study of sewage sludge pyrolysis [5], and is presently being used in this laboratory to study and identify the complex pyrolytic mechanisms of a variety of natural fuels.

ACKNOWLEDGEMENTS

The author would like to express his thanks to Dr. Henry Friedman, past president of NATAS, for sharing his wisdom on the subject of pyrolysis chemistry so freely over the past years.

This research was supported by U.S.D.O.E. under DOE/Battelle Subcontract No. B-C5822-A-Q. The author wishes to thank Drs. B. Berger, D. Stephens, and Mr. S. Friederick for their interest in this work.

REFERENCES

1. J. Sestak, V. Satava and W. Wendtlandt, Thermochim. Acta, 1973, 7, 333.
2. J.H. Flynn, Thermochim. Acta, 1980, 37, 225.
3. M.J. Antal, H.L. Friedman and F.E. Rogers, Comb. Sci. and Tech., 1980, 21, 141.
4. H.L. Friedman, J. Polym. Sci., 1964, C6, 185.
5. D.L. Urban and M.J. Antal, to appear in FUEL.

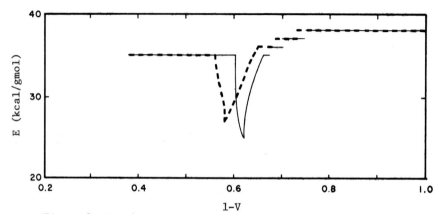

Figure 1 Friedman signature of mechanism 2AB (--- 1,2 and 5°C/min; —— 10,20, and 50°C/min; $n_1=1.$, $n_2=0.5$; $s_1=0.5$, $s_2=0.0$)

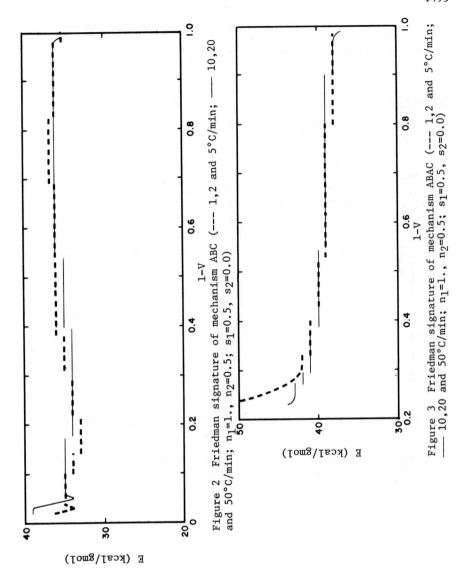

Figure 2 Friedman signature of mechanism ABC (--- 1,2 and 5°C/min; —— 10,20 and 50°C/min; $n_1=1$, $n_2=0.5$; $s_1=0.5$, $s_2=0.0$)

Figure 3 Friedman signature of mechanism ABAC (--- 1,2 and 5°C/min; —— 10,20 and 50°C/min; $n_1=1$, $n_2=0.5$; $s_1=0.5$, $s_2=0.0$)

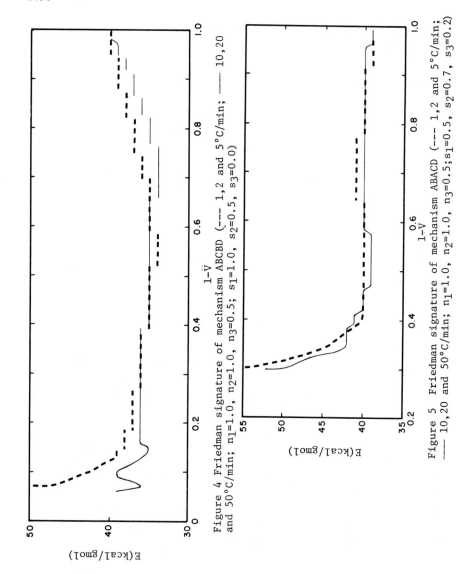

Figure 4 Friedman signature of mechanism ABCBD (--- 1, 2 and 5°C/min; —— 10, 20 and 50°C/min; $n_1=1.0$, $n_2=1.0$, $n_3=0.5$; $s_1=1.0$, $s_2=0.5$, $s_3=0.0$)

Figure 5 Friedman signature of mechanism ABACD (--- 1, 2 and 5°C/min; —— 10, 20 and 50°C/min; $n_1=1.0$, $n_2=1.0$, $n_3=0.5$; $s_1=0.5$, $s_2=0.7$, $s_3=0.2$)

CHARACTERIZATION OF TECHNICAL PRODUCTS
BY AUTOMATED THERMAL ANALYSIS

Hans G. Wiedemann and G. Widmann
Mettler Instrumente AG, CH-8606 Greifensee / Switzerland

Research and industry are making more and more use of
thermal analysis. This is mainly due to the development
of data processing and the wide spread applications of
thermal methods as well as the possibility of combination of methods, such as thermogravimetry (TG), differential scanning calorimetry (DSC), gaschromatography (GC),
masspectrometry (MS), etc.

A short review of the most important thermoanalytical methods is followed by a discussion of a number of applications which have been selected from research, process and
production control in industry.

Of special importance is the recent trend to automation
of thermoanalytical measurement methods. This is comparable to similar developments which took place for chemical analysis. New TA equipment, such as the METTLER
TA3000 system, is ideally suited for routine analysis
in industrial laboratories. Microprocessor control allows
fully automatic analysis including numeric evaluation
and plotting of the data.

The lateness of this development from manually operated
instrumentation towards automation is undoubtedly due to
the complexity of the experimental conditions and the variety of the sample properties.

The new METTLER system TA3000 allows, without any prior
special knowledge in thermal analysis, automatic and rapid determinations of reaction enthalpy, purity, glass
transition, specific heat and parameters of reaction kinetics. Methods supported are differential scanning calorimetry (DSC), thermo-mechanical analysis (TMA) and thermogravimetry (TG).

THERMAL EXPANSION AND KINETIC OF DECOMPOSITION OF CALCITE

The following example shows that the system TA3000 in
connection with the TMA module 40 (TMA = thermomechanical analysis) can be used successfully for investigations of the thermal expansion of minerals. The "layer-type" structure of calcite and dolomite is responsible
for the strong anisotropy of physical properties, in-

	TMA	Heating X-ray
βa =	$-6.1 \times 10^{-6}/°K$ ± 2.3	$-5.7 \times 10^{-6}/°K$ ± 0.2
βc =	$+25.5 \times 10^{-6}/°K$ ± 1.5	$+26.5 \times 10^{-6}/°K$ ± 1.0

Table 1

Fig. 1 TMA cell with calcite crystal inserted in different directions: A) c-axis, B) a-axis

cluding thermal expansion. This means that if calcite is heated up, it contracts in one crystallographic direction whereas it expands in the other.

The "negative expansion" of calcite in the a-direction has been attributed to a rotation of the CO_3- groups out of their planes which leads to a decrease of the C-O distance projected on that plane during heating up. Calcite has the highest c/a ratio (and smallest rhombohedral angle) of all the carbonates. This means less dense packing of the Me and CO_3 layers as compared to the other carbonates and therefore a greater possibility for the CO_3 groups to rotate out of their planes [1].

Table 1 lists thermal expansion of calcite. Results drawn from x-ray photographs can be compared with TMA-mean values. The latter were evaluated automatically by the apparatus after the measurements (see fig. 2).

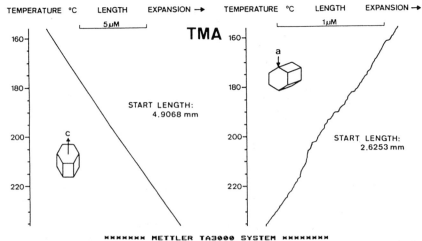

Fig. 2 Thermal expansion of calcite in direction of the c- and a-axis.

The x-ray measurements took about two days whereas the TMA measurements and evaluations were completed in less than three hours.

The decomposition of calcium carbonate measured by thermogravimetry is shown in Fig. 3. The evaluation is performed by means of our multiple linear regression analysis method. The basic equations are those of Willhelmy and Arrhenius. The primary results of kinetic investigations are the kinetic data: order of reaction, activation energy and frequency factor as well as their confidence limits computed for a probability of 95 %. These primary results are not of great practical use but they allow the mathematical description of the chemical reaction, the applied kinetics: degree of conversion as a function of the reaction time at any isothermal reaction temperature (Fig. 4) [2].

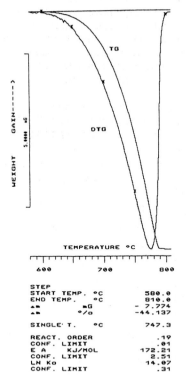

Fig. 3 TG curve of calcium carbonate. Sample weight 17.61 mg, heating rate 10 K/min

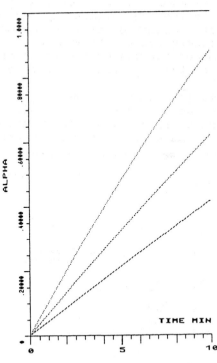

Fig. 4 Conversion as a function of reaction time for 3 isothermal temperatures of 700, 720 and 740°C

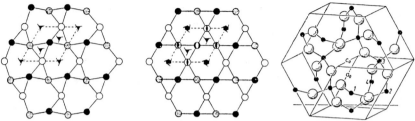

Fig. 5 High and low temperature modification of quartz Si atoms in Fig. 6 High quartz (0001) projection

β-α Quartz [3]

In thermal analysis DSC-measurements of the heat of transformation of the β-α phase transition (see figure 5) are used for the identification of quartz (e.g. β-α quartz in bauxite). The phase transition at 573° can also be used to calibrate the temperature sensors of DSC-apparatus. The structural transition from low quartz to high quartz at 573° C is only due to a change of the bond angle of the linked SiO_4-tetraeder (see fig. 6). This change lowers the symmetry from the hexagonal to the trigonal-trapezoidal class. The corresponding space groups are D_3^4-P3 21 and D_3^6-P3 21, respectively (see figure 5).

The transition can also be observed with TMA-measurements. Figure 7 shows an expansion from room temperature up to 560° C. The expansion stops immediately after the change in the bond angle of the linked SiO_4 tetraeder. These results explain in an unorthodox way why quartz glass is temperature shock proof. Reproducible curves can be obtained only after the samples have been heated up and cooled down for 2 to 3 times.

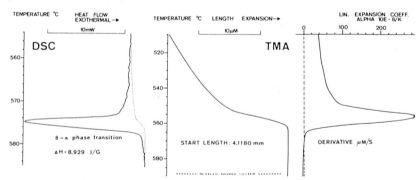

Fig. 7 DSC- and TMA-curves of the β-α transformation of quartz

THERMOANALYTICAL INVESTIGATIONS OF ZIRCON

The relative instability of the zircon structure is sometimes used as an example for the explanation of the decay of a structure under radioactive radiation. This process is called "metamictization". After radiation the structure leaves the orderly state, the material becomes amorph or vitreous [4].

By gemstones of zircon are such metamicte variants known and show sometimes weak radioactivity. Probably radioactive elements, like uranium are built-in, in the structure of zirkon or are deposited as inclusions. The crystal structure of zircon is shown in fig. 8. As in all silicates, silicon is surrounded by a tetrahedron of four oxygen atoms. Zirconium, on the other hand is surrounded by a trigon-dodecahedron of eight oxygen atoms. These SiO_4 tetrahedrons and ZrO_8-dodecahedrons are linked to chains which run parallel to the z-axis.

With the metamictization this structure of zircon is distroyed. This leads to lower specific weight and smaller refraction of light.

Natural zircon can be found in various different states, between normal crystalline structure and complete metamicte. Recrystallization of metamicte mineral (e.g. zircon) can be measured e.g. with a thermobalance, while the sample is heated up linearly (see fig. 9a).

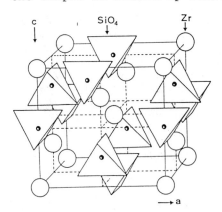

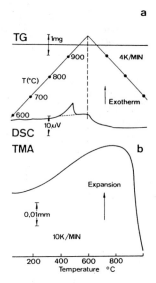

Fig. 8 Crystal structur of zircon

Fig. 9 TG, DSC (a) and TMA curves (b) of metamicte zircon

The weight curve (TG) shows in this case no changes while in the DSC-curve an exotherme peak (between 800 - 900° C) appears which corresponds to the heat of recrystallization. Fig. 9b shows the TMA-curve of the recrystallization process. The crystal was inserted into the measuring head and heated up linearly (10 K/min). Up to 800° C the sample expanded as expected. Then recrystallization lead to a contraction of the crystal which is finished at ~ 1000° C.

Thermoanalytical investigation of Carbohydrates

Thermal analysis is used to investigate the pyrolysis of the main components of hardwood, xylan and cellulose.
F. Shafizadeh [ ] used model compounds consisting of α-D-xylose, substituted phenyl β-D-xylopyronosides, and β-D-glucopyranosides etc.
At lower temperatures these molecules displayed anomerization, loss of water and phase change. Investigation of the 1,6 anhydro-β-D-glucopyranose (s compound) with the Mettler TA3000 DSC (fig. 10) shows a solid state transition of the molecule at 113°C and melting at 180°C. This is followed by pyrolysis at higher temperatures not shown here. Anhydro sugars such as 3,6-anhydro-D-glucose (5), 1,6-anhydro-β-D-glucofuranose (5), and 1,4-anhydro-β-D-glucopyranose (5) melt at 122°, 111°, and 85°C, respectively. In the same region, 1,6-anhydro-β-glucopyranose (levoglucosan) undergoes a solid state transition and melts at a much higher temperature than could be expected.
When a crystalline compound melts, the molecules of the lattice are set free. Parts of the molecule, which were previously restrained in the lattice from reorienting, gain freedom, too. The resulting change in entropy can be measured. Since the entropy of fusion is much lower if the molecules have freedom of rotation and are randomized in the solid state, compounds showing an entropy of fusion of less than 5 cal/mole degree are considered to have a plastic crystal form [6].

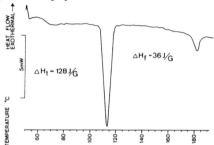

Fig. 10 DSC-curve of 1,6 anhydro-β-D-glucopyranose

SOME MEASUREMENTS ON EDIBLE FATS

An other field of application is the analysis of the melting behavior of fats and oils such as palm oils. Palm oils of different origin and fractionated palm oils like palm olein or palm stearin have a different melting and crystallization behavior. The DSC fusion curves are integrated continuously and the percent integral is plotted as "Liquid Fraction" versus sample temperature (Fig. 11).

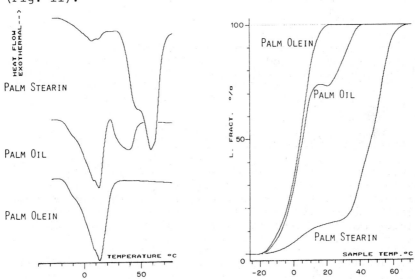

Fig 11 Left: DSC fusion curves of palm oil and fractionated palm oils (sample weight: 30 mg, heating rate: 10 K/min)
Right: Liquid fraction curves computed from fusion curves by integration

REFERENCES

1. G. Bayer, Z. Krist., 1971, 133, 85
2. H. Wyden and G. Widmann, "Angewandte chemische Thermodynamik und Thermoanalytik", Birkhäuser, Basel, (1979), p. 284
3. H. Strunz, "Mineralogische Tabellen", Akademische Verlagsgesellschaft, Leipzig (1970) p. 46
4. G. Bayer and H.G. Wiedemann, Chemie in unserer Zeit, 1981, 15, 88
5. F. Shafizadeh, J. Polymer. Sci.: Part C, 1971, 36, p. 2
6. J. Timmermans, J. Chim. Phys., 1938, 35, p. 331, Phys. Chem. Solids, 1961, 18, p. 1

CRYSTAL TRANSFORMATION IV→III KINETICS OF AMMONIUM NITRATE IN THE LIME AMMONIUM NITRATE

G.Rasulić
Control and Research Department
Chemical Industry"Pančevo"-Pančevo
Yugoslavia

INTRODUCTION

The granulated ammonium nitrate presents, besides urea, the most utilized nitrogen fertilizer nowdays. At storing or transportation the salt loses its flowability, passes into a monolithic mass and becomes, without previous processing, practically useless as a fertilizer.

By many authors the ammonium nitrate crystal transformation from the rhombic/IV/ into monoclinic/III/ form/further in the text: transformation IV→III/ is considered as one of the causes of this phenomenon. It appears at $32,3^oC$ according to the literature cited, and is followed by the specific volume change.

The alternating transformations of rhombic into monoclinic ammonium nitrate cause granules destruction, formation of fine crystal powder that fulfills the space between non-destructed granules, which enables their mutual connection and monolithic mass formation.

In our country, as well as in most European countries, a granulated mixture of ammonium nitrate and calcium carbonate, instead of technical ammonium nitrate is used as a fertilizer. The ground limestone is introduced into ammonium nitrate melt, the mixture of which is granulated after homogenizing.

The subject of these investigations has been the influence of calcium carbonate on the crystal transformation IV→III of ammonium nitrate in the lime ammonium nitrate.

PROBLEM ANALYSIS

Erofeev and Mickevič/1-4/ and Wolf and Scharre/5,6/ dealt with crystal transformation IV→III kinetics of pure ammonium nitrate. All their investigations were carried out under isothermal conditions by dilatometer, the construction of which was different, as well as dilatometer fluid. Both groups of authors used KEKAM equiation for experimental results treatment. The kinetic order of transformation according to their results varied from 2-3 in dependence of the previous preparation of the sample for investigation, more exactly on the new phase centers presence.

Erofeev and Mickevič could not express the relationship between the kinetic constant "k" and temperature T, and they concluded that Arrhenius's equation could not be applied for temperature dependence description of ammonium nitrate crystal transformation IV→III rate constant. Sakovič/7/ has shown, by their experiment results, that if one does not start from absolute null, as Erofeev and Mickevič had done, but from transformation start temperature, as the reaction rate for it is equal to null, Arrhenius's equation becomes applicable. He has determined activation energy and it amounts 36,13 J/mol. Wolf and Scharre have determined activation energy in the same way and they have got 38,94 kJ/mol., although both groups of authors used Kolmogorov-Erofeev's equation for the experimental results treatment.

Sneerson/8/ has investigated the influence of calcium and magnesium nitrate on the ammonium nitrate crystal transformation III→IV, using first mercury and then xylene as the dilatometric fluid. He has remarked the great change of rate constant with the change of dilatometric fluid. According to his opinion, the isothermal investigation of ammonium nitrate transformation by dilatometer and such determined rate constant and activation energy, can present only a relative indicator of the kinetic process.

Novikova at all./9/ used Erofeev's equation for nonisothermal investigations of crystal transformation I→II and II→IV ammonium nitrate by DTA. The determined kinetic order amounts 0,7-1,7 and activation energy was not determined.

We have chosen DTA for ammonium nitrate crystal transformation IV→III investigation. The aim of this paper is not to discuss about DTA application for kinetic study of solid-state processes like this and about real meaning such determined kinetic parameters. We wanted to investigate influence of calcium carbonate on the crystal transformation IV→III ammonium nitrate in the lime ammonium nitrate. For this investigation we used as a criterion kinetic parameters calculated according to the methods or equations proposed for determination of kinetic data from DTA curve.

EXPERIMENTAL RESULTS AND THEIR DISCUSSION

All our investigations of the ammonium nitrate crystal transformation IV→III have been carried out on the Derivatograph, at heating rate of $2,5°C/min.$, DTA sensitivity 1/1, weight of sample about 600 mg. and with small platinic crucible.

In order to explain the influence of calcium carbonate on ammonium nitrate crystal transformation in lime ammonium nitrate the following samples have been prepared: to the 95% ammonium nitrate solution of $140^{\circ}C$ temperature the ground limestone - calcium carbonate- has been added in such amount as to provide ever the mixture with the content of NH_4NO_3 of 77,14% /calculated to the waterless sample/ but of different grain size. The sample has been homogenized for ten minutes on this temperature, then rapidly cooled and dried over conc.sulfuric acid, up to the moisture content of about 0,30%. As the reference, the sample p.a. ammonium nitrate, that had the same treatment,except that calcium carbonate has not been added to it, has been used. The review of the prepared samples has been given in the Table 1.

TABLE 1. THE REVIEW OF THE PREPARED SAMPLES

internal mark	NH_4NO_3 % wt.	$CaCO_3$ % wt.	moisture % wt.	grain size of added $CaCO_3$ μm	
I	76,92	22,80	0,28	- 43	
II	76,92	22,79	0,29	- 74	+ 43
III	76,86	22,78	0,36	-149	+ 74
IV	76,91	22,79	0,30		+149
V	99,70	-	0,30		

The results treatment and kinetic parameters calculation have been carried out according to Borchardt and Daniels/10/; Pilojan/11/; to Davies at all./12/;by applying KEKAM equation/13/ and modified KEKAM equation/14/ According to Kissinger/15/ and Balarin/16/ reaction order from peak shape has been calculated. According to Davies's method we try to determine the kinetic parameters and the mechanism of crystal transformation by supposing topochemical reaction in the solid, which can be expressed by Avrami-Erofeev equation. The calculation has been carried out for n = 1/3; 1/2; 3/4; 1; 4/3; 2 and 3.

All experimental results have been shown in the Table 2. as the average from the two analysis carried out under the same conditions.

According to Kissinger, from DTA peak shape, the determined reaction order of ammonium nitrate crystal transformation IV→III/column 2./ varies from 2,74 for the sample of lime ammonium nitrate with the finest grains of limestone to 1,75 for the sample of pure ammonium nitrate. The reaction order shows a clear decreasing tendency with the added limestone grains increase. If the limestone grains are larger they obviously disturb less the crystal lattice of ammonium nitrate. The reaction order is in accordance with the results of isothermal experiments by Erofeev and Mickevič's for pure ammonium nitrate/1-4/.

According to Balarin the asymmetry of DTA peak shape can be considered in the form of ratio of the areas under the curve above and below the peak maximum. So determined kinetic orders have been shown in column 3. The value for asymmetry varies from 5,66 to 2,07, which is a great discrepancy, but the results have shown the same tendency. We find Kissinger's rule and results in column 2 more correct.

According to KEKAM equation, the value for "n" shows fairly good agreement with the reaction order determined from DTA peak shape according to Kissinger, both in value and in the decreasing tendency with limestone grains increase towards pure ammonium nitrate. The rates constant "k" calculated from the value "n" and "K" shows a distinct decreasing tendency from the finest limestone grains to the pure ammonium nitrate/columns 4,5 and 6/.

Borchardt and Daniels's method application points out that the crystal transformation reaction of "0" or the first order, which is neither in accordance with the previous investigations/1-6/, nor with the reaction order, which we have determined and shown in the columns 2 or 3. The activation energy for n=1, shows a decreasing tendency from 544,55 to 443,16 kJ/mol. with a limestone grains growth and it is the smallest for pure ammonium nitrate/columns 7 and 8/.

For the reaction order 2 and 3 in the plot ln k towards 1/T distinct curved dependences are obtained, which exclude the possibility that the reaction is the second or the third order.

Modified KEKAM equation application according to Criado, shows the rectilinear dependences in the system $\ln/-\ln(1-\alpha)/-2n \ln T$ towards 1/T for n= 0; 1; 2 and 3. The activation energy determined from these dependences, with two exceptions, shows the decreasing tendency with the increase of limestone grains size to pure ammonium nitrate. The activation energy varies from 740,41 to 526,10 kJ/mol. without great difference with the change of reaction order.

The activation energy determined according to Pilojan shows a distinct decreasing tendency with the increase of limestone grains size towards pure ammonium nitrate. According to value, the activation energy determined by this method, is in accordance with the activation energy value determined by Borchardt and Daniels and by modified KEKAM equation /column 13/.

Davies's method application excluded the possibility that reaction order of this transformation is less than the first or about the first order, for rectilinear dependences were obtained only for the second and the third for the supposed reaction mechanism in accordance with KEKAM equation. The activation energy has been determined on the basis of the data processed, excluding the two experimental points with $\alpha < 0,02$ /columns 14 and 15/, and using all experimental points /columns 16 and 17/. The activation energy, with small exceptions, decreases with the limestone grains increase to pure ammonium nitrate, in both treatments.

CONCLUSION

According to the investigations of crystal transformations of rhombic into monoclinic ammonium nitrate kinetics in the lime ammonium nitrate, by applying the methods of Borchardt and Daniels, Pilojan, Davies, Kissinger, Balarin, KEKAM equation and its modification according to Criado, for DTA curve treatment, the following can be concluded:
1. The presence of limestone, that is calcium carbonate, influences on the ammonium nitrate crystal transformation IV→III in the lime ammonium nitrate, by increasing the reaction order of the transformation as well as the activation energy.
2. With the reduction of the added limestone grains the reaction order and the activation energy of ammonium nitrate transformation IV→III increase in the lime ammonium nitrate.

REFERENCES:

1. B.V.Erofeev,N.I.Mickevič, ŽFH,1950, 24,No.10,1235
2. B.V.Erofeev,N.I.Mickevič, ŽFH,1952, 26,No. 6, 846
3. B.V.Erofeev,N.I.Mickevič, ŽFH,1952, 26,No.11,1631
4. B.V.Erofeev,N.I.Mickevič, ŽFH,1953, 27,No. 1, 118
5. F.Wolf, W.Scharre, Zeszyty Naukowe Universytetu im. A.Mickiewicza, Poznaniu,Mat.fiz.Chem.,1967,No.65, 3
6. F.Wolf, K.Benecke, H.Fürtig,Z.Phys.Chem.,DDR, 1972, Bd.249,No.3/4,274
7. G.V.Sakovič, ŽFH,1959, 33, No.3, 635
8. A.L.Sneerson,V.A.Klevke,M.A.Minovič,ŽPH, 1956, 29, No.5, 682
9. O.S.Novikova,Ju.V.Cehanskaja,O.I.Titova,T.I.Sontorenko, ŽFH, 1977, 51, No.1, 257
10. H.J.Borchardt,F.Daniels,J.Am.Chem.Soc.,1957, 79, 41
11. G.O.Pilojan,I.D.Rjabčinikov,O.S.Novikova, Nature, 1966, 212, 1229
12. P.Davies,D.Dollimore,G.R.Heal, J.Therm.Anal. 1978, 13, 473
13. V.M.Gorbaev, J.Therm.Anal. 1978, 13, 509
14. J.M.Criado,J.Morales, Thermochim.Acta, 1976, 16,382
15. H.E.Kissinger, Anal.Chem.,1957, 29, 1702
16. M.Balarin, Thermochim. Acta, 1979, 33, 341

TABLE 2. THE KINETICS PARAMETERS DETERMINED ACCORDING TO DTA CURVE

Intern. mark	Reaction order		KEKAM equation		
			n	K	k
1	2	3	4	5	6
I	2,74	5,66	2,83	0,04987	0,9810
II	2,18	4,17	2,67	0,02990	0,7171
III	2,19	3,58	2,57	0,02472	0,6091
IV	1,99	3,10	2,40	0,03144	0,5678
V	1,75	2,07	2,31	0,02681	0,4822

Intern. mark	Activation energy kJ/mol according					
	Borchardt-Daniels		modified KEKAM equation			
	n=0	n=1	n=0	n=1	n=2	n=3
	7	8	9	10	11	12
I	318,73	544,51	692,88	890,86	740,41	712,69
II	-	540,45	831,47	731,69	731,69	681,81
III	115,52	436,52	631,92	798,21	598,21	598,66
IV	177,39	448,07	634,95	621,34	574,47	526,10
V	175,20	443,45	626,80	592,26	562,26	630,64

Intern. mark	Activation energy kJ/mol according				
	Pilojan	n=2	Davies at all. n=3	n=2	n=3
	13	14	15	16	17
I	653,30	369,54	230,96	426,47	294,17
II	530,06	336,68	228,03	417,14	287,18
III	507,20	284,04	204,09	370,42	255,57
IV	490,56	282,70	205,94	376,15	244,07
V	489,64	273,87	214,63	351,85	245,76

DTA STUDIES ON CHALCOPYRITE - COPPER SULPHATE CONVERSION

M. Aneesuddin, P. Narayana Char and E.R. Saxena, Regional Research Laboratory, Hyderabad - 500 009 INDIA.

INTRODUCTION

DTA curves of sulphide minerals of copper including chalcopyrite indicate the formation of copper sulphate in the normal reaction sequence [1]. DTA studies on chalcopyrite - Copper sulphate conversion were undertaken to: (1) fix the temperature and duration of roasting the ore to get maximum conversion to copper sulphate, (2) ascertain the amount of copper sulphate formed when oxidation is complete, and (3) detect the presence of unconverted sulphide in the roasted ore and thus establish DTA as a good quality control tool in the process of direct conversion of chalcopyrite to copper sulphate by roasting the ore.

EXPERIMENTAL

Chalcopyrite concentrates from Chitradurga, Karnataka State in South India selected for the present work contain Cu 25%, Fe 27.90% and S 26.92%.

DTA

DTA was done using Leeds & Northrup unit with pt-pt 10% Rh thermocouples and Robert Grimshaw type ceramic sample holder. Calcined alumina was used as the thermally inert reference material. Rate of heating was maintained at $12\frac{1}{2}$°C per minute. For each run 0.3 g (-300 mesh BSS) roasted ore was taken.

Determination of the amount of copper sulphate formed in ores roasted at different temperatures and for different durations required the preparation of a caliberation curve.

DTA curve of $CuSO_4$ 5 H_2O (Fig.1) shows three endothermic peaks in the temperature range 90°C-330°C ascribed to melting, boiling and dehydration of the hydrated form. The 4th and 5th endothermic peaks starting at 690°C and ending below 900°C are ascribed to decomposition of copper sulphate to cupric oxysulphate and to cupric oxide respectively[1]. The 6th peak above 1000°C is due to the reduction of CuO to Cu_2O, this peak is reversible as observed in the cooling cycle. The 5th endothermic peak ascribed to the formation of CuO has been selected for preparation of caliberation curve for anhydrous copper sulphate. Peaks between 90°C-330°C are for the hydrated form, whereas copper sulphate formed during roasting is anhydrous, the possibility of the 4th peak at 690°C getting merged with the decomposition peak for ferrous sulphate (Fig.1) and disturbing its magnitude cannot be ruled out, therefore, these peaks have not been considered in the preparation of caliberation curve.

Since Fe_2O_3 is the major insoluble impurity formed during roasting of chalcopyrite, DTA curves of different amounts of $CuSO_4$ 5 H_2O diluted with Fe_2O_3 were taken and a caliberation curve (Fig.2a) was prepared by plotting amplitude of the 5th endothermic peak in inches against amount of $CuSO_4$ 5 H_2O in mg.

Colorimetric determination of copper sulphate

To verify the accuracy of the data obtained through DTA, amount of copper sulphate formed during roasting was also determined colorimetrically and a caliberation curve drawn (Fig.2b) using klett summerson photo electric colorimeter.

Sample preparation

3g (-200 mesh BSS) sample was taken in a gooch crucible with ceramic beads distributed uniformly throughout the sample for accessibility of air and heated in atubular furnace, (25 cm length and 5 cm diameter). The product was ground to pass 300 mesh BSS. 0.3 g of the product was subjected to DTA and colorimetric analysis.

RESULTS AND DISCUSSION

DTA curve of natural sample (Fig.1) indicates oxidation in the temperature range 375°C to 610°C, chalcopyrite ore was therefore roasted at temperature intervals of 50°C right from 350°C to 700°C for durations ranging from 30 minutes to 180 minutes. The temperature range 500°C-700°C was studied in greater detail as the formation of $CuSO_4$ is reported in this range[1]. Amount of $CuSO_4$ estimated both by DTA and colorimetrically are graphically represented in Fig. 3 to 6.

Maximum amount of $CuSO_4$ is formed in ores roasted between 500°C - 650°C for 180 minutes. On an average 87.23% of Cu present in the ore is converted to $CuSO_4$ in this temperature interval.

DTA curves of roasted chalcopyrite (Fig.3 & 4) show exothermic oxidation effects between 300°C to 600°C and edothermic decompositional reactions between 690°C to 850°C. The exothermic effects between 300°C to 600°C are due to the oxidation of chalcopyrite left unaltered during roasting i.e. an indication of formation of an additional amount of $CuSO_4$ during DTA. This is proved by the fact that with increase in duration of roasting, this exothermic effect gradually decreases and is practically absent in samples roasted at higher temperatures for 120 minutes and more, whereas it is maximum in samples roasted at lower temperatures and for shorter durations (Fig.4). Besides, the magnitudes of the amplitude of differential thermal peaks of samples roasted at lower temperatures and for shorter durations indicate greater amount of $CuSO_4$ compared to the corresponding colorimetric values. For longer durations of roasting at 550°C to 650°C there is very good coordination between the amount of $CuSO_4$ estimated by DTA and colorimetry.

During roasting of the samples, the exothermic reaction taking place in the sample, as indicated by the exothermic peaks in the region 300°C to 600°C, would elevate the sample temperature compared to the furnace temperature which is precisely maintained at stated values. This could give rise to slightly erratic values of $CuSO_4$ estimated both by DTA and colorimetry, especially in the case of lower roasting temperatures and shorter durations. In longer durations of roasting there is appreciable chance for the conditions to reach equilibrium.

In general, the amount of $CuSO_4$ formed increases with duration of roasting at all temperatures (Fig.5). It also increases with temperature in case of samples roasted for 180 minutes at 350°C to 500°C, from 500°C to 650°C it is more or less constant and again slightly decreases at 700°C due to decomposition of $CuSO_4$ during roasting itself (Fig.6).

In view of the results reported above it is evident that DTA can be used to:-
1. Fix the temperature and duration of roasting chalcopyrite to get maximum conversion to $CuSO_4$
2. Estimate the amount of $CuSO_4$ when oxidation of sulphide is complete and
3. detect the presence of unconverted sulphide by the presence of exothermic effect between 300°C to 600°C on DTA curves of roasted ore.

DTA can thus be used as a good quality control tool in the process of conversion of chalcopyrite to copper sulphate by direct roasting of the ore.

REFERENCE

E.M. Bollin - "Chalcogenides" - "Differential Thermal Analysis, Vol.I", Edited by R.C. Mackenizie, Academic Press, N.Y. (1970) p 193.

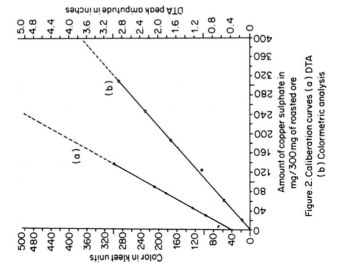

Figure.2. Caliberation curves (a) DTA (b) Colormetric analysis

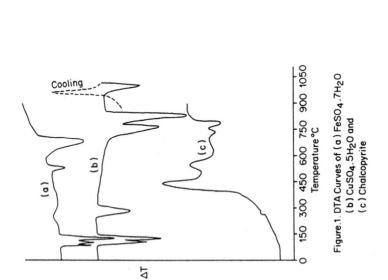

Figure.1. DTA Curves of (a) $FeSO_4 \cdot 7H_2O$ (b) $CuSO_4 \cdot 5H_2O$ and (c) Chalcopyrite

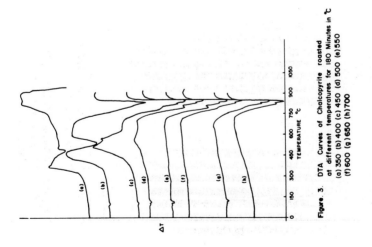

Figure. 3. DTA Curves of Chalcopyrite roasted at different temperatures for 180 Minutes in °C (a)350 (b)400 (c)450 (d)500 (e)550 (f)600 (g)650 (h)700

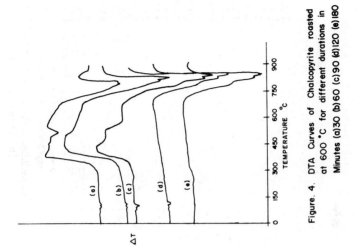

Figure. 4. DTA Curves of Chalcopyrite roasted at 600 °C for different durations in Minutes (a)30 (b)60 (c)90 (d)120 (e)180

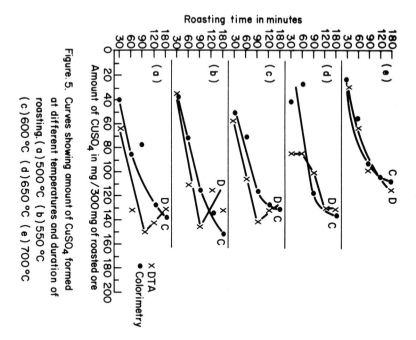

Figure. 5. Curves showing amount of CuSO₄ formed at different temperatures and duration of roasting. (a) 500 °C (b) 550 °C (c) 600 °C (d) 650 °C (e) 700 °C

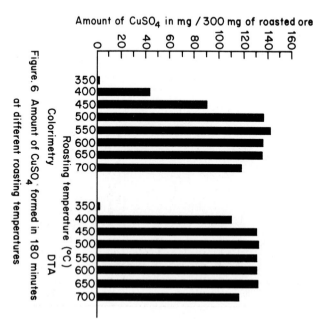

Figure. 6 Amount of CuSO₄ formed in 180 minutes at different roasting temperatures

THERMAL BEHAVIORS OF LONG-CHAIN MONOMERS AND
COMBLIKE POLYMERS

YOSHIO SHIBASAKI AND KIYOSHIGE FUKUDA

DEPARTMENT OF CHEMISTRY, SAITAMA UNIVERSITY
SHIMO-OKUBO, URAWA, 338, JAPAN

INTRODUCTION

The long-chain monomers, such as octadecyl acrylate and methacrylate, crystallize in a layered structure similar to the monolayer assemblies and exhibit the characteristic polymorphisms, depending on the chemical nature of functional group and the length of hydrocarbon chain [1]~[4]. The rate of polymerization of the long-chain monomers in the solid state and the structure of the resultant comblike polymers are influenced by the modes of molecular packing. In the present paper, the thermal behaviors of long-chain n-alkyl acrylates and methacrylates ($C_{20} \sim C_{12}$) and also their comblike polymers have been investigated together with X-ray and infrared analyses, in relation to the mechanisms of the solid-state polymerizations by γ-ray initiation.

EXPERIMENTAL

The materials used are eicosyl to dodecyl acrylates and methacrylates. These compounds were prepared by transesterifications of the corresponding methyl esters with long-chain alcohols, and purified by repeated recrystallizations. Abbreviations and their melting points are given in Table 1.

Table 1. LONG-CHAIN n-ALKYL ACRYLATES AND METHACRYLATES

Compound	Abbreviation	mp (°C)	
Eicosyl acrylate	$C_{20}A$	39.1 ~	39.5
Octadecyl acrylate	$C_{18}A$	32.5 ~	33.0
Hexadecyl acrylate	$C_{16}A$	24.2 ~	24.7
Tetradecyl acrylate	$C_{14}A$	14.0 ~	14.5
Dodecyl acrylate	$C_{12}A$	1.3 ~	1.9
Eicosyl methacrylate	$C_{20}MA$	36.6 ~	37.7
Octadecyl methacrylate	$C_{18}MA$	28.7 ~	29.7
Hexadecyl methacrylate	$C_{16}MA$	20.5 ~	21.1
Tetradecyl methacrylate	$C_{14}MA$	9.6 ~	10.1
Dodecyl methacrylate	$C_{12}MA$	−5.4 ~	−4.9

The thermal behaviors were measured by a Parkin-Elmer model DSC-1 differential scanning calorimeter, with samples of about 5 mg and scanning rate of $2°C \cdot min^{-1}$. The packing modes of the molecules were examined by X-ray diffraction patterns of the powder and infrared spectra of the KBr pellets. Polymerizations were carried out at various temperatures after irradiation with ^{60}Co γ-ray (0.1-2.0 Mrad) at $-196°C$.

RESULTS AND DISCUSSION

Polymorphic behaviors of n-alkyl acrylates and methacrylates. The DSC curves, X-ray diffraction patterns, and infrared spectra of octadecyl acrylate are shown in Figure 1. From these results, it can be deduced that octadecyl acrylate exhibits four crystalline forms, which we refer to as α, sub-α, β_1, and β_2 forms, respectively. The phase transition behaviors is represented as shown in Figure 1. When the molten sample is cooled, at first the α form crystal is obtained, then it transforms into the β_1 form, through the sub-α form. On immediate heating, the β_1 form retransforms into the α form at 19°C and the α form melts at 30°C. All of these crystals are in metastable states. During a storage at room temperature, the β_1 form transforms gradually into the stable β_2 form with higher melting point of 33°C, which can be also obtained from the solution.

The X-ray pattern for the α form crystal shows only one short spacing of 4.16 Å, while the long spacing diffractions correspond to the fully extended molecular length of 30.8 Å. Therefore, the α form is a hexagonal packing with perpendicular orientation. The β_1 form

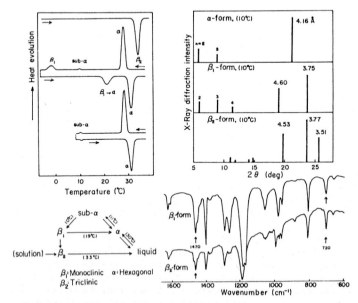

Figure 1. DSC curves, X-ray diffraction patterns, and infrared spectra of octadecyl acrylate.

shows two short spacings of 3.75 and 4.60 Å, and the long spacing is slightly less than that of the α form, whereas the β_2 form shows three short spacings of 3.51, 3.77, and 4.53 Å, but the long spacing diffractions are not clear. Furthermore, in the infrared spectra of the β_1 and β_2 forms the scissoring and rocking bands of methylene groups (1470 and 720 cm^{-1}) are both single and not doublet. These results suggest that the β_1 form is a monoclinic and β_2 form is a triclinic crystal with parallel chain plane.

The molecular arrangements and the subcell dimensions in the α, β_1, and β_2 form crystals of octadecyl acrylate are shown schematically in Figure 2. In the α form, the long-chain molecules are packed hexagonally in the lamellar structure with perpendicular orientation and the rotational motion around the long chain axis is allowed. In the β_1 form, the long axis is slightly inclined to the layer plane and the molecules are in monoclinic packing with parallel chain plane. On the other hand, the long-chain molecules in the β_2 form are tilted considerably and packed in triclinic subcell with parallel chain plane. Molecular packing is most compact in the β_2 form and expanded by about 10% in the α form, that in the β_1 form is intermediate.

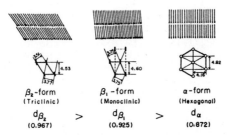

Figure 2. Molecular arrangements and subcell dimensions in the α, β_1, and β_2 form crystals.

Eicosyl methacrylate also exhibits four crystalline modifications (α, sub-α, β_1, and β_2 forms). The phase transition behavior, however, differs from that of octadecyl acrylate. In this case, the metastable β_1 form transforms into the stable β_2 form without retransforming into the α form on the heating process. The X-ray pattern for the α form with hexagonal packing shows one short spacing of 4.18 Å, while the long spacing diffractions can not be observed. The β_1 form (monoclinic) shows two short spacings of 4.52 and 4.77 Å. In addition, some sharp diffraction peaks that correspond to the second, third, and fourth orders of the long spacing of 27.0 Å were obtained. This spacing is fairly less than the length of the fully extended molecule of eicosyl methacrylate (32.4 Å); hence the molecules seem to be inclined to the basal plane (001). All of the diffractions for the β_1 form diminished gradually during the $\beta_1 \rightarrow \beta_2$ transformation, and finally the X-ray pattern of the thermally treated sample coincided with that for the β_2 form obtained from the methanol solution. The β_2 form (triclinic) shows three short spacings of 3.58, 3.85, and 4.65 Å, but the long spacing diffractions are not observed.

The polymorphic behaviors of n-alkyl acrylates and methacrylates with different chain length are illustrated in Figure 3. The α and

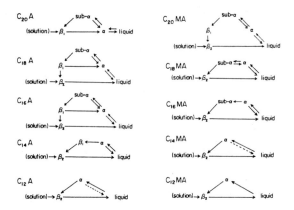

Figure 3. Polymorphic behaviors of n-alkyl acrylates and methacrylates. α:Hexagonal β₁:Monoclinic β₂:Triclinic

β_2 forms appear for all compounds used except for eicosyl acrylate, in which the β_2 form can not be obtained even from the solution. On the other hand, the β_1 form can be observed only for higher alkyl acrylates and eicosyl methacrylate. In both series, when the alkyl chain is shortened the polymorphic behaviors become simpler and the sub-α form disappears.

Effects of the molecular arrangements on the polymerizability. Figure 4-A shows the polymerizabilities in the different aggregation states for tetradecyl acrylate, as an example. All of the long chain monomers used can polymerize quickly in the α form. Since this packing mode of monomer molecules closely resembles that of side chains in the resultant comblike polymers, it is expected that the functional groups are concentrated in the layer plane and situated most effectively for the polymerization. When the polymerization temperature

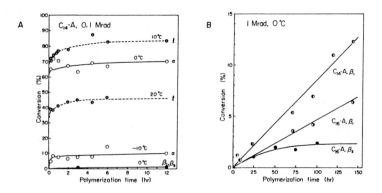

Figure 4. Postpolymerization of tetradecyl and hexadecyl acrylates in the different states.

is raised the polymerizability in the α form increase, but beyond the melting point it decreases markedly as shown by the broken lines. In most cases maximum polymerizability was observed just above the melting point of α form. The monomer molecules in the $β_1$ and $β_2$ forms are hardly polymerizable, while the polymerizability in the $β_1$ form (monoclinic) is only slightly higher than that in the $β_2$ form (triclinic) as shown in the enlarged figure (Figure 4-B).

Effects of chain length on the polymerization mechanism in layered structure and on the aggregation state of comblike polymer. DSC curves of the comblike polymers of the long-chain n-alkyl methacrylates and acrylates together with those for monomers are shown in Figure 5. The melting points of the polymers with longer side chains are higher than those of the corresponding monomers in the $β_2$ form by a few degrees or more, while the melting points of the polymers with tetradecyl are about 10°C below those of their monomers. Furthermore, in the cases of the polymers of dodecyl methacrylate and acrylate any peak for crystallization can not be observed in the cooling curve down to about −50°C. The polymer with sufficiently long side chains ($C_{20} \sim C_{16}$) crystallize in a stiff lamellar structure, in which the long side chains are tightly packed in the hexagonal mode, whereas the polymers with tetradecyl chain solidify in a loosely packed soft crystal similar to the smectic liquid crystal. With shorter chain monomers, such as dodecyl methacrylate and acrylate, the polymer samples obtained in higher conversions have a tendency to gelation. The samples of poly(dodecyl methacrylate) can not crystallize at all. In any case, the comblike polymers of n-alkyl methacrylate and acrylates exhibit no polymorphic behavior.

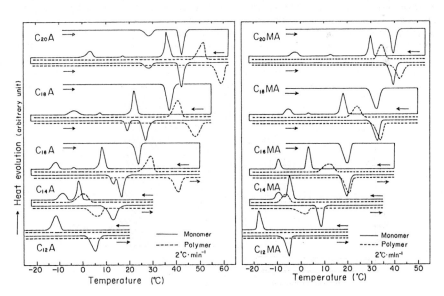

Figure 5. DSC curves for n-alkyl acrylates and methacrylates and their comblike polymers.

Figure 6 shows schematically the arrangement of the monomer molecules in the hexagonal and triclinic packings and also of the resultant comblike polymers. The comblike polymers crystallize in a hexagonal packing with respect to the long side chains, although eight or nine methylene units of the side chains in the vicinity of main chain are amorphous [5],[6]. In the α form with hexagonal packing, the polymerization occurs very fast, because of the most effective arrangement and rotational freedom of the monomer molecules. On the contrary, in the triclinic packing (β_2 form) the polymerization of longer chain monomers ($C_{20} \sim C_{14}$) is practically inhibited, because of difficulties in the molecular rearrangement and in the volume expansion. When the alkyl chain of the monomer is shortened to dodecyl, the side chains in the resultant polymer become amorphous and the polymerization may proceed successively at the interface of β_2 form crystallites, and also propagation across the polymer chains already formed may occur which results in the gelation through strong entanglement of the comblike polymer chains.

On the other hand, it has been found that in the polymerization of dodecyl acrylate two kinds of polymers can be obtained; one is crystalline and the other is noncrystalline. The crystalline poly-(dodecyl acrylate) with melting point of 23°C (about 20°C higher than that of the monomer) is obtained in the solid-state postpolymerization at lower temperature (−20°C), whereas the polymer samples obtained at higher temperatures are amorphous. The DSC curves and X-ray pattern for the crystalline polymers of dodecyl acrylate are shown in Figure 7. This polymer shows relatively sharp melting and crystallization. The X-ray pattern shows two short spacings of 3.75 and 4.20 Å together with clear diffractions corresponding to the long spacing of 40 Å. These results suggest that a kind of stereoregular polymer, such as syndiotactic, is obtained in the postpolymerization of docecyl acrylate at a sufficiently low temperature. The polymer seems to crystallize in the double layer structure, because the length of fully extended monomer molecule is 22 Å, and the side chains may be packed in a monoclinic or orthorhombic mode with parallel chain plane.

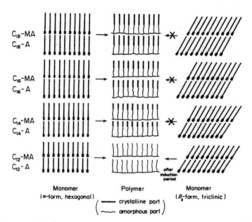

Figure 6. Effects of chain length on the aggregation state of the comblike polymers and polymerizability.

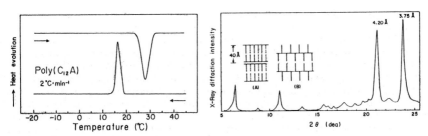

Figure 7. DSC curves and X-ray pattern for the crystalline poly(dodecyl acrylate): α form, 1 Mrad, −20°C.

The configuration of the resultant polymers can be discussed basing on the two-dimensional polymerization mechanism [2]. When the interlayer reaction and also the propagation reaction across the polymer chain already formed in a single layer are excluded, the polymerization occurs in a typical two-dimensional manner. This is the case for the polymerization with sufficiently long chain monomers or at fairly low temperatures. In the hexagonal packing each monomer molecule stands at a distance of 4.85 Å from the others, whereas the distance between every other methylene units in the main chain of the polymer is 2.54 Å. Therefore, when the monomer molecules aligned on one side of a propagating chain are linked successively, the isotactic triad may be formed. On the other hand, if a radical propagates by sewing up alternately the monomer molecules aligned on both sides, the syndiotactic sequence may be formed. The both sequences are combined by the heterotactic triad. Because the main chain of the isotactic sequence is somewhat distorted as compared with that of the syndiotactic sequence, it is expected that the syndiotactic triad may be formed predominantly at lower temperature.

REFERENCES

1. Y. Shibasaki, H. Nakahara, and K. Fukuda, J. Polym. Sci. Polym. Chem. Ed., 1979, 17, 2387.
2. Y. Shibasaki and K. Fukuda, J. Polym. Sci. Polym. Chem. Ed., 1979, 17, 2947.
3. Y. Shibasaki and K. Fukuda, J. Polym. Sci. Polym. Chem. Ed., 1980, 18, 2437.
4. Y. Shibasaki and K. Fukuda, "Thermal Analysis, Proceedings of the 6th ICTA, 1980, Vol.2," Birkhäuser Verlag, Basel, (1980), p.339.
5. E.F. Jordan, Jr., D.W. Feldeisen, and A.N. Wrigley, J. Polym. Sci. A-1, 1971, 9, 1835.
6. H.W.S. Hsieh, B. Post, and H. Morawetz, J. Polym. Sci. Polym. Chem. Ed., 1976, 14, 1241.

A NEW EGA APPROACH FOR THE KINETICAL STUDY OF A COMPLEX THERMOLYSIS
(KEROGENS)
G. Thevand, F. Rouquerol and J. Rouquerol
Centre de Thermodynamique et de Microcalorimétrie du C.N.R.S.,
26, rue du 141ème R.I.A. - 13003 Marseille - France

INTRODUCTION

The *"kerogens"* are defined as the insoluble part (in common organic solvents) of the sedimentary organic matter |1|. They may be considered at one and the same time as potential sources of fossil fuels and as by-products of their genesis.

We had the three following *reasons to carry out the study of the thermal degradation of kerogens* :
a. it could allow to understand the *natural degradation* of kerogens (which is supposed, by most authors to-day, to be essentially thermal) ;
b. it could be easily transferred to the understanding of the artificial thermolysis of *asphaltenes* (either from crude oil or from bitumen) which are of similar structure |2| and of high economical interest ;
c. any methodology developed to give an insight into the mechanism of such a complex thermolysis would be a good tool to study any simpler thermal decomposition.

In this paper we present the thermal analysis method which we developed for that purpose, along with the first results obtained in the case of kerogens.

CHOICE AND DESCRIPTION OF THE THERMAL ANALYSIS METHOD

A recent and comprehensive survey on our present knowledge on kerogen, including its thermal degradation may be found in |3|. It may be seen that, just like in the more general case of coals |4| an extremely careful control of all the parameters of the thermal decomposition experiment (TG, DTA or EGA) is absolutely needed |5|. Actually, although these parameters are apparently numerous (at least : size of sample, ambient atmosphere, flow rate, size and shape of crucible (with or without cover), heating rate) they may be summarized into *two* which are those having a basic influence on the course of the reaction and which are namely the *temperature* and the *atmosphere composition* (or, in other words, the partial pressures) *in the close neighbourhood of the reaction interface*. The safest way to control these two parameters is to keep them uniform, if possible, through out the sample and to analyse the evolved gas. This was achieved by combining, in the method described hereafter, the advantages of Controlled Decomposition Rate Thermal Analysis (which allows to reduce at will the temperature and pressure gradients within the sample during the thermal decomposition |6|) with those of Evolved Gas Analysis (with the use of a quadrupole mass analyser).

The *heating control* follows the general principle of Controlled Decomposition Rate Thermal Analysis |7| : the *heating of the furnace is not programmed* to follow any predetermined temperature path *but is controlled from a parameter directly related with the rate of the reaction*. This parameter can be the total flow of gas evolving from the thermal decomposition but, in the case of a complex thermal decomposition such as that of kerogen (where several reactions may overlap, giving rise to various gases such as H_2O, CO_2, SH_2, SO_2, CH_4, C_2H_6..) the total flow of gas (which could be measured by TG or by EGA) would be nearly useless for the determination of any significant kinetical parameter (just as would be also the total flow of heat as determined by DSC). For these reasons, we decided to choose, as the controlling parameter of the experiment, the *rate of production of one selected gaseous species*. The control loop therefore works in the following way (cf. Fig. 1) :
- sample S is being decomposed under a vacuum of about 10^{-5} torr ;
- *the gas evolved are pumped* (by a turbomolecular pump T Alcatel-Riber model TPM 140 backed with a primary pump P) *through the quadrupole analyser Riber QX 200* ;
- the peak selector is set on the chosen species ;
- *its signal*, which is both *related with* the partial pressure of this species and with its rate of pumping (i.e. with *the rate of production of the gas selected*) is used to control the heating of the sample ;
- the resulting constant rate of evolution of this gas may be kept, if needed, as low as *a few micrograms per hour*.

The *analysis of the gas* remains possible by using a set of several peak selectors. It takes place in extremely good conditions since *the whole flow of gas goes through the quadrupole analyser* (whereas, usually, a small portion of the gas to be analyzed is sampled with the help of a needle - or leak - valve which introduces separation effects, so that the analysis is not safely relevant to the main gas flow |8|). This is made possible because of the permanent control of the pressure and decomposition rate which keeps the pressure in the whole system (including the quadrupole) at a satisfactorily low level, although no constriction at all is used anywhere (the stainless steel tubings and valves are 2 cm bore).

EXPERIMENTAL RESULTS AND DISCUSSION

The three *samples of kerogen* studied here were selected from the same evolution path (type III, following the classification suggested by Tissot et al. |9|) but at increasing depths of burial (and therefore of degradation) in the order A, B, C. They originate from Mahakam (Indonesia). Each experiment was carried out with samples of about 50 mg.

The *calibration* of the quadrupole analyser (in terms of individual flow of the components of the gas stream) was carried out with the help of a one-metal steady leak (0.38 cm^3 STP h^{-1} bar^{-1} for CO_2), of a high precision gas volumetry set up (normally used for gas adsorption determinations on small surface area samples |10|), and of a standard thermal decomposition (sodium hydrogenocarbonate) delivering equimolecular amounts of CO_2 and H_2O and allowing an easy calibration

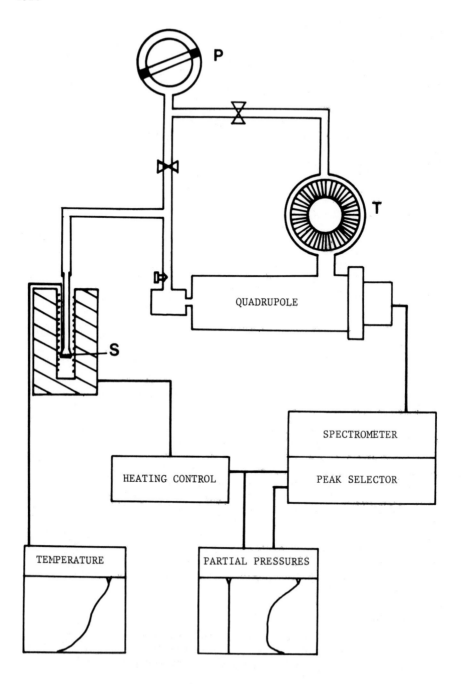

Figure 1 - Flow diagram of the Controlled Decomposition Rate - EGA assembly for the latter (which cannot be easily introduced, in a steady way, through a leak).

The first type of experiments we carried out is the *controlled decomposition rate EGA* of our three kerogens, choosing H_2O as the flow rate (and partial pressure) controlled species. The recorded curves are reported on Fig. 2. The dashed (and straight) line gives the rate of production of H_2O (here : 36 µg per hour for the three experiments) vs time. The dotted lines give the rate of production of CO_2 (as recorded also from the quadrupole analyser). Finally, the solid lines give the *temperature of the sample* vs time, i.e. *vs the amount of water lost*. The latter curves could therefore be considered as *"partial TG curves"* (since they would be obtained from a hypothetical thermobalance able to only take into account the loss of *one* preselected species ...). The whole set of curves clearly shows that the deeper the burial of the kerogen (from A to C) the smaller the residual amount of oxygenated functions giving rise to CO_2. Moreover, for the youngest sample (A) there is an interesting region where both H_2O and CO_2 flows are produced following a steady molar ratio of 2, suggesting that they are produced by the same reaction.

The second type of experiment presented here aims *to determine the activation energy* for the production of *one selected gaseous species*. This type of measurement directly stems from the cyclic two-rate method we suggested earlier |11| and from the Controlled Decomposition Rate EGA experiments presented in Fig. 2. Here, two partial flow rates (in a ratio of about 1 to 4) are chosen for the selected gaseous species (instead of one flow rate for the experiments of Fig. 2). By means of the special control loop of the equipment, *the sample under decomposition is brought to swing* (with a period, here, of 1 to 2 hours) *between both decomposition rates* (r_1 and r_2, cf. Fig. 3). The corresponding temperatures (like T_1, T_2) allow the direct calculation of the activation energy of the reaction producing the selected species. This determination is "pure" (or independent) from any other reaction taking place simultaneously in the same heteregeneous sample and producing different gaseous species. In the case of our kerogen A we could carry out, on a 50 mg sample, between 140 and 370 °C, 15 successive determinations of the activation energy for the formation of CO_2 (controlling here the rate of formation of CO_2 instead of that of H_2O reported in Fig. 2). The values measured steadily increase from 30 to 65 kcal. mol^{-1} as the degradation proceeds. Similar determinations concerning the kinetics of production of the main 4 to 6 gaseous species produced (H_2O, CO_2, SO_2, SH_2, CH_4, C_2H_6) are now needed to clarify the mechanism of this complex degradation. Except the time needed for these extra 50 to 100 determinations of activation energies (per sample) we don't see any extra difficulty for these measurements.

ACKNOWLEDGEMENTS

The authors are indebted to the Institut Français du Pétrole for a research grant (to G. Thevand) and to the CNRS for a financiel help in the scope of the A.T.P. "Energie et matières premières".

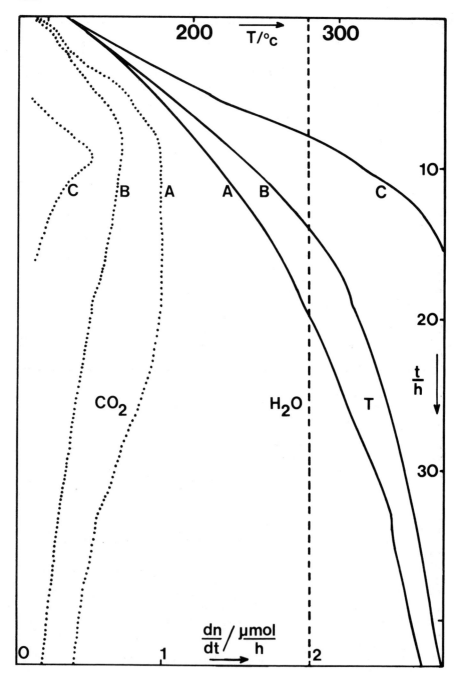

Figure 2 - Controlled Decomposition Rate EGA : temperature (T), CO_2 and H_2O curves vs. time

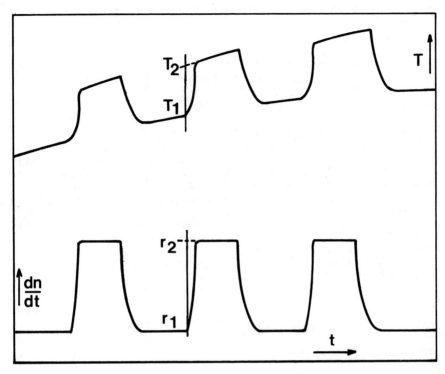

Figure 3 - Principle of the cyclic two-rate method for the determination of the activation energy of *one selected species* produced *under two partial rates* r_1 and r_2

REFERENCES

1. B.P. Tissot and D.H. Welte, "Petroleum Formation and Occurence," Springer-Verlag Berlin (1978), p. 123.
2. Ibid. p. 141.
3. B. Durand, "Kerogen", Technip, Paris (1980).
4. G.J. Lawson, in : "Differential Thermal Analysis", ed. by R.C. Mackenzie, Vol. 1, Academic Press, London (1970), p. 705.
5. Ibid. Ref. 3., p. 143.
6. J. Rouquerol, in : "Thermal Analysis", ed. by R.F. Schwenker,Jr. and P.D. Garn, Academic Press, New York (1969) p. 281 ; J. Rouquerol and M. Ganteaume, J. Therm. Anal., 1977, 11, 201.
7. J. Rouquerol, J. Therm. Anal., 1970, 2, 123.
8. W.D. Emmerich and E. Kaisersberger, J. Therm. Anal., 1979, 17,197.
9. B. Tissot, B. Durand, J. Espitalie and A. Combaz, Am. Assoc. Pet. Geol. Bull. 1974, 54, 499.
10. Y. Grillet, F. Rouquerol and J. Rouquerol, J. Chim. Phys., 1977, 74, 179.
11. J. Rouquerol, J. Therm. Anal., 1973, 5, 203 ; F. Rouquerol and J. Rouquerol, in : "Thermal Analysis", ed. by H.G. Wiedemann, Vol. 1, Birkhäuser Verlag, Basel (1972), p. 373.

APPENDIX

REPORT OF THE COMMITTEE ON STANDARDIZATION

The Committee on Standardization has carried out a preliminary testing of an inorganic glass having a glass transition temperature near 600°C. Preparations are under way for a larger test program.

A purity reference program has been initiated, chaired by E. Marti.

The arrangements for supplying reference materials to the U.S. National Bureau of Standards have been working smoothly. A new batch of silver sulfate has had to be tested and certified.

The Certified Reference Materials now available from the U.S. National Bureau of Standards are:

	NBS Catalog Number
Polystyrene (glass transition)	GM 754
DTA/DSC below 350K	GM 757
DTA/DSC from 125-940°C	GM 758
	GM 759
	GM 760
Thermogravimetry (magnetic materials)	GM 761

Members of the Committee, in addition to the Chairman, are

        E. J. Charsley
        P. K. Gallagher
        G. Lombardi
        R. C. Mackenzie
        E. Marti
        H. G. McAdie
        C. B. Murphy
        F. Paulik
        J. P. Redfern
        O. T. Sorenson
        R. H. Still

                                              P. D. Garn
                                              Chairman

AUTHOR INDEX

(Pages 1-791 refer to Volume 1;   793-1530 refer to Volume 2)

Aartsen, A. G.	457	Brown, G. R.	1030
Alamolhoda, A.	436	Brown, M. E.	58
Ali, M. El Sayed	344	Buri, A.	85
Aneesuddin, M.	1510		
Antal, M. J., Jr.	1490		
Arendt, V. D.	1150	Camino, G.	1137,1144
Argo, J.	470,744	Campero, E.	240
Argyropoylos	1099	Campion, P.	1111
Arnold, M.	214,621	Cardillo, P.	1182
Atanassov, O.	1276	Carr, N. J.	443
Auroux, A.	1189	Cattalani, A.	1217
Aylmer, D.	141,1270	Cejka, J.	713
		Centola, P.	1182
Back, P. S.	106	Cesaro, A.	815
Bair, H. E.	1155,1362	Chan, R. K.	1280
Bajpai, P. K.	558	Chandrasekharaiah, M. S.	220
Balek, V.	371	Char, P. N.	1510
Banerjee, A. C.	769	Charles, R. G.	264
Bappa, M. A.	488	Charsley, E. L.	1440
Baran, G.	120	Chen, G.	629
Barandiarán, J. M.	134	Chen, H.	1303
Barbooti, M. M.	687	Chen, L. Y.	1230
Barham, D.	765	Chen, T.	1016
Baró, M. D.	127	Chipashvili, D. S.	565
Bartels, W.	1050	Chiu, J.	37,979
Barton, T. J.	1440	Chiu, T. H.	793
Barve, S. D.	464	Choquette, D.	614
Bayer, G.	1470	Chow, T. S.	966
Bedford, W. K.	843	Chowdhury, B. B.	999
Behnken, D. W.	98	Choy, C. L.	964
Benoit, P. M. D.	98	Chung, D. D. L.	418,1392
Berte, R.	1224	Clark, G. M.	300
Bissett, F. H.	1332	Clavaguera, N.	127
Blaine, R. L.	254,909,994,1368	Clavaguera-Mora, M. T.	127
Blazer, T. A.	1368	Colmenero, J.	134
Blecic, D.	233	Comel, C.	148
Bossi, A.	1217,1224	Compero, E.	240
Bourrie, D. B.	284	Connelly, R. W.	954
Brar, A. S.	476	Cooney, J. D.	1325
Breakey, D. W.	255	Costa, L.	1137,1144
Brennan, W. P.	255	Cresser, M. S.	325
Breuer, K.-H.	169	Criado, J. M.	99
Bridle, T. R.	843	Crighton, J. S.	809

Dabek, R.	1118	Gabelica, Z.	1203
Darshane, V. S.	731	Gallagher, P. K.	1
Day, M.	1325,1342	Galwey, A. K.	38,58,443
Debras, G.	1203	Gantcheva, T.	1276
Delben, F.	815	Gao, Y.	406,413
Del Rosso, R.	1182	Garbassi, F.	1224
Deniz, K. U.	796	Garn, P. D.	436,899
DeRites, B.	1249	Gaur, U.	950
Derouane, E. G.	1203	Gavande, A. M.	693
Deshpande, D. A.	781,785	Geoffroy, A.	904
Deshpande, G. T.	781	George, T. P.	724
Deshpande, N. D.	781,785	Ghose, J.	706
Desseyn, H. O.	457	Gilbert, A. J. D.	650
Dharwadkar, S. R.	220	Gill, P. S.	994
DiCortemiglia, L.	1137	Gilson, J. P.	1203
DiRenzo, F.	1182	Goebelbecker, J.	608
DiRocco, R.	429	Gokhale, K. V. G. K.	558
DiVito, M. P.	255	Granzow, A.	1150
Dobal, V.	1254	Gravelle, P. C.	1189
Dollimore, D.	636,1111	Grebowitz, J.	1083
Dondur, V.	1209	Greenberg, A. R.	1092
Dyszel, S. M.	272	Gregorski, K. S.	90
		Griffiths, T.	1440
		Grundstein, V. V.	803
Earnest, C. M.	657,1260		
Eisenreich, N.	719		
Emmerich, W.-D.	163,279		
Engel, W.	719	Habash, T. F.	899
Eysel, W.	169	Haglund, B. O.	1402
		Haider, S. Z.	917
		Haines, P. J.	650
Fagherazzi, G.	660	Hakvoort, G.	1175
Farmer, J. B.	650	Halldahl, L.	344
Farrington, G. C.	585,593	Hanack, M.	836
Fellman, J. D.	1350	Hardy, M. J.	876,887
Ferragina, C.	429	Hartung, J.	863
Ferrillo, R. G.	1150	Heiber, O.	904
Fiato, R.	1249	Hemminger, W.	156
Fidler, D.	1209	Herman, M. A.	457
Filsinger, D. H.	284	Hirano, K.	392,399
Fischer, S. G.	979	Hodd, K. A.	1099
Flynn, J. H.	97	Hoffman, D. M.	1029
Frase, K. G.	593	Hofmans, H.	457
Freeman, E. S.	667	Hohne, G. W. H.	955
Frost, C. E.	1286	Hole, P. N.	809
Fu, L.	1386	Holm, J. L.	306,699
Fu, S.	1016	Hoppe, T. F.	1447
Fujieda, S.	176	Houser, J. J.	899
Fukuda, K.	1517	Hunter, J. E. III	909
Fyans, R. L.	255,1260	Hussain, K. I.	687

Ichihashi, M.	258	Locardi, B.	660
Ikoma, S.	450,578	Lønvik, K.	306
Ito, K.	331	Lu, Z.	413
		Luda di Cortemiglia, M. P.	1137
		Lum, R. M.	1349
Jain, K. C.	492		
Jain, P. C.	1024		
Janney, J. L.	1426,1434		
Jasim, F.	687	MacCallum, J. R.	54
Jefferis, S. A.	541	MacKenzie, R. C.	25
Jones, R. P.	1349	MacKnight, W. J.	1078
Jovanovic, S.	1237	Madhusudanan, P. M.	226
		Maesono, A.	258
		Malik, K. M. A.	917
Kaisersberger, E.	163,279	Mani, B.	585
Karasz, F. E.	1078	Manley, T. R.	1380
Karmazsin, E.	148,337	Margomenou-Leonidopoulou, G.	551
Kas, V.	1254	Marjit, B.	870
Kaushik, N. K.	482,492	Marjit, D.	870,1483
Kawalec, B.	899	Markert, C. L.	1362
Kelleher, P. G.	1362	Marotta, A.	85
Kher, V. G.	781,785	Marti, E.	904
Khorami, J.	614	Martin, C. A.	205,829
Kilian, H. G.	955	Martinez-Baez, L.	240
Kimmerle, F. M.	614	Maruta, M.	331
Kishi, A.	258	Maruyama, T.	351
Kobayashi, K.	758	Masri, M. S.	1190
Koch, E.	71	Mathur, A. B.	1064,1071
Kodama, M.	822	Mathur, G. N.	1064,1071
Kolenda, Z.	240	Maurer, J. J.	1040
Kosaka, M.	258	Mazzocchia, C.	1182
Krishnan, K.	226,1197	McCauley, J. A.	893
Kromer, H.	513,526	McCauley, R. A.	1409
Krug, D.	836	McGhie, A. R.	120,585,593
Kuresevic, V.	1078	Mendelovici, E.	533
Kusy, R. P.	1092	Meriani, S.	660
Kuwabara, M.	822	Milanovic, L.	1237
Kvantaliani, L. K.	565	Miller, D. J.	765
Kwakye, K. A.	291	Milodowski, A. E.	642
		Mirza, E. B.	796
		Moll, J.	835
Laczko, M.	1356	Molnar, S. P.	1125
LaGinestra, A.	429	Morales, R.	1118
Lamprecht, I.	849,857	Morgan, D. J.	642
Langer, H. G.	1350	Mosia, D. V.	1419
LaPerriere, D. M.	1373	Mrazek, Z.	713
Laskou, M.	551	Murat, M.	148
Lau, S.-F.	950		
LeParlouer, P.	183,190		
Levy, P. F.	254	Nabar, M. A.	464
Ligethy, L.	1356	Nakamura, T.	365
Liptay, G.	1356	Nakanishi, M.	176
Lin, C. Y.	1230	Nanda, V. S.	1024
Liu, X.	406,413	Negita, H.	737
Livesey, N. T.	325	Ninan, K. N.	226,1197

xxxi

Norwisz, J.	240	Randhawa, B. S.	476
Nya, A. E.	680	Rao, K. V. C.	1197
Nyilas, E.	793	Rao, M. S.	558
Nuzzio, D. B.	1477	Rao, M. S. P.	499,506
		Rao, U. R. K.	751,796
		Rasulic, G.	1237,1504
Ohara, S.	113	Ray, M. N.	693
Ohshima, T.	392,399	Razig, B. E. I. Abdel	571
Okino, T.	331	Rehim, A. M. Abdel	600
Operti, L.	1144	Richard, M.	1249
O'Reilly, J. M.	954	Riesen, R.	1050
Ors, J. A.	1373	Rivacoba, A.	134
Ozawa, F.	450,578	Roberts, F. J. Jr.	973
		Roessler, M.	1318
		Rogers, R. N.	1426,1434
Pal, A.	781	Romand, M.	337
Paoletti, S.	815	Rossi, P.	1151
Papazian, J. M.	385	Rouquerol, F.	70
Paranjpe, A. S.	796	Rouquerol, J.	70
Park, W. R. R.	1057	Rowe, M. W.	141,1270
Parker, K. M.	571	Rudloff, W. K.	65,667
Parvathanathan, P. S.	796	Rumsey, J. A.	1440
Pasteur, G. A.	1155	Rustchev, D.	1276
Patankar, A. V.	796		
Pätel, M.	857		
Patrono, P.	429		
Paulik, F.	214,621	Sagi, S. R.	499,506
Paulik, J.	214,621	Saiello, S.	85
Pavlath, A. E.	90,1190	Saito, K.	258
Pernicone, N.	1217	Saito, Y.	351,758
Petrik-Brandt, E.	1356	Sangari, H. S.	482
Petrini, G.	1224	Sato, T.	450,578
Pfaffenberger, H.	163	Satre, P.	337
Phatak, R. N.	731	Saxena, E. R.	1510
Pickard, J. M.	106	Schaarschmidt, B.	849,857
Piloyan, G. O.	565	Schiraldi, A.	1151
Pithon, F.	183	Scheider, H. A.	1106
Poleski, M. B.	1189	Schneider, O.	836
Porter, R. S.	964	Schönborn, K.-H.	156
Prest, W. M. Jr.	966,973	Schüller, K. H.	513,526
Prime, R. B.	984	Schwartz, E. M.	803
		Scurr, C.	1380
		Sebesta, P.	1254
		Seifert, H.-J.	358
Quan, X.	1349	Seki, S.	822
		Ser, K.	1280
		Sharp, J. H.	571
Raeder, H.	699	Shibasaki, Y.	1517
Rahman, A. A.	1310	Shimada, S.	737
Rahman, M. M.	917	Shyamala, M.	220
Ramachandran, V. S.	1296	Sigua, T. I.	1419
Ramana, K. V.	499,506	Singh, K.	488
Randall, V. G.	1190	Singh, P. K.	1071

Singh, R. P.	482	Vucelic, D.	1209
Smykatz-Kloss, W.	518,608	Vukovic, R.	1078
So, C.	765		
Soled, S.	1249		
Sood, R. K.	680	Wadsten, T.	917
Sood, S.	769	Walters, R. R.	106
Soraru, G.	660	Wang, L.	964
Sørensen, O. T.	344	Wani, B. R.	751
Sparks, S. L.	196	Warczewski, J.	358
Sponseller, S. P.	98	Warne, S. St. J.	1161
Srivastava, S. K.	1064,1071	Weir, E. D.	1447
Stapler, J. T.	1332	Weiss, R. A.	1010
Stepkowska, E. T.	541	Wendlandt, W. W.	320
St-Pierre, L. E.	1030	Whiting, L. F.	1456
Stradella, L.	1244	Widmann, G.	1497
Surinach, S.	127	Wiedemann, H. G.	1318,1470,1497
Svanidze, L. K.	1419	Wight, F. R.	1373
Swindall, W. J.	443	Wiles, D. M.	1325,1342
Szoor, G.	1463	Willis, J. M.	1030
		Won, S. B.	392,399
		Wood C. D.	1350
Takeda, M.	927	Wright, W. W.	1099
Tan, Q.	406,413	Wunderlich, B.	950,1084
Tanaka, H.	737		
Taniguchi, M.	113		
Taylor, T. J.	636	Yadav, Y. S.	1024
Telleria, I.	134	Yariv, S.	533
Thevand, G.	70	Yoganarasimhan, S. R.	680
Thiel, G.	358		
Tindyala, M. A.	1409		
Tou, J. C.	1456	Zagu, T. N.	1419
Townend, D. J.	313	Zheng, F.	406
Tran, H. N.	765	Zivkovic, Z. D.	233
Trögele, P.	955		
Trossarelli, L.	1137,1144		
Tsitsishvili, G. V.	565		
Udupa, M. R.	724,775		
Urbanec, Z.	713		
Van Dooren, A. A.	80		
Van Reijen, L. L.	1175		
Vasičkova, S.	713		
Vankateswarlu, K. S.	751		
Venturello, G.	1244		
Vigo, T. L.	1286		
Villalba, R.	533		
Vitol, I. M.	803		
Vogel, K.	246		
Vratny, F.	1155		